Principles of Geology

A Series of Books in Geology

EDITORS: James Gilluly and A. O. Woodford

Principles of Geology

James Gilluly
FORMERLY OF THE U.S. GEOLOGICAL SURVEY

Aaron C. Waters
UNIVERSITY OF CALIFORNIA, SANTA CRUZ

A. O. Woodford
POMONA COLLEGE

FOURTH EDITION
revised by
James Gilluly

W. H. FREEMAN AND COMPANY
San Francisco

Library of Congress Cataloging in Publication Data

Gilluly, James, 1896–
  Principles of Geology.

  Includes bibliographies and index.
  1. Geology. I. Waters, Aaron Clement, 1905–
joint author. II. Woodford, Alfred Oswald, 1890–
joint author. III. Title.
QE26.2.G54 1975   550   74–23076
ISBN 0–7167–0269–X

Printed in the United States of America

1 2 3 4 5 6 7 8 9 10

Fourth Edition

Principles of Geology

$12.95

JAMES GILLULY, U. S. Geological Survey,
AARON C. WATERS, University of California, Santa Cruz,
and A. O. WOODFORD, Pomona College.
Revised by James Gilluly.

Principles of Geology was first published in 1951. During the lifetime of this widely used and highly acclaimed text, the science of geology advanced more than in any other equal period of its history. These advances have been reflected in the second edition of 1959, the third edition of 1968, and this, the fourth edition.

The fourth edition of *Principles of Geology* offers an expanded coverage of plate and global tectonics, seismology, mineral resources, and a general updating of all other subject areas. New illustrations have been added. As in earlier editions, the text is not dogmatic: the authors supply evidence as well as conclusions, and the emphasis throughout is on scientific method, not merely conclusions. Each chapter ends with a set of study questions and a list of recommended readings.

Principles of Geology is a sound choice for a textbook for introductory geology and physical geology courses, especially those offered to majors in the earth sciences, engineering, and the life sciences.

JAMES GILLULY has had a long teaching and professional career in geology. He is a former Research Geologist with the United States Geological Survey and, though retired, continues to be active in research and lecturing. Dr. Gilluly studied at the University of Washington and Yale University, receiving his Ph.D. from Yale in 1926. In 1959, he received an honorary Sc.D. from Princeton University. Dr. Gilluly is a member of the National Academy of Sciences, Past President of the Geological Society of America (1948), Foreign Member of the Geological Society of

London, and Fellow of the American Academy of Arts and Sciences. In 1958 the Geological Society of America awarded him its Penrose Medal, in 1962 the Geological Society of London named him William Smith Lecturer, and in 1969 he received the Bucher Medal of the American Geophysical Union. Dr. Gilluly is the author of numerous United States Geological Survey Professional Papers and other research studies.

AARON C. WATERS received his bachelor's and master's degrees from the University of Washington and his Ph.D. from Yale University (1930). He has served on the faculties of Yale University, Stanford University, The Johns Hopkins University, and the University of California, Santa Barbara. Currently Professor Emeritus of Geology at the University of California, Santa Cruz, he received that institution's Citation for Distinguished Achievement in 1973. In 1973, Professor Waters was honored with a Special Commendation from the Geological Society of America for his work in astronaut training. He is a member of the National Academy of Sciences and the American Academy of Arts and Sciences. Professor Waters is the author of more than 70 scientific publications.

A. O. WOODFORD, Professor of Geology, Emeritus, Pomona College, has had a teaching career spanning nearly forty years. He received his bachelor's degree from Pomona College (1913) and his Ph.D. from the University of California, Berkeley (1923). Professor Woodford retired from active teaching in 1955. He was associated with the United States Geological Survey from 1943 to 1960. He is a Fellow of the Geological Society of America (Second Vice President, 1948), and Past President (1962) of the National Association of Geology Teachers. Professor Woodford is a member of the Mineralogical Society and the Association of Petroleum Geologists. He is the author of *Historical Geology* (W. H. Freeman and Company, 1965).

GEOLOGY, RESOURCES, AND SOCIETY
An Introduction to Earth Science
H. W. MENARD, Scripps Institution of Oceanography,
University of California, San Diego

Intended for all introductory courses in the earth sciences, this text combines a solid treatment of geological principles with a discussion of how modern geological knowledge relates to many public concerns. Some of the examples included are the seismic effects of underground nuclear testing, the environmental impact of the Alaska oil pipeline, and the rate of resource depletion. A helpful appendix on rock and mineral identification is included.

1974, 621 pages, 219 illustrations, 27 tables, 28 boxes, $12.95

EARTH
FRANK PRESS, Massachusetts Institute of Technology,
and RAYMOND SIEVER, Harvard University

Here is an introductory earth science text that focuses on the concepts needed to understand our planet. Plate tectonics and the latest discoveries in geophysics, geochemistry, oceanography, and space science provide a framework in which the subject matter of classical geology can be related to modern scientific thinking and student concerns.

1974, 945 pages, 729 illustrations, 14 boxes, 28 tables, $13.95
Study Guide by Donald W. Newberg, Colgate University ($2.95)
Instructor's Guide by Roger D. K. Thomas, Harvard University

Please order through your bookstore, or, if this is not possible, use the order card below.

Please send me the following book(s). If not completely satisfied, I may return the order for full refund within 10 days after receipt.

Title(s)	Price	Quantity	Total

Subtotal _____

☐ I enclose full payment.   California residents add 6% sales tax _____
  (Publisher pays postage.)

(Plus ½% for residents of BART counties) _____

TOTAL COST OF ORDER _____

Name _____

Address _____

City _____ State _____ Zip _____

PLANET EARTH
Readings from SCIENTIFIC AMERICAN
With Introductions by FRANK PRESS,
Massachusetts Institute of Technology, and
RAYMOND SIEVER, Harvard University

"The main emphasis in this volume," write the authors, "is on topics in the geosciences that represent new directions. . . . " Nineteen of the twenty-four articles have been published since 1970; they give the reader an exciting view of recent advances in the study of plate tectonics, environmental geology, energy resources, geomorphology, and geophysics.

1974, 303 pages, 258 illustrations (182 with color), cloth $12.00, paper $6.95

GEOLOGY
Principles and Concepts
A Programmed Text
DENNIS P. COX, U.S. Geological Survey, and
HELEN R. COX

This is a programmed text in physical geology based on the same authors' *Introductory Geology: a Programmed Text* (Preliminary Edition). New features of this book include: chapters on the latest advances in plate tectonics and related topics; many more drawings and photographs; and introductions and behavioral objectives for each chapter. Sections in this book are keyed to corresponding sections in several popular geology texts in current use.

1974, 463 pages, 400 illustrations, 32 tables, paperbound, $6.95

Contents

Preface

This is the fourth edition of a textbook that first appeared in 1951. The advances in the science of geology during the twenty-four intervening years have been greater than in any equal period in history. While incorporating some of these advances, advantage has also been taken of the opportunity to correct the few large and many small errors that seem always to creep into this work.

I have imposed on many friendly colleagues for advice and criticism: Paul Averitt, R. R. Doell, Carl Hedge, Arthur Lachenbruch, Stanley Lohman, Charles Naeser, John Obradovich, Louis Pakiser, Zell Peterman, Arthur Pierce, John Saaa, and Robert Smith have each read anywhere between a paragraph and a chapter and eliminated many errors. I am grateful to them all.

Professors Woodford and Waters have not participated actively in this revision but have both painstakingly read the entire typescript and called my attention to errors of fact and infelicities of organization, most of which I hope I have corrected. However much rewritten, the book still retains large sections to which they supplied the leading thoughts in the earlier editions and I am delighted that both have agreed to remain as authors of record. Much of any merit the book may have is due to their sharp pencils.

October 1974   *James Gilluly*

CHAPTER 1

The Science, Geology

Geology, the science of the earth, is concerned with the systematic study of rocks and minerals, in which there is preserved a record of the changes produced by processes that have long been at work on the surface of the earth and deep within it. In seeking information about the earth's origin and place in the solar system, the science of geology obviously impinges on astronomy; in attempting to understand the earth's dynamic evolution, it must make use of physics and chemistry; and in its inquiry into the origin and evolution of life, geology must both embrace and contribute to biology. In short, geology is an eclectic science in which every available tool is applied in the effort to understand the planet on which we live. Investigations of the physical properties and internal processes are called **geophysics**, those that concern chemical processes are **geochemistry**, and those that deal with ancient biology are **paleontology**. Geology has powerful tools of its own, primarily geologic mapping (Chapter 5), but in large part it draws upon its sister sciences in investigating the origin and history of the earth.

Geologic information is gathered by systematic observation, measurement, and analysis of rocks and minerals, of soils and fossils,* of the diverse forms of the land surface and the sea floors, and of the various processes that produce and change them.

The beginnings of geology are lost in antiquity. Although a few of the ancients made sound geological observations, their work was lost sight of during the Middle Ages, when appeal to the dogma of Aristotle largely substituted for observation and experiment. Virtually all of geology has developed within the past three centuries — the name itself is less than 200 years old. In this short time geology has compelled a revolution in thought. Just as astronomy has proved the immensity of space, geology has demonstrated the immensity of time (Chapter 6) and thus provided the background, sufficient and

---

*Fossils are the remains or imprints of animals and plants that lived in the past, naturally preserved by burial in sediments or in sedimentary rocks. Sediments are accumulations of such materials as dust, sand, and volcanic ash from the atmosphere; of stream gravel, sand, and mud on the lands; and of gravels, sand, clay, and organic remains on the sea floor. Sedimentary rocks are consolidated sediments, discussed in Chapter 4.

compelling, to demonstrate the reality of organic evolution—a principle that revolutionized man's ideas of his past and of his place in nature.

Over the past century, geology has become the basic tool of western civilization in its application to supply of energy and mineral resources. The way in which its applications have revolutionized our culture are deferred to Chapter 18; here we continue with the historical development of the science.

THE GEOLOGIC CYCLE

In 1785 James Hutton shocked his fellows of the Royal Society of Edinburgh with the statement that the history of the earth consists of a series of cycles so long and so numerous that he could see no vestige of a beginning nor any prospect of an end. Rejecting the notion that the earth had been created only about 6000 years ago—a date arrived at from the biblical genealogies—he reasoned that the time required to produce the readily discerned complexities in the rocks was so much greater as to be truly immeasurable.

What was his evidence? In essence, he noted that many rocks exposed high in the mountains contain fossils so like the hard parts of living marine creatures as to indicate that they had been deposited in the sea. Layering in those rocks so closely resembled that observed in a pit dug in a sandy beach that he inferred that the rocks had been similarly formed, even though now so well cemented as to be very resistant to deformation. But the layering in the rocks, which we now call *bedding*, is no longer nearly horizontal, as is that in sandy beaches, but stands at all inclinations to the horizon. Hard and resistant to bending as the rocks now are, they had obviously once been plastic enough to be greatly contorted, and though deposited in the sea, had been lifted high above it to form mountains. But the cycle of change does not end there: the uplifted rocks do not form "everlasting hills"; at every rain, rills form and wash away some soil, the rills unite to form muddy streams that carry the soil to the sea, and there deposit new sediments.

A most striking example of the cyclic nature of geologic history—one that deeply impressed Hutton—is seen on Siccar Point, Berwickshire, east of Edinburgh (Fig. 1-1). Here, tightly folded rocks, including sandstones that contain

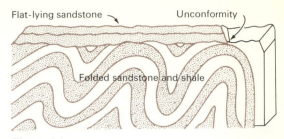

Figure 1-1
The relations exposed at Siccar Point. The folded rocks contain marine fossils. They are cut off by a surface of erosion, an *unconformity*, above which a new series of beds is superposed.

marine fossils, are overlain above a nearly horizontal surface by flat-lying sandstone. The interpretation was obvious to Hutton: earlier sediments had been consolidated to rock, folded and uplifted, eroded by streams or waves to a nearly flat surface that was then buried by younger sediments, and again uplifted. Two complete cycles are seen on one cliff: uplift followed by erosion, then by sedimentation, and finally by renewed uplift. How many similar cycles have run their course in the history of the earth? Surely, he concluded, the mountains we now look upon have not existed throughout the history of the earth; they are merely the latest in a series indefinitely long.

This book analyzes such geological cycles and their various sequences. We shall find incontrovertible evidence that, although sedimentary and volcanic rocks are formed at the earth's surface, other rocks have been formed deep within it. All rocks exposed at the surface are attacked by the weather, wind, rain, streams, and glaciers, broken down commonly to soils, and washed down to lower elevations. The cycle is illustrated diagrammatically in Figure 1-2.

THE UNIFORMITARIAN PRINCIPLE

Hutton summarized his observations in the *Uniformitarian Principle*, which was elaborated on by his friend John Playfair and given such wide currency by the brilliant British geologist Charles Lyell during the next century that it became one of the main guidelines of geology. It is generally somewhat enigmatically stated as "The present is the key to the past," but this

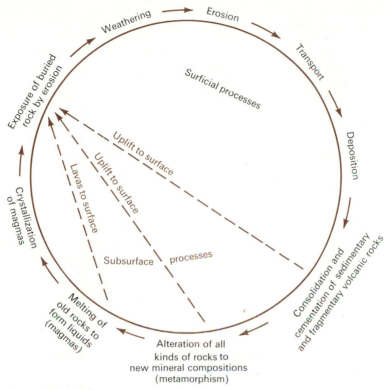

Figure 1-2

The geological cycle. Surficial processes are diagrammed in the upper part, deep-seated in the lower.

demands elaboration. It does not mean, as some disputants have claimed, that conditions now prevailing have been invariant through time. What is postulated is that those laws of nature that now prevail have always prevailed, and that, consequently, the products of processes now active are like the products of similar processes of the past; the earth changes, but only in accordance with unchanging physical laws.

Thus water has always flowed downhill and washed sand and silt toward the sea; cold and motionless rocks identical in mineral content and texture to lavas we see flowing from modern volcanoes were once themselves flowing lavas; objects having the form, growth patterns, and chemical composition of shells of living clams are indeed remains of formerly living creatures, even though now entombed in solid rock. In short, rocks having features identical to those of rocks we now see forming must also have been formed by similar processes operating in the past.

In Chapter 19 strong evidence is presented that early in decipherable earth history the atmosphere contained little or no oxygen (today it is 21 percent oxygen), but we assume that the causes of the winds that circulated this early atmosphere were the same as they are now — solar radiation and the rotation of the earth.

Throughout this book are innumerable examples confirming Hutton's thesis. The earth is indeed most active; new rocks have formed and are now forming, both in the earth's interior and by consolidation of fragments of former rocks deposited in the sea.

As shown in later chapters, the record is clear that rocks have been squeezed together and uplifted into mountains that were subsequently either rejuvenated in place or replaced by new ranges forming elsewhere. We call attention to this, perhaps prematurely, but so that the student can follow the various parts of the cycle as they are discussed. As Hutton said, the record is clear that cycle upon cycle of mountain-building and destruction by erosion have followed one another for as far back in time as we can decipher the record. Along with the geologic cycle goes a continual cycle of mineral changes as the materials of the earth's crust adjust themselves to the changes of temperature, pressure, and chemical environment that they encounter. These manifold changes, as we shall see, demand no catastrophic alterations in the earth processes with which we are familiar or in those that are compellingly inferred from observations. The premise that "the present is the key to the past" is nowhere contradicted in the almost inconceivably long history of the earth.

LIMITATIONS ON STUDY METHODS

For more than a century and a half geologists, physicists, and chemists have attempted to duplicate Nature's geological experiments in order to elucidate the factors involved in the formation of rocks and ores, in the flow of streams and glaciers, and in many other geologic processes. Much insight has been gained by these inquiries, but because of the almost unimaginably long time scale of earth history (Chapter 6) and because of the extremely high pressures and temperatures prevailing deep within the earth (Chapter 7), many earth processes are beyond the capability of the experimenter. Where is the calorimeter that will measure the thermal energy of a volcano? Where is the pressure bomb that can duplicate the pressure and temperature deep within the earth and maintain them for millions of years?

As with the life sciences, geology has now and always will have problems that can never be experimentally tested or stated in terms of meaningful mathematics; they will always remain in the domain of "natural history." But this is one of the most challenging and therefore fascinating features of the science. From circumstantial evidence and clues, the geologist must reconstruct, as logically as possible, the probable environments and events of the distant past. Laboratory studies have thrown much light on many geologic problems and have established limitations on the many conceivable possibilities, but for most geologic problems experiments can furnish only analogies, not final answers. In 1952, Nobel laureate Harold Urey wrote, in a discussion of the origin of the planets: "It should be noted that most phenomena in nature cannot be described in mathematical ways, and this is true of most chemical phenomena, like that presented here."

That being the situation in chemistry, how much more applicable it is to geology! For this reason, the focus of this book is on the natural history aspects of geology. Attention is given to the quantitative studies that seem pertinent, but no attempt is made to do this consistently. Such detailed and quantitative studies are beyond the broad introduction that is the objective of this book.

Scientific American Offprints

The *Scientific American* Offprints listed below and after the other Suggested Readings in this book are available from your bookstore or from W. H. Freeman and Company, 660 Market Street, San Francisco, California 94104, and 58 Kings Road, Reading, England RG1 3AA. Please order by the number preceding the author's name. The month and year in parentheses following the title of an article is the issue of the magazine in which the article was originally published.

846. Loren C. Eisley, "Charles Lyell" (August 1959).

CHAPTER 2

The Earth's Size, Shape, and Surface Features

That the earth is almost spherical was familiar to us long before the remarkable satellite photographs were published; the ancient Greeks had deduced this centuries ago from observations—the shadows cast on the moon during eclipses and from the way in which a ship coming over the horizon seems to rise out of the water.

EARLY MEASUREMENTS

In the third century B.C., Eratosthenes, a Greek geographer and astronomer, first measured the earth. Though techniques have since been refined, his reasoning is still basic to modern geodesy. (**Geodesy** is the science of measuring the earth and determining directions from point to point on its surface and distances between them.) Eratosthenes heard that at Syene (now Aswan), in Upper Egypt, the sun shines vertically down a well only at noon on the longest day of the year (summer solstice). At Alexandria, 5000 stades to the north, a part of every well is always in shadow. (The ancient Greek stade equals about 185 meters.) Eratosthenes measured the angle between a vertical line (plumb line) and the edge of the shadow cast by the noon sun in a well at Alexandria on the day of the summer solstice (Fig. 2-1). He then calculated the size of the earth on these premises:

a. That the sun is so distant that its rays to Syene and Alexandria are virtually parallel.
b. That Alexandria lies due north of Syene, so that a plane passing through Syene, Alexandria, and the noon sun would also pass through the center of the earth.
c. That the plumb line points directly to the earth's center.
d. That the east and west walls of the well in Alexandria are north-south planes.
e. That the earth is a sphere.

On these assumptions, the angle between plumb line and shadow edge equals the arc of the earth's curvature between the two points (Fig. 2-1). Accordingly:

$$\text{Circumference of earth} = \frac{360°}{\substack{\text{Angle of sun's rays to} \\ \text{vertical at Alexandria}}} \times 5000 \text{ stades.}$$

Eratosthenes' result was, in modern units, about 45,000 km. For reasons partly explained in Figure 2-1, this is about 14 percent larger than the measurement now accepted. About a century later, Poseidonius measured another arc but had less luck with compensating errors; his earth was a quarter too small, which may have led Columbus to mistake America for India.

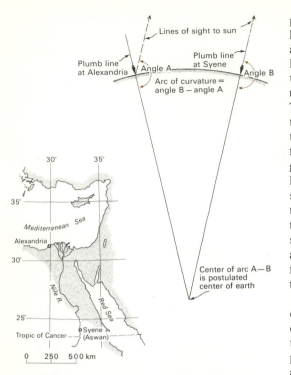

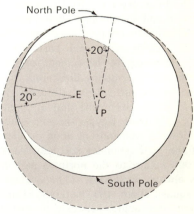

Figure 2-1

Eratosthenes' method for measuring the earth. Note that Syene does not lie due south of Alexandria, so the distance between them was not measured along a meridian, as assumed. And Syene is not precisely on the Tropic of Cancer but a little to the north; the measured arc was too small. For both these reasons the calculated size was too large. Angle A in the diagram to the right is greatly exaggerated.

In these early crude measurements, it was assumed that the earth is a sphere. Eventually, more accurate measurements made necessary successive modifications of this assumption. The history of science affords few better examples of scientific method and modification of theory in accord with improved observations than the refinement of the figure of the earth.

MODERN MEASUREMENTS

In the seventeenth and eighteenth centuries, expanding navigation demanded more accurate charts; Eratosthenes' method was resurrected. North-south lines, "arcs of meridian," were measured in different latitudes—in Finland, France, and Peru—and the angles of arc com-

puted. These showed that a degree of latitude is longer in Finland than in Peru. If we retain the assumption that the earth is a sphere, we would have to conclude that the sphere measured near the pole is somewhat larger than that measured near the equator. Figure 2-2 shows the problem. The broken circle with the center P represents the size of the earth as deduced from polar arcs; the dotted circle with center E, that deduced from equatorial data. The difference in size is greatly exaggerated but clearly shows that the hypothesis of a spherical earth is wrong. The simplest way to reconcile the measurements is to assume that the earth is slightly flattened at the poles; that is, it is an oblate spheroid, the solid figure obtained by revolving an ellipse about its shorter axis. The solid line in the figure, an ellipse with center C, reconciles all the data.

Satellite orbits other than equatorial ones are of course modified by the attraction of the equatorial bulge. Utilizing satellite observations, the International Union of Geodesy and Geophysics recommended (1967) international adoption of the following dimensions as the best available:

Equatorial radius	6,378,160 meters
Polar radius	6,356,775 meters
Difference	21,395 meters
Flattening	1/299.75

Figure 2-2

Diagram showing how the differing lengths of equal arcs of meridian measured in high and low latitudes suggest that the shape of the earth is ellipsoidal rather than spherical.

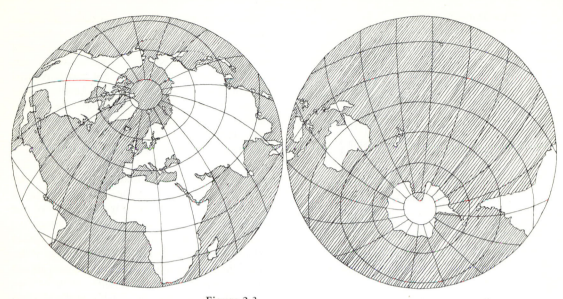

Figure 2-3
The land and water hemispheres.

THE EARTH'S RELIEF

Measurements from the best available maps show that more than two-thirds of the earth's surface is covered with water:

Area of the seas	361,059,000 km²	70.8 percent
Area of the lands	148,892,000 km²	29.2 percent

Land and sea are very unevenly distributed: the earth can be divided into hemispheres such that four fifths of all the land lies in one, while the other is nine-tenths water (Fig. 2-3). Note the extent of the dissymmetry: nearly 45 percent of the sea lies diametrically opposite sea, but less than 2 percent of the land is diametrically opposite land.

The range in relief of the surface is nearly 20 km: Mount Everest towers 8884 m above sea level, and the Nero Deep in the Mariana Trench lies 11,055 m below. Great as are these heights and depths by human standards, they are trivial compared with the earth's radius. If the largest ellipse that can be drawn on this page represented the earth, a normal pencil line would include all the mountains and ocean deeps, from highest to lowest. At this scale, the earth is relatively as smooth as a billiard ball, though it might be difficult to persuade a Tibetan of this.

From maps and charts, careful estimates have been made of the areas of the earth's surface standing between various altitude limits (Fig. 2-4). This graph shows that the extreme heights and deeps are trivial in area but that two dominant altitude ranges are conspicuously represented. Though the average oceanic depth is 3730 m, nearly a fourth of the earth's surface is ocean floor with depths between 4000 and 5000 m. The average height of land is about 875 m. About a fifth of the earth's surface lies between 500 m above and 200 m below sea level. The submerged part of this upper range is part of the continental shelves.

Neither the mountains nor the ocean deeps are distributed at random (Fig. 2-5). Most of the world's high mountains are within two main groups: one, the circum-Pacific group, virtually rings the Pacific Ocean in a belt generally 300 km or less wide, though in North America it widens to well over 1600 km, with some high plateaus included. A second group extends from Indonesia through Burma to the Himalayas and thence westward through the Hindu Kush, Zagros, Elburz, Caucasus, Carpathian, Toros, Peloponnesus, Dinaric Alps, Alps, Apennines, Betic Mountains of Spain, Pyrenees, and Atlas of North Africa. To the north of the Himalayas stands the high Tibetan Plateau and a whole series of roughly east-west trending mountains in Mongolia. The circum-Pacific and the Indonesian-Himalayan-Mediterranean chains include all the

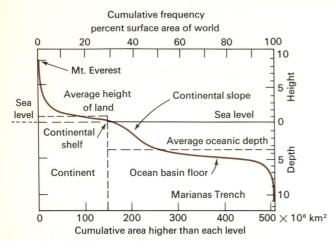

Figure 2-4
Graph showing percentages of the earth's surface
lying between various levels above and below sea
level. Vertical scale in kilometers. Note the two
broad flat areas of the graph, one near sea level, the
other between 4 and 5 km below it. [Data from
Kossina and from Menard and Smith.]

notably high mountains of the earth, except for
such isolated blocks as Ruwenzori on the Uganda-
Rwanda border and the volcanoes Kilimanjaro in
Tanzania and Mauna Loa in Hawaii. As Figure
7-26 (p. 112) shows, the two principal belts are
the loci of most earthquakes on land.

Much more subdued mountains (which we
will find to have been formed much earlier and so
have lost much of their bulk by erosion) constitute the Appalachians and Ouachitas in North
America, the Caledonian mountains of Scotland
and Scandinavia, the Urals, and the Great
Dividing Range of Australia.

The upper of the two conspicuous, dominant
altitude ranges includes the low-lying plains that
abound on both sides of the Atlantic and most of
the Arctic shores, the south shore of the Baltic,
the west shore of the Black Sea and the north
and east shores of the Caspian Sea. The largest
of the plains areas is the one that extends from
the Gulf of Mexico to the Arctic in the central
parts of Canada and the United States. Much of
Australia, a large part of the Amazon Basin, and
the Pampas of the Argentine also are plains. Such
plains are missing from the Pacific Basin, except
in China, behind the Japanese arc.

Most large plains extend beneath the sea to
form the **continental shelves**: the edges of the continents are submerged because the ocean basins
are overfull. Many shelves slope seaward very
gradually, say 1 m per km, but most average
three or four times this gradient. At an average
distance of about 50 km from shore, the shelves
steepen abruptly at the top of the **continental
slope**. This break in slope is of irregular depth,
but is generally about 150 m deep. In the Barents
Sea and off Norway and Labrador, however, it is
as deep as 450 m. Around the Antarctic continent it varies from 300 to 700 m, averaging
about 450 to 500 m. Nor is the width of the
shelves uniform: off Chile and southwestern
Alaska there is no shelf at all, whereas off Siberia
a shelf extends 1350 km into the Arctic. Alto-
gether the shelves cover about 7 percent of the
sea floor, so that the continents really occupy
about 34 percent of the earth's surface, even
though only a little more than 29 percent is land.

The continental slopes fall seaward steeply, in
places with a gradient as high as 1:6 and an
average of perhaps 1:20, notably steeper than the
shelves. As Figure 2-4 shows, they cover about
5 percent of the earth's surface. Off most shores
the slope extends to depths of about 2000 or
3000 m, where it joins the much gentler gradient
of the **continental rise**, which is generally between
1:50 and 1:700. At depths of 4000 to 5000 m are
abyssal hills with gentle contours or **abyssal
plains**, the flattest extensive surfaces on earth,
with slopes commonly less than 1:1000 and some
as low as 1:7000.

Notable oceanic features are the **abyssal cones**
that lie offshore of the Mississippi, Congo,
Ganges, Indus, and Magdalena rivers. With
apices at the river mouths, they extend across
continental slopes and rises to merge at their
bases with abyssal plains. They are obviously
piles of sediment derived from the rivers.

Most continental slopes and terraces are cut
by sporadic **submarine canyons** of a great range
in size, some deeper and more precipitous than
the Grand Canyon of the Colorado. Many can
be traced to great depths. A few such canyons
begin off the mouths of large rivers, such as the
Hudson and Congo; others, equally impressive,
begin at the shoreline, far from the mouth of any
large river, and many others begin as notches in
the continental shelf far from land. The origin of
such canyons is considered in Chapter 16.

The Ocean Ridges and Rises

Among the many surprising discoveries of oceanography since World War II has been the finding of the longest mountain range on earth — the Mid-Atlantic Ridge. A mid-ocean ridge was known in the Atlantic for more than a century, since the time the first trans-Atlantic cable was laid, but its great extent was not suspected (Fig. 2-5). The ridge, which occupies about the middle third of the Atlantic, halfway between the Old World and the New, rises between 2 and 3 km above the ocean floor for much of its length and projects above the sea in Iceland, the Azores, St. Pauls Rocks, Ascension, Saint Helena, Tristan da Cunha, and Bouvet Islands. Although repeatedly offset, in places for several hundred kilometers, by straight transverse fractures, the ridge is otherwise continuous for many thousands of kilometers.

The ridge is almost precisely in mid-ocean, and it extends — at about the same distance from shore — part way around the Cape of Good Hope and then veers northeastward as far as the middle of the Indian Ocean. Here the ridge splits, with one branch trending north and northwestward as the Carlsberg Ridge and extending to the mouth of the Gulf of Aden; the other, the Mid-Indian Ridge, trends southeastward, passing between Australia and Antarctica and then becoming the Pacific Antarctic Ridge. This is separated from a considerably smoother and lower rise, the East Pacific Rise, by a west-northwest-trending lineament. The East Pacific Rise does not follow a mid-ocean course but, with several offsets, trends generally northward, finally becoming obscure or dying out off the west coast of Mexico, several hundred kilometers south of the Gulf of California. Several shorter ridges are found off Vancouver Island and in the southwest Pacific.

The longest straight ridge on earth is the Ninety East Ridge, which extends almost precisely

Figure 2-5
The distribution of ocean ridges, rises, and trenches, and the Circum-Pacific and Mediterranean-Himalayan-Indonesian mountain chains.

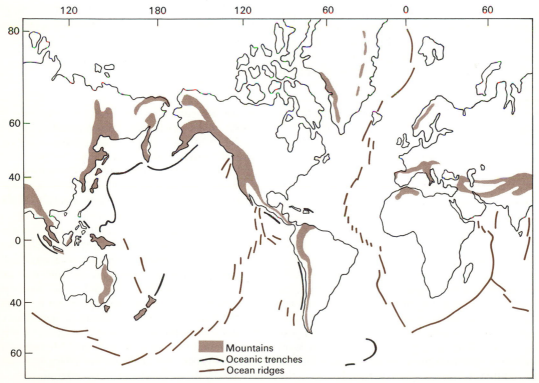

Mountains
Oceanic trenches
Ocean ridges

along the meridian of 90°E from south of the Tropic of Capricorn to well north of the Equator, more than 2500 km. A comparable meridional ridge extends from just west of Bombay to a junction with the Mid-Indian Ridge at about 20°S.

Island Arcs and Ocean Deeps

Among the most striking features of the Pacific are the great festoons of **island arcs** that fringe its northern and western sides. These include the Aleutian, Kurile, Japanese, Ryukyu and Philippine Archipelagos. Island-crowded arcuate ridges separate comparatively shallow seas—the Bering, Okhotsk, Japan, Yellow, and East China —from the Pacific Basin proper. Other comparable arcs branch southward from Japan. The seas within these outer arcs are not, however, shallow; the Philippine Basin on their shoreward side is as deep as the Pacific. The Indonesian islands form a comparable arc bounding the Indian Ocean on the northeast; the Antilles and Southern Antilles are the only similar features in the Atlantic.

Closely associated with the island arcs are the deepest parts of the oceans. Depressions deeper than 7 km are **deeps**. The deepest so far surveyed is the Nero Deep in the Marianas Trench, but there are several others deeper than 10 km in the Pacific. The only deeps not associated with an island arc are the Chile Deep and the Middle America Trench.

In the Atlantic, the deepest trenches are the Puerto Rico Trough, north of Puerto Rico, and the South Sandwich Trench, on the convex side of the Southern Antilles arc, each more than 8 km deep.

We will find, in Chapter 8, some remarkable connections between most of these major features of earth relief.

Gravity and a Level Surface

In making a topographic map, or in setting the floor joists of a building, what is meant by "a level surface"? The surface of a quiet pond is level and appears to be a plane, but in fact it is curved, just as we know the surface of the sea to be. A level surface is therefore not a plane but a surface so curved that it is everywhere perpendicular to the direction of the local plumb line. All plumb lines

converge toward the central part of the earth, although as we have seen, not toward a single point at the center. The direction of the plumb line is vertical; its overhead projection is toward the *zenith*, the point directly above us among the stars.

The Law of Gravitation

The attraction of the earth for a plumb bob is simply one illustration of the universal **Law of Gravitation**, first formalized by the great English scientist Isaac Newton (1642–1727). It is usually stated: *Every particle in the universe attracts every other particle with a force directly proportional to the product of their masses and inversely proportional to the square of the distance between them.*

The mathematical statement of the law is:

$$F \propto \frac{M' \times M''}{D^2},$$

in which F is the force of attraction, M' is the mass of one body (in our example, the plumb bob), M'' the mass of the other (the earth), and D the distance from the bob to the center of the earth. The symbol $\propto$ means "varies as."

Weighing a Mass

Precise measurements of the force of gravity at various places yield valuable information about the figure of the earth and something of its interior. The **weight** of a mass is the measure of the gravitational attraction between the earth and itself.

Two common weighing instruments are the spring scale and the beam balance (Fig. 2-6). A spring balance measures weight by the stretching of a spring. We assume that Hooke's Law holds—that the spring is perfectly elastic and stretches in exact proportion to the pull of gravity on the object weighed. The scale is calibrated by marking the stretching produced by standard weights. By international agreement, the standard kilogram mass is preserved at the International Bureau of Weights and Measures at Sèvres, France. We calibrate the spring scale by comparing it with the standard kilogram or one of its duplicates; a second mass that stretches the

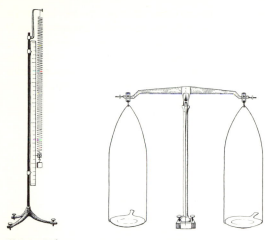

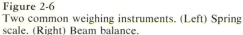

Figure 2-6
Two common weighing instruments. (Left) Spring scale. (Right) Beam balance.

spring equally is another kilogram mass. One kilogram mass precisely balances another when placed in opposite pans of a beam balance. But if the equipment is moved to some other point, we find that not all of these relations may hold precisely.

EFFECT OF ALTITUDE

If we take kilogram masses, scale, and balance to a peak in the Alps, 3 km above sea level, we find that although one kilogram mass will still balance another, as at Sèvres, neither will weigh

a kilogram on the spring scale, but slightly less. The only different factor is altitude. According to Newton's law, the attraction of the earth for the kilogram mass varies directly with the product of the two masses and inversely as the square of the distance between them. The product of masses is of course unchanged during the move, but the distance to the earth's center is enough greater in the Alps that a kilogram weight on the spring scale weighs about a gram less than it does at Sèvres. Thus the distance to the earth's center must be taken into account when weighing on a spring scale, though not on a beam balance, for the pull of gravity is equally lessened for both masses.

EFFECT OF THE EARTH'S ROTATION

Because the earth spins on its axis, any object not on the axis tends to be thrown off by centrifugal force, just as mud is thrown from a spinning automobile wheel. This force of course weakens the pull of gravity. It is very small, even at the equator, where it amounts to only about 0.33 percent of gravity, and it steadily decreases with latitude as the parallels of latitude diminish in radius (Fig. 2-7).

Thus centrifugal force acts, along with the differences in radius, to decrease the effect of gravity at the equator as compared with that at the poles. The sum of the two effects amounts to about 0.5 percent of gravity. Thus if a polar bear weighing a metric ton at the pole were transported to the equator, it would weigh only 995 kg.

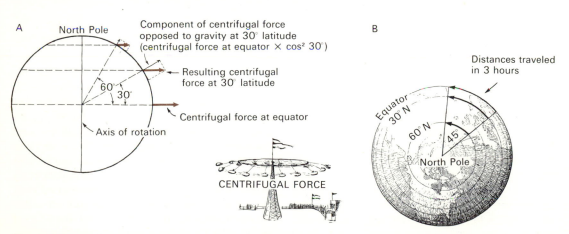

Figure 2-7
Diagram showing a familiar demonstration of centrifugal force. (A) How the centrifugal force due to rotation varies from equator to pole. (B) How the linear velocity varies with latitude.

EFFECT OF LOCAL VARIATIONS IN
ROCK DENSITY

The distribution of rocks of different densities also alters the local pull of gravity. For example, imagine that the United States gold reserve buried at Fort Knox, Kentucky, forms a single block of pure metal (Fig. 2-8). A cubic meter of gold weighs more than 21 tons. The inverse-square relation in the Law of Gravitation makes obvious that an object immediately above the gold must weigh a little more than it would if weighed on the roof of one of the many limestone caverns nearby. Although at the same altitude and latitude, the pull of the nearby gold would surely exceed that of an equal volume of air. We know that the rocks near the surface vary measurably in density from place to place, though not, of course, to the extreme of our artificial example. Thus measured gravity must everywhere vary with the density of the immediately underlying rocks.

This principle is of demonstrated economic value even where the density contrasts are far less than in our example. In many lands, precise gravity measurements have led to the discovery of dense ores concealed by stream deposits. More nearly parallel to our example is the situation at Carletonville, on the West Rand gold-field, South Africa. The ore lies at depth beneath a thick cavernous formation. During mining, great volumes of water were pumped from the caverns, leaving many empty of the water that had formerly helped to support their roofs. A good many caved in, some with considerable loss of life; many miners refused to face the risks, and for a time it appeared that the mine must be abandoned. The problem was solved by making systematic, precise measurements of gravity over a closely spaced network. Notably low values of gravity marked the roofs of dangerous caverns; houses and roads were relocated onto areas of higher gravity. Though the collapses still go on, they no longer threaten life and property; the hazardous areas have been pin-pointed by the gravity measurements.

MEASURING GRAVITY

We have seen that the local pull of gravity depends on (1) altitude, (2) latitude, and (3) the density of the immediately underlying rocks. Each effect is relatively small; very sensitive instruments are needed to measure the differences—instruments far more sensitive than even the best spring scale. The gravity pendulum is a relatively simple instrument that meets the need.

A free-swinging pendulum—that is, one not driven by clockwork or other external means—oscillates because of the force of gravity. If we pull such a pendulum aside and release it, the weight falls toward the earth, following the arc constrained by the length of the pendulum. The pull of gravity accelerates the pendulum; at the bottom of its swing, its inertia carries it against the pull of gravity up the similar arc beyond until its inertia is overcome and it reverses its swing. Because of friction with the air, the swings shorten and eventually cease; the pendulum becomes a plumb bob.

The great Dutch scientist Huygens (1629–1695) discovered the law governing the *period of oscillation*—the time for one complete to-and-fro movement—of a pendulum: it varies inversely with the square root of the local acceleration of gravity and directly with the square root of the length of the pendulum,

$$P \propto \sqrt{\frac{l}{g}}$$

where P is the period, l the length, and g the acceleration of gravity.

Newton showed how this law explains why pendulum clocks systematically gain or lose time when moved from place to place and why an allowance must be made for adjusting pendulum length. A pendulum clock that keeps good time in Paris would run slow in the high Alps because gravity is smaller there. We are thus able to measure the force of gravity by timing the swings of a pendulum of known length. Swings are

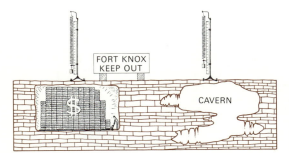

Figure 2-8
Gravity effects of density of nearby masses.

counted through a known time interval, and gravity is computed from Huygens' Law.

Modern gravity pendulums hang on knife-edge jewel bearings, swing in vacuum chambers and are timed by precision chronometers accurate to 1/10,000 second; they measure gravity to within a few parts per million. Such pendulums are unwieldy to transport and the observations long and tedious, so most practical gravimetry is done with somewhat less accurate but readily transported *gravimeters*. These compare the distortion of a quartz fiber at an observation station with that at an accurately known pendulum station. The principle is the same as that of a spring scale, but because the distortion must be read to 1 part in 10 million, many refinements are necessary in the construction. Observations require only a few minutes, but because of "drift" —fatigue of the quartz fiber—the instrument must be calibrated with respect to pendulum stations every few days.

By means of these instruments the variations of gravity have been measured over large areas with considerable accuracy: gravimeters have recently been modified to operate in airplanes in flight and on surface vessels at sea, where ordinary gravity pendulums cannot work because of ship motion. Before this improvement, the only gravity data from the seas were obtained with multiple pendulums developed by the Dutch geodesist F. A. Vening-Meinesz, which had to be used in submarines submerged deeply enough to minimize wave motion. These instruments used two pendulums swung in opposite phases and were operable in slight waves. The new gravimeters have increased gravity measurements by the millions from both land and sea, and the data provide insight into many important geologic questions, as noted later in this and other chapters.

DEFLECTIONS OF THE PLUMB LINE

Supplementing and confirming the gravity data are many surveying observations. **Triangulation** (described in Appendix I) is a method of fixing the position of a third point by sighting it from each of two known positions. The position of a point can also be determined by measuring its latitude and longitude astronomically, as is done in navigation by reading angles between the

horizon and lines of sight to the sun or stars. Such observations determine the *zenith*—the direction of the vertical—at the observation point, for it is normal to the horizon. If the plumb line invariably pointed to the earth's center, position of a point determined by latitude and longitude should coincide with that determined by triangulation from two known points. (This would be strictly true only if the earth were a sphere, as noted in Figure 2-2; the slight modification due to polar flattening is here disregarded.)

As we saw, however, other factors than latitude affect the plumb line. The Law of Gravitation tells us that the plumb bob is attracted by a given mass 1 km away with a force 100 times that exerted by a like mass 10 km away. Thus the plumb line should be deflected from the vertical toward a nearby mountain, and in fact it generally is.

Consider the situation in a deep Norwegian fjord, nestled between massive cliffs rising 2000 meters above. A plumb line suspended near one side of the fjord is deflected toward the adjacent mountain. The sea surface is also tilted slightly upward toward the mountain mass (Fig. 2-9, *Bottom*). Such deflections are of course extremely small, generally only a few seconds of arc, but they can lead to appreciable errors in astronomical position. Moreover, we now see that an oblate spheroidal figure of the earth, though closer to fact than a spherical figure, is still somewhat imprecise. Differential local attractions make any level surface of wide extent vastly more complex than the surface of a simple oblate ellipsoid, even though each part is exactly normal to the plumb line. We turn now to the discovery of a fundamental relation in earth structure—the principle of isostasy.

The Trigonometrical Survey of India

In the mid-nineteenth century, the Trigonometrical Survey of India was organized under Sir George Everest (for whom Mount Everest was later named) to establish the control for mapping that huge subcontinent. Many points were located by careful triangulation (see Appendix I) and distances between them calculated. Many were also fixed astronomically.

If latitude, longitude, and *azimuth* (the angle between a line of sight and true north) of a line to another station are known, it is relatively easy to

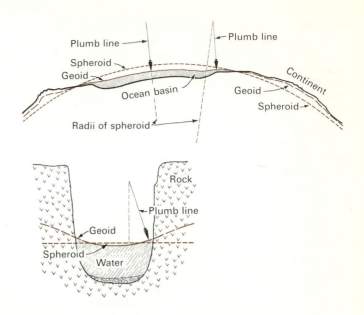

Figure 2-9
The relation between spheroid and geoid, greatly exaggerated to show the qualitative effect of irregular topography. (Top) Broad view of their relations as between continent and ocean. (Bottom) The hypothetical relations (greatly exaggerated) in a Norwegian fjord.

calculate latitudes and longitudes of all the other stations. Such calculations assume a certain figure of the earth, but the only astronomical observations required are at the station of origin. If latitude and longitude of a second station are independently determined astronomically, the two fixes can be compared.

It soon became obvious that geographic positions found by triangulation from a base line did not agree with those found astronomically. Careful checks revealed no errors in either triangulation or astronomical observations. Several of the discrepancies were far too large to be accounted for by errors in so careful a survey.

Two stations, Kaliana and Kalianpur (Fig. 2-10), are among those studied by Archbishop Pratt, a British cleric and amateur mathematician, who examined these discrepancies. Kaliana is on the Indo-Gangetic plain, close beneath the towering Himalayas; Kalianpur lies far to the south, near the center of the peninsula. Pratt surmised that the plumb line at Kaliana would be deflected northward by the mass of the mountains, so that the latitude difference between the stations would be less than that computed from triangulation. This proved true:

Difference in latitude measured by triangulation	5°23′ 42.29″
Difference in latitude measured astronomically	5°23′ 37.06″
Discrepancy	5.23″

The discrepancy of 5.23″ corresponds to a distance of about 160 meters, far more than could be attributed to surveying error.

Pratt estimated the approximate mass of the Himalayas above sea level and, assuming that the mountains rest on an otherwise uniform crust, he computed how much the plumb line should have been deflected at each of the stations. His results were at first surprising: the northward deflection of the plumb line at Kaliana should have been 27.853″ and at Kalianpur 11.968″. The difference, 15.885″, is more than three times the 5.23″ actually measured and far greater than could be explained by errors in surveying or in estimating the mass of the mountains.

Pratt's Hypothesis of Isostasy

Pratt saw that one or more of his assumptions must have been wrong. A basic one was that the crust beneath both mountains and plains is identical. If the rocks beneath the Himalayas were less dense than those beneath the plains, the discrepancy would be explained. This suggested to him that both mountain and plain are "floating" on an underlying layer of denser material and the heights at which their surfaces stand are inversely related to the densities of the two blocks. In other words, the high-standing Himalayas might be "compensated" by a corresponding deficiency of mass in the underlying rocks.

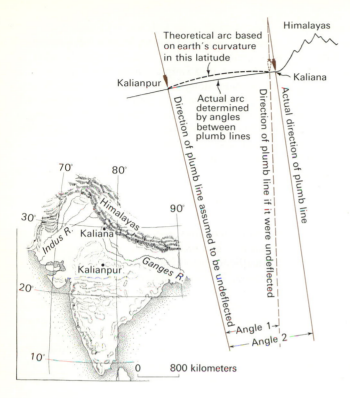

Figure 2-10
The effect of the Himalaya Mountains
on the computed distance between
Kaliana and Kalianpur. Note the
difference in arc distance between the
two cities, produced by the deflection
of the plumb bob, so that angle 1 is
greater than angle 2. The angles are
greatly exaggerated.

A simple illustration of Pratt's idea is shown in Figure 2-11. In the upper diagram, blocks of four different metals, each weighing exactly the same and with identical cross sections, float in a pan of mercury, a liquid of very high density (13.6 g/cm³). Since the weights and cross sections of the blocks are identical, all will sink to the same depth; their surfaces stand at heights inversely

proportional to the ratios of their densities to that of mercury. For ease of computation, let us assume that each block weighs 136 g and has a cross section of 10 cm². Each sinks until it displaces 136 g of mercury; that is, 1 cm. The lead block (density 11.4) is 13.6/11.4 cm long and thus stands with its surface 0.19 cm above the mercury; antimony (density 6.6) rises 1.06 cm above it. Pratt assumed that mountains, plains, and ocean floors have similar relations.

Pratt's was the first formulation of a theory of isostasy, and his scheme of "compensating" for differences of elevation by variations in density of crustal rocks is known as **Pratt's Hypothesis of Isostasy**. Pratt did not use the word "isostasy"; it was coined thirty-four years later by the American geologist C. E. Dutton, from Greek roots for "equal standing."

Airy's Hypothesis of Isostasy

The same volume of the *Transactions of the Royal Society* (1855) that contains Pratt's formulation contains a brief note by G. B. Airy, the Astronomer Royal of Great Britain. Airy

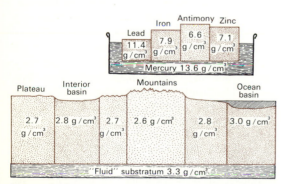

Figure 2-11
Pratt's theory of isostasy. The densities shown in the lower diagram were not specified by Pratt, but are based on modern estimates. [Modified from William Bowie, *Isostasy*, E. P. Dutton, 1927.]

said that Pratt's conclusions should have been anticipated because it could be shown that no rocks are strong enough to sustain loads as great as plateaus and high mountains; they would crush and flow laterally until balance were restored. Mountains thus must be floating masses, as Pratt thought, but Airy suggested a different mechanism of flotation. He saw no reason why rocks underlying a mountain should differ from those beneath a plain. If the rocks have the same density but unequal thicknesses and float on a denser substratum, they would also stand at unequal heights. The height of the mountain is compensated by a "root" of its own material that projects downward into the "fluid" layer and displaces it. Airy's is the **Roots of the Mountains Hypothesis** of isostasy. It accounts for the "errors" in the Survey of India just as well as Pratt's, and accords much better with what we know of rocks in general. It operates more widely in minor crustal adjustments, but, as we shall see, Pratt's hypothesis must be called on to explain the differences between continents and ocean basins, and in many other places. Neither mechanism operates to the exclusion of the other.

A simplified view of the Roots of the Mountains hypothesis is shown in Figure 2-12. In the upper diagram, several blocks of copper of identical density and cross-sectional area but differing heights float in a pan of mercury. The thickest block stands highest and also extends deepest into the mercury.

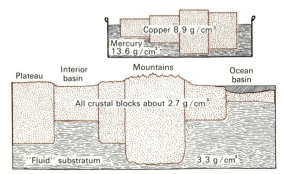

Figure 2-12
Airy's Roots of the Mountains Theory of Isostasy. The densities shown in the lower diagram were not specified by Airy, but are modern estimates.
[Modified in part from C. R. Longwell, *Geological Review*, 1925.]

The diagrams in Figures 2-11 and 2-12 are obviously greatly oversimplified. Good evidence presented later shows that the earth's crust is not divided into simple blocks free to move past one another along frictionless boundaries. Nor is the substratum a fluid, although it does respond by plastic flow to loads of long duration, reacting virtually as a viscous fluid. But the crust is not without considerable strength; loads must exceed a threshold value before it yields.

THE GEOID AND THE SPHEROID

We have seen how arcs of meridian measured at different latitudes forced abandonment of Eratosthenes' spherical earth in favor of an oblate spheroid. But the facts just discussed show that further refinement is needed.

Though mountains do not deflect the plumb line as much as we might expect, they nevertheless do deflect it. Even the surface of the sea cannot be a perfect oblate spheroid. It is warped by the pull of the continents, islands, and submerged ridges. This irregular surface, which is of course everywhere normal to the plumb line, is called the **geoid** (Figs. 2-9, 2-13). The geoid may be thought of as the surface of the ocean and, in continental areas, as the surface of the water in a system of narrow sea-level canals that we imagine to be cut through the land.

The geoid—the imaginary level surface corresponding to sea level over the whole earth—obviously does not correspond to any regular mathematical figure. Satellite orbits are measurably disturbed by irregularities in the geoid. With the great improvement in accuracy of measurement made possible by the laser, the flattening at the poles, mentioned on page 6 as being internationally accepted in 1967 (a value of 1/299.75), has been corrected to 1/299.255 by a team from the Smithsonian Institution (Fig. 2-13). The Smithsonian Standard Earth shows upward bulges near New Guinea (81 m), near the Gold Coast (31 m), in the North Atlantic (61 m), in Peru (30 m), and south of Madagascar (56 m). Downward dimples exist off the coast of Baja California (45 m), south of India (113 m), and southeast of New Zealand (73 m), with many minor deviations. The North Pole stands about 25.8 m above a truly ellipsoidal surface with the above flattening, and the South Pole about 25.8

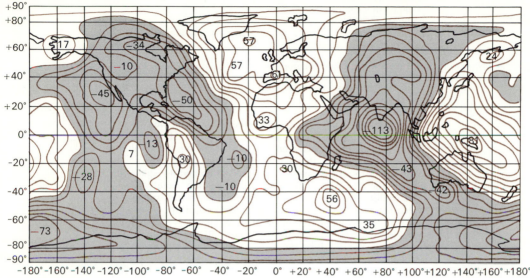

Figure 2-13
Geoid heights in meters relative to an ellipsoid of flattening 1/299.255. [Smithsonian Standard Earth, 1969.]

m below it, so that the earth's shape is somewhat like that of a pear. Furthermore, the equator is not a true circle, but bows a bit outward at 10°W and 140°E longitude and bows a bit inward at 120°W and 80°E.

Although these deviations are trivial compared to an earth radius, they are much too large to permit using the geoid as a geodetic reference surface. Geodesists have therefore adopted, as their reference surface for mapping, an oblate **spheroid** that is as close to an average geoid as possible. Though the deviations of the geoid from the spheroid are indeed small, they are highly significant in our understanding of the Earth's interior. We return to them in Chapter 8.

GRAVITY MEASUREMENTS AND ISOSTASY

At any point on the spheroid, the force of gravity theoretically depends only on the latitude. Both distance to the earth's center and the centrifugal force of rotation are identical everywhere on any parallel of latitude of a true oblate ellipsoid. With pendulum or gravimeter we can measure the actual force of gravity, but to compare this with the theoretical value, we must take several factors into account. These are (1) the altitude above or

below sea level; (2) the gravitational pull of the rock between the station and sea level (if the station is above sea level) or the lack of pull of the rock (if the station is below sea level); and (3) the attraction of the nearby topography, as we saw in the Indian example. Clearly, many adjustments and computations must be made if we are to compare the measured and theoretical values of gravity in any meaningful way (Fig. 2-14).

Thousands of measurements on land and sea have brought out a most significant fact: *on the average, the materials of the earth beneath high land are less dense than those beneath low, if the computation allows for enough depth.* This is what Pratt and Airy found from the deflections of the plumb line—a completely independent method.

The procedure is as follows: The value of gravity measured at a land station is compared with the theoretical value at a point vertically beneath it on the spheroid. By convention, the theoretical value is subtracted from the measured value as corrected for disturbing factors. The difference is the **gravity anomaly** for the station. In computations of each of the several kinds of gravity anomalies, one or more disturbing influences is taken into account. The first correction is

18

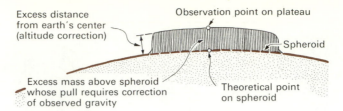

Excess distance from earth's center (altitude correction)

Observation point on plateau

Spheroid

Excess mass above spheroid whose pull requires correction of observed gravity

Theoretical point on spheroid

Figure 2-14
Factors involved in correcting observed gravity for comparison with the theoretical value by the Bouguer method. Height above the spheroid, taken alone, diminishes the observed value as compared with the theoretical (greater distance to earth's center), but the rock mass between the station and the spheroid adds to the attraction. When both corrections are made, most land stations yield negative anomalies (that is, the measured value, after correction, is less than the theoretical; the more negative the value, the higher the station), and most sea stations yield positive anomalies. This is so commonly (though not invariably) true that it becomes strong evidence for general isostasy.

for altitude: the measured value is increased to the computed amount it would have if it were possible to make the measurement on the spheroid, closer to the earth's center. Subtracting the theoretical value from this corrected value gives the "**free-air anomaly**." It is equivalent to assuming that the rock between station and spheroid has no mass—thus the "mountains are empty eggshells" assumption. But of course, the rock does have mass and adds its pull to that measured. Accordingly, when corrected in this artificial way, values measured nearly everywhere above sea level are greater than the theoretical. Free-air anomalies are thus positive for nearly all land stations; it would be amazing if they were not.

When, more realistically, we correct the measured value by subtracting the theoretical pull of the rocks between station and spheroid (having already made the free-air correction), we have a **Bouguer anomaly**, named for the French geodesist who first suggested this after measuring an arc of meridian in Peru. Bouguer anomalies offer strong support for the theory of isostasy: they are chiefly negative on land, and, in general, the higher the station, the greater the negative Bouguer anomaly (Figs. 2-15, 2-16). Conversely, sea stations generally have slightly positive Bouguer anomalies. The correlation is by no means perfect, but the tendency is unmistakable.

Neither of the two corrections applied in arriving at the Bouguer anomaly can be seriously in error. We can measure altitude more accurately than gravity; even though an average rock density is conventionally used rather than one specific for the region, the possible error here is also small. So too with the disturbing effect of nearby topography, except in rugged mountain areas.

The theoretical value of gravity on the spheroid is based on the average of the whole earth. That Bouguer anomalies are generally negative in mountainous areas, and more strongly negative the higher the mountains, indicates that the rocks that underly mountains are on the average less dense than those that underly lowlands. As Pratt and Airy found from the Indian Survey, *the excess mass of the mountain rocks above sea level is in general compensated for by a deficiency of mass below*. Figure 2-15 shows this correlation for the Alps; Figure 2-16 shows it for a very much larger region. There are, of course, deviations from the rule, but its generality is obvious.

Conversely, the positive anomalies at sea may be explained by the existence of underlying material denser than the average of that beneath the continents. In Chapter 8 we shall discuss many other reasons for accepting this interpretation.

Deflections of the plumb line and gravity measurements both show that the larger segments of the earth are roughly in floating equilibrium (isostatic balance) one with another.

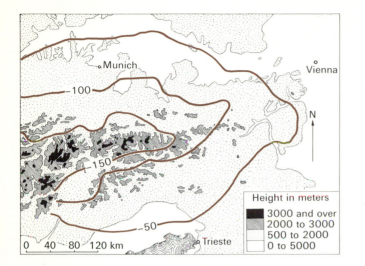

Figure 2-15
The topography and Bouguer anomalies of the Alps. Note the general relations of the topography (shown by shading) to the lines of equal anomaly (shown in brown). The highest negative anomalies are in the mountains. [After Paavo Holopainen, 1947.]

In the image: Munich, Vienna, Trieste, N

—100

—150

—50

0 40 80 120 km

Height in meters
3000 and over
2000 to 3000
500 to 2000
0 to 5000

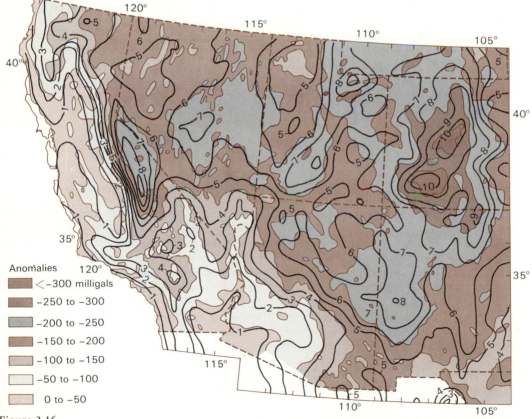

Anomalies
< −300 milligals
−250 to −300
−200 to −250
−150 to −200
−100 to −150
−50 to −100
0 to −50

Figure 2-16
Bougeur anomalies of the southwestern United States (shown by patterns) superposed on a generalized topographic map with altitudes averaged over about 6000 km². Contour interval approximately 300 m. One milligal equals 0.001 gal (named for Galileo). One gal corresponds to an acceleration of 1 cm/sec. A milligal is thus about one millionth of the normal value of gravity at sea level. [Bougeur anomalies from map by American Geophysical Union and U. S. Geological Survey, 1964.]

Large areas of high land stand above large areas of low land because they are underlain by materials of lower density (Pratt), by materials of greater thickness and equivalent density (Airy), or by materials in which both factors are involved—probably the most common case for large mountain chains.

STRENGTH

The isostatic tendency suggests that the earth has a plastic interior that buoys up crustal irregularities by flotation. But the rocks with which we are familiar behave not at all like fluids. Mountain peaks do not seem to be flattening out under their own weight, and as we see in Figures 2-15 and 2-12, good evidence shows that some areas are not in isostatic equilibrium and are being sustained in nonequilibrium positions by the strength of the earth's crust or by forces operating on them. Must we conclude from these evidences of surface strength and bodily weakness that the earth has a strong solid crust floating on a fluid interior? We shall see that there is cogent evidence that the subcrust is only liquid in very small part and that under short-term stresses it behaves as though it were twice as strong as steel.

What is meant by "strength"? Airy long ago stated that a solid or rigid earth's interior could not possibly support large masses, no matter how strong we think the rocks to be; the subcrust must behave plastically. Moreover, where deep erosion exposes metamorphic rocks, their contorted parting surfaces (Chapter 5) and fantastically complex fold patterns show that they have flowed and recrystallized, and now retain patterns of the sort we can duplicate by deforming putty or toothpaste. The folds in many unmetamorphosed rocks (Chapter 8) show that large masses of rock can deform plastically even near the earth's surface.

Definition of Strength

The **strength** of a body is defined as the force (load) per unit area that is required to break it under some specified temperature and pressure. Thus if a cube of granite one centimeter in diameter breaks under a weight of 2200 kg, it has a *compressive strength* of 2200 kg/cm². A steel cable with a cross section of 1 cm² that breaks under a load of 10,000 kg has a *tensile strength* of 10,000 kg/cm². Every solid material requires a definite force that must be reached to make it break. This is its *rupture strength*.

Fluids, both liquids and gases, have no strength; they yield continuously under the slightest stress, though they vary greatly in the speed of yielding. At first glance, tar seems "stronger" than water; an iron bar will not sink into it so quickly. Yet the bar does sink, and will eventually reach the bottom; the tar is simply more viscous than water—it has no strength.

Most solids are not brittle; they begin to deform at lower stresses than the rupture strength, and some yield greatly before breaking. If the stress drops below a minimal value—the *yield stress*—they cease to deform. Yield stresses of all substances differ at high temperatures and pressures from those at room conditions.

EFFECT OF TEMPERATURE

Temperature greatly influences yield stress. At red heat, a bar of iron is solid but will flow at stresses much lower than those needed to make it bend at room temperature—a fact utilized in forging iron. The general rule is that all solids are weaker at high temperatures.

EFFECT OF CONFINING PRESSURE

All solid substances become stronger when placed under confining pressures. If the pressure is equal on all sides, it is *hydrostatic pressure*. Laboratory tests show that the crushing strength of the fine-grained limestone from Solnhofen, Germany, is about six times as great under a hydrostatic pressure of 10,000 kg/cm² as it is at room conditions—10,200 kg/cm² instead of 1700. Such a pressure is equal to the weight of a column of granite about 36 km high. Presumably this is the hydrostatic pressure at a depth of 36 km in the earth. Despite the increased strength at high pressures, the limestone should yield at greater stress.

From temperatures measured in drill holes and mines, we know that the earth is hotter inside than at the surface. Clearly, factors of opposite tendency affect rock strength at depth: the increased pressure tends to increase strength, the increased temperature to diminish it. There is nothing really inconsistent in an earth whose rocks are "rigid and strong" enough to sustain large relief features at its surface, yet so "weak" as to react plastically under large loads. We must think to scale.

EFFECT OF SIZE, OR SCALE

In everyday life we rarely think of how size affects strength of materials. A dramatic illustration of the effect of scale has been given by M. King Hubbert in the quarry operation illustrated in Figure 2-17. Suppose we are to quarry a single block of flawless granite the size of the State of Texas. It is about 1200 km across, and we want to hoist a piece about a quarter as deep as it is broad, say 300 km. Grant that we have a crane capable of hoisting it and that the rock has a crushing strength of average granite, about 2500 kg/cm². Will the block hold together during the hoisting? Obviously we cannot test this directly, but we can investigate the properties of the block by using a scale model.

To make a convenient model we could reduce the length of 1200 km to 60 cm; the model would be on a scale of 1 to 2,000,000. The thickness would be 15 cm. Because both original and model are at the earth's surface, the force of gravity is unchanged. Our model can be of the same density as granite, roughly 3 g/cm³. If the model is to behave the same mechanically as the original, the strength of its material must be reduced in the same ratio as the size; the material must have a strength of 1/2,000,000 that of granite, or about 1 g/cm². It is hard to envisage so

Figure 2-17
Quarrying the State of Texas. [After M. King Hubbert; reproduced by permission, American Association of Petroleum Geologists, 1945.]

weak a solid. With crushing strength of 1 g/cm² and a density of 3 g/cm³ any column higher than 1/3 cm would collapse of its own weight. Yet the model planned is 15 cm thick and its weight 81 kg. The pressure on the base would be about 45 g/cm² or 45 times the crushing strength of the model material!

Thus if we tried to lift the model in the way shown in the figure, the eyebolts would pull out; if we tried to support it on a pair of saw-horses, its middle would collapse; were we to place it on a table, the sides would fall off. To lift it at all would require a scoop shovel. That this is not unreasonable can be verified by calculation on the original block—the State of Texas. Hubbert pointed out that "the pressure at its base would exceed the crushing strength of its assumed material by a factor of 45. The inescapable conclusion, therefore, is that the good State of Texas is utterly incapable of self-support."

Questions

1. What evidence can you give that the earth is approximately an oblate spheroid, and not shaped like a football?

2. What assumptions underlie Eratosthenes' measurement of the Earth? Neglecting measuring errors, are there reasons for doubting the validity of these assumptions?

3. When we sight through an accurately leveled telescope, are we sighting parallel to (a) the geoid, (b) the ellipsoid, or (c) the spheroid? Where, in general, would you expect the three surfaces to be most nearly parallel?

4. How would you expect the plumb line to be deflected from a line at right angles to the spheroid at Denver, just east of the high Rockies? If surveying failed to show such a deflection, what reason could account for this?

5. Geologists commonly assume that the hydrostatic pressure at depths of a few kilometers is due to the weight per unit area of the overlying rocks. How can this assumption be justified?

6. There is a very large negative isostatic anomaly at Seattle, nearly at sea level. Within less than 30 km east and west of Seattle, there is practically no anomaly. What suggestions can you offer to explain these facts?

7. Would a gravity pendulum swing faster or slower at the surface than at the bottom of a deep mine shaft?

8. In a certain valley in Turkestan, the plumb line is deflected toward the center of the valley rather than toward the bordering mountains. Suggest a possible explanation.

Suggested Readings

Daly, R. A., *Strength and Structure of the Earth,* Englewood Cliffs, N. J.: Prentice-Hall, 1950. [Especially the Introduction and Chapter 1.]

Hubbert, M. K., *Strength of the Earth.* American Association of Petroleum Geologists Bulletin, v. 29, 1945, pp. 1630–1653.

Poynting, J. H., *The Earth.* Cambridge, England: Cambridge University Press, 1913.

Scientific American, *The Planet Earth.* New York: Simon and Schuster, 1957.

Scientific American Offprints

268. Marcus Reiner, "The Flow of Matter" (December 1959).

273. George Gamow, "Gravity" (March 1961).

812. W. A. Heiskanen, "The Earth's Gravity" (September 1955).

873. Desmond King-Hele, "The Shape of the Earth" (October 1967).

CHAPTER 3

Materials of the Earth– Minerals and Matter

SYSTEMS AND PHASES

Chemists and physicists deal with **systems**— arbitrarily designated parts of the physical world that can be analyzed in terms of their component **phases.** *A phase is a homogeneous part of a system, separated from other parts by physical boundaries*, such as the interfaces between a gas, a liquid, a solid or between two solids. For example, a bottle containing air, water, and a floating chunk of ice is a system that comprises three phases. If the system is heated, so that the ice melts, it is reduced to two phases; if it is further heated, so that the water evaporates, it is reduced to a single phase—a homogeneous gas composed of air and water vapor. In this kind of phase change, **change of state**, we see that the stability of the phases at surface pressures depends on temperature.

Besides change of state, other kinds of phase change take place. One is the **phase transition**, a change in internal structure of a substance in the solid state. The American physicist Percy Bridgeman compressed ice in a pressure chamber and converted it successively into half a dozen denser phases, each stable at different temperatures and pressures. Another kind is **solution**. A crystal of common salt in a beaker alongside a cup of water is a stable system of two phases—salt and water. But if the water is turned into the beaker, the salt dissolves and the system consists of a single phase, the brine.

These simple illustrations can be generalized to apply to all phases in nature: *each is stable only under certain ranges of temperature, pressure, and chemical environment, called its field of stability*. Many phases can exist outside of their fields of stability because of the sluggishness of many phase changes; these are called *metastable phases* because they tend to change, and are not in *equilibrium* with their environment.

Examination of soils and nearly all rocks shows them to be composed of one or more distinct phases we call **minerals**. In the common rock granite, for example (Fig. 3-1), the naked eye can commonly recognize three or four distinct phases and under a microscope a few more can usually be seen. Obviously, if we are to understand either soils or rocks, we must learn something about their constituent phases. We defer the formal definition of a mineral, though, for we must turn to a more fundamental subject, the properties of matter in general.

ATOMS

In 1805 the English chemist John Dalton noted that different substances react chemically with each other only in constant ratios of their weights.

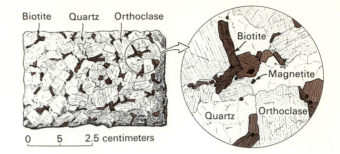

Figure 3-1
The mineral constituents of granite.
Note that the mineral magnetite is
visible only under considerable
magnification.

He suggested that all matter is made up of minute individual particles that he called **atoms**. Further work in chemistry, physics, and mineralogy has verified the atomic theory completely. We will not review the advances in detail but merely summarize some elements of atomic theory now generally accepted.

Atoms are extremely minute, ranging from 2 to 5 Ångströms* in diameter.

Forty million atoms in line would only extend 1 cm; a cube 1 cm on a side contains about 6×10^{22} atoms. No one has ever seen a single atom, yet ingenious experiments have shown that, small as an atom is, it is made up of still smaller particles. Three of the many subatomic particles now recognized determine the mass and chemical behavior of atoms: the **proton**, the **neutron**, and the **electron**.

Subatomic Particles

Electrons, neutrons, and protons are all of about the same extreme minuteness—about 10^{-12} cm (a millionth of a millionth of a centimeter) in diameter. Every electron carries an identical negative electric charge whose amount has been agreed upon as the international unit of electricity. The proton has a precisely equal positive charge; the neutron, as the name implies, is electrically neutral. Neutrons and protons also differ in weight: the proton weighs 1836 times as much, and the neutron 1837 times as much, as the electron.

Each of these three particles is constant in properties from one atom to another. All atoms

*An Ångstrom, named after the Swedish physicist A. J. Ångstrom (1814–1874), is 1×10^{-8} cm, or 10^{-10} m.

have protons and electrons in equal numbers and are thus electrically neutral. The chemical behavior of an atom depends on the number of protons in its nucleus—the **atomic number**. The number of neutrons, though, is not generally fixed. Most elements consist of several **isotopes**; that is, all their atoms have identical numbers of protons but varying numbers of neutrons. Thus each isotope of an element has chemical properties identical with those of every other isotope, but their masses differ.

Structure of the Atom

Every **atom** has a minute nucleus that contains one or more protons and, except in the most abundant isotope of the simplest element, hydrogen, one or more neutrons. The nucleus contains more than 99.9 percent of the mass of any atom, but occupies only about a billionth of its effective volume; the outer parts of the atoms are mostly empty space.

Atoms behave as though their electrons move around the nucleus in concentric shells. Hydrogen has one proton, around which an electron revolves in complex orbits at a constant distance; its isotopes, deuterium and tritium, also have single protons and electrons but have, respectively, one and two neutrons. They therefore weigh nearly twice and thrice, respectively, as much as an atom of hydrogen. Helium, the next simplest element, has a nucleus with two protons and two neutrons. Its two electrons move in complex orbits of uniform radii (Fig. 3-2).

More complex atoms, whose larger nuclei contain more protons, retain an inner orbital shell of two electrons like that of helium; additional electrons—enough to balance the number of

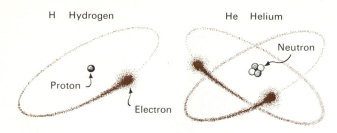

H Hydrogen

He Helium

Proton

Electron

Neutron

Figure 3-2
Schematic drawing of hydrogen and helium atoms. The orbits are diagrammatic, as the positions of electrons cannot be located except as to orbital radii.

protons—are situated farther out, in one or more additional shells. The heaviest atoms have as many as seven such shells.

Each element is assigned a definite symbol, such as H for hydrogen and Pb for *Plumbum,* Latin for "lead." This shorthand is used in chemical formulas. The elements, together with their symbols, atomic numbers, and atomic weights, are listed in Appendix IV.

For some reason, *the elements whose outermost electron shells contain eight electrons are chemically the most stable.* These, the *inert* or *noble gases,* neon, argon, krypton, xenon, and radon, are never found in nature combined with each other or any other elements. Helium, with only two electrons, is also an inert gas. (Chemists have recently been able, under rigorous laboratory conditions, to make some inert gases combine with other elements, but no natural compounds of any are known.)

Except for helium, any atom whose outermost electron shell contains less than eight electrons can combine with others to form **compounds.** The chemically most active atoms are those whose outermost shells contain either one or seven electrons. For example, sodium (Na) has only one electron in its outer shell; chlorine (Cl) has seven. Both are chemically very active. Sodium has only to lose and chlorine to gain one electron for each to attain the stable number of eight. Combined, they form the *compound* sodium chloride (NaCl), common table salt. An electron from the sodium atom is transferred to the chlorine atom. The compound differs greatly from its component elements. After losing an electron the sodium atom is no longer neutral, in fact is no longer an atom! Having one more proton than it has electrons, it has a positive electric charge. Charged atomic nuclei are **ions.** Because *unlike charges attract* and *like charges repel each other,* the sodium ion (positive) is

attracted to the chlorine ion (negative) to form the electrically neutral compound NaCl.

We can now define minerals: *A mineral is a naturally occurring solid phase possessing a characteristic internal structure determined by a regular arrangement of the atoms or ions composing it, and with a chemical composition and physical properties that are either fixed or that vary within definite ranges.* These properties characterize all specimens of any individual mineral, whether one came from France and another from Peru, or whether one crystallized on a coral reef and another in a hot spring. Minerals are *natural objects,* not man-made. Synthetic laboratory products, even though identical in physical and chemical properties, are not minerals. Synthetic topaz and synthetic diamond are not minerals.

The definite character of a mineral is its internal structure, the systematic arrangement of its component atoms or ions. This is the core of the definition, as we shall see in discussing diamond and graphite.

FORM AND STRUCTURE OF MINERALS

Geometric Form

Some mineral crystals are familiar to all: garnet, quartz (rock crystal), and ice. Perfect crystals have strikingly regular geometry. Their surfaces (crystal faces) are planar and meet at sharp angles (Fig. 3-3). But such regular crystals are rare. Most snowflakes show marvelous six-fold symmetry, but hailstones show no crystal faces, merely concentric structure. But enough perfect crystals have been studied along with the much commoner imperfect ones to enable mineralogists to determine the internal symmetries of

26

Figure 3-3
Common minerals that show good crystal form.
[Photos by Alexander Tihonravov.]

hundreds of minerals (and consequently the shape of their building units) long before physicists and chemists had proved the atomic theory and worked out the internal details of crystal structure.

Constancy of Interfacial Angles

Nicolaus Steno (1631–1687), a Danish monk living in Florence, pioneered several aspects of geology, including the first important study of crystal form. With only crude instruments, Steno was able to show that the long faces of all elongate quartz crystals always meet at angles of 120°, regardless of the gross shape of the specimen (Fig. 3-4). Guglielmini, an Italian, demonstrated a few years later similar constancy for other minerals, though the characteristic angles differ from one mineral species to another. Most specimens of halite, for example (Fig. 3-7), have interfacial angles of 90°, thus forming cubes or other box-like bodies.

Later workers improved on these early works and showed that the constancy of angles was a general phenomenon. From it, the French student René Haüy inferred that this could only mean that every crystal is built of minute particles, each of which has similar angular shape (Fig. 3-5). The packing determines the interfacial angles and is identical in all specimens of a species, though the gross form may vary because

the particles do not pack uniformly in all directions. To their immense credit, the mineralogists of the early nineteenth century were able to deduce, from interfacial angles alone, all the internal patterns of crystals now recognized with the aid of much more elaborate equipment.

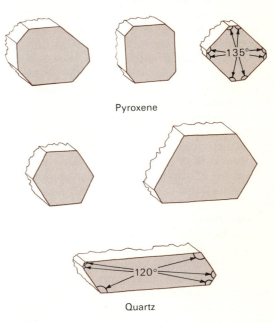

Pyroxene

Quartz

Figure 3-4
Outlines of crystals, showing constant interfacial angles in crystals of varying gross form.

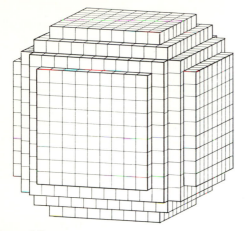

Figure 3-5
Haüy's idea of the structural units that compose a cube of halite or galena, showing how each unit has the characteristic angles.

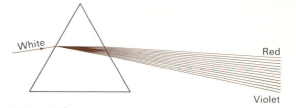
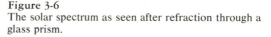

Figure 3-6
The solar spectrum as seen after refraction through a glass prism.

RADIANT ENERGY

Since the time of J. C. Maxwell (1831–1879), physicists have considered the radiation of light to be an electromagnetic phenomenon in which energy is transferred by wave-like vibrations moving in all directions at right angles to the direction of propagation. (The medium in which the vibrations take place used to be called "the ether," but no such material exists; in the theory of relativity, time has become part of the fourth dimension called "space-time." But our intuition does not generally encompass such a concept, and even physicists retain an intuitive affection for the word "ether" as the vibrating medium.) Wavelengths in electromagnetic phenomena range continuously over a vast *spectrum*, from "gamma rays" of less than 0.1 Ångstrom to electric currents of several millions of Ångstroms; those between 3,800 Å (violet) and about 7700 Å (red) are able to excite the human retina as visible light.

From high school physics, we remember that a triangular prism refracts (bends) the different wavelengths in sunlight at different angles (Fig. 3-6). With a sensitive thermometer it is possible to show that waves other than those visible to the eye also exist, both at longer wavelengths (*infrared*) and shorter (*ultraviolet*). Other methods are of course necessary to detect radiations

of different wavelengths. We have mentioned these features here because of the importance of the concepts in further advances in mineralogy.

Optical Properties of Crystals

Long before Haüy, the Dutch physicist Christiaan Huygens (1629–1695) had discovered the "double refraction" of the mineral calcite (Fig. 3-7). If a clear rhomb of calcite is placed over a dot on a surface, two dots are seen. Revolving the crystal makes one dot trace a circle about the other. More than a century later, physicists found that this could be explained by division of the light ray into two by the internal structure of the crystal. Instead of vibrating in all directions, as does ordinary light, each of the divided rays vibrates in a single plane—its *plane of polarization*—which is at right angles to the plane of polarization of the other. Clearly, this can be explained only in relation to the systematic arrangement of the internal structure of the crystal.

It is now known that each nonopaque crystal has its own individual internal structure and, consequently, its own unique *optical properties*,

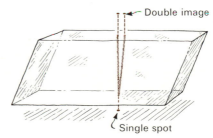
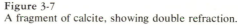

Figure 3-7
A fragment of calcite, showing double refraction.

which permit its identification with a properly equipped *petrographic microscope*. The techniques involved are beyond the scope of this book but are powerful means for the study of minerals and rocks.

X-ray Study of Crystals

X-rays, another form of electromagnetic radiation, are produced in an evacuated glass tube in which a high-voltage direct current impinges on a metal plate (Fig. 3-8). The rays radiate in all directions and, because of their extremely short wavelength, can penetrate to some extent objects that are opaque to ordinary light. A minute hole permits a bundle of nearly parallel X-rays to emerge from the tube and strike a target. Rays diffracted by the target are recorded on photographic film.

Max von Laue (1879–1960), a German physicist, suggested in 1912 that since crystals were thought to be composed of particles arranged in closely packed parallel planes, and if X-rays are like light rays except for their apparently much shorter wavelengths, a thin crystal might diffract X-rays in much the same way that a narrow slit diffracts visible light. Two of his students placed a crystal of copper sulfate pentahydrate in an apparatus much like that shown in Figure 3-8 and produced the first "Laue X-radiogram," thus verifying von Laue's surmise.

Two British physicists, W. H. Bragg (1862–1942) and his son, W. L. Bragg (born 1890), promptly followed up this lead and showed that,

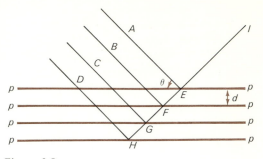

Figure 3-9
Interference and reinforcement of X-rays in a crystal lattice.

unlike light rays, which reflect from a mirror at all angles of incidence, X-rays have such short wavelengths that they are reflected only at angles that depend on the pattern and spacing of the particles composing the crystal.

A simplified 2-dimensional model is shown in Figure 3-9. (The principle is the same for real 3-dimensional crystals, though much more complex.) Layers of regularly spaced particles in a crystal are represented by the lines labeled p, spaced at uniform distances d from each other. The parallel X-rays A, B, C, and D impinge on layers at points E, F, G, and H, and all are reflected along HI, thus reinforcing each other. This reinforcement can take place only if the reflected rays are precisely *in phase*. Thus rays $D(DHG)$, $C(CGE)$, and $B(BFE)$ must be an integral number of wavelengths long: $n\lambda = DHG$, $m\lambda = CGE$, $l\lambda = BFE$, in which λ is wavelength and n, m, and l are integers.

If the path of the reflected ray is not precisely an integral number of wavelengths long, the reflected waves will interfere and tend to cancel each other; if they are in phase, a strong reflection reaches the photo film. The path BF is obviously related to the interplanar spacing, d, and the glancing angle θ (theta) by the relation

$$\sin \theta = \frac{n\lambda}{2d},$$

where n is an integer.

This equation is Bragg's Law, derived later the same year in which von Laue's experiment laid the foundation for the analysis of crystals by X-rays. These techniques are invaluable in mineral studies, for they can be applied to grains so small as to be hardly visible microscopically.

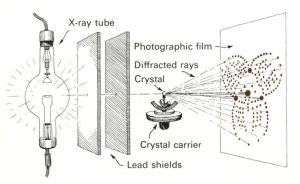

Figure 3-8
Sketch of apparatus used in obtaining an X-ray diffraction pattern. The crystal is calcite.

Figure 3-10
The cubic form (right) and internal structure of halite. The lattice diagram (left) shows the relative positions of the Na and Cl ions. They can be distinguished because the heavier Cl ions reflect more strongly than the lighter Na ions, thus producing more intense photographic imprints. The packing arrangement is depicted in the center. [Photo from the Smithsonian Institution.]

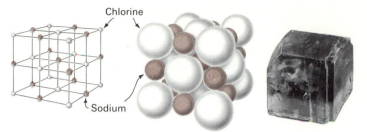

Internal Structure of Crystals

X-ray studies give us an accurate picture of internal packing arrangements. For example, they show that crystals of halite (NaCl) have the structure illustrated in Figure 3-10. We have already seen that in forming halite the lone electron in the outer shell of the Na atom is transferred to the Cl atom, thus giving each nucleus a stable outer shell of eight electrons and converting the former atoms to ions, held together by their unlike charges. In a liquid or gas, the chlorine and sodium ions would be drawn together to form **molecules**. (Molecules are distinct groups of ions closely bound together.) But the close packing in the crystal fixes each ion in its place; no two can be considered as making a discrete molecule.

The geometrical packing maximizes the attractions of ions of unlike charge and the repulsions of those of like charge. Thus each positively charged sodium ion is equidistant from six symmetrically placed chlorine ions; each negatively charged chlorine ion is surrounded by six symmetrically placed sodium ions. Each ion is said to have *six-fold coordination*. Most minerals are held together by similar **ionic bonds**, though most have internal structures far more complex and less easily visualized than that of halite.

Some minerals—diamond, for example—are held together by shared electrons. Diamond is pure carbon, an element with four electrons in the outer shell. Each carbon atom is bonded to four others in such a way that each electron of the outer shell is shared with an adjacent atom (Fig. 3-11). Thus each atom can be considered to have a complete outer shell of eight electrons, though each electron is oscillating from the orbit of one atom to that of its neighbor. Such bonds

are **covalent**; there are no ions. Covalent bonding of the carbon atoms in diamond is so strong that diamond is the hardest natural substance known. The bonding in some solids is intermediate between ionic and covalent bonds, partly one and partly the other.

Diamond and graphite illustrate that the fundamental distinction between minerals is not chemical but structural. Both are pure carbon; their differing crystal structures account for their dramatically contrasting physical properties. Diamonds are superlatively hard, and most are transparent; all graphite is soft, greasy, and opaque. Diamond is used as an abrasive and in cutting tools; graphite as a lubricant whose fine flakes glide smoothly one over another.

Ionic Radii

As in halite, each ion in a crystal tends to be surrounded by ions of opposite charge. The number of these surrounding ions—the **coordination number**—is determined partly by the charge on the ion but more importantly by the *sizes* of the several ions involved.

An atom's size depends largely on the number of its electron shells. When an atom becomes ionized, the gain or loss of an electron upsets the electrical balance and changes the radius. A positive ion, having lost an electron, pulls the electron shells a little closer because of its excess nuclear charge. The extra electron on a negative ion allows the shells to expand. Ionization that changes the number of electron shells acts similarly: Neon, which is electrically neutral, has the same number of electrons in the same shells as the sodium ion, but the excess charge on the sodium ion pulls its shells in and makes it

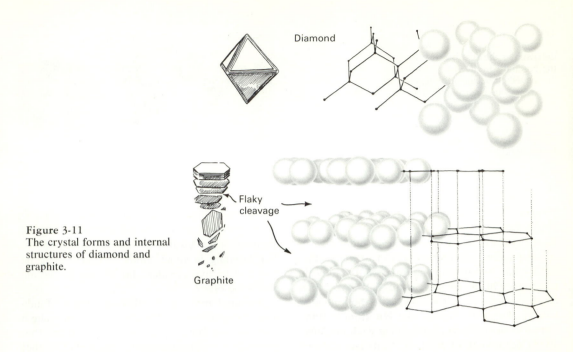

Diamond

Flaky
cleavage

Graphite

Figure 3-11
The crystal forms and internal
structures of diamond and
graphite.

smaller. Table 3-1 gives the atomic and ionic
radii of the nine most abundant elements of the
earth's crust. Sizes of atoms and ions are measured by X-ray methods based on Bragg's Law.
Note the consistent differences between atomic
and ionic radii.

The abundance of the elements tabulated has
been estimated as follows: Many thousands of
chemical analyses have been made of the different rocks of the earth's crust. Geologic maps
(Chapter 6) permit estimates of the areas occupied by the various rocks and thus the relative

Table 3–1
Most abundant elements in the earth's crust.

Atomic number	Element	Atomic size and ionic radii in Ångströms		Abundance in crust	
		Atom	Ion	Weight percent	Volume percent
8	Oxygen	O 0.60	O⁻⁻ 1.40	46.60	93.77
14	Silicon	Si 1.17	Si⁺⁺⁺⁺ 0.41	27.72	0.86
13	Aluminum	Al 1.43	Al⁺⁺⁺ 0.51	8.13	0.77
26	Iron	Fe 1.24	Fe⁺⁺ 0.74 Fe⁺⁺⁺ 0.64	5.00	0.43
20	Calcium	Ca 1.96	Ca⁺⁺ 0.99	3.63	1.03
11	Sodium	Na 1.86	Na⁺ 0.95	2.83	1.32
19	Potassium	K 2.31	K⁺ 1.33	2.59	1.83
12	Magnesium	Mg 1.60	Mg⁺⁺ 0.65	2.09	0.29
22	Titanium	Ti 1.46	Ti⁺⁺⁺ 0.76 Ti⁺⁺⁺⁺ 0.68	0.44	0.03

SOURCES: Data on abundance from Brian Mason, *Principles of Geochemistry*, John Wiley &
Sons, New York, 1966 (3rd edition). Data on atomic and ionic radii from L. H. Ahrens, *Geo-
chimica et Cosmochimica Acta*, v. 2, pp. 155–169, 1952; from Jack Green, *Geological Society of
America Bulletin*, v. 64, pp. 1001–1012, 1953; and from Linus Pauling, *The Nature of the Chem-
ical Bond and the Structure of Molecules and Crystals* (3rd edition), Cornell University Press,
Ithaca, New York, 1960.

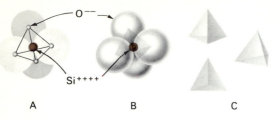

Figure 3-12
Three representations of the silica tetrahedron.
(A) Lattice diagram. (B) Atoms drawn to scale.
(C) Geometric tetrahedra.

abundance of each variety and the elements composing it.

Table 3-1 shows that oxygen and silicon make up nearly three-fourths of the mass of the crust, and so must predominate among crustal minerals. Most minerals are *silicates*, composed of oxygen, silicon, and one or more of the metals listed. Volume relations are still more disparate: though oxygen forms only forty-seven percent of the mass of the crust, its ion (O^{--}, 1.40 Å) is so large that oxygen fills 92 percent of the volume. Thus more than nine-tenths of the crustal volume is filled by an element we usually think of only as an atmospheric gas!

The silicon ion (Si^{++++}) has the much smaller radius of 0.42 Å. In silicates, four oxygen ions and one silicon ion form a four-cornered, four-sided configuration, the *silica tetrahedron* (Fig. 3-12). The nucleus of the silicon ion is at the center of the tetrahedron and the nucleus of an oxygen ion at each corner. The silica tetrahedron is an ion, not a molecule, for the association of one silicon ion (4 positive charges) and four

oxygen ions (8 negative charges) leaves the tetrahedron with four unsatisfied negative charges. In an electrically neutral crystal, the tetrahedron must be bonded with enough positive ions for neutrality—for example, with two ions of magnesium (Mg^{++}) or of iron (Fe^{++}), as in the mineral olivine (Fig. 3-14).

More than nine-tenths of the earth's crust is made of silicates, and except for the extremely rare stishovite (Chapter 19), all siliceous minerals are made of silica tetrahedra bonded to one or more other elements. *The silica tetrahedron is the most important "building block" in the architecture of the earth's crust.* The tetrahedra are linked in different patterns in different minerals. In quartz, for example, every tetrahedron is linked by its corners to another identical tetrahedron. Every oxygen ion is thus shared between two Si ions; the oxygen-silicon ratio is 2:1, and the formula is SiO_2. In some minerals the silica tetrahedra do not join but are held together by ions of metals, such as Mg or Fe, which bond the corners of two separate tetrahedra, as in olivine (Fig. 3-13).

In some minerals two tetrahedra share a single oxygen ion and thus form a group of composition $Si_2O_7^{6-}$, whose unshared O^- ions are linked to metallic ions. Mica has rings of tetrahedra joined at common O corners and linked by metal ions between the rings (Fig. 3-13,B). The common mineral pyroxene consists of tetrahedra strung in long chains (Fig. 3-13), and in olivine and monticellite (Fig. 3-14) they form a framework tied together by ions of Mg, Fe, and Ca.

The German mineralogist Hugo Strunz has classified the silicates according to the arrangements of the silica tetrahedra (Table 3-2).

Table 3-2
The classes of silicate minerals.

Class	Arrangement of tetrahedra		Representative minerals
Nesosilicates	Separate tetrahedra	(SiO_4)	Olivine
Sorosilicates	Double tetrahedra	(Si_2O_7)	Melilite (rare)
Cyclosilicates	Ring arrangements	(SiO_3)	Beryl
Inosilicates	Single chains	(SiO_3)	Pyroxene
	Double chains	(Si_4O_{11})	Amphibole
Phyllosilicates	Sheets	(Si_2O_5)	Mica, Kaolinite
Tectosilicates	Frameworks	(SiO_2)	Feldspars

SOURCE: After Hugo Strunz.

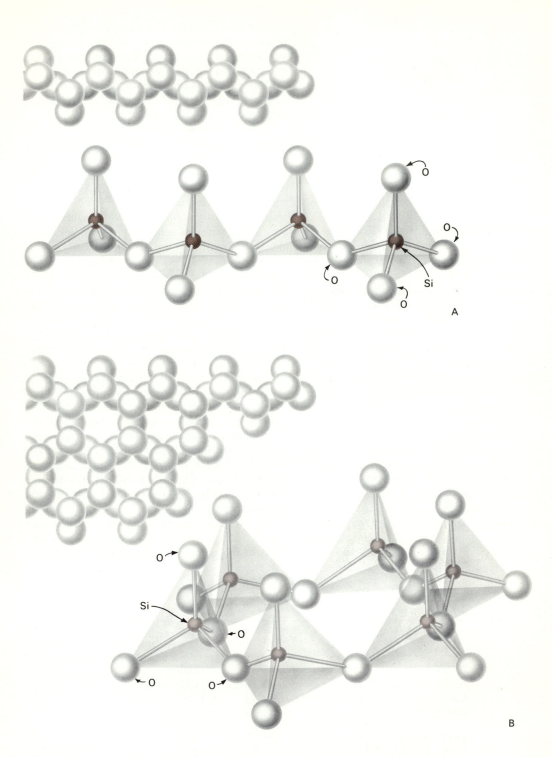

Figure 3-13
Diagrams showing how the silica tetrahedra form chains in pyroxene and double plates whose unshared electrons face each other in mica. (A) Pyroxene (side view). (B) Mica (top view).

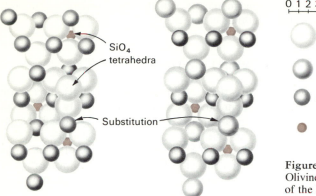

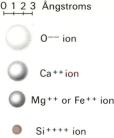

0 1 2 3 Ångstroms

◯ O^{--} ion

◐ Ca^{++}ion

● Mg^{++} or Fe^{++} ion

● Si^{++++} ion

SiO$_4$
tetrahedra

Substitution

Olivine — (Mg, Fe)$_2$SiO$_4$ Monticellite — MgCaSiO$_4$

Figure 3-14
Olivine and monticellite, showing how substitution
of the larger calcium ion for an iron or magnesium
ion requires an expansion of the crystal lattice.
[After H. W. Bragg, 1928.]

VARIATIONS IN CHEMICAL COMPOSITION OF MINERALS

Some minerals, diamond (C), sulfur (S), copper
(Cu), and gold (Au), are elements. Others, ice
(H$_2$O), quartz (SiO$_2$), calcite (CaCO$_3$) and kaolinite (H$_4$Al$_2$Si$_2$O$_9$), are compounds whose composition can be expressed in simple formulas.
Others, as noted in our definition, vary within
specified limits in their composition: their compositions cannot be expressed in simple formulas because one element may substitute for
another in the crystal lattice, which remains
virtually unchanged. Such substitution is **solid
solution**.

A simple example of solid solution is the olivine group of minerals. The formula, written
(Mg,Fe)$_2$SiO$_4$, states that different specimens
may have compositions anywhere between one
end member, pure Mg$_2$SiO$_4$ (forsterite), and the
other, pure Fe$_2$SiO$_4$ (fayalite). The proportions
of Mg and Fe may vary widely; those of Si and O
remain constant. The intermediate members are
considered solid solutions of the end members,
for they are all single phases without physical
boundaries between the components. Solid solutions are ionic structures, as now discussed.

Mechanism of Substitution in Solid Solutions

Joint chemical and X-ray studies show how one
element substitutes for another. The major factor
is not, as might be thought, the number of electrons in the outer shells — the chemical *valences* —
but their ionic radii. Iron and magnesium can
readily substitute for each other in olivine, as
they each have valence +2, but more important,
closely similar ionic radii. In many mineral
groups sodium (valence +1) readily substitutes
for calcium (valence +2) because their ionic radii
are very similar (0.95 and 0.99 Å, respectively),
although their valences differ. But sodium cannot
substitute to nearly the same extent for potassium, even though they have identical valences,
because the radius of the K$^+$ ion (1.33 Å) is so
much larger, despite the close similarity in chemical properties of the two elements.

In plagioclase feldspar a little K$^+$ may indeed
substitute for Na$^+$, the lattice warping slightly to
accept the larger radius, but beyond a very small
warp the lattice becomes unstable; it breaks up
into tiny interlocking crystals of two distinct
minerals: K-feldspar and plagioclase. Warping of
the crystal lattice is seen in Figure 3-12, where
the substitution of Ca^{++} for about half the Fe^{++}
and Mg^{++} ions in olivine produces the slightly
different structure of monticellite, which is rather
similar chemically.

Clearly, an ion of valence 1 cannot substitute
for another of valence 2 without destroying electrical neutrality; a concurrent substitution in the
opposite sense must take place. Thus in the
plagioclase solid solutions, the gradation from
pure albite (NaAlSi$_3$O$_8$) to pure anorthite
(CaAl$_2$Si$_2$O$_8$) takes place by the simultaneous
exchange of both Ca^{++} and Al^{+++} for Na (valence
+1) and Si^{++++} (valence +4). The valences
balance, as $3 + 2 = 4 + 1$. The ionic radii of Na$^+$

and Ca^{++} are nearly the same; that of Al^{+++} is close enough to that of Si^{++} (Table 3-1) that the slightly distorted lattice is not unstable. Though the structural changes are slight, they produce readily recognizable changes in plagioclase crystals of differing compositions.

MINERALOIDS

Some natural substances that lack orderly internal structure and so are not true minerals are nevertheless commonly grouped with them as **mineraloids**. A familiar example is opal, most specimens of which lack orderly structure; another is amber.

IDENTIFICATION OF MINERALS

More than 2,000 different minerals have been recognized, but only about twenty are at all abundant. Most of these can be identified at sight by anyone who carefully observes their physical properties. Appendix II describes the methods of sight identification and lists the properties of twenty common minerals and of ten others of economic importance.

Questions   *(Based in part on Appendix II)*

1. What is the essential difference between a mineral and an animal? Between a mineral and a rock? Between a mineral and a chemical element?

2. What controls the external geometric form of crystals? How was this discovered?

3. Name three subatomic particles and give their chief characteristics.

4. State the distinctions between elements, ions, atoms, crystals, molecules.

5. What is the difference between a crystal and a molecule? Why does the concept of molecules fail to apply to crystals?

6. Define solid solution; what factors are involved, and which dominate the phenomenon?

7. How do the differences of internal structure account for the different physical properties of diamond and graphite: cleavage, hardness, specific gravity?

8. Why is the specific gravity of quartz definite (2.65) whereas that of pyroxene varies (3.2 to 3.6)?

9. Why is the streak (the color of finely powdered material) of some minerals more diagnostic than the color of a large piece?

10. Describe at least three various arrangements of silica tetrahedra in different minerals.

Suggested Readings

Bragg, W. L., Claringbull, G. F., and Taylor, W. H., *Crystal Structure of Minerals.* Ithaca: Cornell University Press, 1965. [Chapters 1 and 2.]

Ernst, W. G., *Earth Materials*. Englewood Cliffs, N. J.: Prentice-Hall, 1969.

Mason, Brian, *Principles of Geochemistry*. New York: John Wiley & Sons (3rd edition), 1966.

Scientific American Offprints

249. G. H. Wannier, "The Nature of Solids" (December 1952).
260. R. L. Fullman, "The Growth of Crystals" (March 1955).
262. A. M. Buswell and W. H. Rodebush, "Water" (April 1956).

CHAPTER 4

The Materials of Geology–Rocks

To a geologist, any aggregate of minerals, whether so loosely coherent as to crumble between the fingers or so tightly cemented together as to require an explosive to disintegrate it, is a rock. By somewhat arbitrary convention, soils (Chapter 10) are generally treated separately. Though soil may conceal the underlying rocks over wide areas, a well boring, a foundation excavation, or a roadcut almost anywhere is likely to reach rock at no great depth.

We have stressed the physical and chemical properties of minerals. Much of **petrology**, the study of rocks, is also concerned with these matters. But rocks are much more than mineral aggregates; they are also historical documents that record events far back in earth history. Although the mere composition of some rocks is enough to betray some of the conditions of their origin (as with coal, Chapter 18), and the physical properties of others alone suffice (as with glacial ice), the geologist usually needs a combination of several or many attributes of a rock and its relations to other rocks in order to define the conditions of its origin and thus interpret another bit of earth history.

SEDIMENTARY ROCKS

Everyone has seen muddy rills form on a steep slope during a rainstorm and deposit conical piles of mud, sand, and gravel on gentler slopes below. Be it rill or giant river, every stream sweeps such debris downstream, intermittently dropping much of it in sand and gravel bars, only to pick it up again at the next high water and carry it farther—eventually to the sea or to a closed basin.

The ancient Greeks recognized that beds of sand and gravel high above flood level resulted from such action by similar streams at earlier times.

Noting the presence of clam shells in loose sand high above the strand, they concluded that either the land must have risen or the sea fallen. Perhaps Omar Kayyám was the first to assert, late in the twelfth century, that *firmly cemented sandstones* containing sparse shells were marine deposits, even though they were found on high mountains in Iran, far from the sea. Omar was severely punished for heresy; the Koran teaches of an unchanging world. The learned world of the West at this time was also dominated by churchmen of similar convictions; to them fossils were not of organic origin but were "sports of nature" emplaced by the devil to deceive mankind. Even Georg Bauer (Agricola, 1494–1555), the brilliant student of medieval mining in Saxony, denied that fossil shells were remains of animals, although he did recognize the organic origin of fossil leaves and fish skeletons.

The same Nicolaus Steno who discovered the Law of Interfacial Crystal Angles observed that the "tongue stones" found in the cliffs of Malta and regarded as "collector's items" for centuries were virtually identical in shape and composition with the teeth of a modern shark he had dissected. He took the common sense view that they were indeed shark's teeth, and that the rocks forming the Maltese cliffs were of marine origin. The loose beach sand and the firm sandstone differ only in *cementation*, the filling of the interstitial voids by calcite, which holds the grains together. More than a century later, when geologists began to make geologic maps (Chapter 5), they found many examples of the lateral gradation of weakly consolidated sand into firm, well-cemented rock.

Cementation

Only slight magnification enables us to see that the individual grains of a fossiliferous sandstone (Fig. 4-1,C) have the same shapes and about the same mineral composition as the sands of a modern beach (Fig. 4-1,A). Though fossil shells in the sandstone differ from those of living animals, they are so similar in detailed structures as to compel the conclusion that they are indeed organic. Although some amino acids are probably of inorganic origin, several others that are almost surely formed only by living organisms have been found in fossil shells several hundred million years old (Chapter 6).

Sandstone is representative of the **sedimentary rocks**—rocks formed at the earth's surface, either by accumulation and cementation of fragments of rocks, minerals, and organisms or as chemical precipitates and organic growths in water. Many rocks other than sandstone are sedimentary; equally surely, many other rocks have very different origins.

Characteristics of Sedimentary Rocks

SORTING

Swift water currents can move larger pebbles and sand grains than slow ones. Accordingly, when a current slows down as a flood recedes or a wave retreats, the coarser material is first to be left at rest and the finer last. In short, the sediment being carried by a stream or current is *sorted according to grain size* as the current slows. But streams are not the only carriers of sediment. Winds carry dust and sand, and "dust devils" carry material high into the air. When these currents slow and stop, a similar sorting takes place as the wind-borne materials settle to earth.

Wind whisks the dust from sand but cannot move coarse boulders; the grains of sand in a dune are thus very well sorted according to size. Water currents are generally not so efficient in sorting sediment, but glaciers are even less so; the sediment dumped at the melting terminus of a glacier is almost completely unsorted, though of course the parts reworked by the meltwater streams immediately downslope become somewhat sorted.

ROUNDING

As material is rolled along a stream bed or bounced along the ground before the wind, the friction and impact of particles colliding with each other and scraping against the stream bed or the ground wear away sharp corners and round the particles into smooth sand grains and rounded pebbles. The coarser material in streams accumulates in gravel bars that slowly migrate downstream with each successive high-water stage. The farther the travel, the smoother and rounder the pebbles. Thus the degree of rounding gives a clue to the distance the sediment has traveled.

STRATIFICATION

Most sedimentary rocks are distinctly layered. The layering—called **stratification**, or **bedding**—divides the rock into individual **strata**, or **beds**,

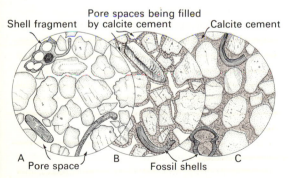

Shell fragment | Pore spaces being filled by calcite cement | Calcite cement

A   Pore space   B   Fossil shells   C

Figure 4-1
Stages in the cementation of sand, as seen under the microscope. (A) Loose sand from an Oregon beach. (B) Partly cemented sandstone from near a Brazilian coral reef. (C) Completely cemented sandstone from Ohio.

which record variations in current velocity. Beach material is modified by every storm with the varying power of the waves. A pit dug by a child in a beach nearly always reveals distinct layers of varying grain size. A great storm may sweep away a layer of sand and hurl a sheet of coarse gravel in its place. Volcanic ash is sometimes erupted and spread widely over the sea floor, forming a distinctive bed, and severe storms may leave a layer of mud over a shelly bed. Such interruptions in rate and kind of deposition are preserved by burial and give rise to obvious bedding. Because of the irregularity of natural processes, some beds are many meters thick and others less than a millimeter.

Stratification is an important guide to conditions of sedimentation. A lake deep enough so that wave disturbance does not reach the bottom generally has thinly bedded, fine-grained mud, reflecting either annual or seasonal variations in stream inflow. Cuts through sand dunes show fantastically complex lamination (Fig. 4-2); lower sets of curving layers are cut off and overlain by others at different inclinations and directions—a feature repeated many times. Stream bars of sand and gravel also show steeply inclined bedding though not usually in such irregular patterns. Cross-bedding of many kinds is preserved in most sandstones (Figs. 17-5, 17-7, 17-8).

Figure 4-2
Cross-bedding in dune sand of the San Luis Valley, Colorado, exposed by excavating wet sand. [Photo by E. D. McKee, U. S. Geological Survey.]

Steno also formalized the common-sense **Law of Superposition**: *In any pile of sedimentary strata not strongly folded or tilted, the youngest stratum is at the top, the oldest at the bottom.* Obviously, each stratum had to have a base on which to come to rest.

Both of these laws are obvious truisms, but they are not trivial; our understanding of the make-up of the earth's crust, the history of the earth, and the evolution of life depends almost entirely upon them.

Laws of Sedimentary Sequences

Some sediments come to rest on steep alluvial cones built where mountain torrents spread out upon reaching a valley flat. These cones slope as much as 15°, and their crude bedding therefore has a corresponding slope, like that of some storm-deposited beach gravels. But most sediments are deposited on much gentler—indeed, virtually imperceptible—slopes. This generalization was another enunciated by the brilliant Steno in his **Law of Original Horizontality**: *Waterlaid strata are deposited nearly parallel to the surface on which they accumulate, and thus nearly horizontally.* This applies to all waterlaid sediments with the minor exceptions of those that exhibit marked cross-bedding. The term cross-bedding refers to the internal layering that crosses the main strata of stream-laid deposits at various angles.

Classification of Sedimentary Rocks

Sedimentary rocks are classified on the basis of **texture** (size and shape of component particles) and **composition** (of both particles and cements). Sandstone is cemented sand; similarly cemented gravels are **conglomerates**. Very fine-grained **siltstones** and **mudstones** are compacted silt and mud, respectively; if they part readily along surfaces nearly parallel to bedding, they are **shales**. Mudstones, siltstones, and shales are collectively called **pelites** (from *pelos*, Greek for "clay"). Conglomerates, sandstones, and pelites are **clastic** (from *klastos*, Greek for "broken") rocks, as they are mainly made up of broken and worn fragments of older minerals, rock particles, or shells deposited from water, wind, or ice and then cemented or compacted. Clastic particles range from many meters down to microscopic in size; they may be sharply angular to nearly spherical. Though other minerals abound, the coarser monomineralic particles are generally quartz, a mineral more resistant to both chemical and mechanical attack than most others. The cements that bind most coarse clastics together are calcite, clay, quartz, and limonite; zeolites, gypsum, and halite are less common cements. Many of the finer-grained rocks have no cement but are held together by interlocking clay particles.

Some sedimentary rocks are made up of reef-forming organisms, such as coral and associated algae. Deposits of marine animal shells broken by waves and later cemented together are called coquinas. Some limestones are neither organic nor clastic but are **chemical deposits**, precipitated directly from solution (though organic activity may have altered the composition of the water to bring this about). The most abundant chemical sediments are **evaporites**, precipitated by evaporation of waters in landlocked basins. The principal products are halite, anhydrite, and gypsum. Most organic and chemical sedimentary rocks contain more or less clastic material, washed or blown into the evaporating basin.

The reader should refer to Appendix III for more details about sedimentary rocks. Incidental information follows in almost every chapter of this book, but knowledge of the contents of Appendix III is hereafter assumed.

IGNEOUS ROCKS

Volcanic Rocks

Late on Christmas Eve in 1965, fissures opened on the southeast flank of Kilauea volcano, Hawaii. While the ground shuddered from swarms of small earthquakes and a few large ones, lava gushed from some of the new fissures, formed a pool about 15 meters deep in one of the satellite craters, and flooded wide areas of forest. Fissuring and quaking continued for several hours until early Christmas morning when a huge area of the south flank of the volcano dropped two or three meters and much of the still-fluid lava began to drain back into the source fissures; lava that remained at the surface chilled to form the rock basalt (Appendix III). By 4 A.M. the eruption was over, though for several days earth-quakes continued and clouds of vapor rose from fissures and deposited sulfur on their walls. Evidently the whole south side of the volcano had slid southward about a meter or two with respect to the summit, opening fissures into which the lava drained.

Early on a raw November morning in 1963 a fishing crew off the southwestern coast of Iceland smelled sulfurous fumes and, when dawn came, saw two or three dark columns rising like smoke from the sea. The columns were soon identified as volcanic ash—fine, porous fragments of lava quenched in the sea. By noon they rose more than 4 km and soon united into a single curtain more than 400 m long. The next day an island rose 10 m from the sea. The island must have been composed entirely of unconsolidated ash, for a single storm cut it back fully 100 meters. In April lava emerged and flowed intermittently for more than 15 months. By this time, the island (named Surtsey in honor of a Norse mythological hero who hailed from the south) was 170 m high and covered more than 2 km². Two other neighboring islands appeared before activity ceased in late 1965; both must have been built entirely of ash, for they were soon destroyed by the waves, leaving the ash and lava isle of Surtsey as the sole survivor of the eruption.

Many other volcanoes emit great volumes of fragmental lava, improperly but generally called ash, though it is not a product of combustion. (Interestingly, the term "tephra"—not herein used—has recently been unnecessarily introduced as a synonym for ash; but since this word is itself derived from the Greek for "ash," the long-entrenched error is perpetuated rather than rectified.) Pumice (highly vesicular lava) and ash exploded from Vesuvius in A. D. 79, overwhelming the nearby cities of Pompeii and Herculaneum so catastrophically that whole families perished together in almost life-like postures.

The Mediterranean volcanoes taught the ancients that lavas congeal and ash compacts to form solid rocks, but for many centuries such rocks in areas distant from active volcanoes were not distinguished from the associated sedimentary rocks. Ash from later eruptions has washed down the slopes of Vesuvius, and in places stream-laid beds have buried the ash that destroyed Pompeii. In Samoa, lava has been seen to spread over coral reefs; today, corals grow on the congealed lava. It is thus not surprising that volcanic rocks interbedded with ordinary sedimentary rocks in areas of inactive volcanism were once thought by many students to be sedimentary.

Actually, the proof of volcanic origin had been presented in the 1760's, before this erroneous suggestion was made. Desmarest, a French amateur geologist, working in the historically inactive volcanic field of the Auvergne in the Plateau Central of France, noted that the soil beneath one basalt layer was hardened and that the massive basalt grades upward into a coarsely porous rock with many small spherical holes—a **scoria**. Scoria is common at the base and still more common at the top of basalt flows. It has been seen to form in flowing lava as expanding bubbles of steam rise and are trapped by the congealing liquid. Desmarest attributed the hardening of the soil beneath the flow to baking by the molten lava, and correctly inferred that the scoria was due to steam bubbles. He traced the flow to its source in a conical hill that, though much eroded, still retains a crater-like depression in its summit. We now know from microscopic studies that the interlocking relation of the component minerals of lava is wholly different from the texture of any sediment; there is no longer a problem in their discrimination (see Fig. 4-3).

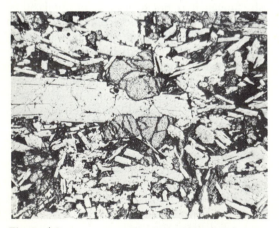

Figure 4-3
A much magnified thin slice of basalt. Compare with Figure 4-1; the contrast in texture permits ready distinction of a lava from a sedimentary rock. [Photomicrograph by Warren Hamilton, U. S. Geological Survey.]

Figure 4-4
Dolerite dike cutting sedimentary rocks on Alamillo Creek, near Socorro, New Mexico. [Photo by N. H. Darton, U. S. Geological Survey.]

Plutonic Rocks

Basalts have been seen to form from lava; they are **igneous rocks** (from *ignis*, Latin for "fire," in error, for no fire is involved in their formation). Molten rock beneath the surface is called **magma**. When it reaches the surface it flows as lava or explodes to form ash; both are **volcanic rocks**, newly formed from subterranean melts.

Where erosion has cut through lava flows it is possible in many places to find the conduits through which they arose—conduits like those that have been seen to open during volcanic eruptions. Some flows grade down into pipe-like masses (**volcanic plugs**) or long, narrow, filled fissures (**dikes**) that cut through the neighboring rocks (Fig. 4-4). A few plugs and dikes have been traced down valley walls into larger bulbous or irregular masses, **plutons**. Some plutons are several scores of kilometers across. Dikes, plugs, and plutons that formed beneath the surface are intrusive bodies. Magma thus crystallized deep within the earth forms **plutonic rocks**. Although a few plutons can be traced into lavas,

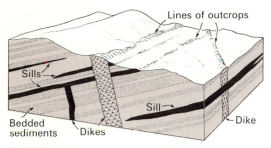

Figure 4-5
Sills are concordant, dikes are discordant tabular intrusions.

most of them appear to have congealed entirely beneath the surface.

Most intrusive rocks are coarser grained than lava and have crystals large enough to be identified with the naked eye or a hand lens. The rocks overlying an intrusive act as a blanket that permits only slow cooling. Laboratory measurements show that rocks conduct heat much more slowly than convection currents in air. The slower the cooling, the more time for ions to migrate within the magma toward centers of crystal growth, and thus the larger the grains.

Many intrusive bodies have features that prove that they formed by crystallization of magmas:

their borders are finer grained than the rest of the mass, suggesting quicker cooling; tongues and stringers of the intrusive cut the bordering rocks, which are commonly changed in mineral content and, almost invariably, in texture by heat from the intrusive. Strata at a distance are not affected. We infer that these *contact metamorphic changes* were brought about by magmatic heat and that the chemical changes locally demonstrable in the contact zone were due to fluids emanating from the magma as it congealed.

Gas bubbles, so abundant in lava selvages, are very small or absent in plutonic masses, because the weight of the roof rocks tends to keep the gas in solution in the magma and the slower cooling allows seepage of the gases through the wall rocks, bringing about the contact changes mentioned. Large bubbles are trapped only under light load, at or near the surface.

Plutons are more fully described in Chapter 9, but one intrusive form, the **sill**, should be discussed here because of its seeming breach of the Law of Superposition. Most intrusives transgress the bedding of associated strata, but sills are virtually concordant with strata above and below, having congealed from magma squirted between two strata like grease from a grease gun (Figs. 4-5, 4-6). A sill low in a sedimentary sequence

Figure 4-6
Sill—lower, dark-gray body—cut by a diagonally sloping mass of dolerite (upper left to lower right) in the Beacon Sandstone, South Victoria Land, Antarctica. [Photo By Warren Hamilton, U. S. Geological Survey.]

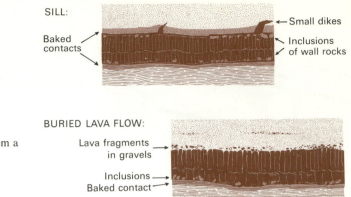

SILL:

Baked contacts → ← Small dikes

← Inclusions of wall rocks

BURIED LAVA FLOW:

Lava fragments in gravels →

Inclusions →
Baked contact →

Figure 4-7
Criteria for distinguishing a sill from a buried lava flow.

may be younger than any of the overlying strata, so that the Law of Superposition cannot be applied to intrusive rocks. How is a sill distinguishable from a lava flow that has been buried by later strata, and to which does the law apply?

A lava flow may contain fragments of the underlying stratum; so may a sill; both may bake the rocks beneath. A stratum overlying a flow may contain fragments broken from the lava, but it obviously cannot be baked by it. A sill may not only bake the overlying bed but also send dike-

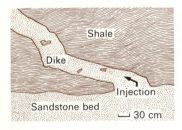

Shale

Dike

Injection

Sandstone bed

⌐ 30 cm

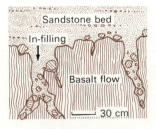

Sandstone bed

In-filling

Basalt flow

⌐ 30 cm

Figure 4-8
(Top) Sandstone dike in the Modelo Formation, Santa Monica Mountains, California. (Bottom) Sandstone infiltrating crevices in an underlying lava flow.

lets into it (Fig. 4-7). Furthermore, an eroded lava flow that has been covered by sand is likely to contain sandstone dikes filling crevices that extend downward from its surface (Fig. 4-8, Bottom); these are the converse of those that sills commonly send into beds above and below. Not all sandstone dikes, however, are made of material younger than the beds they penetrate. Some are made of older material squeezed upward in a semi-liquid state from poorly cemented beds into younger but well-consolidated strata (Fig. 4-8, Top).

Baked contacts are not always readily recognized. Percolating water may stain and impregnate the rocks, making microscopic study necessary for discrimination. Although Desmarest recognized baking beneath the Auvergne lavas, this is exceptional. Many flows creep forward beneath a cooler rind that breaks up and rolls down the front like the tread of a tractor, making a pavement cool enough so that it does not alter the soil significantly. Intrusive masses, though, invariably alter their wall rocks somewhat.

Classification of Igneous Rocks

Although studies with the petrographic microscope have permitted the discrimination of literally hundreds of varieties of igneous rocks, most varieties are very rare and the great bulk of igneous rocks can be included in about 15 major groups (Appendix III; Table III-2). This book ignores most of the rare varieties; most rocks considered here can be identified with only a hand lens of low magnification.

Comparison of the mineral and chemical compositions of many thousands of igneous rocks has shown that mineral composition depends primarily on composition of the magma: silica-rich (*salic*) magmas yield much feldspar and quartz; silica-poor (*mafic*) magmas yield much pyroxene and olivine.

Variations in amounts of glass and crystals and in size and arrangements of the crystals determine a rock's texture, important in classification (Appendix III). Grain size depends chiefly on cooling rate, though chemical composition is a factor. Field studies suggest and laboratory experiments confirm that a high content of water and other volatile substances promotes the growth of large crystals. In most plutons, most crystals are large enough for identification with a hand lens, but lava flows generally cool too quickly to produce large crystals. Many flows and dikes contain sporadic large crystals that formed on the way to the surface; a mixture of such large crystals (**phenocrysts**) in a surrounding **groundmass** of smaller crystals or glass is called **porphyritic**. Very rapidly cooled magma chills to glass, with few or no crystals.

The Granite Enigma

For nearly two centuries, spirited controversy has gone on over the origin of the very common rock **granite**; it still goes on, though it is generally conceded today that the rock can be formed in more than one way.

Mapping has shown that granite and its close relative, granodiorite, are among the commonest plutonic rocks. Many granite masses are hundreds of square kilometers in extent. Many are overlain by pebbly sandstones containing particles weathered and eroded from them (Fig. 4-9). Because the sandstones are clearly younger than these granites, some early geologists erroneously deduced that granite is the earth's oldest rock—part of an "original crust." Careful study of the nonsedimentary contacts of granite masses with

Figure 4-9
Contact between granite and a younger sandstone, El Paso County, Colorado.
[Photo by N. H. Darton, U. S. Geological Survey.]

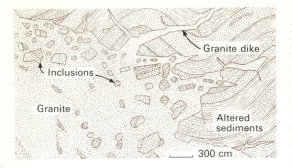

Figure 4-10
Intrusive relations at a granite contact.

their country rocks commonly shows that these masses are finer grained at their borders than in their interiors and that the country rocks have been altered and penetrated by granitic dikes. Granites commonly enclose fragments of the surrounding country rocks (Fig. 4-10). These are unmistakable evidences of magmatic emplacement as intrusives. Diked contacts and traces of magmatic flowage show that many granite bodies are *composite*—formed by successive invasions of slightly different magmas separated by quiet intervals, just as most volcanic rock masses are formed by episodic eruption.

By no means, however, do all granite masses show such contacts. Many contacts are neither erosional nor clearly intrusive, but are gradational, showing gradual changes in mineralogy and texture in distances of hundreds of meters— from rocks that are truly granitic to others that are almost equally coarse-grained but which nevertheless retain unmistakable remnants of sedimentary or volcanic features.

Some granitic rocks retain faint, nebulous patterns resembling stratification, outlines of pebbles, and other sedimentary structures. Other masses contain thin layers of recrystallized limestone or sandstone only a fraction of a meter thick, projecting from an indefinite wall rock for many meters into the pluton. It is inconceivable that magmas could shatter and engulf wall rocks as shown in Figure 4-10 and yet leave such relatively delicate layers unbroken. We conclude that some originally sedimentary rocks have been transformed into granite by slow recrystallization and replacement without distortion—and thus in the solid state. The beds least susceptible of replacement are relict. Some granites thus appear to be *metamorphic* (from the Greek for "change of form"). Their distinction from magmatic granites is often very difficult, hence a cause of controversy.

METAMORPHIC ROCKS

We have seen that many rocks are readily identified as of sedimentary or igneous origin. Many others, the **metamorphic rocks**, may lack features diagnostic of either origin or their heritage may be obscured by the overprinting of younger structures and new minerals. For example, a rock may show the distinct forms of pebbles and layering that can be seen in a normal conglomerate, but the pebbles appear to be stretched into lozenge shapes and lie in a matrix of neither sand nor clay but of clear quartz and mica. In places long, delicate needles of tourmaline may penetrate both pebbles and matrix (Fig. 4-11).

Such a rock was obviously once a conglom-

Figure 4-11
Metamorphosed conglomerate
with tourmaline needles
cutting other minerals (enlarged
about 4 times).

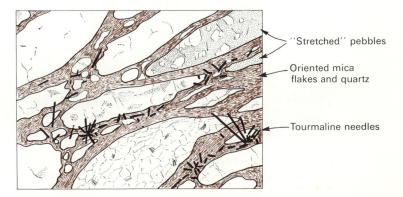

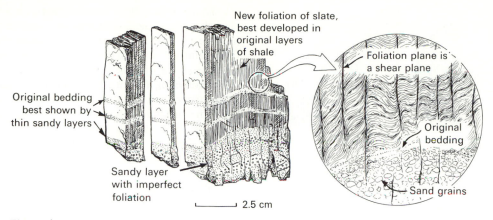

Figure 4-12
Fragments of slate (left) showing foliation (vertical lines) and traces of bedding. The enlargement shows small offsets of the bedding along the surfaces of foliation, the *cleavage*.

erate, but equally obviously, neither the mica nor the tourmaline were present then; they must have grown not only after the pebbles had been squeezed and stretched but during recrystallization of the matrix of sand and clay. The relict structures show that the rock did not melt during recrystallization; it must have remained virtually solid throughout these drastic changes. We must infer that either during the deformation or later, through the action of solutions permeating the original conglomerate, the larger sand grains of the matrix were able to grow at the expense of the smaller, and that the original clay minerals of the matrix became unstable and converted to mica under the new conditions.

The mineral phases that were stable under the conditions of formation of the original rocks became unstable under the new conditions and altered to other phases. As metamorphic rocks commonly contain within themselves unmistakable evidence of deformation and recrystallization, we infer that the metamorphism is due to heat, pressure, shearing, and the chemical effects of contained fluids. Innumerable gradations exist between the parental igneous and sedimentary rocks and the great variety of metamorphic rocks derived from them.

Foliation, and the Origin of Slate

Many metamorphic rocks exhibit a closely spaced layering called **foliation**—a textural feature due to the parallel orientation of some minerals. The layers may be slightly wavy and as much as a millimeter or so thick in a **gneiss**, somewhat thinner in a **schist**, or paper-thin and nearly planar, as in the familiar rock **slate**. Foliated rocks that have not been thermally metamorphosed after the foliation was formed (see p. 191) split readily along the foliation. Foliation seems to record the breaking of original grains by a slow pervasive movement within the rock mass, causing them to be streaked out or rotated into a parallel alignment. These newly oriented "seed crystals" then grew into interlocking fabrics.

Metamorphic foliation has often been mistaken for stratification. Slates commonly, but not always, show two layered structures, of which the more nearly planar set displaces the less regular (Fig. 4-12). The older set parallels changes in grain size, color, and composition, as well as bedding in adjacent, less-foliated rocks. This set is the true stratification; the younger is metamorphic foliation.

In many slates the relict stratification is broken and offset by very small, generally microscopic distances (Fig. 4-12). Some fail to show such offsets, but in most of these the foliation nearly parallels the bedding, so that any displacement would be hard to detect. In a few slates the foliation seems to have formed simply by compaction normal to it, collapsing randomly oriented flaky minerals into parallelism.

Slates split readily along the foliation surfaces but not along the bedding, as do most sedimentary rocks. Where foliation and bedding coincide,

Figure 4-13
The fossil at the left was collected from undistorted limestone in Idaho; the deformed fossil at the right (the identical species), from slate in the Inyo Mountains, California. [Identifications by Professor S. W. Muller, Stanford University.]

recognizable but distorted fossils are occasionally found on cleavage surfaces. Compared with the fossils from unmetamorphosed sedimentary rocks, these are greatly flattened parallel to the cleavage (Fig. 4-13). Where foliation crosses bedding at high angles, most fossils are so broken and displaced by microscopic slips along foliation as to become unrecognizable. Vestiges of stratification, the presence of fossils and the evidence from chemical analyses show that most slates are derived from fine-grained sedimentary rocks, such as shale, but field and microscopic studies and chemical data also show that some were originally tuff.

Slates are extremely fine-grained; many of their mineral particles are ultramicroscopic and identifiable only by means of X-rays; the main constituent is usually muscovite, the white mica, in minute crystals in nearly parallel orientation. Since mica has only one direction of cleavage, this alignment of crystals accounts for the excellent cleavage of the rock. Evidently slate is merely shale or tuff that has been heated and sheared enough to cause the original particles of clay or glass to recrystallize into flakes of mica.

Deep-seated Origin of Metamorphic Rocks

Some slates have been traced along the surface into more coarsely crystallized schists, which retain no vestige of either stratification or fossils. The minerals of these rocks differ from those in slate, gradually replacing and obliterating them

just as the micas in a slate replaced the clay minerals of a shale. We therefore conclude that this schist is merely a more altered sedimentary rock. (Not all schists, however, are *metasedimentary*, as we will see in Chapter 9.)

The common association of highly metamorphosed rocks with plutons that congealed well below the surface under high pressures and temperatures leads us to infer that many metamorphic rocks have been formed at depth. Other evidence lies in the structure of mountains (Chapter 8).

Most of our insight into metamorphic processes comes from studying transitions. We have noted that slates locally grade into more coarsely crystalline schists. Detailed mapping of metamorphic rocks commonly shows fantastically complex patterns of both foliation and compositional banding that record flowage much like that seen in taffy. How could this come about? At the surface these rocks are rigid and brittle. But as we will see in Chapter 13, very similar complex flowage patterns develop in glaciers. Ice, too, is brittle in daily experience: strike a piece with a hammer and you have crushed ice! Yet glacial crevasses are open only to a depth of about 60 m; at greater depth they close by flowage beneath the weight of the overlying ice. We infer that rocks, at some unknown depth, also flow in the solid state. But at what depths and under what conditions—temperature, pressure, permeating solutions—are the different kinds of metamorphic rocks formed?

Experimental Study of Metamorphic Processes

Experiments are increasingly employed to gain evidence on the physical conditions under which the various kinds of metamorphic rocks form. Ingeniously devised presses can squeeze minerals under pressures equivalent to the rock load at a depth of more than 80 km, which is probably deeper than any rock seen by man ever has been (Chapter 8). The pressure is therefore more than adequate to simulate metamorphic conditions. Rocks have also been sealed in "bombs" and subjected to high temperatures and pressures, both in the presence and absence of water vapor. Under such conditions many minerals become unstable and convert to new phases, either by reacting with other minerals or by undergoing internal rearrangement. Even diamonds have been made artificially.

Geochemists, who draw on the techniques of physical chemistry and solid-state physics, use such experiments to work out some of the chemical reactions that go on during metamorphism. But minerals are complex: many are solid solutions with widely varying properties, and despite the brilliant beginnings that have been made in this difficult field, knowledge of the chemistry and physics of solid silicates lags far behind that of the chemistry of fluids. It was not until 1953 that the common mineral muscovite was synthesized from clay minerals in the presence of water vapor—the major reaction in the transformation of shale to slate and schist.

Experiments have shown that under surface pressures, ordinary quartz ("low" quartz) inverts from one space lattice to another at a temperature of 573°C. At 870°C the new form, called "high quartz," inverts to a still different space lattice—that of the mineral *tridymite*, which is stable up to the temperature 1470°C. Above that temperature it inverts to another mineral, *cristobalite*, which is stable up to the melting point of silica at 1713°C. On cooling, the truly stable forms in the various temperature ranges are the same as those found on heating, but the inversions are so sluggish that in nature we find, in different specimens, glassy silica (really stable only above 1713°C), cristobalite, tridymite, and high and low quartz all persisting at room temperature. Except for low quartz, these forms are metastable at room temperature and pressure; they are outside their stability ranges. One of the major difficulties in experimental work with silicates is that of distinguishing metastable phases from stable ones, because many reactions are incalculably sluggish.

At different pressures the inversions from one form of silica to another take place at different temperatures. At pressures of 30,000 bars (equivalent to the rock load at a depth of 90 km in the earth) and temperatures appropriate to this depth (Chapter 7), the stable form of silica is the mineral coesite, with a density of 2.93; at the extreme pressure of about 120,000 bars (equivalent to that at a depth of about 360 km in the earth) and at temperatures thought to prevail there, the stable form is stishovite, a mineral with a density of 4.35. Stishovite is unknown and coesite ex-

tremely rare except in certain craters, which are known from other features to have been formed by meteorite impact (Chapter 19).

The physical properties of minerals also vary with the environment. For example, quartz in the presence of water vapor under moderate pressures flows readily at 600°C even though, when dry, its strength at that temperature is comparable to that of steel. The micas, which are hydrous minerals, occur widely in metamorphic rocks, strongly suggesting that water vapor is commonly present during metamorphism. Experiments show that water vapor is highly conducive to the formation of coarse crystals, even of minerals whose crystal lattices embrace no hydrogen.

Temperature, pressure, and chemical environment are thus all important in fixing not only the stability ranges of the various minerals but also their physical properties within these ranges. We are learning much about these matters, but, more than anything, what we have learned in the laboratory about the variability of solid solutions, the sluggishness of many reactions, and the marked effects of even very minor constituents in changing both stability ranges and physical properties serves as a constant reminder of how far we are from a complete understanding of the broad spectrum of igneous and metamorphic reactions that take place deep within the earth. As one winner said in acceptance of the Day medal,* "unless laboratory results are constantly checked against field studies, they may be leading us into a thorough knowledge of a world that never existed!" The precautionary note is valid, but laboratory studies so checked are essential to a real understanding of geological processes.

Thermal Metamorphism

Many limestones contain nodules of chert, an extremely fine-grained rock composed of quartz. Most people are acquainted with flint, the dark variety of chert. Where such a limestone has been highly heated, as near a pluton, the flint nodules have commonly disappeared, and their place is taken by sheaves of the white, fibrous

---

*Awarded by the Geological Society of America for the application of physics and chemistry to the solution of geological problems.

mineral wollastonite. Wollastonite has the composition $CaSiO_3$. It seems clear that the following reaction has taken place:

$$SiO_2 + CaCO_3 \rightleftharpoons CaSiO_3 + CO_2$$
chert limestone   wollastonite carbon
                                      dioxide

In the laboratory, where carbon dioxide is free to escape as a gas, wollastonite forms at 500°C. But in nature the gas is more or less confined, depending on the rock pressure and permeability, so that the reaction is hindered and goes on only at a somewhat higher temperature. We cannot, therefore, accept the presence of wollastonite as a safe guide to the exact temperature. Another obstacle in the use of this reaction as a "geologic thermometer" is that other minerals than silica and calcite may be present and participate in reactions whose temperatures depend not only on pressures but on the composition of the reactants.

The Significance of Metamorphic Rocks in Geologic History

Hundreds of examples could be added to those we have discussed, but they would all show that minerals are stable only within limited ranges of temperature, pressure, and chemical environment. A rock tends to adjust to changes in conditions, though reactions may be so sluggish as to be long delayed. New minerals such as wollastonite may form, but some minerals may simply recrystallize into coarser grains as marble recrystallizes from fine-grained limestone and quartzite from sandstone.

Geologic mapping (Chapter 5), microscopic studies, and laboratory experiments combine to show that most metamorphism takes place deep in the crust, far below depths accessible to direct observation. Where large areas of metamorphosed rock are exposed at the surface, we can therefore be sure that deep erosion has taken place. The working out of depth relations by methods discussed in later chapters shows, however, that slates and nonfoliated thermally metamorphosed rocks near plutons do not require extremely high pressures for their formation. A depth of one or two kilometers may suffice, though they may also form at much greater depths.

Any rock may be metamorphosed in any of several ways; hundreds of varieties have been recognized. For example, basalt (as identified by chemical composition) has been metamorphosed into at least five different kinds of rock, each with its own mineral composition and texture (Chapter 9). The thorough understanding of metamorphic rocks demands a sound knowledge of physical chemistry as well as thorough field and microscopic studies. Nevertheless, the more common metamorphic rocks can be identified at sight, as indicated in Appendix III, Table III-3.

Questions   (Based in part on Appendix III)

1. The bedding in most sand dunes is curved, and some parts slope more than 15°. How do you reconcile this with the Law of Original Horizontality?

2. The long controversy over the origin of basalt took place before the invention of the petrographic microscope. How would microscopic studies, if available, have helped in a solution?

3. Of two intersecting sets of structures in a rock, how can you tell which is the older?

4. The surface temperature of lavas is as high as or higher than that of most intrusive magmas. Evidence of metamorphism is widespread along intrusive contacts but trivial beneath lava flows. Suggest at least three reasons why this is so.

5. How can limestone be distinguished from sandstone? Basalt from granite? Phyllite from shale? Coal from slate?

6. In examining a contact between a granite and an overlying sedimentary rock, what would you look for in trying to decide whether the contact is erosional or intrusive?

7. What is the origin and significance of porphyritic texture? (Refer to Appendix III.)

8. What holds the sand and other mineral fragments together in a sandstone? What holds the mineral grains together in a granite?

9. How can you tell shale from schist?

Suggested Readings

Geikie, Sir Archibald, *The Founders of Geology*. Baltimore: Johns Hopkins Press, 1901. [A fascinating account of the early history of geology.]

Hutton, James, *Theory of the Earth* (2 vols.). Edinburgh, 1788. (Reprinted 1959 by Stechert-Hafner, Inc., New York.)

Huxley, T. H., *On a Piece of Chalk*. London, 1893. (Reprinted 1965 by Scribners, New York.)

Mather, K. F., and Mason, S. L. (eds.), *Source Book in Geology*. Cambridge: Harvard University Press, 1970.

Playfair, John, *Illustrations of the Huttonian Theory of the Earth*. Edinburgh, 1802. (Reprinted 1956 by Dover Publications, New York.)

Scientific American Offprints

101. Philip H. Abelson, "Paleobiochemistry" (July 1956).

803. P. H. Kuenen, "Sand" (April 1960).

819. O. Frank Tuttle, "The Origin of Granite" (April 1955).

846. Loren C. Eiseley, "Charles Lyell" (August 1959).

CHAPTER **5**

The Tools of Geology–
Geologic Maps

Conspicuous in many valleys of the Sierra Nevada are flows of black basalt that rest on the prevailing light-gray granodiorite; similar contrasts of basalt against gleaming white coral rock are found on some Samoan beaches. The abrupt contrast allows such boundaries of rock masses—the geologic **contacts**—to be followed readily. By following the contacts and carefully plotting them on a topographic map, a **geologic map** is made. Even so simple a map as one that depicts only the distribution of basalt may be economically useful. Crushed basalt may be needed for road metal in the Sierra; a good geologic map not only shows where it may be found, but also the relations between contacts and topographic contours, permitting an accurate estimate of the quantity available.

Geologic maps are used in predicting where petroleum, water, coal, iron ore, and other valuable materials buried beneath soil and rocks may be found. The accuracy of such predictions has been tested, and again and again confirmed, by tunneling and drilling. But they have more than economic value; maps are basic to all geology—to deciphering the stages in the development of a mountain range, the steps in the evolution of fossil organisms (Chapter 6), and the climatic fluctuations and other vicissitudes that are part of the long and eventful history of the earth.

DIFFICULTIES OF GEOLOGIC MAPPING

Rock contacts are seldom as obvious as those on the barren High Sierra or the wave-washed shores of Samoa. Indeed, on the fertile prairies of Illinois or the Ukraine, no rock of any kind may be naturally exposed (**crop out**) for many kilometers. Only in the walls of creeks or in man-made excavations can bedrock be found. Geologic mapping may require digging pits or drilling holes to obtain samples of the underlying rocks. Nevertheless, geologic maps were first made in parts of western Europe where outcrops are few.

EARLY GEOLOGIC MAPS

Among the earliest maps to delineate the geology of a considerable area are two of the Paris region, published jointly by the French naturalists Georges Cuvier and Alexandre Brongniart in 1810 and 1822. French scientists had long known that the rocks near Paris are gently tilted strata of limestone, clay, gypsum, and sandstone seen both in natural outcrops and in the many pits dug for porcelain clay. As early as 1782 the great chemist Lavoisier noted that many nearby

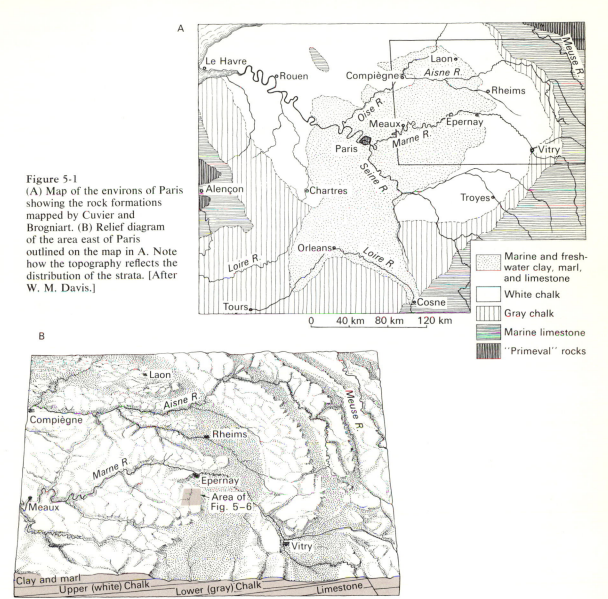

Figure 5-1
(A) Map of the environs of Paris showing the rock formations mapped by Cuvier and Brogniart. (B) Relief diagram of the area east of Paris outlined on the map in A. Note how the topography reflects the distribution of the strata. [After W. M. Davis.]

quarries exposed the same succession of strata, with chalk at the base, overlain in turn by beds of clay, sandstone, limestone, gypsum, impure limestone, and, at the top, siliceous limestone. Cuvier and Brongniart went much further: they noted that some beds contain characteristic fossils, that the color and composition of some beds change gradually as they are traced along their trends, and that the rocks can be classified into recognizable groups that we now call **formations**, distinctive enough to be mapped for considerable distances.

The Rock Succession near Paris

East of Paris low ridges and narrow lowlands lie in a broad arc, partly encircling the city. Each ridge slopes gently toward Paris, and steeply on the side away from the city (Fig. 5-1). Beyond the outermost ridge lies a soil-covered lowland where scattered outcrops and excavations all expose chalk.

This chalk lowland almost completely encircles Paris. The chalk of the outer lowland is gray, with many thin layers of pale green sandstone. The chalk nearer Paris is white and porous; much of it encloses potato-shaped nodules of black flint, but interbedded layers 5 or 6 m thick contain no flint.

The bedding of both the gray and the white chalk slopes gently downward toward Paris, showing that the white chalk rests upon the gray; a well started in the white chalk enters the gray at depth. Cuvier and Brongniart recognized the two as distinct formations, the Lower Chalk, chiefly gray chalk and greensand (a sandstone containing the green iron-bearing mineral glauconite), and the Upper Chalk (a massive white chalk containing flint nodules and scores of varieties of fossil shells that resemble, but are not identical with, shells of living animals). The Upper Chalk forms the "White Cliffs of Dover" across the channel, and there, too, wells started in it penetrate the gray chalk at depth.

Gullies in the steep hills on the Paris side of the chalk lowland expose plastic clay that rests on the Upper Chalk and is overlain in turn by limestone and marl (clayey limestone); all these strata slope gently toward Paris. Cuvier and Brongniart could find no fossils in the clay and concluded that it was deposited in fresh water. The marine animals that flourished in the sea while the chalk was being deposited could not survive in fresh water. In some places, they found pebbles of chalk in the basal bed of clay, showing that the chalk was coherent before the clay was deposited; they inferred a considerable time interval between deposition of the two rocks.

Cuvier and Brongniart eventually worked out the whole succession of rocks between the chalk lowland and Paris: limestones, sandstones, clay, and gypsum. Some beds contain marine shells, others bones of land mammals, birds, fish, and impressions of leaves. Each formation shows characteristic physical features, and many contain distinctive fossils.

The larger formations retain consistent features throughout the area shown in Figure 5-1, but of course not every bed, nor even every group of beds composing a formation, can be traced con-

tinuously. In places, some beds had been eroded before the next younger was deposited, and others were originally lenticular and discontinuous, like modern sandbars. The record of interbedded marine and land-laid rocks is sure testimony to vertical crustal movements, both up and down. The fact that both rock facies are exposed on the present hilly surface shows that on balance the uplifts have exceeded both the down-sinkings and the rate of land erosion in the vicinity of Paris.

Fossils as Stratigraphic Markers

Cuvier and Brongniart found that they could not always distinguish one limestone from another merely on the basis of color, details of bedding, thickness, or other physical characteristics. But when they considered the fossil content of the strata, they found that: *Each group of closely related strata contains its own characteristic assemblage of fossils.*

This generalization, checked again and again by the Law of Superposition, has proven fundamental in the correlation of isolated outcrops, even across the oceans. For example, virtually the same successions of fossils are found in the sequential strata of England, France, and Morocco. The reason for this relationship was not understood, however, till 1859, when Darwin published *The Origin of Species.*

William Smith's Geologic Map
of England

William Smith worked as a surveyor on many of the canals that were built in England just before the railroad age. He observed the rocks exposed in many miles of canal excavations and elsewhere, and recorded their characteristics, including their fossil content. As a surveyor, he took great pains to locate formation contacts accurately on his map. Starting from a contact exposed in a canal or stream cut, he would follow its approximate position across a soil-covered hill, using as guides the rock fragments in the soil or at the mouths of rabbit holes, until he could again determine the contact's exact position in a natural or artificial exposure. After twenty-four years of such observations, he completed and published a colored geologic map of England—the greatest individual feat in the history of geology as a science.

For many years Smith had been able to predict, from his knowledge of strata and their succession, the kind and thickness of rock that would be encountered in an excavation. His map made such information widely available. But its greatest impact was to show the **stratigraphy**—the sequence and nature of strata—throughout an entire country, thus proving the continuity of individual formations over considerable areas. Smith proved conclusively that if a particular bed lies above another in one exposure, it is never found below it anywhere else unless a structural disturbance has affected one of the beds. This is, of course, simply a consequence of the Law of Superposition. Eventually, the formations Smith and other pioneers established were extended, refined, and grouped into the Standard Geologic Column to which we now refer sedimentary formations the world over (see Table 6-1).

This brief account of early geologic maps hints of the methods and principles used in geologic mapping to this day. We proceed to illustrate them further.

FOUR FUNDAMENTAL POSTULATES OF GEOLOGIC MAPPING

Geologic mapping is based on four basic postulates. We have already discussed two: the Law of Superposition and the Law of Original Horizontality (Chapter 4). A third, the **Law of Original Continuity**, is merely a common-sense deduction from the others, and was also first stated by Steno: *A water-laid stratum, when formed, must continue laterally in all directions until it thins out as a result of nondeposition, or until it abuts against the edge of the original basin of deposition.*

An important corollary of this law, not stated by Steno, was nevertheless widely recognized by the late eighteenth century: *A stratum that ends abruptly at a point other than its depositional edge must have had its original continuation removed by erosion* (Figs. 5-1, 5-2), *or else displaced by a fault* (Fig. 5-10).

Figure 5-2
The Grand Canyon of the Colorado River, in Arizona. [Photo by L. F. Noble, U. S. Geological Survey.]

Underlying all of our stratigraphic inferences are four simple principles—(1) *superposition* (the higher bed is the younger), (2) *original horizontality* (bedding is formed roughly parallel to the earth's surface), (3) *original continuity*, and (4) *truncation only by faulting or intrusion*. They are not absolute rules to be applied perfunctorily. Some beds once horizontal have become highly tilted—even overturned—by movements of the earth's crust, so that a stratum originally beneath another may now lie upside down upon it. Other strata, as at the front of a delta, may have original steep inclination; landslides may end abruptly and not thin gradually. There are other exceptions, but most can readily be recognized by a trained geologist.

FORMATIONS

Problems do arise in applying these postulates, even though, as we have stated, they appear to be truisms. The most important is the selection of mappable contacts. Obviously not every bed can be separately shown, even in areas of completely barren rock; a map that attempted this would be a mass of black unless its scale were unreasonably large. The basic unit of the geologic map is the **formation**. A formation must satisfy two criteria: (1) both top and bottom of a sedimentary or volcanic formation must be recognizable and capable of being traced in the field, and (2) the formation must be large enough to be shown on the map. Thus, on a map of 1/10,000 scale, considerably more (and thinner) formations may be mapped than on a scale of 1/50,000, for example (see Appendix I).

Cuvier and Brongniart could recognize many bedding surfaces within the Lower Chalk, but so similar are the beds above and below them that no bedding surface could be identified in even a nearby outcrop. They therefore could not establish mappable subdivisions of the Lower Chalk. But the contact between the Lower and the Upper Chalk is readily recognizable by the differences in color, flint content, and fossils; it is a mappable contact. The Upper Chalk also could not be subdivided because the beds within it were not distinctive, but the contact with the overlying Plastic Clay is readily recognized and mappable. The Upper Chalk is thus a mappable unit and meets the requirements for its recognition as a formation at the map scale used.

The strata above the Plastic Clay present a more difficult problem. They include many kinds of rock in thin and lenticular beds—limestone, shale, sandstone, gypsum, and clay. Only on an absurdly large scale could these units be separately mapped. Cuvier and Brongniart could do nothing but group them all into a single formation until, on going to higher strata, "rising in the section," another distinctive and truly mappable contact could be identified.

The scale of the map, the abundance and quality of exposures, the distinctiveness of the beds, the intended use of the map, and, by no means least, the discriminating abilities of the geologist determine the selection of map units. Any characteristics that permit a particular stratum or group of strata to be identified in scattered outcrops are adequate to justify recognition as a formation if both top and bottom can be separately mapped. Many formational contacts are gradational through several or even some tens of meters; that is, rocks characteristic of the thicker units of contrasting lithologies above and below, may be interbedded. So long as the transitional strata can be recognized over a considerable area and are not so thick as to require discrimination of their top and bottom at the scale of the map, the transitional zone may itself be depicted as a formation boundary.

In America a geologic formation is nearly always named from a geographic locality near which it was first identified, followed by the name of the dominant rock of which it is composed, or, if various kinds of rock are inseparably interbedded, by the word "Formation." Examples: Austin Chalk, Yakima Basalt, Chattanooga Shale, Denver Formation, both words always capitalized. In Europe, the practice is less formal, many names having survived from the infancy of geology. Some are named from a characteristic fossil (Lingula Flags, a thin-bedded sandstone containing abundant fossils of the brachiopod genus *Lingula*), from some economic characteristic (Millstone Grit), or even from a folk name (Norwich Crag).

Mapping of Poorly Exposed Formations

On the barren walls of the Grand Canyon (Fig. 5-2) stratigraphic details can readily be traced for many kilometers; in most areas, though, soil and vegetation cover the surface to greater or lesser degree, and natural outcrops are scarce. On William Smith's map of England, the contacts between the various formations are shown for distances of hundreds of kilometers. Yet, in tracing an individual contact for 100 km it is doubtful whether Smith found as many as 50 exposures of the actual contact. How, then, was he justified in presenting the contact as demonstrated? A geologist can see no farther through the soil than anyone else. How can he make and map inferences about the position of the buried strata that will withstand the tests of wells drilled or mine shafts dug?

The stratigraphy must be pieced together from scattered outcrops. Though actual contact between two formations may be seen only rarely, there are generally many more exposures of either the upper or the lower stratum between which the contact is bracketed. Even with no clean outcrops at all, loose fragments of rock are common in the soil on a slope. Inasmuch as such fragments cannot creep uphill, their source must be upslope from the point where they are found. These fragments, called **float**, are thus very useful in fixing the *upper* contact of their source rock. Where float characteristic of the lower formation is no longer present on going up a slope, the upper contact is nearby. Just how near, of course, depends on the thickness of the creeping rock waste; a little digging or examination of animal burrows or ant hills usually makes a reasonable judgment possible.

Correlation of Strata

Modern geologists use essentially the same methods as Cuvier, Smith, and Desmarest. In a ravine on a grassy hillside, we may see a bed of clay with well-marked horizontal stratification; we assume it continues horizontally into the hill (How else can its stratification be projected?). If we follow a contour along the hillside before seeing, in another ravine, a similar clay, we may suppose it to be the same bed; if we have seen some clay float on our traverse, the likelihood is greater. If both show concretions (nodular harder lumps), and especially if the concretions are about the same in size and spacing, our confidence is further heightened. If farther on we find a deep ravine that exposes scores of meters of strata in which there is but a single clay bed, we feel virtually certain that our correlation between outcrops is correct. Mapping the clay is simple; as its bedding is horizontal, it will map along the contour we traversed. The deep ravine may show us that the clay bed is overlain by a brown sandstone 20 meters thick and that it rests on a black limestone with many fossils.

If we now go to a more distant outcrop and find a clay bed a little thinner than the first one seen, resting on a mottled black and gray limestone and overlain by a red-brown sandstone, we might question the correlation with the first bed, but would not necessarily reject it. We might look for intervening exposures of both the overlying and underlying strata to check them for chances of gradations in **lithology**—the sum total of such features as composition, grain size, color, and kind of bedding. Nearly all strata do change laterally to some degree, and the possibility must be tested.

This example illustrates the use of three important elements in rock correlation:

1. *Lithology*—The closer the resemblance between rocks in scattered exposures, the more likely is their correlation. Over large lateral distances, however gradual, changes in lithology are to be expected.

2. *Sequence*—Similar successions of strata suggest correlation, and the more numerous the comparable items in identical sequence, the more likely it is that the correlations are correct.

3. *Fossil content*—William Smith found that "Each stratum contained fossils peculiar to itself, and might, in cases otherwise doubtful, be recognized and discriminated from others like it by examination of them." Cuvier and Brongniart reached the same conclusion. Fossil assemblages are indeed excellent keys to correlation. Smith used fossils as though they were merely distinctive pebbles; we will find in Chapter 6 that their irreversible changes in time make them much more significant guides to correlation than pebbles or other inorganic features.

A still more reliable method of correlation is by actually *tracing a continuously exposed contact from one area to another*. But only where rocks are nearly bare of soil and vegetation can this be done for any considerable distance, and of course it cannot be used across a sea or beneath a cover of younger rock. Correlation by similarities in lithology, in sequences, and in fossil content—all of which call for the exercise of judgment—are everywhere required in geologic mapping because of the generally sporadic outcrops.

Thus sound geologic mapping can never be reduced to a mechanical routine; regardless of the apparent simplicity of stratal arrangement, an element of judgment is invariably involved. Some correlations are certain, others tentative, others doubtful. Two geologists may disagree about the doubtful ones, just as equally qualified physicians may differ in diagnosing pathological symptoms. But nearly all such differences concern minor features; most thick sequences of strata embrace enough diagnostic features to lead two careful observers to identical conclusions.

GEOLOGIC SECTIONS

Geologic sections are vital to the extrapolation of surface information to depth, the first step in the application of geology to economic problems. A **geologic section**, also called a **structure section**,

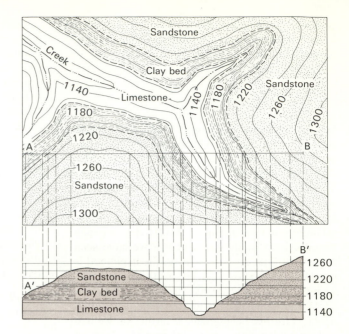

Figure 5-3
Construction of a geologic section from a geologic map.

shows how the rocks would appear on the side of a trench cut into the ground. The edge of a slice of layer cake is a section through the cake; a geologic section is a picture of what a like cut into the earth would disclose. As an example, we prepare a section through the clay bed shown in Figure 5-3. The section is shown below the map, together with the construction lines used in preparing it.

The section shows how the strata would appear in the wall of a trench dug along the line A-B on the map. First the topographic contours were used to develop the topographic outline, using the same scale for both horizontal and vertical measurements. By extending perpendicular lines from the points at which A-B crosses each contour to the appropriate depths in the section, we can prepare the outline of the hills and valleys crossed by the section—the **surface profile**. Similarly, perpendiculars were extended to the surface profile from the points where A-B intersects a geologic contact on the map. Connecting points on each of the contacts below the surface completes the section.

Such a section may be of practical value. If the clay is suitable for brickmaking, we can measure on the section the volume of useless soil and rock that must be removed to get to the clay. If a tunnel

is to be driven through the hills, the clay may pose problems of roof support or water disposal. Mines, tunnels, and wells severely test the validity of geologic maps and sections, and of the assumptions underlying them.

ROCK STRUCTURE AND GEOLOGIC MAPPING

In most places, strata are no longer horizontal, as they were when deposited; they have been more or less warped and folded, as we shall see later in this chapter. The beds near Paris, we recall, are tilted gently toward the city, forming what geologists call the Paris Basin. In mapping tilted beds and plotting them on sections, the same principles apply as with horizontal ones, but the projections are controlled by the amount and direction of tilt.

Dip and Strike

The attitude of a tilted plane is fixed if we know the compass trend of a level line drawn on it—the **strike** (measured with a compass)—and the inclination of the steepest line that can be drawn on it—the **dip** (measured with a clinometer; see Fig. 5-5). The simplest model is that of a roof. The trend of the ridge pole is the strike of the sections of roof to either side. Rainwater flowing from the roof runs down the steepest slope following a map direction at right angles to the strike of the ridgepole. The water runs in the **direction of dip**. The angle between the horizontal and the path of flowing water is the steepest line that can be drawn on the roof: it is the **angle of dip**.

In nature, though, these angles may not be so readily found. The first problem is to find the compass direction of a level line on the stratum concerned. Again we use a simple example. Figure 5-4 illustrates the improbable assumption that the lake shore to the left of the figure is formed for some distance by the top surface of an inclined bed. Here, as the lake surface is horizontal, the geographic trend of the straight segment of shore is the **strike** of the stratum. The projection of the stratum into the distant hill must be along this same trend until it is interrupted or diverted (Steno's Law). Strike is measured with a *geologist's compass*, equipped with a *level bubble* for determining the horizontal and with a *clinometer* for determining dip (Fig. 5-5). Normally a geologist measures strike by maneuvering until his eye lies as nearly as possible in the plane of the surface to be measured. The plane then appears as a line. With level bubble set to zero, his eye sweeps laterally until his level line of sight intersects the edge of the bed; this is the line of strike whose compass bearing he then reads with the compass face held horizontally. With the clinometer parallel to the bed and the bubble level (Fig. 5-5), the dip is read as the maximum acute angle between the bed and the horizontal.

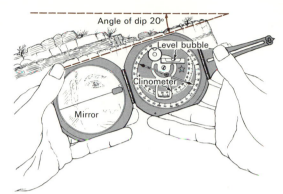

Figure 5-5
Determining dip with a Brunton geological compass. The clinometer is rotated by a lever on the back of the compass until the bubble is centered, while the ruling edge of the compass is held parallel to the dip of the stratum. The dip is then read on the inner graduated arc.

In recording strike and dip on a map, a symbol like a capital T with a long crossbar is used. The trend of the strike is shown by plotting the crossbar in the correct geographic direction. The stem of the T points in the direction of dip, and the angle of dip is plotted alongside; the symbol ₄₀ indicates a strike of N45°E (identical with S45°W) and a dip downward at an angle of 40° to the southeast. It is unnecessary to record the strike in degrees; this can always be found by comparing the trend of the long bar with the north arrow, which, by convention, points to the top of nearly all maps.

Topography, Inclined Beds, and Geologic Mapping

In contrast to a horizontal bed, a dipping contact projects along a topographic contour only when this parallels the strike. If it does not, the trace of the contact will rise (see Fig. 5-4) or fall, depending on the direction in which it is followed. This relation is the basis for determining the attitude of most beds with low dips, for a clinometer is difficult to read to a degree or less, though it may be easy to find two points on a contact at the same elevation on opposite sides of a narrow valley. The compass bearing of a line joining such

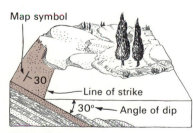

Figure 5-4
The strike and dip of an inclined bed exposed along a lake shore.

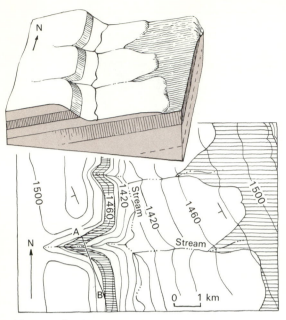

Figure 5-6
Relief diagram and geologic map of the small area south of Epernay (Fig. 5-1, B). Note the strong V in each contact as it crosses a stream valley. The strike of the beds can be determined by joining points such as A and B, where a contact intersects a contour line. (Why?)

shown on this section is projected to depths far below the observed surface; we are justified in this projection by the Law of Original Continuity, which is as valid for tilted as for horizontal strata. Projection must take account of dip.

Relation of Topography to Structure

The relation just noted between topography and structure of the underlying bedrock is a general one, though not invariable. Weathering and erosion reduce the less resistant rocks to lowlands while the more resistant remain as hills, reflecting in the landscape the differences in position and structure of the underlying rocks. Thus the curving ridges near Paris, shown in Figure 5-1, mark outcrops of resistant rocks such as sandstones, but porous chalk and other easily eroded rocks are found only in the lowlands. The pattern of both hills and lowlands and the centripetal dips show that the formations have been warped into the shallow saucer of the Paris Basin—a pile of strata each shaped like a giant plate or saucer, and all stacked into a shallow basin with Paris near the center of the uppermost and smallest saucer.

The conformity of topography with rock structure in the Paris Basin is by no means perfect; it is closest where soil is thin or absent (Fig. 5-2). In most areas, despite thick soil, the topography nevertheless reflects trends in the underlying rocks. Sandstone is generally more resistant to erosion than shale, and where the two are interstratified (Fig. 5-7) it forms ridges. It is also more permeable to water, so that where it rests on shale there is likely to be a line of springs or a strip of flourishing vegetation. Thus minor features, both topographic and vegetational, nearly everywhere give clues to the kinds and trends of underlying bedrock. If firm conglomerate crops out on a ridge crest at one point, it is likely to underlie the whole ridge; this is readily checked by a ridge traverse. A drainage ditch may expose chalk in a lowland; do wells in the lowland also cut chalk? Piecing together scattered data generally enables the geologist to establish the succession and distribution of even poorly exposed rocks.

points is of course the strike of the contact. This relation also gives the basic clue to reading the succession of beds and their direction of dip from a geologic map.

In the Paris Basin most streams drain to the Seine. East of Paris (Fig. 5-1) they flow westward, crossing the arcuate ridges nearly at right angles. We note that any particular bed, when traced westward along a valley wall, gradually descends to the stream; on the opposite wall of the valley, it reverses trend and rises. Each contact forms a V with the apex directed downstream (Fig. 5-6).

The Upper Chalk crops out farther west in the valleys of the Marne and Aisne than on the intervening ridgetops; (note especially the V pointing downstream at Epernay); its contact with the overlying Plastic Clay must therefore dip westward—the basis for the section in the upper part of Figure 5-6, B. Each formation

Removed by erosion

Figure 5-7
Erosion of tilted sedimentary rocks. The conspicuous ridges are formed of sandstone strata; the more subdued topography is underlain by shale and sandy shale. The projections above the structure section suggest only a small part of the rocks that have been eroded away. [Photo by U. S. Air Force.]

LIMITATIONS OF SCALE

Geologic maps, like all others, demand rigorous selectivity; they necessarily emphasize some features at the expense of others (see Appendix I). The smaller the scale, the fewer the details that can be shown. Figure 5-6 shows detail impossible to portray on the scale of Figure 5-1. The larger the scale and the better the exposures, the greater the number of mappable formations in a given area, though the number of lines per square unit of map is generally about the same, as it is on the two examples cited.

The geologic map, our most valuable tool, is a shorthand record that requires interpretation. Many a map has been misunderstood, sometimes even by the man who compiled it; a more experienced or more imaginative geologist may derive more information from a map than its maker. But any map that has been properly done is a summary of the geologic history of an area and of the mutual relations and structures of its rocks. But to determine the ages and environments of deposition of the various strata and intrusive rocks requires still further work — particularly, a consideration of the fossils in the rocks and their significance as to both age and environment.

SOME BASIC FINDINGS FROM GEOLOGIC MAPPING

As the maps by Smith and by Cuvier and Brongniart show, marine rocks are exposed at the land surface over wide areas — convincing evidence that either the land has risen or the sea has fallen. We will see later that both processes have taken place; we stress here only that the saucer shape of the rock structure around Paris indicates clearly that strata originally almost horizontal have been deformed into a basin. The rocks of England have also been deformed, and there a much thicker series of strata has been laid bare by erosion.

On the broad plains bordering the Mississippi, nearly every visible stratum appears to the naked eye to be horizontal, but careful observation reveals clear evidence of deformation. Oil wells in southern Illinois penetrate an unusual shale bed at a depth of about 1400 m below sea level, but the bed rises gradually eastward and appears at the surface near Louisville, Kentucky. It also rises, though less steeply, to the north, south, and west: southern Illinois thus lies in a rock basin, even though the land surface slopes gently outward toward the bounding rivers.

If we are mining a coal bed by open-pit methods, stripping away the overburden of barren rock and soil above the coal, as is widely done in Illinois, even very low dips may be critical to the success of the operation. If the coal bed lies at a depth of 8 m at one point and dips only 1°, within a kilometer in the down-dip direction the overburden will have increased to about 25 m, perhaps a prohibitive thickness for economic mining. A few hundred meters up dip the coal will intersect the surface, where it will have been eroded away.

Structure Contour Maps

For many economic purposes, both in mining and oil-field development, a **structure contour map** is almost indispensible. As on topographic maps (see Appendix I), contours are used to represent lines of equal elevation, here not of the ground surface but of the surface of a stratum. If we could remove all the overburden from a coal bed and then make a topographic map of the newly exposed surface of the coal, the result would be a structure contour map of the coal bed.

Figure 5-9, a map of part of the Wasatch

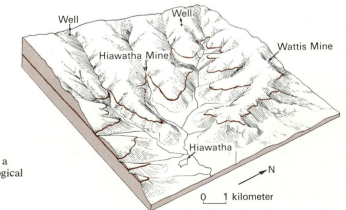

Figure 5-8
Relief diagram of the Wasatch Plateau coal field, Utah. [Based on a map by E. M. Spieker, U. S. Geological Survey.]

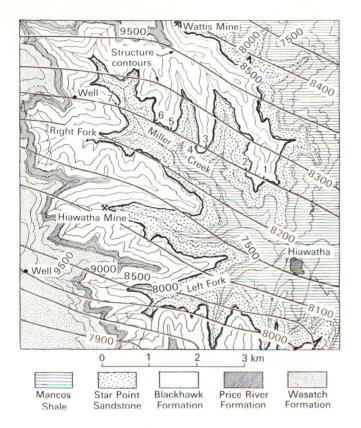

Figure 5-9
Topographic, geologic, and structure contour map of the area shown in Figure 5-8. (Topographic contours black, structure contours brown.) [Modified from E. M. Spieker, U. S. Geological Survey.]

Legend:
- Mancos Shale
- Star Point Sandstone
- Blackhawk Formation
- Price River Formation
- Wasatch Formation

Plateau coal field sketched in Figure 5-8, illustrates how control is obtained. It is a topographic map on which the formations and structure contours have been plotted. The outcrop of the Hiawatha coal bed is shown as a heavy line. Note that the topographic contours do not parallel the coal; the beds have been tilted. At seven places (numbered) in the ravines north of Miller Creek, the coal bed crosses the 8250-foot topographic contour, thus fixing the 8250-foot structure contour. On the south wall of Left Fork Canyon, the bed crosses the 8000-foot topographic contour; this fixes the 8000-foot structure contour.

Borings and wells give other points. A well on the ridge north of the right fork of Miller Creek penetrates the coal at a depth of 750 feet; as the well head is at 9000 feet, the coal is at 8250, giving another point on that structure contour. The coal is overlain by the Blackhawk Formation, several measurements of which show that it averages 750 feet in thickness. Thus wherever the top of the Blackhawk formation crosses a contour, we can assume the coal to lie 750 feet below, fixing another point on the appropriate contour.

Such a structure-contour map makes it easy to find the thickness of overburden: at any point, we need only subtract the elevation of the structure contour from that of the topographic contour.

FOLDS

In many places, most notably in the mountain chains, the rocks have been much more intensely deformed than in our examples from the Paris Basin, Illinois, and the Wasatch Plateau. Folds are the most common structures in mountains, and they even deform parts of plateaus and plains. They range from small features (Fig. 5-10) through folds a kilometer or two across (Fig. 5-11), to gigantic arches and troughs 100 km or more across. Upfolds, or arches, in the strata are called **anticlines** (Figs. 5-10, 5-11); downfolds, or troughs, are **synclines**.

Figure 5-10
Anticline in limestone and chert, Barranca de Tolimán, Hidalgo, Mexico. [Photo by Kenneth Segerstrom, U. S. Geological Survey.]

Figure 5-11
Air view of an eroded plunging anticline in the Zagros Mountains, Iran. The plunge is away from the observer. [Photo by Aerofilms, Ltd., through courtesy of John S. Shelton.]

Figure 5-12
Air view of an eroded, plunging syncline in northwest Africa. The plunge is to the left. Note how the asymmetric ridges show gentler slopes toward the interior of the arc, steeper to the exterior, a common landform in areas of moderately dipping beds. [Photo by U. S. Air Force.]

On geologic maps, eroded anticlines show older rocks along their central (axial) parts and younger and younger rocks on either side; conversely, synclines show younger rocks along their centers and older and older rocks dipping under them from either side (Fig. 5-12). A **monocline** is a fold formed by a linear steepening of an otherwise uniform dip (Figs. 5-13, 5-14).

The terms "anticline" and "syncline" have no reference to the topographic form; an anticline may underlie a topographic valley or form a ridge; so may a syncline, depending entirely on the relative resistance of the various strata to

erosion and the length of time since the fold was formed. The original forms of most folds have been obscured by erosion (Fig. 5-15), though the differential erosion of weak and resistant beds may clarify their structural relations.

Structure symbols Structure contours record more details about a fold than any other graphic device (Fig. 5-16), but if the information necessary to draw them is lacking, then **structure symbols** alone suffice to reveal the essential features of most structures. Figure 5-17 shows the more common symbols used for this purpose.

The structure symbols on the map shown in Figure 5-18 imply the structure illustrated in the block diagram. The structure symbols are plotted at the points where the recorded observations were made. The map shows two beds of sandstone and two of shale; the shales show innumerable small puckers, whereas the sandstones lie in more regular folds. We say that the shale shows **incompetent folding** and that the sandstone shows **competent folding**. The dip of the shale in the minor puckers does not reflect the general structure unless they are viewed in the ensemble.

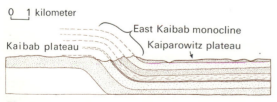

Figure 5-13
Cross section of the Kaibab monocline, Utah. [After H. E. Gregory and R. C. Moore, U. S. Geological Survey.]

Figure 5-14
The northward-plunging end of the Comb Ridge monocline, near Blanding, Utah. Note the nearly horizontal attitude of the rocks on the skyline to the left, the downward bend as the strata are followed to the right. Out of view in the foreground to the right, the strata are again horizontal. View is north. [Photo by Tad Nichols, Tucson, Arizona.]

Figure 5-15
Deeply eroded limb of an anticline, Flaming Gorge, Utah. [Photo by W. H. Jackson, U. S. Geological Survey.]

Figure 5-16
Structure contour map and cross section
of folds in the Big Horn Basin,
Wyoming. The bed contoured is the top
of a sandstone of Cretaceous age.
[After D. F. Hewett and C. T. Lupton,
U. S. Geological Survey.]

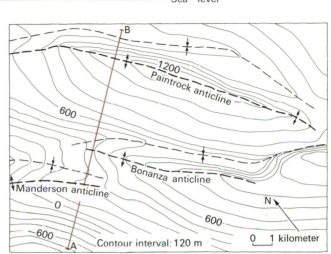

 20
Strike and dip
of beds

 40
Strike and dip
of joints

Strike–slip fault,
showing relative movement

4
Plunging anticline (top)
and upright syncline

 60
Fault, showing dip

21
Plunge of fold axes and
other linear structures

Figure 5-17
Examples of structure symbols.

The surface that divides a fold as symmetrically
as possible is the **axial surface**, often mistakenly
called the **axial plane**. The line along which the
axial surface intersects a bed is the **axis** of the
fold. If the dip of the axial surface is steep, as in
Figure 5-18, the axis is also the line on the ground
where each bed reaches its highest level as it
arches over the anticline, but if the axial surface
dips at a lower angle, as in Figure 5-15, the highest points on the beds—which form the **crest
line**—may be far removed from the axis. (This
distinction is not always made; many maps show
the crest line as the axis, even where the axial
surface dips gently.) The crest line of an anticline
is shown as a line with diverging arrows (Fig.
5-18); the trough line of a syncline, as a line with
converging arrows.

No fold continues indefinitely; all must end by
plunging out, as seen in both Figures 5-14 and
5-18. The anticline in Figure 5-18 plunges northward, as indicated on the map by the arrow with
the number 8 at its point. The plunge is the dip of
a bed exactly on the fold crest or trough.

Forms of folds Where a crest line is horizontal (that is, where a fold shows no plunge), a map
of level country will show the beds to be about
parallel on either side of the fold. Figures 5-12
and 5-18 show that the beds of a plunging fold
curve smoothly to join with those of the sides,
commonly in a canoe-shaped arc. Erosion acts on
each stratum differently, the pattern of ridges and
valleys produced being characteristic of plunging
folds.

A **dome** is an anticline that plunges in opposite
directions from a high point, so that it is about as
wide as it is long. A doubly plunging syncline
forms a **basin**, like the one near Paris.

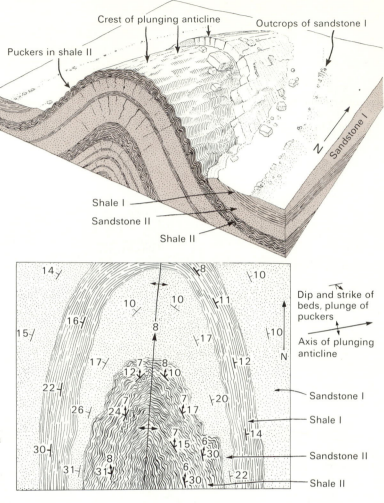

Figure 5-18
Geologic map and relief diagram
of a small fold near Amity,
Arkansas. [Based on map by
A. C. Waters, U. S. Geological
Survey.]

Many folds, like the Paintrock anticline of Figure 5-16, are **asymmetrical**—the strata of one limb dip at steeper angles than those of the other. Though asymmetrical, such folds are still considered **upright** if their limbs dip away from each other, but in regions of intense deformation many folds are **overturned**—that is, the beds of one limb have been bent beyond the vertical (Figure 5-15), so that both limbs dip in the same direction (not necessarily at the same angle). Other folds are **recumbent** (Figure 5-24), with their axial surfaces virtually horizontal. In recumbent anticlines, the lower limbs are completely overturned; in recumbent synclines, the upper limbs. If the limbs of the fold are parallel, the fold is **isoclinal**, whether upright, overturned, or recumbent. Several folds in Figure 5-24 are recumbent.

FAULTS AND JOINTS

Most rocks show some degree of fracturing. If the fractures have merely opened, without one side being appreciably displaced relative to the other, they are **joints**; if displacement is observable, they are **faults**. Small fault displacements are usually ignored in mapping, but if the offset is large enough to be mappable the fault is shown. Many faults have displacements measured in kilometers.

Because of their abundance and general irregularity, most joints are not mapped, but if there is some reason for showing them (for example, some ore veins follow joints), their attitude is commonly plotted by the upper-right symbol shown in Figure 5-17. Figure 5-19 shows un-

Figure 5-19
Rectangular joint pattern in massive sandstone, Colorado Plateau. [Photo by V. C. Kelley, University of New Mexico.]

Figure 5-20
Air view of a fault near Great Bear Lake, Northern Territory, Canada. The fault extends for 130 km. Deformed sandstone strata on the left, granite on the right. Note the small faults and joints that cut the granite at acute angles to the fault. These suggest horizontal movement of the right-hand block toward the reader or the left-hand block away, or both. [Photo by Royal Canadian Air Force; data from A. W. Jollife, Queens University.]

usually regular sets of joints. Others are shown in Chapter 7.

Joints and small faults are formed by erosional unloading, dessication, thermal contraction, and other processes, but large faults and many joints traceable for hundreds of meters evidently record crustal movements (Fig. 5-20).

Varieties of faults Faults may be found anywhere but are especially abundant in mountain ranges. The several varieties are distinguished by the direction of *apparent* movement along fractures separating two earth blocks. A **dip-slip fault** is one along which the apparent movement has been predominantly parallel to the dip (Fig. 5-21,A and D). A **strike-slip fault** is one whose movement has been predominantly parallel to the strike (Fig. 5-21,F). Most faults show components of both strike and dip movement; where these are both considerable, the fault is an **oblique-slip fault** (Fig. 5-22).

Most faults have appreciable dips; from early times, miners have recognized such faults, calling the lower block, on which they might be standing, the **footwall** and the overlying block, which might collapse on them, the **hanging wall**. A **normal** fault is one whose hanging wall has *apparently* moved down with respect to the footwall (Fig. 5-21,A and D). A **reverse fault** is a fairly

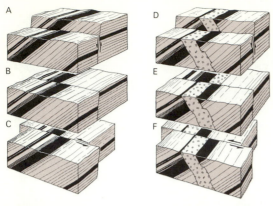

Figure 5-21
Diagram showing how dip-slip and strike-slip faults can produce identical outcrop patterns after erosion has worn away the fault scarp (A, B, and C). Under ideal conditions, where a second structure cuts the bedding, they can be distinguished (D, E, and F).

Figure 5-22
Cliff (fault scarp) formed during the Mino-Owari earthquake, Japan, in 1891. The displacement, as measured on the offset road, was 6 m vertically and 4 m horizontally. [Sketch based on photo by Professor Koto.]

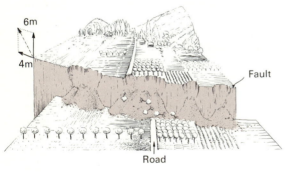

"with respect to," "apparently," and "relative movement." These cautions are necessary, for we do not normally see the actual fault movement, but only the geometry of the rock masses after some erosion. The **offset** of a stratum along a fault can be measured, but we cannot usually tell whether the hanging wall of a normal fault actually dropped with respect to the earth's surface, whether the footwall rose, or whether both blocks moved. We can normally determine only the apparent movement with respect to the other block. The bending of strata as they approach the fault, called **drag**, commonly records some components of the relative movement, but is not reliable, for some faults apparently reverse their movement at different times, and some of them rather surprisingly retain drag from the smaller of successive displacements.

Where strata are uniformly tilted, it is commonly impossible to tell whether the movement was dip-slip or strike-slip, as the same pattern is left by each after erosion (Fig. 5-21,A, B, C). If, however, another feature, such as a dike, differs in attitude from that of the strata, and can be identified on both sides of the fault, the true direction and amount of displacement can be determined (Fig. 5-21,D, E, and F). Very exceptionally the attitude of a drag fold permits unambiguous determination of the direction of fault movement. Close to the San Andreas fault in the Mecca Hills, California, the weakly consolidated alluvium has been folded into vertically plunging isoclinal folds (Fig. 5-23). The folds die out away from the fault, which shows that they are related to the fault motion—clearly a virtually horizontal (strike) slip.

The whole classification of faults is necessarily artificial: not only are there all angles of diagonal movement between dip-slip and strike-slip, but some faults, when traced along strike, change from normal to reverse (**scissors faults**) or change in dip, as does, for example, the great Uinta Fault, on the north side of the Uinta Mountains, Utah. At the west end it dips gently south (a thrust fault); along strike it steepens to vertical, and toward the east end of the range it dips north and is a normal fault, though the strata to the south are everywhere hundreds or thousands of meters higher than the same strata to the north.

Fault classification is further complicated by the fact that in some places the faults have been later deformed. For example, the contorted fault

steeply dipping fault in which the hanging wall has *apparently* moved upward with respect to the footwall. If the fault dip is relatively low, such a fault is called a **thrust fault** (Fig. 8-19).

Strike-slip, or **lateral**, faults are characterized as **left lateral** or **right lateral**, depending on whether the continuation of the rock on which one stands while facing the fault is displaced to left (Fig. 5-21,F) or to right (Figs. 8-2, 8-3) on the opposite side of the fault.

Note the cautious words in these definitions:

Figure 5-23
Weakly consolidated alluvium, folded isoclinally about vertically plunging axes. The view is northwestward, parallel to the San Andreas fault, just out of sight to the left. Mecca Hills, California. [Photo by Warren Hamilton, U. S. Geological Survey.]

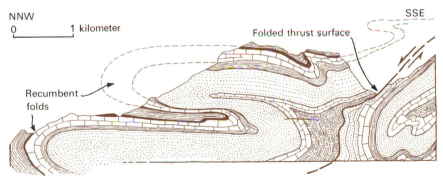

Figure 5-24
Recumbent folds and a folded thrust fault, east end of Lake Lucerne, Switzerland. [After Albert Heim, 1922.]

at the right side of Figure 5-24 would at first be considered a normal fault, as the rocks to the left have clearly moved downward with respect to those of the footwall; observations over a considerable area surrounding this locality show, however, that the fault originated as a low-angle thrust fault, moving from right to left, but has since been folded into its present conformation. Such intense deformation is not uncommon in some mountain ranges.

Folds and faults may grade into each other: some normal faults die out into monoclines along strike, and some thrusts are merely broken anticlines. Nevertheless, many faults, both normal and reverse, show no evidence of having grown from folds.

UNCONFORMITIES

Figures 4-9 and 5-25 illustrate significant relations between rock masses. In the first the well-bedded sandstones of the cliff rest on granite, with fingers of sand extending down into the

Figure 5-25
An angular unconformity, Wyoming. Tilted and eroded beds of sandstone, shale, and coal, overlain by flat-lying sandstone. [Photo by C. J. Hares, U. S. Geological Survey.]

Figure 5-26
Picture Gorge, Oregon. Angular unconformity between the Mascall Formation and the Rattlesnake Tuff. [Photo by Oregon State Highway Department.]

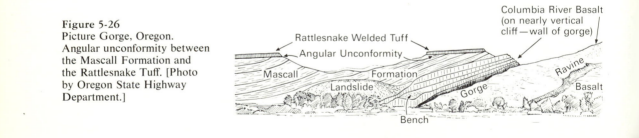

Figure 5-27
Disconformity between thin-bedded sand below and massive silt above, near Apaxo, Tlaxcala, Mexico. [Photo by Kenneth Segerstrom, U. S. Geological Survey.]

granite, showing that the sand was unconsolidated when it came in contact with its floor. As granite does not form at the earth's surface, it is clear that deep erosion took place after the granite congealed and before the sandstone was deposited. The surface of deposition is thus a buried erosion surface, an **unconformity**. Of the several kinds of unconformity, one like this, in which younger sedimentary rocks rest directly on plutonic rocks, is a **nonconformity**.

Figure 5-25 shows an **angular unconformity**. The rocks below the flat-lying sandstone at the top of the cliff had obviously been tilted to angles approaching 45° and beveled by erosion before the upper sandstone was deposited. Figure 5-26 shows a somewhat similar relationship; the dark basalt flows that dip to the left (right center of

picture) were tilted after the Mascall Formation had been deposited on them. Both were then eroded to a nearly level surface before the Rattlesnake Tuff covered the whole unconformably. Even the tuff has been slightly tilted to the left, adding to the dips of the underlying rocks.

An unconformity records a reversal of the local process of sedimentation and erosion; the time intervening between the formation of the youngest rocks beneath the unconformity and that of the oldest of the overlying rocks is a lost interval in the local geologic record. Most unconformities show that some tilting or other disturbance of the underlying rocks took place before deposition of the younger beds, but many do not, and are merely irregular erosion surfaces cut on the lower beds; these are **disconformities** (Fig. 5-27).

BASAL CONGLOMERATES AND
REGRESSIVE SANDSTONES

When a broad erosion surface of low relief is warped beneath the sea, the rate of submergence is generally so slow that the shoreline advances only a few meters per century, as has been proved by methods discussed in Chapter 6. Waves are powerful agents of erosion (Chapter 16), and during the slow advance of the sea they strip away any soil and generally some of the under-lying bedrock, wash out the finer particles produced during their attack on the land, and thus generally leave thin veneers of pebbles and sand —**basal conglomerates**—as records of their transgression across the landscape. Thus many conglomerates lie at the base of unconformable stratigraphic sequences, recording a sea encroaching over a land surface. When the sea retreats, it generally winnows out the finer silt and clay and leaves a regression sandstone to mark its departure.

Questions

1. What differences in topography would you expect if the rocks of the Paris Basin had been warped into a dome, rather than a saucer shape?

2. Give a general rule relating contours and outcrop patterns of horizontal strata.

3. If you traced a thick lava flow and came to the termination of its outcrop, what four possibilities might account for its termination?

4. How would a dike of basalt that dips vertically and strikes north appear on a map? What difference, if any, would the map pattern show where the dike crosses a sandstone ridge trending due east?

5. If you traced a shale-sandstone contact across flat country for several kilometers and found that it leads in an elliptical path back to the starting place, what inferences could you make concerning the dips in the area?

6. Some buttes in the New Mexico desert are capped by flat-lying basalt flows; others are volcanic plugs. How would the patterns differ on a geologic map if both kinds of forms had several steep gullies on the slopes?

7. Draw a hypothetical geologic map showing a series of tilted beds that include a lava flow and a sill. Where on the map would you expect to find lava fragments included in a sedimentary bed?

8. A geologic contact that V's downstream invariably indicates a downstream dip. But if the stream falls steeply, a contact that dips downstream may actually V *upstream*. Show how this might happen, and frame a general law expressing the relation between the angle of dip of a contact and the gradient (slope) of the stream bed.

9. Draw a geologic map showing two anticlines and an intervening syncline, all eroded to a nearly flat surface. The folds all plunge north and are asymmetrical. Four formations are exposed.

10. In the field how could you tell an angular unconformity from a thrust fault?

11. By what criteria could you differentiate a nonconformity from the top of a sill?

12. Draw one cross section showing all the following features:
   a) A series of folded marine sedimentary rocks overlying a nonconformity.
   b) A series of nearly flat lava flows that rest with angular unconformity on the folded strata.
   c) Two thrust faults that are older than the lava but younger than the sedimentary rocks.
   d) A normal fault younger than the lavas.
   e) A dike younger than the thrusts but older than the normal fault.

Suggested Readings

Adams, F. D., *The Birth and Development of the Geological Sciences*. Baltimore: Williams & Wilkins, 1938. [Chapters 7 and 8.]

Babbage, Charles, "The Temple of Serapis," *in* K. F. Mather and S. L. Mason (eds.) *Source Book in Geology*. Cambridge: Harvard University Press, 1970. Reprinted from *Quarterly Journal Geological Society of London*, v. III, pp. 186–217, 1847.

Blackwelder, Eliot, "The Valuation of Unconformities." *Journal of Geology*, v. 17, pp. 289–299, 1909.

Harrison, J. M., "Nature and Significance of Geological Maps," *in* C. C. Albritton (ed.) *The Fabric of Geology*. San Francisco: Freeman, Cooper & Company, pp. 225–232, 1963.

Hutton, James, *Theory of the Earth* (1795). New York: Stechert-Hafner, 1960.

Smith, William, "The Strata of England," *in* K. F. Mather and S. L. Mason (eds.), *Source Book in Geology*. Cambridge: Harvard University Press, 1970. Reprinted from *Memoir to the Map and Delineation of the Strata of England and Wales* (1815).

CHAPTER 6

Fossils, Strata, and Time

The discovery that fossils lend identity to strata and thereby permit correlation over wide distances—even across the sea—was fundamental to geology. Cuvier, one of the first systematic biologists, soon recognized consistent relations among the fossils: each fossil assemblage differs from that of noncontemporaneous strata—and the older the assemblage, the less its fossils resemble the shells or bones of comparable living creatures. This is broadly true of vertebrate, invertebrate, and plant remains the world over.

CORRELATION AND FAUNAL SUCCESSION

That assemblages of older fossils differ more from living creatures than younger ones implies that old forms of life have died out and that they have been succeeded by new ones. As we shall see later in this chapter, this generalization (to which a few exceptions are known) was made the basis by Lyell for subdividing the stratigraphy of the latter part of the geologic column. Neither idea was generally accepted before Cuvier. In fact, the great Swedish naturalist Linnaeus, founder of the systematic biological classification basic to the one we use today, had

written only a generation before that "existing species of animals are now as they were created in the beginning."

Cuvier's discoveries not only disproved this but also posed a tantalizing question: How did new species arise? Cuvier's answer—by a succession of new Creations, each following a universal catastrophe that had destroyed all earlier life—was soon proved wrong, but his discovery of the extinction of old species and the rise of new ones was basic to the great strides in biology and geology that led, half a century later, to Darwin's demonstration of evolution.

Geologic Chronology

Geologists in many lands soon turned up thousands of additional fossil species as well as many specimens of those already known from England and the Paris Basin. And, following Cuvier, they confirmed his discovery of a secular change toward a modern fauna. As discussed more fully in Appendix V, biologists and *paleontologists* (students of ancient life) group individual organisms (or their fossil remains) into *species*. Species considered closely related are grouped into *genera* (singular, *genus*), genera into families, and so on to higher and higher categories. It was soon found that many widespread species are found only within narrow stratigraphic intervals and, together with similarly limited genera, imply the **Law of Faunal Assemblages** upon which stratigraphic correlations, both local and world wide, are everywhere based: *Like assemblages of fossils indicate like geologic ages for the strata containing them.* Worldwide demonstrations of this law form the basis for the standard geologic column.

THE STANDARD GEOLOGIC COLUMN

Simple Early Column

By the middle of the eighteenth century, Italian and German geologists had classified rocks into three categories that they considered chronological: Primary (rocks such as granite and gneiss, without bedding or fossils), Secondary (firmly cemented sedimentary rocks, generally found in the mountains associated with granite and gneiss), and Tertiary (commonly weakly consolidated rocks, generally confined to lowlands). Though fossils were known in both Secondary and Tertiary rocks, they were not utilized in the classification.

Standard Column

The present **standard geologic column** was pieced together in Europe during the nineteenth century by expanding the stratigraphic sequences of Smith and of Cuvier and Brongniart into the rest of Europe. The fossil assemblages in these strata are those to which all others are ultimately compared, the world over, and theirs is the standard sequence. Many correlations are rendered approximate by the limited distribution of certain species that lived in narrow environmental niches, either climatic or ecologic. On the other hand, though, some marine species are both abundant and cosmopolitan; their numbers allowed many shells to be preserved even though the vast majority have been dissolved away or crushed during earth movements.

The fundamental subdivisions of the geologic column are the rock **Systems**, most of which were represented on Smith's map. Smith grouped his oldest, widely recognizable British strata as the Old Red Sandstone. These rocks contain few fossils and exhibit many clues to deposition in a widespread desert; but in Devonshire, richly fossiliferous marine limestones thicken southward between strata of the Old Red desert sandstones, showing their essential contemporaneity. Because marine fossils are far more abundant than others, they are generally the most useful in correlation: the old Red Sandstone is therefore included in the *Devonian System*, named from the fossiliferous limestones there. Smith called the next higher beds the Mountain Limestone. These are overlain by a sequence of sandstone, shale, iron ore, and coal—the Coal Measures of Smith. Both Mountain Limestone and Coal Measures are grouped in the *Carboniferous System*, so called because of the abundant coal it contains.

We shall not go into the details of the construction of the standard column. Suffice it to say that more than a generation after Smith, three fossiliferous systems were recognized beneath the Devonian, in descending order: Silurian, Ordovician, and Cambrian. The Cambrian strata are the oldest that contain abundant fossils, though sedimentary rocks many thousands of feet thick crop out beneath them in some regions, and many of these contain primitive fossils (Fig. 6-1). Such fossils, however, are sparse and nearly devoid of characteristic features useful in correlation, though this is a field of very active research in which some broad correlations of algal remains have been recently suggested and are now being tested. In this book all rocks older than Cambrian are classified simply as Precambrian. As noted in Table 6-1 and discussed later in this chapter, the Precambrian strata represent far more of geologic history than do Cambrian and younger rocks (grouped together as *Phanerozoic*, Greek for "plainly evident life"), and are vastly more voluminous. Although the Precambrian rocks can be mapped and their local relations quite as well determined as those of fossiliferous rocks, their lack of fossils renders correlation over considerable distances far more difficult. Dating by radioactivity has begun to facilitate correlations (see pp. 80–85), but many correlations are rather uncertain, although rapid progress is being made.

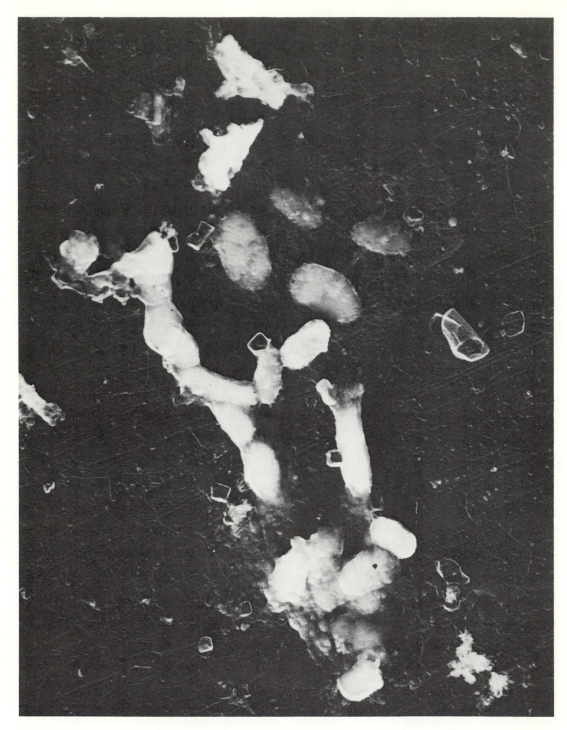

Figure 6-1
Fossil bacteria from the Gunflint Formation of Ontario, north shore of Lake Superior. The Gunflint Formation has been dated as about 2 billion years old by radiometric methods. [Electron photomicrograph by courtesy of Professor Elso Barghoorn, Harvard University.]

Table 6-1.
Geologic column and time scale.

Era	System or Period (rocks) (time)	Series or Epoch (rocks) (time)	Approximate age in millions of years (beginning of unit)
Cenozoic *(recent life)*	Quaternary (an addition to the 18th-century scheme)	Holocene *(entirely recent)*	0.01
		Pleistocene *(most recent)*	1.7 to 2.0
	Tertiary (Third, from the 18th-century scheme)	Pliocene *(very recent)*	5 to 6
		Miocene *(moderately recent)*	25 to 27
		Oligocene *(slightly recent)*	37 to 39
		Eocene *(dawn of recent)*	53 to 54
		Paleocene *(early dawn of the recent)*	63
Mesozoic *(intermediate life)*	Cretaceous *(chalk)*		136 to 138
	Jurassic (Jura Mountains, France)		190 to 195
	Triassic (from three-fold division in Germany)		225
Paleozoic *(ancient life)*	Permian (Perm, a Russian province)		270 to 280
	Carboniferous (from abundance of coal)		345 to 350
	Pennsylvanian		320 to 325
	Mississippian		345 to 350
	Devonian (Devonshire, England)		395 to 420
	Silurian (an ancient British tribe, Silures)		440 to 450
	Ordovician (an ancient British tribe, Ordovices)		ca. 500
	Cambrian (the Roman name for Wales, Cambria)		ca. 570
Precambrian	Many local systems and series recognized, but no well-established world-wide classification has yet been delineated.		

Sources: Approximate ages from Holmes, 1964; Evernden, Savage, Curtis, and James, 1964; Folinsbee and others, 1964, 1973; Obradovich and Cobban, 1973; Phanerozoic Time Scale of the Geological Society of London, 1964; Rubinshteyn and Gabuniya, 1973.
Notes: Terms in italics indicate the Greek derivations of some names. Many provincial series and epochs have been recognized in various parts of the world for Mesozoic and older strata. Most of the systems have been divided into Lower, Middle, and Upper Series, to which correspond Early, Middle, and Late epochs as time terms.
Pennsylvanian and Mississippian Systems, named for States of the U.S.A., are not generally recognized outside of North America; elsewhere the Carboniferous System is regarded as a single system.

GEOLOGIC TIME SCALE

The standard geologic column is the basis of the **geologic time scale**. The names of the systems (strata) are also used for the periods (time intervals during which the respective systems were deposited). Thus the Carboniferous Period was the interval of time during which the Carboniferous System was laid down. **Systems** and **Periods** are farther subdivided into **Series** and **Epochs**, respectively. Thus the Comanche Series of Texas was deposited during the Early Cretaceous

Epoch. The fundamental elements in the classification are the strata—tangible objects. The periods and epochs are abstractions—units of time derived from them.

More subdivisions are recognized among younger than among older strata. Just as in human history, the records are more complete for recent than for ancient times. Obviously, the older a rock, the greater chance it has had either to be eroded away or be buried from sight beneath younger deposits. In fact, if the durations of the several periods are as shown in Table 6-1, the

Cretaceous rocks are 40 times as well exposed over the American continents, per million years of period length, as are the Cambrian rocks. The Cenozoic rocks are nearly three times as well exposed proportionately as those of the Cretaceous. In substance this means that we have at least 100 times as good a record of the Cenozoic as of the Cambrian. No wonder that many geologists, not realizing the magnitude of the disparity in the record, have inferred that geologic processes are speeding up with time, contrary to the laws of thermodynamics. We shall see that this is not true; the illusion is an artifact of "geologic perspective."

The systems are grouped into larger units, formally called *Erathems*; the corresponding time intervals are **Eras**: Paleozoic Era, Mesozoic Era, etc. The word "erathem" is little-used; instead, the usual practice is to use the names of eras as adjectives: for example, Paleozoic rocks, Mesozoic rocks.

Gaps in the Standard Column

The divisions of the standard column into systems and series are based on abrupt changes in the fossil assemblages in the European strata. The boundaries were naturally selected at stratigraphic levels at which rocks, and especially fossil assemblages, above and below, contrast most conspicuously—levels that commonly corresponded to times of local erosion or nondeposition. The longer such interruptions lasted, the more drastic the fossil changes were likely to be. As stratigraphic work was extended, however, fossil assemblages intermediate between those of adjacent European "type systems" were found; these had lived during the times of the gaps in the European deposition.

Near many systemic boundaries are beds whose fossil assemblages pose "boundary problems"—that is, uncertainty whether they should be referred to the upper part of an underlying system or the lower part of an overlying one. Work on these boundary problems has helped to shrink many intersystemic gaps. Even the gap between the Paleozoic and Mesozoic strata, long thought to be unrepresented by any marine beds, is bridged by virtually complete sequences in Nevada, Greenland, and the Himalayas. Even in Europe, where the divisions were established, further research has narrowed the gaps. For example, the Paleocene Series was established to include rocks not certainly referable either to the typical Eocene or to typical Cretaceous. Now we find that different paleontologists refer the same beds to Paleocene or Cretaceous, depending on their subjective appraisals of the similarities of the fossils with one or the other European type section. Two new zones of uncertainty have been introduced, but the precision of correlation has improved.

The uncertainties posed by "boundary problems" are thus concerned more with form than with fact. Such disputed strata must at least partly fill an intersystemic gap in the type section: their position in the *sequence* is rather precisely fixed, and the systemic assignment relatively unimportant. The ambiguity is rather a tribute to the precision of paleontologic correlation than a criticism of its uncertainties. It seems certain that the intersystemic gaps now recognized will nearly all eventually be filled when the vast areas of little-studied strata scattered over the earth are known as well as those of Europe. The systemic divisions are purely arbitrary, not natural, and would quite certainly have been differently placed had the standard section been built up in Africa or western America instead of in Europe.

Intercontinental Correlation

Thus, although the original systemic boundaries were based largely on gaps in physical stratigraphy, *correlations of distant strata must be based on comparisons of fossils*—on biostratigraphy, not on physical stratigraphy. A stratigraphic break in Texas, for example, cannot be cited as evidence that the overlying rocks are Permian and the underlying Carboniferous (though this claim has been made), for there is no reason to think that an interruption in sedimentation took place simultaneously in Texas and in Russia, where the type Permian was established. The age assignment of the Texas strata must be based on their fossil assemblages, not on their relation to interruptions of deposition.

BIOSTRATIGRAPHIC DIFFICULTIES—
LONG-RANGING SPECIES

The labors of many paleontologists have shown that some fossils, even from very ancient rocks, differ very little from shells of organisms now

Figure 6-2
The use of fossils in stratigraphy. The long-ranging species on the left is of only general use in dating the packet of limestone beds indicated between the horizontal lines. But the short ranges of the next three fossils make them accurate time markers, and the presence of the second, fourth, and fifth shows that the beds in which they occur are not at the very base of the zone indicated by the range of the third fossil species. The specimens were collected from Jurassic rocks in England. [Data from S. W. Muller, Stanford University.]

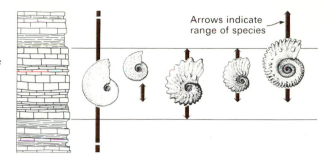

Arrows indicate range of species

living. Such species, fortunately for geology, are few. Though a long-lived species may occur throughout a thick sequence of strata, associated species generally range through far lesser thicknesses; they change from bed to bed, old ones dying out and new coming in until, finally, though an old and persistent species may continue, none of its associates at lower levels do. Individual species range through very different thicknesses, even in lithologically uniform rocks (Fig. 6-2).

Obviously, short-ranging species make possible more precise correlations than long-ranging ones. But finding such species is not the usual thing; most often one must be content with less precise stratigraphic fixes than that suggested in Figure 6-2. Obviously, the fossils most suitable for long-distance correlation must be abundant, readily preserved, have short time ranges and wide distribution.

SEDIMENTARY FACIES AND
FACIES FOSSILS

Biostratigraphic correlation also is limited by lateral changes in environment. Depositional environments are many and varied: river flood-plains, sheltered estuaries, open beaches, coral reefs, and hundreds more. Each has its own characteristic plants and animals living contemporaneously with quite different floral and faunal groups in different environments. Thus we cannot expect to find fossils of all contemporaneous organisms in a particular deposit of a given time; antelopes don't live on coral reefs, nor do clams inhabit desert dunes! Any penguins in Cuba are in zoos. The fossil breadfruit in the Cretaceous of Greenland are more likely to signify non-Arctic climate rather than a world-wide expansion of breadfruit distribution.

Analogy with living organisms suggests that some marine fossils were free-swimming and others floating; these might be found in any part of the sea. Fragile shells seldom survive the breakers and hence would be rare in near-shore deposits. The remains of organisms that lived only on muddy bottoms would be rare in limestone or conglomerate. Thus fossil species found in ancient rocks, like modern organisms, were restricted to special **sedimentary facies**, of which many surely coexisted at any one time.

Most organisms have (and had) restricted ecologic niches; fossils practically limited to particular kinds of sedimentary rocks are *facies fossils*. Thus one assemblage may characterize limestone reefs; another of the same age may be found only in shale, and the two might have few or no species in common. This may occasion considerable error in correlation. In Figure 6-3 a fossil of a long-ranging species (Fossil A) occurs *above* one of a shorter-ranging species (Fossil B) in one region but *below* it in another. This may appear to contradict the statement that once a species dies out it never reappears, but it actually results from differing facies and accidents of preservation. Similarly, a species may be missing from one bed in which another is found while they may occur together in older and younger beds. This may be because one species could adapt to slightly changing conditions of deposition while the other one could not.

Despite these and many other complexities, brought to light by the work of hundreds of paleontologists over nearly two centuries, Cuvier and Brongniart's generalization has been thoroughly justified: the older the rocks the less their fossils resemble living forms. Data from many parts of the world now permit reference of almost any collection of more than a few species—and even some individual fossils—to a fairly restricted part of the standard column.

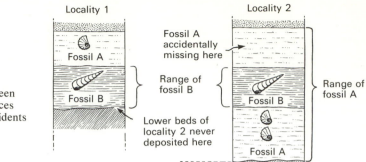

Locality 1

Locality 2

Fossil A

Fossil A accidentally missing here

Fossil B

Range of fossil B

Fossil B

Range of fossil A

Lower beds of locality 2 never deposited here

Fossil A

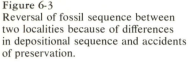

Figure 6-3
Reversal of fossil sequence between two localities because of differences in depositional sequence and accidents of preservation.

EARLY ESTIMATES OF GEOLOGIC TIME

When Cuvier discovered the sudden disappearance of some species and the abrupt rise of new ones, he jumped to the conclusion that there had been a series of major catastrophes in the history of the Paris Basin, in each of which all life was destroyed and after each a new creation populated the region. This notion of Catastrophism, as it is called, was surely suggested by the Biblical story of the deluge. Buffon, a French naturalist, had earlier estimated from some crude (and as we now know, wholly inapplicable) experiments that it would have taken the earth about 75,000 years to cool to its present temperature from the white heat he assumed it must once have had, and that it would have been habitable only for the last 40,000. Cuvier accepted this estimate and thought he could recognize four distinct faunas, thus three catastrophes and four creations in that time. The creation of *Genesis* was the most recent, having occurred about 6,000 years ago — a date obtained from biblical genealogies.

D'Orbigny, like Cuvier, believed that catastrophic destruction of all life was followed by new creations, but recognized not three catastrophes but dozens. Although he observed that some fossil species were found in more than one of his sequences, he attributed this, not to survival through a catastrophe, but to recreation of an identical species in more than one creation! The error of Cuvier and d'Orbigny was due partly to the fact that in the Paris Basin a continental environment commonly succeeded a marine one. They thus confused facies faunas with new creations. The greater part of their error, however, was probably due to the human predilection to accommodate novel data into the ideas current at the time — a weakness still with us.

Catastrophism was short-lived. Oppel, a German student, recognized many zones of strata characterized by particular fossil associations but found also that several or many of the species abundant in one zone persist through lower and higher zones, though in lesser abundance — in short, that each species had its individual span of life, as recorded in the fossil assemblages. Lyell soon showed that in the Paris Basin itself, some fossil forms of the Cenozoic formations are those of species still living, and that these are more abundant in younger than in older strata — in fact, he made the percentages of living forms the basis for subdividing the Cenozoic strata as Eocene through Pleistocene.

MODERN ESTIMATES OF GEOLOGIC TIME

Many scientists have estimated the age of the earth, starting from various postulates: Buffon, 75,000 years; Lord Kelvin, 40,000,000; Sollas, 34,000,000, to 75,000,000; Joly, 99,400,000; to name a few. It is not worthwhile to review these attempts; all have been shown to be fantastic underestimates because of false starting premises.

Radiometry

The way to more realistic estimates was opened by Becquerel's discovery of radioactivity (1896). A few elements, among them uranium and thorium, spontaneously disintegrate into lighter elements by changes in their nuclei that give rise to radiation of three kinds: alpha, beta, and gamma. Alpha radiation is the emission of helium ions from the nucleus at speeds of thousands of kilometers per second, converting the original atom into another with atomic weight four units less. The emitted ions collide with those in the surroundings to produce considerable heat (Chapters 8 and 9). Beta emission consists of electrons, derived from the breakdown of neutrons in the nucleus to form protons and the ejected electrons. Though the electron is ejected at an even higher speed than the alpha particle, its mass is so small that the heat produced by its collisions is negligible. Gamma rays are short X-rays, emitted with the speed of light.

The emission of either alpha or beta particles from the nucleus of an unstable atom converts it into a different element. Thus an atom of ^{238}U decays slowly through a series of seven intermediate daughter elements, all themselves radioactive, until a stable, nonradioactive isotope of lead, ^{206}Pb, is produced.

The rate of disintegration is constant for each radioactive isotope, but rates differ greatly from one isotope to another. Although the rate of disintegration of some radioactive elements (^7Be, for example) is slightly affected by chemical factors, no experiment with those used in geochronology has ever shown a deviation in rate to be caused by variations in temperature, pressure, or chemical environment. Disintegration rates are expressed in terms of the *half-life* of a radioactive substance—the time required for half of its atoms to disintegrate. The half-life of some members of the ^{238}U series is only a fraction of a second, but ^{238}U itself has a half-life of 4,468 million years (abbreviated to m.y.). Thus, of an initial gram of ^{238}U, only half a gram is left after 4,468 m.y.; after another 4,468 m.y., a quarter gram; and so on. The rest has changed into lead, helium atoms, electrons, and small amounts of intermediate elements in the decay series.

Uranium has two radioactive isotopes: ^{238}U (constituting 99.28 percent of natural uranium), with a half-life of 4,468 m.y., and ^{235}U (making up only 0.72 percent), with the much shorter half-life of 704 m.y. The proportion of ^{235}U was thus much greater in the distant geologic past than it is today. The isotope ^{235}U also disintegrates ultimately to a stable isotope of lead, ^{207}Pb. Similarly, thorium, ^{232}Th, with a half-life of 14,010 m.y.,

disintegrates to another stable isotope of lead, ^{208}Pb. Natural lead thus consists of a mixture of the three radiogenic isotopes, ^{206}Pb, ^{207}Pb, and ^{208}Pb, together with a fourth isotope of nonradiogenic origin, ^{204}Pb.

Many minerals—most comparatively rare—contain measurable amounts of uranium, thorium, or both. By analyzing such minerals for the ratio of radiogenic lead isotopes to radioactive parental isotopes (Fig. 6-4), the age of the mineral can be computed, provided certain conditions are fulfilled. Obviously, the minerals analyzed must be absolutely fresh, for circulating solutions might have leached out lead and its parental isotopes at different rates, thus producing great errors in the calculated age. Metamorphism of the mineral would drive out lead faster than either uranium or thorium, thus invalidating the ratio as a measure

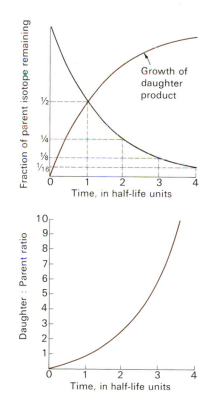

Figure 6-4
Graphs showing disintegration of a radioactive element, growth of the stable daughter isotope, and the ratio of daughter to parent. The small amounts of intermediate elements in the disintegration series are neglected; in the decay of ^{40}K to ^{40}Ar and of ^{87}Rb to ^{87}Sr, there are none.

of age. And if the radioactive mineral happened to be closely associated with a lead mineral containing chiefly common lead, it might be difficult to sort out the several isotopes, especially if the nonradiogenic isotope greatly predominated.

These conditions are stringent, but many minerals seem to meet them and have been carefully analyzed. The analyses are difficult and require the use of mass spectrometers and other complex instruments. The proportions of all the isotopes involved must be determined, as well as that of the nonradiogenic isotope, ^{204}Pb. Despite meticulous care, it is still unfortunately true that few analyses give completely consistent results when each daughter-parent pair is considered. Ages computed from $^{208}Pb/^{232}Th$ ratios commonly differ from ages derived from the ratios of uranium-derived leads to their parents. Lead differs greatly in chemical behavior from either uranium or thorium; even slight leaching by permeating solutions produces errors in the ratio of daughter to parent. Probably the best measure, at least for the older rocks, is the ratio $^{207}Pb/^{206}Pb$, for the isotopes of lead are chemically identical and should be leached in proportion to their abundance; their *ratio* should be unaffected by alteration, though their *quantities* may be drastically reduced. It is clear from the many concordant analyses that many uranium and thorium minerals are many hundreds of millions of years old.

Uranium and thorium are relatively rare elements, found in but few minerals in quantities adequate for analysis. Radiometric dating — radiometry — began with them, and suitable materials containing them are still avidly sought. In 1948, however, L. T. Aldrich and A. O. Nier, physicists at the University of Minnesota, found that the very common element potassium, also known to be radioactive, can be used in geochronology; K-bearing minerals are abundant in many rocks and consequently radiometric dating has advanced briskly in the past 25 years.

Potassium, the seventh most abundant element in the earth's crust, consists of three isotopes, ^{39}K, ^{40}K, and ^{41}K. The first and last of these are stable, but ^{40}K is radioactive and disintegrates mainly by two modes of a branching process. In one, an electron is ejected from the nucleus, thereby converting a neutron into a proton and the original K atom into ^{40}Ca. This conversion is of no value in geochronology because ^{40}Ca is the most abundant isotope of a very common ele-

ment, and the radiogenic part cannot be segregated from the same isotope of nonradiogenic origin. The useful mode of decay is the capture of an electron by a nuclear proton, which converts the proton into a neutron and the K atom into one of argon of the same atomic weight. The half-life of the combined process is 1.31 billion years, with 11 percent of the ^{40}K converting to ^{40}Ar.

Argon is an inert gas that enters no chemical compound. It may readily enter or leave some crystal lattices by diffusion, but in others it seems to be retained indefinitely, notably in those of biotite, muscovite, hornblende, and sanidine (the high-temperature form of potassium feldspar found in many volcanic rocks) unless the minerals are heated above about 250°C. Determination of ^{40}Ar : ^{40}K ratios requires painstaking and skillful analysis, but recently improved techniques yield consistent results with probable errors between 1 and 2 percent. Under favorable circumstances it is possible to date minerals as young as 50,000 years, in which radiogenic argon is sparse, indeed. The techniques of measuring K/Ar ratios are beyond the scope of this book, but the results of many careful workers on geologically dated materials underlie many of the younger dates given in Table 6-1.

Rubidium is so similar in chemical behavior to potassium that any K-bearing material is virtually certain to contain some rubidium. One of rubidium's two isotopes, ^{87}Rb, is radioactive and disintegrates to Sr of the same atomic weight, offering still another geochronologic tool. The half-life is somewhat uncertain, but if it is assumed to be 50 billion years the computed ages agree very well with the best U-Pb ages on associated rocks. Common Sr consists of four isotopes: ^{84}Sr, ^{86}Sr, ^{87}Sr, and ^{88}Sr; not all ^{87}Sr is derived from the breakdown of Rb, so that it is necessary to correct for the nonradiogenic ^{87}Sr in the mineral analyzed. Because of this correction to a mass already small, Rb-Sr age determinations are extremely sensitive to error in relatively young rocks but are much more reliable in Mesozoic and older rocks. As Rb is not a gas, like Ar, these ages are less subject to alteration by slight metamorphism than are those of the K-Ar method; most of the older ages given in Table 6-1 are based on Rb-Sr determinations.

Most minerals used in radiometric dating are from magmatic rocks, though the mineral glauconite, which is known to form on the sea floor before the accompanying sediment is cemented, contains potassium and has been successfully dated. Most geochronologists think that argon diffuses more readily from the crystal structure of glauconite than from the lattices of the micas, thereby reducing accuracy somewhat. Many lava and ash flows of known stratigraphic position have been dated radiometrically, but to use radiometry for estimating ages of sedimentary rocks, it has usually been necessary to bracket them between the dates from older and younger radiometrically dated intrusives. Suppose, for example, that a thick stratal sequence is bounded by unconformities. If one intrusive cuts the lower unconformity and another cuts both, the radiometric dates of the intrusives would give respectively, the maximum time range and upper age limits of the sequence, perhaps an entire system or even more. Such bracketing is represented in many of the dates listed in Table 6-1, and accounts in large part for the spreads shown by most of them. We can thus be sure that many of them are subject to change, but the corrections are unlikely to be significant—no more than a few percent.

Thousands of rocks have been dated radiometrically, many by more than one method, with consistent results. But even inconsistent results may reveal significant events in geologic history. For example, a K-Ar analysis of the minerals in a given rock may give an age of 750 m.y. for hornblende and only 440 m.y. for biotite. Many such discrepancies, checked by U-Pb dating of the highly retentive mineral zircon from the same rock, have shown that the lattice of biotite is less retentive of Ar than is that of hornblende. The discrepancy may indicate that the rock was formed 750 m.y. ago but was reheated 440 m.y. ago or even later. This, of course, is not the only possible interpretation, but one that should be tested. Another example is also instructive. Biotite crystals from some highly metamorphosed rocks in the southern Alps give Rb/Sr dates of about 11 m.y., whereas muscovite crystals from the same specimens give dates of about 12 m.y. Experiments show that biotite heated to temperatures higher than 350°C no longer give stable ages, whereas muscovite must be heated to about 550°C before its lattice permits loss of Sr. Accordingly, this apparent age difference between the muscovite and the biotite may represent the time it took for the rock to cool from 550°C to 350°C. If the increase of temperature with depth beneath the Alps was the same 12 m.y. ago as it is today, this temperature difference amounted to a depth difference of about 1 km. The cooling rate was thus brought about by the erosion of 1 km of rock per million years—roughly the present rate in the high Alps.

Advances in geochronology have been great; errors of 10 percent are unusual today, and comparisons of analyses done in different laboratories generally show that results agree within two or three percent. Cretaceous ash falls (Chapter 17) can be dated with a precision of less than one percent error—comparable to that of paleontologic dating. Moreover, radiometry offers the only means of interregional correlation within the vast time span represented by Precambrian rocks. Because argon is lost during reheating and slow cooling, the K-Ar method is far less reliable for older rocks than younger ones. It has nevertheless been widely used as the basis for subdividing the Precambrian—a classification explicitly rejected in this book. Fortunately, neither the Rb-Sr nor the U-Pb method is so readily susceptible to "resetting of the atomic clock" by reheating, and both are far superior for use with older rocks. The U-Pb method is especially reliable when applied to the highly retentive mineral zircon. In fact, recent progress in scrupulously lead-free laboratories has been so great that age determinations of rocks as much as 2600 m.y. old are repeatable within an error of 1 m.y. Because of the necessary correction for ordinary lead, this does not mean that the true age is quite so accurately known.

The tremendous sweep of geologic time is no longer debatable: it is now probable that the earth's crust is at least 4580 m.y. old; rocks of this age have been reported in eastern Siberia. The next oldest yet measured, 3750 m.y. old, are in West Greenland. The oldest rocks yet found in North America are gneisses from the Minnesota River valley, dated at about 3550 m.y. Simple organisms such as bacteria and algae (Fig. 6-1) have been found fossil in rocks more than 3000 m.y. old, and multicelled organisms have existed since late in Precambrian time, more than 600 m.y. ago. Some meteorites have given ages as great as 4600 m.y.—an age that many workers think from this and other evidence is the approximate age of the earth as a planet (Chapter 19).

It is noteworthy that the estimated age of the earth is about equal to the half-life of ^{238}U and more than seven times the half-life of ^{235}U. The earth must have originally contained twice the present amount of ^{238}U and more than forty times that of ^{235}U. Moreover, its internal energy must then have been far greater than at present, and many geologic processes correspondingly much more active.

Fission-track Dating

Atoms of uranium break down not only by the process previously outlined, but also by fissioning into two roughly equal-sized nuclei, at the fantastically sluggish rate of 1 atom in 69×10^{16} atoms per year. The fragments fly forcefully apart and knock many ions out of their normal positions in the crystal lattices of surrounding minerals. The lattice imperfections so produced are extremely small—about 50 Å wide and 10 to 20 microns long, far below the resolving power of ordinary microscopes, but they are less stable than the undamaged parts of the lattices, and therefore much more susceptible to chemical attack (which enlarges them to visibility) than the undamaged lattice alongside.

Virtually all rocks contain a few uranium atoms as impurities, either within minerals or at grain boundaries. Nearly all silica and silicate minerals are soluble in sodium hydroxide or hydrofluoric acid, and the radiation-damaged volumes especially so. Brief etching thus reveals microscopic pits and cones that mark the paths of fission fragments.

These etched surfaces are being used to date rocks. If the rock is crystalline, minerals known to concentrate uranium are segregated from it and embedded in plastic so they can be held rigidly for polishing before being lightly etched with sodium hydroxide solutions. If the rock is glassy, hydrofluoric acid is applied to a polished plate. The pits and cones of the lattice imperfections are counted in a measured area. The specimens are then subjected to a measured flux of neutrons in a nuclear reactor, producing a new crop of fission tracks by fissioning of ^{235}U; the etching is repeated and the new tracks counted. As the neutron flux is known, the increase in number of tracks is a measure of the number of U atoms present. From these data it is possible to compute the time required to form the tracks first counted. The method is relatively new and has not been tested as thoroughly as others, but it seems to yield results consistent with them.

Radiocarbon, Chronometer of the Holocene

Another radiometric clock is furnished by carbon. Carbon in coal—"old carbon"—consists chiefly of the isotope ^{12}C (generally about 98.9 percent) and a small admixture (usually about 1.1 percent) of ^{13}C. Neither of these isotopes is radioactive. But carbon in atmospheric CO_2 also contains small amounts of a third isotope, ^{14}C, formed in the upper atmosphere by the collision of cosmic rays (neutrons) with nitrogen. This collision adds an electron to one of the nuclear protons, converting it into a neutron and the ^{14}N atom into one of ^{14}C. Radiocarbon decays to ^{14}N with a half-life of 5,730 years, a time so short that after 40,000, or at most 50,000 years (7 to 9 half-lives), so little is left that it cannot be accurately measured. It is uniquely valuable, however, for use on materials less than 50,000 years old, as this time span is too short for the more slowly disintegrating elements to produce measurable amounts of daughter products. Radiocarbon dating has thus completely revolutionized archeology, and gives us the best dating for the history of the Holocene and late Pleistocene.

The method depends on the assumption that the isotopic ratio of carbon in the cells of living things is identical with that in air, because of rapid equilibration through photosynthesis and respiration. When an organism dies it ceases to exchange carbon with the air, and the ratio of ^{14}C to ^{12}C and ^{13}C begins to decline. A comparison of the ^{14}C ratio in a bone or log with that in the atmosphere is thus a measure of the time since the organism died.

Several sources of error impair the value of this method. Algae in springs whose CO_2 is partly derived from dissolved limestone—"old carbon"—have been found to deposit sinter with a much lower ^{14}C content than the air: the sinter is "born old." Molluscan shells in areas of upwelling ocean water are also hundreds of years "old" when formed. Conversely, porous reef corals, long dead, are continually "rejuvenated" by exchange of ions where they are constantly drenched in sea spray.

A more severe problem is the inconstant supply of ^{14}C, *both in time and in latitude*. Radiocarbon ages from tree-ring series cut from living bristlecone pines (*Pinus aristata*) fail to agree with the ages obtained by ring counting. It is clear from comparing ^{14}C ages of parchment (animal carbon) with those of counted tree rings, that each ring is active for only one year and does not re-equilibrate with the next succeeding ring. Egyptian artifacts known from history to date from the second century B.C. give radiocarbon dates higher than the known calendar dates. Worse still, the rate of ^{14}C production has varied, so that the measured ^{14}C date for a log or a bone might correspond with that of several different calendar ages. The possible confusion because of the variable rates of ^{14}C production is fortunately not more than two centuries, and commonly less than one.

We know, of course, that since the beginning of the Industrial Revolution combustion has added large amounts of "old carbon" from coal and petroleum—as much as 10 or 15 percent of the total now found—and the ratio increases yearly by about 0.2 percent of that present today. But the age discrepancies go far back of the rise of industry and must be due to other causes. We know that the flux of cosmic rays varies inversely with the strength of the earth's magnetic field and directly with solar flares. Incontrovertible evidence shows that the magnetic field has varied both in strength and polarity many times in the geologic past (Chapter 8), so that variations in the rate of production of ^{14}C are what should be expected.

A further weakness is that measurements of the ^{14}C content of the atmosphere, when compared with tree-ring ages, show that the ^{14}C content of air in the southern hemisphere is lower than in the northern, despite rapid atmospheric mixing. "Ages" of contemporaneous events, therefore, are somewhat "older," measured by radiocarbon, in the southern hemisphere than in the northern. This anomaly has been attributed to the far greater extent of the southern oceans and the greater solution of CO_2 in the boisterous seas of the "Roaring Fifties" than in the northern hemisphere. Paradoxically, plants in the area of Kamchatka also seem poor in ^{14}C; some live wood gives a date of about 6000 years.

Still other difficulties include the contamination of tree roots by bacteria and of shells by permeating solutions, so that many geochronologists think it useless to measure ratios of shells that have lost their original lustre.

Thus radiocarbon dates cannot precisely correspond with astronomical dates—they measure time in "radiocarbon years." Yet, except for the latitudinal effect mentioned, the variations in supply of radiocarbon should be virtually simultaneous the world over, because the time of mixing of the atmosphere is not more than two or three years. Comparable sequences can therefore be established in various localities by radiocarbon, just as they have been by fossil assemblages, though neither method measures time in calendar years. Consequently, radiocarbon assays have enormously expanded our knowledge of prehistory and of geologic events of late Pleistocene and Holocene time.

Dating by Varves

In regions of protracted winter freezing and open summers, lake sediments generally show a distinctive pattern. At spring breakup, turbid streams pour into the lakes, dropping their coarser sands in deltas but carrying finer silts and clay well out over the lake floor, there to settle out, the coarser grains more quickly than the fine. As summer progresses, occasional storms supply additional sediments, and algae, diatoms, and other microorganisms flourish in the photic zone above. With the onset of the winter freeze the supply of clastic sediment is cut off, and in the quiet waters beneath the ice the dying and dead microorganisms slowly settle to the bottom. The result is that each year a pair of contrasting sedimentary layers is deposited: the lower one composed of coarser clastic material at the bottom, grading into finer upward; the upper one, an organic-rich, extremely fine-grained layer (Fig. 6-5). These annual deposits, first studied thoroughly in Sweden, are known by the Swedish word *varve*.

The thicknesses of both layers varies from year to year because of climatic fluctuations; some years of major floods produce a thicker clastic layer, some unusually hot summers produce more organic growth. Inasmuch as climatic fluctuations are likely to be similar over a considerable area, the deposits in the lakes of a region—say a hundred kilometers across—are all likely to show similar fluctuations in varve thickness and other characteristics. Some varves are extremely thin. One 5-m core from Lake of the Clouds, Minnesota, contains nearly 10,000 annual layers, but in lakes that receive more sediment, varves are

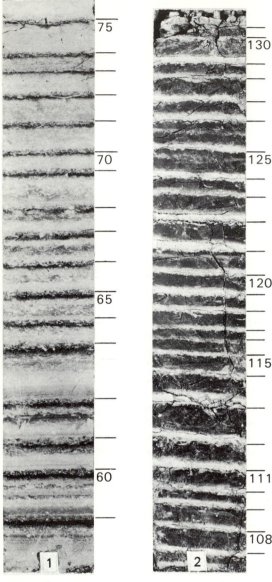

Figure 6-5
Varved glacial clay from Steep Rock Lake, Ontario.
The samples are from drill cores, each 30 cm long.
Lines alongside the cores mark the boundaries of each
varve. The light-colored laminae are summer deposits;
the dark, winter. [Photo by Ernst Antevs.]

much thicker. Nevertheless, the sequence of
varves, whether thick or thin, should be closely
parallel in deposits from nearby lakes.

By comparing these varve records in different
lakes, it has been possible to build a correlation

of lake deposits for many thousands of years
back into the Pleistocene over fairly large contiguous regions such as New England, but
climatic variances from place to place have been
too great to permit correlation across the oceans
or even, without considerable uncertainties,
between Sweden and Finland. Yet much of our
knowledge of Pleistocene and Holocene time is
based upon varve correlations in the north
temperate zone (Chapter 13).

Varves are not, however, confined to lakes
that freeze over, though this is their most familiar
locale. They evidently form in other environments in which the supply of sediment varies
greatly from season to season and where the
deposits are protected from waves. Though difficult to establish as annual alternations—a requirement for true varves—pairs of alternating
layers of sediments in thick marine Cretaceous
shale in Wyoming so resemble the known varves
of Minnesota lakes that most observers think
they were similarly formed. In the Green River
Formation of Eocene age in Wyoming—a lake
deposit of widely ranging salinity—W. H. Bradley
of the U. S. Geological Survey recognized similar
alternating pairs of strata. If these are annual,
they represent about 8 million years. Bradley
estimated the Green River Formation to represent about a third of Eocene time; if the deposits
are indeed varves, this would make the Eocene
about 24 m.y. long—a fair agreement with our
best, but uncertain, estimates from radiometric
dates of about 15 m.y. (Table 6-1). Many saline
deposits, such as those of Great Salt Lake and
the Permian of West Texas (Chapter 17), are
exceptionally well varved.

DISTRIBUTION OF ROCKS OF VARIOUS AGES

Significantly, the various Systems of the geologic column are not distributed at random over
the face of the earth (Fig. 6-6). Rocks of Precambrian age are found in many mountain ranges,
but the great bulk of their exposures are in continental areas of low relief. Precambrian rocks are
exposed in Fennoscandia and most of India
and Korea. The same is true in large parts of
Australia west and south of the Great Artesian
Basin, most of Africa south of the Sahara, much
of eastern Brazil, the northern United States
bordering Lake Superior, and much of eastern

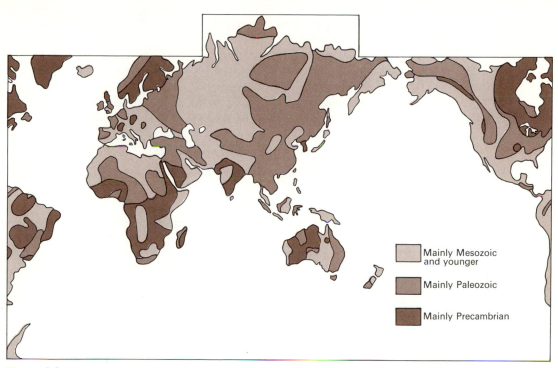

Figure 6-6
Geologic map of the earth, greatly generalized. [From many sources, generalization by the author.]

Canada to the Arctic and, on the coast, to Labrador. These great expanses are nearly all of low local relief, though some parts of Labrador and Greenland reach altitudes of nearly 1 km. They are called **Precambrian Shields**.

The shields are overlain at their borders by younger rocks, but they extend beneath them, generally forming broad surfaces of low relief— the **cratons** of the continents. The mountains— belts of deformed rocks that rise above the cratons—differ widely in age. Except for those of Australia, the mountain ranges immediately surrounding the Pacific Ocean are made up of rocks ranging from Triassic to Pleistocene, and so is the chain that extends from the western Mediterranean through the Alps, Carpathians, Causasus, Elberz, Himalaya, Burma, and Indonesia. Older ranges such as the Appalachians, the Scandinavian Mountains, and the Great Dividing Range of eastern Australia include only Paleozoic and older rocks. The mountain-making process is the theme of Chapter 8.

Questions

1. Most of the Standard Geologic Column is based upon marine strata. What relations between these and land-laid strata permit correlations between widely separated land-laid strata?

2. In so far as all biologists agree that new species arise by sudden mutations, why do they reject Cuvier's notion of Catastrophism?

3. In assessing the possibility of correlating a limestone in Kansas with another in Pennsylvania, would their physical characters or their fossil assemblages be the more useful? Why?

4. Most radioactive mineral samples used in dating are separated from intrusive rocks. How can these give clues to the age of the associated sediments?

5. Many obviously sedimentary rocks contain no fossils. How is their age determined? State some reasons for the absence of fossils.

6. Why is it that fossil correlation over the whole earth is based chiefly on marine organisms?

Suggested Readings

Berry, W. B. N., *Growth of a Prehistoric Time Scale*. San Francisco, W. H. Freeman and Company, 1968.

Dalrymple, G. B., and Lanphere, M. A.: *Potassium-Argon Dating*. San Francisco: W. H. Freeman and Company, 1969.

Eicher, D. L., *Geologic Time*. Englewood Cliffs: New Jersey, Prentice-Hall, 1968.

Faul, Henry, *Ages of Rocks*, *Planets* and *Stars*. New York: McGraw-Hill, 1966.

Geological Society of London, *Phanerozoic Time Scale*: Supplement to v. 129, pp. 13–44, 1964.

Olsson, Ingrid U. (ed.), *Radiocarbon Variations and Absolute Chronology*. Stockholm: Nobel Institute, 1970.

Simpson, G. G., *The Life of the Past*. New Haven: Yale University Press, 1953.

Woodford, A. O., "Catastrophism and Evolution": *Journal of Geological Education*, v. 19, pp. 229–231, 1971.

Scientific American Offprints

102. Harrison Brown, "The Age of the Solar System" (April 1957).

811. Edward S. Deevey, Jr., "Radiocarbon Dating" (February 1952).

832. Robert Broom, "The Ape-men" (November 1949).

837. Martin F. Glaessner, "Pre-Cambrian Animals" (March 1961).

838. Charles T. Brues, "Insects in Amber" (November 1951).

842. H. B. D. Kettlewell, "Darwin's Missing Evidence" (March 1959).

844. J. E. Weckler, "Neanderthal Man" (December 1957).

867. Norman D. Newell, "Crises in the History of Life" (February 1963).

CHAPTER 7

Earthquakes and the Earth's Interior

EFFECTS OF EARTHQUAKES

Almost at dawn on All Saint's Day, 1755, the churches of Lisbon, thronged with worshippers, heaved and shook as the earth shuddered violently. Within six awesome minutes, the great stone arches and roofs collapsed, trapping thousands beneath their rubble. The sea withdrew, exposing the bar at the harbor mouth, and then returned in a wall of water a dozen meters high, drowning hundreds who had taken refuge from falling walls by running to the open docks. A shorter, but intense, shock cascaded slides of rock from the mountains onto the city, raising dust clouds so dense that many thought them to be volcanic ash. Two hours later a third shock struck, and fires completed the devastation; by nightfall more than a fourth of the city's 235,000 people were dead, and thousands of the survivors were maimed or mad from terror.

The Elbe at Lübeck, more than 2300 km away, rose and fell a meter or more, and Loch Lomond, nearly as distant, was thrown into waves more than half a meter high. Though possibly reinforced by a nearly simultaneous but independent shock farther south, the quake virtually demolished the cities of Fez and Mequinez, 600 km away in Morocco, plus scores of other towns in Spain and North Africa. An area at least five times that of the United States was affected.

The great earthquake that jolted southern Alaska in 1964 produced widespread deformation in Prince William Sound. Southeast of a curving line from Port Nellie Juan to Valdez, a belt of crust more than 50 km wide and 200 km long was uplifted variably to a maximum of about 10 m, and a corresponding belt to the northwest was depressed by several meters (Fig. 7-1).

Lisbon is in a hilly region; southern Alaska is mountainous. Most earthquakes center near regions of high relief, either terrestrial or marine (Fig. 7-26), but some of the greatest have taken place in areas of low relief, far from any hills, mountains, or submarine ridges. In 1819 a great earthquake struck the Rann of Kutch, India, at the mouth of the Indus, flooding much of the delta and raising a scarp to the north, the "Allah Bund," or "Dam of Allah," as much as 6 m high and 80 km long.

Perhaps the greatest earthquake ever to strike the North American continent was that of December 11, 1811, near New Madrid, Missouri, on the banks of the Mississippi. At 2 A.M. the quake threw people from their beds; in a few seconds many of their log cabins had toppled, and sand, mud, and water gushed from widespread fissures, burying cleared fields for scores of kilometers in all directions. Undercut banks caved off into the Mississippi, islands sank, others rose, boats were swamped and hurled ashore. During the next 15 months minor shocks continued by the thousand, along with two quakes comparable to the first. A broad tract of swampland was uplifted and drained, and a still larger area nearly 250 km long and 60 km wide sank — in

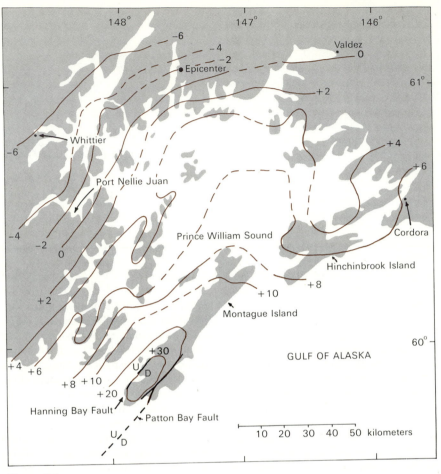

Figure 7-1
Level changes brought about during the earthquake of 1964 in Prince William Sound, Alaska. Contour interval 2 feet between −6 feet and +10 feet (175 cm to 300 cm); 10 feet between +10 and +30 feet (3 m between +3 m and +9 m). [After George Plafker, U. S. Geological Survey, 1965.]

places as much as 3 m—to form new lakes and swamps. Reelfoot Lake, so formed, is 7 km across. The great shocks stopped pendulum clocks and rang church bells as far away as Boston, and cracked plaster in Virginia. A century later the loss of life would have been appalling, but in those pioneer days few were endangered.

Many other major earthquakes may be cited. One struck eastern Sicily and the Calabrian coast across the Straits of Messina on December 28, 1908, destroying the cities of Messina and Reggio and killing about 100,000 people. The coast at Messina dropped more than 60 cm.

Two great earthquakes rocked Kansu, in western China, one in December 1920, the other in May 1927; each was reported to have killed about 100,000 people, chiefly by collapse of dwellings dug into weakly consolidated loess (Chapter 13). A great earthquake struck Assam, India, in 1897, raising a scarp more than 11 m high and destroying buildings in Calcutta, more than 300 km away. During a great earthquake in Mongolia on December 4, 1957, an oblique-slip reverse fault more than 250 km long raised a scarp nearly 10 m high and hurled riders from their camels for several meters; sheep rolled "like balls" down the hills, and houses almost 300 km away were destroyed. A month later travel was still hazardous because of persistent avalanches and landslides.

Even though the earthquake that struck the San Fernando Valley suburban area of Los Angeles on February 9, 1971, was a modest one in terms of energy released, it did kill 64 people when hospitals, homes, and apartments collapsed. Its timing—early enough in the morning so that traffic was light on the freeways—was fortunate, for many of the overpasses were immediately toppled. During the quake a wedge of rock several kilometers long and thickening northward was thrust southward for about 0.8 m with a left-lateral component as great as 1.6 m. The surface of the wedge was raised nearly 2 meters in some places. This earthquake is notable as having the highest acceleration rate ever measured, as much as 1.25 G at times. (Similar accelerations, though unrecorded by instruments, must have occurred on several other faults, for some have actually ejected stones vertically from the surface.)

The Wellington earthquake of 1855 and the Hawkes Bay earthquake of 1931 in New Zealand were each accompanied by notable land tilting and uplift of 2 to 3 m. The great Chilean earthquake of 1960 affected an area more than 1000 km long and 200 km wide, dropping a narrow belt of country as much as 2 m and raising some of the islands as much as 6 m. The first shocks came at the north on May 21 and gradually extended southward. During a great shock on the evening of the 22nd, some of the continental slope was raised near the southern end of the disturbed zone, generating a giant sea wave (tsunami) that cost hundreds of lives in Chile and traveled across the Pacific to drown hundreds more in Hawaii, Japan, and the Philippine Islands. A quake on the floodplain of the Hoang Ho in the sixteenth century is said to have killed more than 800,000 people—almost certainly the greatest catastrophe in history.

The San Francisco Earthquake

The great earthquake that struck San Francisco a little after 5 A.M. on the morning of April 19, 1906, is of special interest because its study led to many of the current ideas about faulting. Many buildings collapsed, especially those on marshy or filled ground; others on solid rock were little damaged by the shock. Most water mains were severed, however, and fires raged for days, destroying many of the buildings left standing. Though the loss of life could only be estimated, it probably exceeded 700; material loss was more than $400 million (in 1974 dollars probably 7 or 8 times that sum). San Francisco suffered most, but towns more than 150 km away were also severely damaged.

A great fault zone, long known as the San Andreas rift, cuts obliquely across the California Coast Ranges for more than 1000 km—from the ocean at Point Arena on the north through southwestern San Francisco and thence far to the southeast of San Bernardino, where the main strand is lost in the alluvium of the Colorado Desert. Not only are geologic formations cut off and the rocks much sheared along this fault, but it stands out in the landscape, as many streams follow it for various distances before resuming transverse trends. Small ponds are strung out along it, some on steep slopes—strikingly anomalous features in arid southern California.

During the 1906 earthquake the ground was rent open along this old fault from Point Arena to San Juan Bautista, more than 450 km, with the ground to the west moving northward relative to that on the east—a right-lateral fault (Fig. 7-2). The greatest displacement, more than 6.5 m, took place about 50 km north of San Francisco, whence it diminished in both directions, though not regularly. Locally, especially toward the north, slight vertical displacements of less than a meter took place. Figure 7-3 shows the offset that occurred in 1940 on one of the San Andreas family of faults.

CAUSES OF EARTHQUAKES

The ultimate cause of earthquakes is still debated (we return to some speculations later), but the immediate cause must be the sudden movement of rock masses along faults, for too many visible displacements have been associated with earthquakes for their coincidence to be accidental. Even so, most earthquakes are unattended by visible displacement, though some must have occurred beneath the surface. Most mine workings reveal faults that do not reach the surface; many oil wells have been cut off during earthquakes by faults that did not rupture the

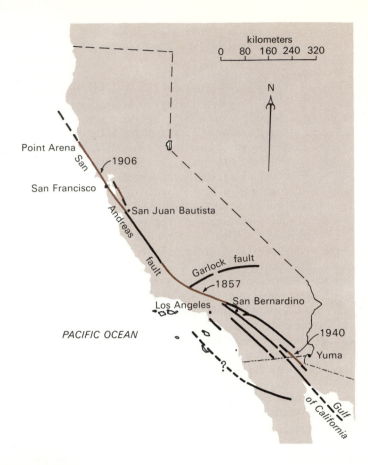

Figure 7-2
The San Andreas and associated fault zones in California and northern Mexico. Brown line segments show where the ground was broken by historic quakes in 1857, 1906, and 1940. [After Clarence R. Allen, 1957.]

Figure 7-3
Orange grove in Imperial Valley, California, displaced by a fault in December 1940. A kilometer to the south, the displacement was more than 4 m. [U. S. Army photo, courtesy of E. Marliave.]

surface. Every fault is of course of finite length and dies out eventually; for example, no movement was determinable on the San Andreas fault in 1906 south of San Juan Bautista. Nearly all students accept faulting as the cause of earthquakes, even though many cause no surface displacement.

Elastic Rebound Theory

The displacement that accompanied the 1906 earthquake was nearly horizontal, and therefore could not be directly due to gravity, as are landslides. The region is not volcanic, and in any event no volcanic eruption has shown energy remotely comparable to that expended in moving thousands of cubic kilometers of rock for several meters along a break 450 km long. What, then, was the source of this tremendous energy?

The most widely accepted suggestion was made by the American geophysicist H. F. Reid. Most crustal movements take place slowly. Though fault displacements are generally sudden, the energy released had probably accumulated slowly as the crustal block west of the fault slowly drifted northward relative to the block to the east. Slip was long delayed by the interlocking of irregularities on either side, allowing the slow accumulation of elastic strain energy — the kind stored when a bow is bent. Ultimately a critical value was reached, the friction along the fault surface was overcome, and the two sides of the fault "snapped past each other."

Reid's suggestion came from his study of the precise surveys by the U. S. Coast and Geodetic Survey in 1851–1865, 1874–1892, and, after the earthquake, in 1906–1907. Though the first was less accurate than the later surveys, it was consistent with them. If the triangulation stations Diablo and Mocho, both between 50 and 60 km east of the fault, are considered not to have moved, the points nearer to and across the San Andreas fault show notable systematic relative movements. Reid grouped the stations according to their distance from the fault and determined their average displacement, north or south (Table 7-1).

The relations may be visualized from Figure 7-4. The line AOC represents a straight line crossing the fault at a time when there was no elastic strain in the region. As regional distortion began, the point A moved relatively northward to A' and C south to C'. Because the fault did not slip, the line AOC became bent into A'OC'. Just before the earthquake, A had advanced to A" and C to C", the distances being such that the total A-A" plus C-C" amounted on the average to about 4.5 m, with a maximum of 6.4 m. Then the elastic strain overcame the friction on the fault, and the stored energy was released with explosive violence. Line A"O straightened out to A"O' and C"O to C"O". It is clear that lines such as A"O' and C"O", which had been straight just before, would be bent during the faulting into curves A"B and C"D, identical with A"O and C"O but

Table 7-1
Average displacement of points between the surveys of 1874–1892 and 1906–1907, relative to the Diablo-Mocho line, held fixed.

Group	Number of stations	Average distance from fault of stations in group, in kilometers		Displacement, in meters	
		East	West	Northward	Southward
A	1	6.6	—	—	0.58
B	3	4.3	—	—	0.85
C	10	1.4	—	—	1.52
D	12	—	2.	2.96	—
E	7	—	5.8	2.38	—
F	1	—	37.1	1.77	—

Figure 7-4
Diagrammatic map showing how the 1906 offset along the San Andreas rift is explained by the "elastic rebound" theory. [After H. F. Reid, 1911.]

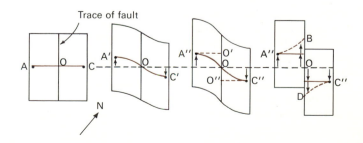

bent in the opposite directions. Reid assumed that when the fault slipped, the elastic strain was almost completely relieved; the fault displacement is merely the "elastic rebound" of the rocks — their recovery from the strain accumulated over a considerable time. The data in Table 7-1 strongly support this view.

Since 1907 the Coast and Geodetic Survey — now a part of the National Oceanic and Atmospheric Administration — has systematically resurveyed the region, confirming a continued strain in the same sense as that before the 1906 earthquake. The average rate of movement is about 3 to 5 cm a year. If we assume that distortion equal to that recovered by the 1906 earthquake is necessary for a repetition (and there is no reason at all to think so), about a century or two must elapse before another great earthquake occurs. Such an estimate is valueless, for we cannot be sure that the binding of the fault is the same as it was before 1906 or that another fault of the system may not break instead of the main fault. But continuance of the movement is strong support for the "elastic rebound" theory.

Seismic Sea Waves (Tsunami)

The great waves that overwhelmed Lisbon and southern Chile were doubtless caused by sudden displacement of the neighboring sea floor, like the tilting of the floor of Prince William Sound during the 1964 earthquake. Some of these waves were caused by submarine slides. Continuous-recording tide gages reveal many much smaller disturbances, probably also caused by fault displacements of the sea floor.

Sea waves caused by faulting are called seismic sea waves, or **tsunami** (both plural and singular, Japanese). Although 60 percent are caused by dip-slip reverse faults (p. 67), so that shoreward water motion would be expected, in most tsunami the first water movement is actually a withdrawal, followed, after a few minutes, by a great inrush of the sea. Some seismic sea waves are truly gigantic when they reach shore: perhaps the greatest on record is one that reached a height of 64 m as it broke on the south tip of Kamchatka in 1737. During the Alaska earthquake of 1964, a tsunami in Valdez Arm washed debris nearly 52 m above sea level and threw barnacle-covered boulders

weighing nearly a ton to heights of 26 m. A tsunami 28 m high struck the city of Miyako, Japan, in 1896; another in 1868 carried the U.S.S. *Watersee* far inland from its Chilean anchorage and left it high and dry. Waves generated by an earthquake off Peru were more than 2 m high when they reached Japan after traveling more than 16,600 km. These heights depend greatly on the configuration of the shores (Chapter 16); in the open ocean, the waves are never noticed. The wave speed depends only on the depth of water; it reaches as much as 700 km per hour in the Pacific but much less in the shallower Atlantic. Damage and loss in the Pacific borders have been so great that tsunami warnings are now issued regularly when seismographs indicate a potentially dangerous earthquake.

EARTHQUAKE WAVES AND THEIR TRANSMISSION

If we break a bat hitting a baseball, our hands are stung by the elastic vibrations. Similarly, when two huge crustal blocks shear past each other along a fault, the elastic vibrations set up travel through the earth in all directions. The nature of these vibrations is diagrammed in Figure 7-5, left.

We assume the point P in the figure is a particle in a uniform mass of perfectly elastic material, and that it is pressed toward P′ by some outside force; obviously the material on its right is compressed (squeezed) while that to its left is rarified (stretched). In a perfectly elastic solid these volume changes mean that elastic energy is being stored; the compressed material tends to expand and the rarified material to contract. Consider the line APB. If the point P is moved to P′ by an external force, the line must bend to some position, such as A′P′B′, thereby setting up a distortion or shear in the line. The nearer each point on APB is to P, the further it tends to slip parallel to PP′ from its original position, as the line becomes A′P′B′. Thus a differential shear arises between P and R, for P-P′ is larger than R-R′, etc. This means that some energy is stored up that can be released in shear.

Now let the outside force be removed. Point P′ will not simply return to its original position but will have a momentum, like a pendulum, that will

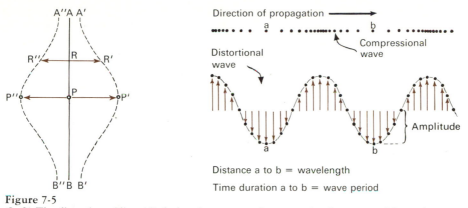

Figure 7-5
(Left) The distortion of line AB during the passage of compressional waves at right angles to AB. (Right) The effect of compressional and distortional waves on the grouping of a series of equally spaced points. Note how the compressional wave affects the spacing in the direction of propagation, whereas the distortional wave offsets them from its path from left to right. [After J. B. Macelwane, *When the Earth Quakes*, Bruce Publishing Company, 1947.]

carry it past P to a point such as P″, reversing the fields of compression and rarification and the direction of shearing tendency. At P″ the kinetic energy that carried the point past the center will have again been transformed to elastic energy, and the point will again oscillate through P to P′. The distance from P′ to P″ is the amplitude of the oscillation.

Rocks are of course not perfectly elastic; some vibrational energy is consumed as frictional heat, and the amplitude dies gradually away. But the principle is identical: if we consider the effects of the vibration of a rock particle at P on its neighbors, and of theirs upon their neighbors, it is clear that two types of waves, one compressional, the other shear, must emanate through the rock in all directions from P.

In the *compressional* wave, the rock particles vibrate back and forth in the line of wave progress as pulses of compression and rarification alternate through the rock. It is a sound wave in rock, and because it travels nearly twice as fast as the shear wave, it is called the primary, or **P wave**. The *shear wave* is transverse; that is, the particles vibrate at right angles to the direction of progress (Fig. 7-5), and because it travels slower than the P wave, it is called the secondary, or **S wave**. The particle motion in the S wave is like that set up in a loosely hanging rope when one end is given a sharp flip. Each particle of the rope moves vir-

tually at right angles to the length of the rope, but the wave travels from one end of the rope to the other. The S waves are not the visible ground waves reported from some great earthquakes; their wavelengths are measured in thousands of meters, and their speeds are far too great for the human eye to see.

In an elastic solid the speed of the P waves depends directly on the resistance of the material to change in volume and shape (shear), and inversely on its density; that of S waves depends directly on resistance to change in shape, and inversely on density. The formulas are:

$$\text{Velocity of } P \text{ wave} = \sqrt{(K + \tfrac{4}{3}\mu/\rho}$$

$$\text{Velocity of } S \text{ wave} = \sqrt{\mu/\rho}$$

in which K = bulk modulus (a measure of resistance to change in volume), μ = modulus of rigidity (a measure of resistance to change in form, shear), and ρ = density. It is assumed that the medium is isotropic and "perfectly" elastic. It should be pointed out that although the density is in the denominator of both equations, the bulk and rigidity moduli both increase much faster with increase in density than does density itself; both P and S waves are speedier in denser than in less dense rocks. The formulas make it obvious that the P wave is the faster.

The incompressibility (bulk modulus) and rigidity (shear modulus) of rocks can be measured in the laboratory.* Assuming the wave theories correct, they can also be computed from the measured speeds of waves produced by accurately timed explosions in precisely located drill holes (artificial earthquakes). With precise chronometers, the travel time from drill hole to seismograph can be known to 1/1000 second. Such field measurements fail to agree with the laboratory measurements, but the differences are generally small and readily rationalized as due to the obvious inhomogeneity of all rocks in nature (joints, foliation, bedding surfaces, grain-size variations, etc).

Explosion seismology — measurement of artificial earthquake waves — is extremely valuable as a tool of exploration (Chapter 18), even though several kinds of rock may fortuitously have very similar elastic properties. Empirical field measurements of wave velocities have given the following data:

Table 7-2
Velocities of P waves in various materials.

Material	Wave velocity kilometers/ second	Density
Unconsolidated sediments	0.3–2.5	1.5–2.2
Consolidated sandstone	1.5–5.4	2.0–2.6
Sedimentary rock in deformed belts	3.0–5.4	2.5–2.8
Metamorphic rocks	4.5–6.0	2.7–3.0
Limestone	3.0–6.0	2.4–2.7
Igneous rocks	4.5–7.5	2.4–3.0

Thus P waves in bedrock are from twice to twenty times as fast as in surficial deposits. In water-saturated soil their speed is two or three times that in dry soil. This contrast is such that it is generally easy to determine the depth to the water table (Chapter 14) by seismic experiments.

The P and S waves (body waves) travel in all

*The bulk modulus, K, is the force per unit area times the volume divided by the change in volume of the body: $K = pV/v$, where p is stress, V volume, and v the change in volume. Similarly, the shear modulus, μ, is the shearing stress, T, divided by θ, the shear produced: $\mu = T/\theta$.

directions through the elastic body of the earth. In passing through rocks whose elastic properties are identical in all directions (isotropic), waves travel in straight lines. But waves that pass through rocks whose properties are not uniform in all directions (anisotropic) or pass from one kind of rock into another, such as from limestone to shale, are bent, or refracted, like water waves that cross a shoaling beach diagonally (Fig. 16-11). They may also be reflected at such a boundary, just as sound is echoed from a cliff. Furthermore, where either a *P* or an *S* wave strikes a sharp boundary — a *discontinuity* in the elastic properties along its path — it sets up two new *P* waves and two new *S* waves. Moreover, several kinds of "surface waves" are formed, so called because they travel along or close to the discontinuity. Such waves are set up along the surface of the earth, where the discontinuity consists of the contact between rock and air or water, as well as within the earth at geologic contacts separating rock masses of differing elastic properties.

The surface waves set up along discontinuities travel much more slowly than the body waves and are far more complex: they are the *L* **waves** (long). Those following the earth's surface produce disturbances that diminish with depth.

From this outline, it is obvious that earthquake waves are extremely complex. Because they are generally felt only near the source, delicate instruments are needed for recording them at long distances. We briefly discuss these **seismographs** before proceeding to the deductions they permit as to the structure of the earth.

SEISMOGRAPHS

We all know that the bottom book in a stack lying on a table can be knocked out so quickly that the books above fall almost vertically, without tipping over. The inertia — tendency to resist acceleration — of the upper books allows them to be decoupled from the bottom book. The seismograph is an instrument designed to measure the ground motion with respect to a mass that, like the upper books, is as independent as possible of motions of the support.

Many ingenious seismographs have been built for this purpose. Perhaps the most readily under-

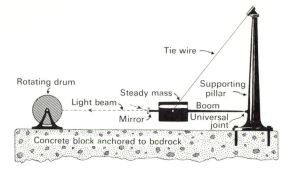

Figure 7-6
Diagram of a horizontal-pendulum seismograph. Modern instruments are much more complex than the old model illustrated, but the principle involved is identical.

Figure 7-7
A seismogram, with letters indicating the arrival of the various earthquake waves. [After L. D. Leet, *Practical Seismology and Seismic Prospecting*, Appleton-Century-Crofts, 1938.]

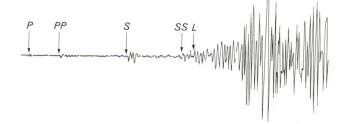

stood (though it is no longer made as shown in Fig. 7-6) is the so-called **horizontal pendulum**. In this device, a heavy weight at the end of a boom is supported by a wire from the supporting pillar at a level slightly lower than the joint that attaches the boom to the pillar.

The weight thus tends to remain at rest while an earthquake vibrates the supporting pillar. A beam of light reflected from a mirror on the weight to a motor-driven strip of photographic paper records the relative motion of weight and foundation. Clearly, this record shows only the motion normal to the boom; to record all horizontal movements completely, two such pendulums are oriented at right angles, usually one with the boom north-south, the other east-west. A third mass supported by springs is needed to record vertical motion. Time is marked automatically by an electric clock suitably linked to the photographic record—the **seismogram**.

Inferences from Seismograms

By comparing seismograms of the same earthquakes at several stations and of different earthquakes at the same station, seismologists can recognize the many waves that elastic theory predicts. Most seismograms show several kinds of waves, each related to a definite wave path within the earth. For example, the arrivals of *P*, *S*, and *L* waves are labeled on the seismogram in Figure 7-7, as well as the reflected *P* wave, *PP*, and the reflected *S* wave, *SS* (see also Fig. 7-10). (Modern electronically recorded seismograms do not closely resemble these outmoded records, but their interpretation is similar.) Note that with the onset of the surface *L* waves, the record becomes much more complex than during earlier phases because of mutual interference of the waves.

REFRACTION AND REFLECTION AT DEPTH

By tabulating the **travel times** of the waves from earthquakes of known source and comparing the records of many stations, **time-distance tables** have been compiled. **Time-distance curves** (Fig. 7-8) are made by plotting distances from the sources (in degrees of the earth's circumference) against travel time. Significantly, the curves for the *P* and *S* waves are concave toward the axis of distance; the greater the distance of travel, the faster the speed of the waves. We infer that the

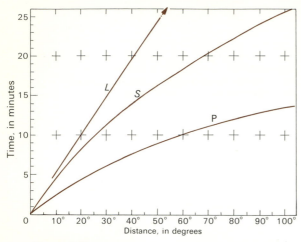

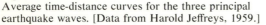

Figure 7-8
Average time-distance curves for the three principal earthquake waves. [Data from Harold Jeffreys, 1959.]

waves travel faster as they penetrate deeper into the earth, for the increase in speed is far greater than the difference (increasing with depth) between *surface distance* (arc) and *straight-line* (chord) distance. The *L* waves, which travel along the surface at constant velocity, have straight time-distance curves.

At points far from the source, the *P* and *S* waves

emerge at higher angles than they would if they followed straight lines. Upon reaching the great discontinuity at the surface, they are reflected and proceed again in curved paths, thus accounting for the long train of waves recorded at distant stations (Fig. 7-10).

We know from the earth's mass and dimensions that its average density is very nearly 5.516 g/cm³ (p. 116), whereas the average density of the surface rocks is only about half that. The material in the earth's interior must therefore be far denser than that at the surface. As the formu-

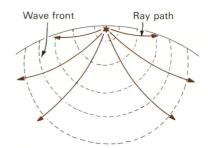

Figure 7-9
The curving paths of earthquake waves due to increase of wave velocity with depth. The waves progress along lines normal to the wave front. Only six of the infinite number of wave paths emanating from the source are shown.

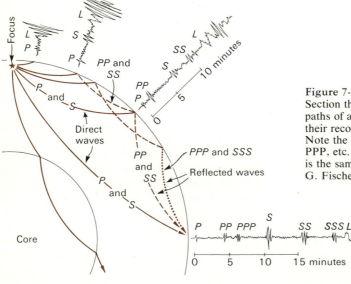

Figure 7-10
Section through a part of the earth, showing the paths of a few of the many earthquake waves and their records on seismograms at four stations. Note the reflected waves from the surface, PP, PPP, etc. The time scale of all the seismograms is the same. [After A. Sieberg, *Erdbebenkunde*, G. Fischer, 1923.]

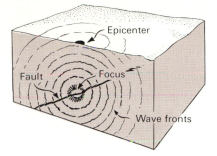

Figure 7-11
Diagram showing the positions of the focus and epicenter of an earthquake caused by deep-seated movement on an inclined fault.

las for wave speed show, if only the density of the rocks increased with depth and their elastic properties remained unchanged, the waves should travel more slowly at depth rather than more swiftly. This proves that rigidity and incompressibility increase with depth even faster than density. We return to this point later.

EPICENTRAL DISTANCE

The sharply pulse-like records on seismograms suggest that most earthquakes begin in very small areas, even though the faults may extend for hundreds of kilometers, like the San Andreas in California. The Chilean earthquake of 1960 was well recorded at many stations. From the point of origin the rupture traveled more than 1200 km southward at a speed of more than 3.5 km per second. The point at which the disturbance begins is the **focus**; the point vertically above the focus is the **epicenter** (Greek, "above the center"). (See Fig. 7-11.)

The time lag between arrivals of *P* and *S* waves allows the distance from epicenter to seismograph to be read from the time-distance curves in degrees of arc. (Modern instruments are so accurate that account must be taken of the earth's ellipticity in locating the epicenter.) The locus of the epicenter lies on a circle of radius equal to the arc read from the time-distance curve drawn around the station (with the slight correction noted). Accordingly, adequate records from three stations will locate the epicenter: it is where three such circles intersect (Fig. 7-12).

The time-distance curves for most stations more than a few degrees distant from an epicenter are much the same, regardless of the direction of approach of the seismic waves. The deep earth must therefore be nearly homogeneous

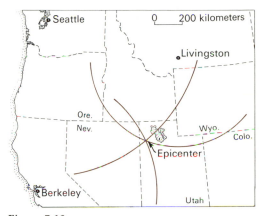

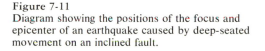

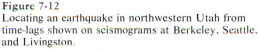

Figure 7-12
Locating an earthquake in northwestern Utah from time-lags shown on seismograms at Berkeley, Seattle, and Livingston.

(or similarly variable with depth) at depths greater than a few score kilometers. A few regions differ from the average, however. For example, the deep earth beneath Hawaii shows slower speeds than below most regions.

ISOSEISMAL LINES

Few epicenters coincide with seismograph stations, so that direct measures of accelerations connected with an earthquake are rare, though indirect measurements are becoming more and more accurate (p. 91). Qualitative information, such as degree of destruction or perceptability of the shock by people, is normally available. Although many "scales of intensity" have been suggested, the one now most widely used is the "Modified Mercalli" scale, 1956, adapted from a proposal of the Italian seismologist Mercalli.

Modified Mercalli Intensity Scale (1956)
(Slightly reworded from Don Tocher, 1964.)

I. Not felt.

II. Felt by persons at rest, especially on upper floors or in similarly favorable places.

III. Felt indoors but may not be recognized as an earthquake. Vibrations resemble those from a passing light truck.

IV. Felt indoors like the vibrations from a passing heavy truck or the jolts of a heavy ball striking a wall; hanging objects swing; standing objects rock; windows, dishes, doors rattle; glasses clink; walls and frames may creak.

V. Felt outdoors; sleepers awaken; liquids move and some spill; small unstable objects move or fall; doors swing; shutters and pictures move; pendulum clocks stop or change rate.

VI. Felt by all; many run outdoors in fright; people walk unsteadily; windows, dishes, glassware break; knicknacks, books, dishes fall from shelves, pictures from walls, furniture moves or overturns; weak plaster and poor masonry crack; small church and school bells ring; trees, bushes shake visibly and rustle.

VII. Noticed by automobile drivers. Walkers have difficulty keeping balance; weak chimneys break at roof lines; furniture breaks; poor masonry cracks; plaster, loose bricks, stones, tiles, cornices fall; small slides and caving develop along sand and gravel banks; water becomes turbid with mud; large bells ring; concrete irrigation ditches are damaged.

VIII. Affects steering of motor cars; damage to good unbraced masonry, with partial collapse; some damage to good, somewhat reinforced masonry but none to masonry reinforced against horizontal stresses; walls of stucco and some of masonry fall; chimneys, factory stacks, monuments, towers, and elevated tanks twist and fall; frame houses not bolted down move on their foundations; loose panel walls thrown out; decayed pilings break off; branches break off trees; flow and temperature of springs and wells change; wet ground and steep slopes crack.

IX. Causes general panic; poor masonry destroyed, good unbraced masonry heavily damaged; reinforced masonry seriously damaged; general damage to foundations; frame structures not bolted shifted off their foundations; frame cracks; serious damage to reservoirs; underground pipes break; alluvial areas cracked conspicuously, ejecting sand and mud; earthquake fountains and sand craters develop.

X. Destroys most masonry and frame structures; some well-built wooden structures and bridges destroyed; serious damage to dams, dikes, embankments; large landslides; water thrown on banks of canals, rivers, and lakes; sand and mud shift horizontally on beaches and flat lands; rails bend slightly.

XI. Puts underground pipe lines completely out of service; rails bend greatly.

XII. Distorts lines of sight and level; damage nearly total; large rock masses displaced; objects thrown into the air.

Local foundation conditions vary so much—for example, from marshy or filled ground to bedrock—that these crude criteria are neither quantitative nor comparable for different localities. They nevertheless give much valuable information to engineers, insurance underwriters, architects, and civic planners. After any considerable earthquake in the United States, the U. S. Geological Survey circulates questionnaires to persons in the affected areas, asking them to check various items according to their experience of the shock. The returned cards are tabulated by locality, and a map is drawn showing lines of equal intensity according to the Modified Mercalli scale.

Lines drawn through points of equal intensity plotted on a map are isoseismals (Fig. 7-13). The isoseismals generally form crude ovals about a

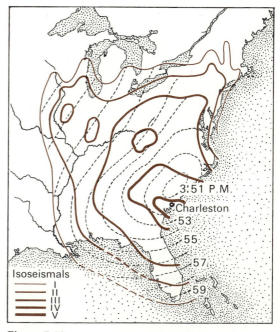

Figure 7-13
Isoseismals of the Charleston earthquake, 1886. The fine broken lines (called coseismals) connect points where the earthquake struck at the same time; the numbers indicate their arrival times. [After C. E. Dutton, U. S. Geological Survey.]

center, which either coincides with the epicenter as determined by seismograms or is close to it. That the instrumentally determined epicenter does not precisely coincide with the "field epicenter" may be accounted for by differences in soils or near-surface geology. Local rock differences not only influence the distribution of earthquake damage (the isoseismals) but also affect the speeds of elastic waves toward nearby seismographs, and thus the position of the epicenter. The isoseismal pattern of the San Francisco earthquake is complex because of the abrupt boundaries between hills of strong rock and small basins filled with weakly consolidated muds and silts. The present-day hazard in the San Francisco area is greater than in 1906 because of the continued building on bay fill.

EARTHQUAKE MAGNITUDE

Intensity, being only a measure of local ground shaking, is not very useful in evaluating the true energy of shocks. A more significant measure, developed by seismologist C. F. Richter of the California Institute of Technology, is based on earthquake **magnitude** (M). This is measured by the amplitude of the largest trace recorded by a standard seismograph located 100 km from the epicenter of an earthquake of normal focal depth. Empirical tables that express the variations in this measure with distance and focal depth permit the magnitude of any earthquake to be calculated from any seismogram. Though agreement as to magnitude as measured at different stations is far from perfect, the range in magnitudes is so tremendous that even a rough estimate is useful.

Semi-empirical studies have led to various estimates of the relations between seismic energy and magnitude; for a 1-unit increase in magnitude, the increase in energy released is about 30 times. Nearly 1000 shocks of magnitude 5 would be required to release as much energy as a single shock of magnitude 7.

Magnitude is always reported in Arabic numerals, intensity in Roman.

The San Francisco earthquake of 1906 had a magnitude of about 8.25; the Tokyo (Kanto) earthquake of 1923, about 8.1. The greatest magnitudes yet measured are those of an earthquake in Colombia (1906) and one in Assam, India (1950), both about 8.6. The Alaska earthquake of 1964 had a magnitude variously estimated at 8.4 to 8.6 and was thus at least twice as energetic as the San Francisco earthquake. Normally a magnitude of 2 corresponds to a shallow shock barely perceptible near the epicenter; a magnitude of 7 is about the lower limit of a major earthquake, but we have already noted the tremendous damage that a shock of only 6.4 did to San Fernando, California (1971).

There is no direct relation between magnitude and ground acceleration: the great Assam earthquake of 1950 produced ground acceleration of about $0.5\,G$, whereas the much less energetic San Fernando earthquake produced a horizontal acceleration of $1.25\,G$ at the Pacoima Dam.

DEPTH OF FOCUS

If the focus of an earthquake lies deep, the lag of the S wave behind the P wave is greater than it is from a shallow source; this makes possible an estimate of the depth of focus. We thus know that the foci of most earthquakes lie at depths between 3 and 30 kilometers. About 4 percent of recorded quakes differ from most in having small, if any, surface waves, and also in arriving at distant stations earlier than the normal quake with like epicentral distance. This and the fact that the isoseismals of such shocks form very erratic patterns, with the felt intensity falling off much more slowly than for normal earthquakes, is interpreted to mean that the depth of focus lies much deeper than usual (Fig. 7-14).

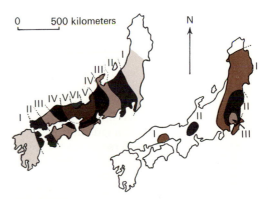

Figure 7-14
Maps of Japan, showing the isoseismals of the normal North Tazima earthquake (left) and of the deep-focus earthquake of March 29, 1928 (right). [After Wadati; from J. B. Macelwane, *When the Earth Quakes*, Bruce Publishing Company, 1947.]

Laboratory studies of rock strength under high pressures have shown a great increase in strength with hydrostatic pressure—an increase so great that many students of rock mechanics had thought faulting must be impossible at such depths. They suggested that the mechanism for such deep-seated faults might be a sudden phase change among the minerals. In favor of this view is the fact that the flowage implied by the general condition of isostasy takes place at much shallower depths than the deepest foci, so that it seems unlikely that strain could accumulate slowly at greater depths, as is implied by the elastic-rebound mechanism; an explosive phase change would seem more likely.

But the seismographic records are like those of shallower earthquakes. If a phase change were involved, the first motion at all stations should be either a compression (if the phase change were a decrease in density) or a rarification (if the change were an increase in density). Instead, as with shallower shocks, stations in some directions record compressive first motions while other stations record rarification. Perhaps the long time scale involved in geologic processes makes the laboratory measures of rock strength misleading.

The systematic relation of most, though not all, deep-focus earthquakes to the ocean deeps, first recognized by the late Professor H. H. Hess of Princeton University, is a strong argument for the interpretation of deep-focus earthquakes as tectonic, as further discussed later in this section and in Chapters 8 and 9 (Fig. 7-15).

FIRST MOTION

From detailed studies of seismograms, the attitudes of fault surfaces and the direction of slip of earth blocks have been determined. In the example of Figure 7-16, we assume a northeast-trending vertical fault along which left-lateral strike slip takes place. Before the earthquake the stress pattern would be as shown in black; on beginning of faulting the compression and rarifaction would obviously reverse, given the elastic rebound mechanism. The pattern of first motion then would be as shown by the brown arrows. The interpretation is ambiguous, however, for a right-lateral movement on a plane at right angles to the fault assumed would give precisely the same pattern of stress release. If the regional geology is well known, it is commonly possible to decide which is the active fault and which the

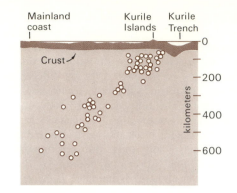

Figure 7-15
Cross section of the outer earth, at right angles to the Kurile Trench and island chain, showing the foci of recorded earthquakes projected to a common plane. There is no vertical exaggeration. [After H. H. Hess, 1948.]

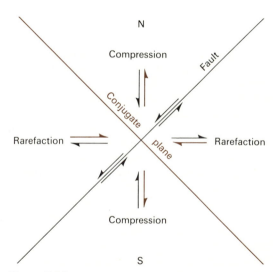

Figure 7-16
First motions as they would occur on the hypothetical vertical fault trending NE–SW if left-lateral movement took place along it. In black, the stress situation before faulting; in brown, the first motion at the time of faulting, due to release of the prior stress pattern. The broken line at right angles to the fault is complementary to the fault, and right-lateral movement along it would produce the same pattern of stress release.

conjugate surface. If the fault surface is inclined and three-component seismograms are available, it is possible to determine all three components of displacement. This has become a most important clue to the major tectonics of the earth (p. 104).

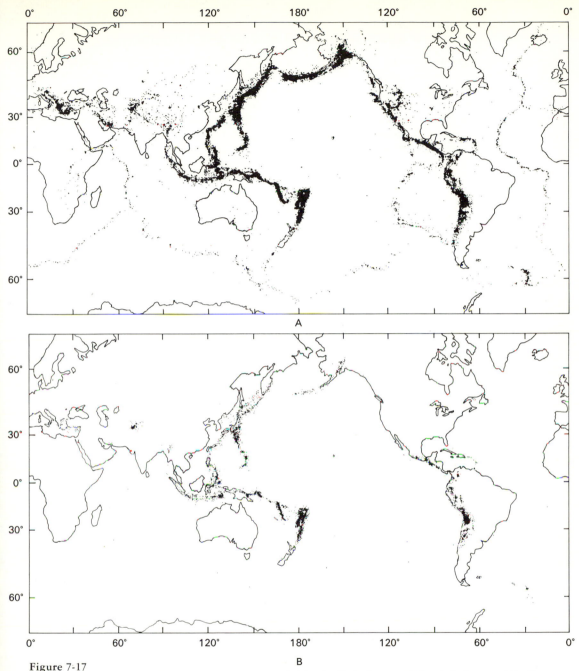

Figure 7-17
Seismicity of the earth, 1961–1967. Epicenters recorded by the U. S. Coast and Geodetic Survey, (A) all epicenters, (B) epicenters whose foci lie between 100 and 700 km. Inasmuch as locating an epicenter requires records from several stations, many minor shocks are not shown. [From M. Barazangi and J. Dorman, *Seismological Society of America Bull.*, 1969.]

DISTRIBUTION OF EARTHQUAKES

Seismologists estimate that every year more than a million earthquakes shake the earth strongly enough to be felt (M = 2), though relatively few are strong enough to be widely recorded. About 220 shocks greater than M = 7.75 and about 1200 strong ones (7.0 < M < 7.75) occur per century. Figure 7-17 shows the distribution of earthquakes recorded between 1961 and 1967; the distribution pattern over several decades would be very similar.

These maps omit innumerable earthquakes not widely recorded. Small, shallow-focus shocks

(M = 5 or less) apparently occur nearly everywhere over the earth, but the larger shocks are nearly, though not all, restricted to narrow belts.

Of the major, geographically restricted earthquakes, the shallow ones, whose first motions indicate normal displacement, abound along all the ocean ridges—the East Pacific Rise, the African Rift Valleys, the northern Mediterranean, and parts of western Mexico and southwestern United States, though they are not restricted to these belts. Others with reverse or thrust first motion are common all around the Pacific "Ring of Fire," and extend in a belt of scattered shocks through Indonesia, Burma, the Himalayas, Iran, and Turkey to the Mediterranean. Scattered shocks of this kind also occur in China and Mongolia. Shocks deeper than 100 km, chiefly with a thrust or reverse sense of motion, are largely restricted to the landward sides of ocean deeps. Their relation to the trenches is nearly everywhere like that shown in Figure 7-15. Others occur beneath the Tibetan Plateau, which may be thought of as inward from the belt of thrust faults at the south foot of the Himalayas.

Along the lines where the mid-ocean ridges are offset as shown by magnetic data given in Chapter 8, many vertical faults show strike-slip motions of moderate magnitude, precisely opposite to the direction of apparent ridge offset. The San Andreas Rift is considered by many to belong to this class of faults, as is the very similar Alpine Fault of New Zealand. Following a suggestion of the Canadian geophysicist J. T. Wilson, these are called **transform faults**, discussed later.

The distribution of the various kinds of faults is extremely significant in the broad picture of earth tectonics, but it is not the whole story; it must not be forgotten that the very powerful Charleston (1886) and Mississippi Valley (1811–1812 earthquakes)—perhaps the greatest ever to strike North America—were in no way related to ocean ridges, trenches, or significant boundaries of crustal blocks. A few great deep-focus earthquakes beneath Spain are also aberrant in the world pattern.

THE CRUST OF THE EARTH

The term "crust of the earth" has been freely used earlier in this book, without definition, as a convenient lay expression for the near-surface part of the earth. But the word is a technical one in geology, as we will now note: In 1909 the Yugoslav seismologist Mohorovičić noted features in the seismograms of continental wave paths that indicate a layered structure in the shallow part of the continents: seismograms recorded at stations less than 800 km from an epicenter show two compressional and two shear waves instead of only one of each. By comparing travel times to more distant stations, he was able to identify the smaller pair as the normal *P* and *S* waves long known from distant shocks. The larger pair traveled more slowly but seemed to have started earlier. It was received first at stations within about 150 kilometers but lagged farther and farther behind the other pair with distance, finally becoming unrecognizable at about 750 to 1000 kilometers from the source.

Mohorovičić showed that this could be explained by assuming a layered structure of the earth: an outer layer—the **crust**—characterized by slow speeds, and resting on the deeper body of the earth—the **mantle**—in which speeds are higher. A shock originating in the crust sends waves directly to the nearby stations—powerful waves that arrive before those that penetrate the mantle, even though mantle waves are faster. But at more distant stations the mantle waves more than make up for the time lost in travel from the focus down to the crust-mantle interface; they overtake the crustal waves and arrive first at distant stations.

Seismic velocities change abruptly nearly everywhere at this interface; clearly, the **Mohorovičić** ("Moho" or "M") **discontinuity** is a fundamental feature of earth structure—a sharp boundary between the contrasting rocks of crust and mantle.

The Continental Crust

Explosion seismology, which gives depths to the M Discontinuity within about 10 percent, has revealed much about the thickness of the crust and the elastic properties of both crust and upper mantle. It was formerly thought that the crust was fairly uniform—say about 35 km thick under lowlands and about 60 km under the mountains—as might be expected from Airy isostasy (Chapter 2). In fact, however, crustal thickness and elastic properties vary widely and abruptly from region to region of the United States, as is shown in Figure 7-18.

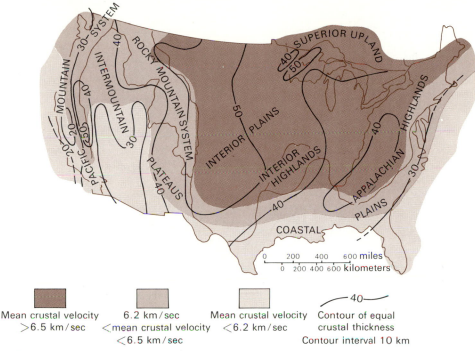

Figure 7-18
Mean crustal thickness and mean velocity of *P* waves within the crust of the United States. [After L. C. Pakiser and J. Steinhart, 1964.]

Interestingly, crustal thickness bears no relation to altitude. The east and west coasts, as Figure 7-18 shows, differ in crustal thickness by 10 km. Much of the crust in the inter-mountain plateaus and the Great Basin is less than 30 km thick; the Great Valley of California, only about 20 km. The Sierra Nevada has a root about 50 km deep. The southern Rocky Mountains have about the same crustal thickness, but it can hardly be considered a root, for the crust is equally thick beneath the high plains of eastern New Mexico and Colorado and western Kansas and Oklahoma. Indeed, the Rocky Mountains in Montana have a thinner crust than the Montana and North Dakota plains. In the southwest, crustal thickness locally changes so abruptly that the M Discontinuity slopes as much as 30°. In general, under the continents the *P* wave in the upper mantle is about 1 km/second faster than in the crust just above the M Discontinuity—a somewhat smaller contrast than is common beneath the oceans.

Mean velocity of crustal *P* waves also varies greatly, from less than 6.2 km/sec to more than 6.5/sec, tending to be higher where the crust is thick and lower where it is thin. In Chapter 2 we noted the remarkable correlation between topographic height and negative Bougeur gravity anomalies; we now see that this is only partly due to greater crustal thickness beneath highlands than elsewhere. The crust beneath many highlands must actually be less dense than that beneath lowlands. Thus isostatic compensation is accomplished only in part by the Airy mechanism and in part by Pratt's; it involves density variations in the upper mantle as well as in the crust.

That the upper mantle also varies notably from region to region is shown by the wide variation in velocity of P_n—the compressional wave in the mantle immediately beneath the M Discontinuity, as seen in Figure 7-19. The range is from less than 7.8 km per second near the Utah-Nevada border to more than 8.3 km per second beneath southern Oklahoma, a range of about 6 or 7 percent of the lower figure, and so great that it must be considered in establishing local time-distance curves for some individual seismograph

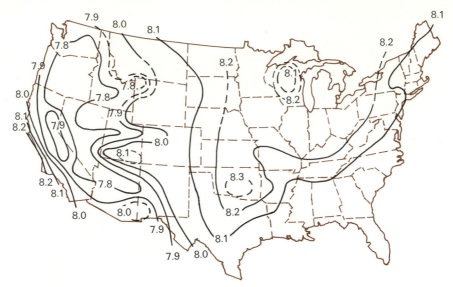

Figure 7-19
Speed of P_n (the compressional wave immediately beneath the M-Discontinuity) in the United States. [After E. Herrin, 1969; based on E. Herrin and J. Taggart, *Seismological Society of America Bull.*, 1962.]

stations. Such great variations in the elastic properties of the upper mantle can only reflect real differences in the rocks composing it. In a rough way, P_n tends to be low where the crust is thin and high where it is thick, but the correlation is far from perfect.

Of what is the crust composed? Geologic mapping, seismic and gravity studies all show that sedimentary rocks are locally as thick as 15 km in narrow zones, discussed more fully in Chapter 8. Over the crust as a whole, the sedimentary layer is much thinner; a recent careful estimate for the area of the conterminous United States gave about 4.5 km for all Phanerozoic rocks, a volume considerably greater than estimated earlier. Of course very large volumes of sedimentary rocks have been metamorphosed to schist, gneiss, and other metamorphic rocks, or melted into magma (Chapter 4). We see, in the great shields of all the continents, highly varied gneisses, schists, marbles, and plutonic and volcanic rocks of many kinds—all in the most diverse structural arrangements and with complex trends. It is not surprising, then, that the crust is far from homogeneous, either in thickness or elastic properties, when studied in detail. The really surprising thing in view of the complexity of basement rocks, is that the *average* elastic

properties of the crust over considerable areas in Europe, Asia, and North America are so nearly the same that roughly the same time-distance curves are widely applicable.

The average velocities of the seismic waves in the upper continental crust are similar to those measured in granite in the laboratory. Granite and gneiss of roughly the same composition are indeed the dominant rocks in the shields. They consist chiefly of minerals rich in silicon and aluminum. Therefore, following a suggestion of the great Austrian geologist Eduard Suess, the material of the upper crust is called **sial** (*si*, for silicon; *al* for aluminum). Laboratory measurements have shown that the elastic properties of rocks largely depend on the average atomic weight of the minerals composing them.

Earthquake speeds increase with depth at a rate greater than can be explained solely by increasing pressure, even within the crust, so that the average composition must change with depth. In a few places the change is abrupt, and a so-called Conrad Discontinuity has been postulated, but generally the change is gradual; certainly nothing in the geology of the shields suggests a sharp horizontal layering. The wave speeds in the lower crust are variable from place to place—it is far from homogeneous—but they are generally appropriate to gabbro or a similar rock. Though, the lower crust is often called the "basaltic layer," the rocks composing it are almost certainly not basalt, though probably of the same average composition; in Suess's terminology this is **sima** (*si* for silicon, *ma* for magnesium). The crust as a whole is thus less siliceous and richer in iron, aluminum, calcium, and magnesium than the average of the exposed rocks (Appendix IV, Table IV-2). The evidence is of course indirect, and the figures must be taken as approximate.

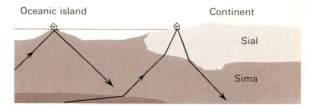
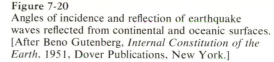

Figure 7-20
Angles of incidence and reflection of earthquake waves reflected from continental and oceanic surfaces. [After Beno Gutenberg, *Internal Constitution of the Earth*, 1951, Dover Publications, New York.]

The Oceanic Crust

Oceanic crust differs fundamentally from continental, as determined by both seismic and gravity studies. The seismological evidence is of three kinds: two from seismograms of natural earthquakes, the third from explosion seismology.

The earthquake evidence was first recognized by the brilliant Beno Gutenberg: the higher the angle at which a wave strikes the earth's surface, the greater the proportion of its energy that is reflected back into the earth. Significantly, the *PP* waves from two earthquakes equal in intensity and epicentral distance, one of which is reflected from the ocean floor and the other from a continental segment, differ greatly in strength. Those reflected from the ocean floor are the weaker, suggesting that they strike the surface at lower angles. This, Gutenberg thought, means that the oceanic crust lacks a sialic layer (Fig. 7-20). As the continental waves slow down in going from sima to sial, they are refracted to strike the surface at a higher angle than they otherwise would; their reflected energy is correspondingly greater.

The surface waves tell a similar story: their speeds, unlike those of the body waves, vary with the wavelength. As a very crude rule of thumb, a surface wave of a given wavelength travels at a speed that depends on the elastic properties of a layer about as thick as the wave is long: a wave 3 km long travels at a rate dependent on the properties of the upper 3 km of the crust; a wave 200 km long travels at a rate dependent on the properties of all the material to a depth of 200 km.

Short surface waves travel at different speeds in different crustal segments, and much faster

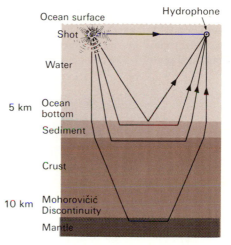

Figure 7-21
Refraction and reflection paths for explosion seismology at sea. [After Maurice Ewing and Frank Press, "Structure of the Earth's Crust," in *Handbuch der Physik*, Springer Verlag, 1956.]

across segments of the ocean floor than across continents. But longer waves, being more dependent on properties of the mantle than of the crust, have nearly the same speeds everywhere. Gutenberg showed that these facts are consistent with the absence of sial from oceanic crust.

Explosion seismology has verified the inferences from earthquake records. An explosion or the discharge of an air gun under water sets up shock waves that reach the sea bottom and are partly refracted and partly reflected, precisely like earthquake waves (Fig. 7-21). Pressure sensors towed by ships record the returning waves, and their records can be analyzed like seismograms. Timing the direct wave in water gives the distance from shotpoint to receiver; timing the waves returning from the bottom gives a measure of the elastic properties of the suboceanic crust.

Immediately beneath the pelagic sediments, most ocean floors are composed of rocks with elastic properties like those of the continental sima. The oceanic sima layer is only about 5 or 6 km thick—in many places even less—above the Moho, which thus generally lies only 10 or 12 km below sea level, in great contrast with its depth beneath the continents.

Over most of the ocean floor the sedimentary cover is only 1 or 2 km thick, but both sedimentary and crustal thickness vary greatly. Although the average may be represented by the left column in Figure 7-22, crustal sections in the

southern Gulf of California are considerably thinner, those on the East Pacific Rise (column 2) are still thinner, and those on the Mid-Atlantic Ridge axis show no sharp discontinuity equivalent to the Moho. Seismic velocities in rocks immediately beneath the thin ridge-axis sediments (P = 5.1 km/sec) increase without a discontinuity from about 7.2 km/sec to 7.8 or higher, characteristic of the normal mantle.

Over wide areas three suboceanic layers can be recognized: (1) unconsolidated sediments averaging about 1 km thick in the Atlantic and half that in the Pacific; (2) consolidated sediment about 1.7 km thick, with a *P*-wave speed of about 5 km/sec and (3) a little less than 5 km of probable basalt or gabbro, with a *P*-wave speed of about 6.7 km/sec. Beneath the M Discontinuity the upper mantle has speeds of 7.8 to 8.3 km/sec. In general the speed of *P* waves in the oceanic

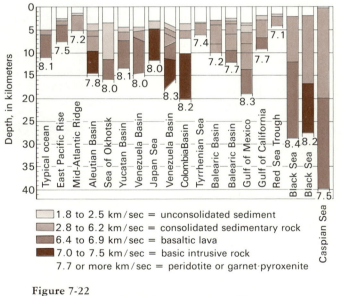

Figure 7-22
Seismic cross sections of oceans and seas. Numbers correspond to velocities of *P* waves from explosions. Oceanographic seismologists generally interpret the velocities as follows:

    1.8 to 2.5 km/sec, unconsolidated sediment,
    3.8 to 5.4 km/sec, consolidated sedimentary rock,
    6.5 to 6.9 km/sec, basaltic lava,
    7.1 to 7.5 km/sec, mafic intrusive rock,
    more than 7.7 km/sec, garnet pyroxenite or peridotite.
[Data from C. L. Drake and others, 1964; J. I. Ewing and M. Ewing, 1959; R. W. Raitt, 1956; J. D. Phillips, 1964; and H. W. Menard, 1967.]

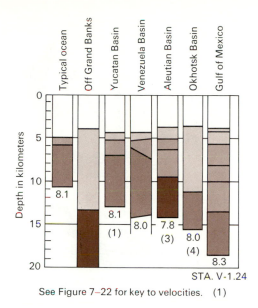

See Figure 7–22 for key to velocities.   (1)

Figure 7-23
Oceanic crustal segments arranged in order of their approach toward continental thicknesses of sial. [Data from C. L. Drake, M. Ewing, and G. H. Sutton, 1959, *Physics and Chemistry of the Earth* (v. 3) and from Menard, 1967, *Science* v. 157, pp. 923–924, copyright by American Association for the Advancement of Science.]

crust and upper mantle differ by about 1.5 km/sec, a somewhat greater contrast than that between continental crust and mantle.

At the foot of the continental slope off the eastern United States, the American oceanographer Drake and his colleagues have found exceptional oceanic crust: elongate basins several scores of kilometers wide and hundreds of kilometers long, locally more than 9 km deep and filled with sediments underlain by a probable basaltic layer. In the Gulf of Mexico and the Black Sea, into which great rivers pour vast amounts of sediment, thicknesses as great as 10 km have accumulated. In the seas behind some island arcs, sedimentary accumulations are also far thicker than the average on the ocean floor (Fig. 7-23).

In some of the thicker oceanic sections shown in Figure 7-23, the crust is about as thick as some of the thinner parts of the continental crust. The American oceanographer Menard has suggested the possibility that these segments of the sea floor

are being converted to continental masses, as many other segments have apparently been converted in the geologic past (Chapter 8). Sedimentary facies changes have shown that many areas, such as those of the Alboran Sae, east of Gibraltar, and the Tyrrhenian Sea, west of Pisa, were formerly continental but are now deeply drowned, their crusts nearly oceanic in character; some of the sections in Figure 7-23 may be evolving in the opposite direction.

Thus though we note considerable variation in sedimentary thickness in many oceanic environments, the mafic part of the crust seems relatively uniform, differing greatly in this respect from the simatic part of the highly variable continental crust. Except beneath the mid-ocean ridges, little oceanic crust has elastic properties yielding *P*-wave speeds between 7.2 and 7.8 km/sec, whereas the lower continental crust contains many such bodies.

Crust Transitional from Continent to Ocean

The great contrast between the crustal structures of continents and oceans renders the transitional zone between them very significant. The transitions are very different on the Atlantic side of North America from those on most Pacific shores. Figure 7-24 shows the depth to the mantle over the western Atlantic. The crust thins from the 2000-m isobath seaward, with shallow mantle beneath the Bermuda Rise and nearby areas. But the thinning is by no means rigidly normal to the coast.

The cross sections in Figure 7-25 show the gradual thinning of the continental crust for distances of 300 to 400 km from the coast before reaching normal oceanic dimensions.

No such areal coverage of sea-floor crustal structure is available for the Pacific shores, but a representative cross section across the Peru-Chile Trench is given in Figure 7-26. No such trenches exist along the Atlantic shores. The medial continental crustal material of inferred density 2.8 wedges out within less than 200 km of the coast, and does not extend west of the trench. Again, the agreement of gravity as inferred from seismic records with independent gravity measurements is convincing.

110

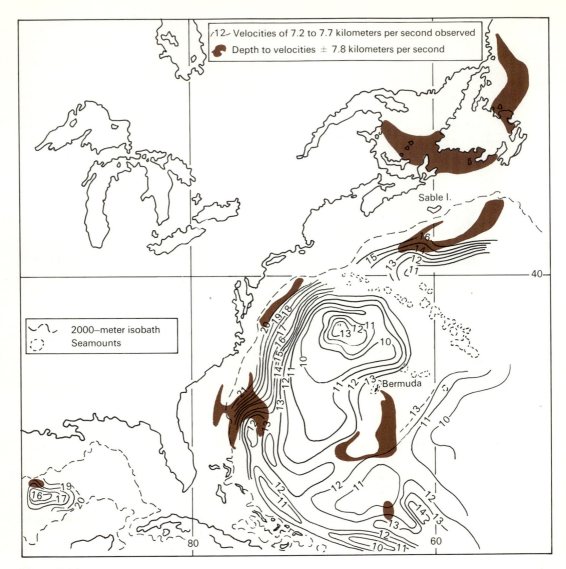

Figure 7-24
Depth to the mantle off eastern North America. Stippled areas indicate P_n velocities between 7.2 and 7.7 km/sec, which are lower than the usual 7.8 to 8.3 km/sec. [After C. L. Drake, J. I. Ewing, and Henry Stockard, 1968, *Canadian Journal of Earth Sciences*, by permission.]

THE MANTLE AND CORE OF THE EARTH

The Low-Velocity Zone

Just below the M Discontinuity, P_n waves generally travel a bit faster beneath the oceans than beneath the continents. Speeds of 8.6 to 9.0 km/sec are commonly reached at 20 km beneath the oceans, but such speeds are not reached until about 100 km beneath the continents. Gutenberg long ago suggested that seismograms of distant earthquakes were best interpreted as evidence for a zone in the upper mantle in which both P and S speeds were somewhat lower than in both overlying and underlying layers—a Low-

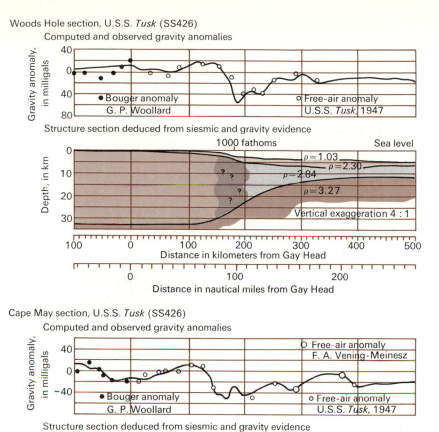

Figure 7-25
Cross sections transverse to the shore, off Woods Hole, Mass., and Cape May, New Jersey. Inferred crustal structure derived from both seismic and gravity observations. Note that the dots on or close to the line of gravity anomalies are the anomalies that should be observed if the crustal densities have been correctly inferred from the seismic records. The close agreement gives confidence in the interpretation. [After J. L. Worzel and E. L. Shurbet, 1955, *Proceedings of the National Academy of Sciences* v. 41, pp. 458–469.]

Velocity Zone (LVZ). For many years this idea was treated skeptically, but with the development of explosion seismology and nuclear bomb tests whose timing and location are accurately known, the existence of the Zone was demonstrated. The LVZ does not have a uniform depth; the top varies from about 70 km to perhaps 150 km in depth and the bottom from 200 to as much

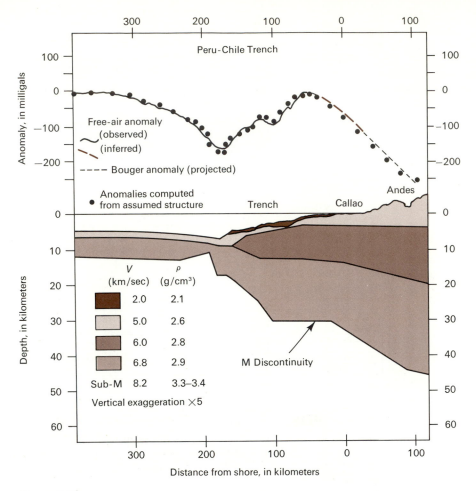

Figure 7-26
Section across the continental-oceanic interface on the west coast of South America near 13°S. Note the wedging out of the layer with *P*-wave velocity of about 6, density about 2.8, before reaching the Peru-Chile Trench. [After D. E. Hayes, 1966, *Marine Geology* v. 4, pp. 309–351.]

as 360 km. Neither top nor bottom is sharply defined; and very locally a top is lacking—there is no M Discontinuity.

The LVZ is generally attributed to the existence of hotter than normal rocks or even to pockets of magma. Here may be the locus of magma generation and of much of the lateral flowage involved in isostatic adjustment. It extends beneath both continents and oceans. The rocks above the LVZ are collectively called the **lithosphere**—the rock shell of the earth. The LVZ is the **asthenosphere**—the weak zone of the earth. We return to this contrast in properties many

times in later chapters, as it is thought to play an important role in many geologic processes.

Below the Low-Velocity Zone the wave velocities again increase downward, indicating that the rigidity and incompressibility of the mantle rocks continue to increase with depth. The rate of increase, however, is not uniform. In studying a Montana earthquake of 1925, Perry Byerly of the University of California found a marked increase, both in *speed* of *P* waves and in the *rate of increase of speed* with depth, beginning at epicentral distances of about 20°. It has since been found that these changes are fairly consistent, though not uniform, the world over and that *S* wave velocities also change greatly at similar distances. Waves emerging at this distance have penetrated to depths of 360 to 400 km. Both rigidity and incompressibility increase rapidly from this depth downward.

The abrupt rate of increase in velocity of *P* waves beginning at about 400 km depth soon lessens somewhat, though the velocities themselves continue to increase, until at about 650 to 700 km, the rate of increase with depth again becomes nearly uniform and less than that at shallower depths (Fig. 7-27). The rate at which *S*-wave velocity increases with depth is modest between 400 and 650 km, much higher in the next 50 km, and somewhat lower at greater depths.

Waves that have penetrated to a depth of 2752 km are the only ones below the Low-Velocity Zone to show decreasing speed with depth. They slow from a velocity of 13.63 km/sec to about 13.32 km/sec at a depth of 2898 km, as found in a careful study by Professor Bruce Bolt of the University of California.

At epicentral distances of about 102° to 104°, emerging *P*-waves follow a curved path whose deepest penetration is about 2898 km (some seismologists favor a slightly shallower depth). A short distance beyond, both *P* and *S* waves suddenly fade—the *S* wave completely, except for reflections from discontinuities, and the *P* wave becomes so weak beyond an epicentral distance of 118° as to be recorded only on superior seismographs. But at epicentral distances of about 143° the *P* wave reappears and is very powerful. Had it traveled through the earth's center at the same speed it had at a depth of 2898 km, it should have reached the antipodes about 16 minutes after the earthquake. Instead, it arrives about 20 minutes after. This delay can only mean that the speed through the **core** of the earth—the part below the Gutenberg-Wiechert Discontinuity at 2898 km—is far lower than in the mantle just above.

Moreover, since the speed is lower in the core, we see the reason for the weak records between epicentral distances of 102° and 143°, the so-called **shadow zone**. When a wave emerges from a medium in which its velocity is high and enters another in which its velocity is lower, the wave is refracted and penetrates deeper into the medium of lower speed, just as a reading glass

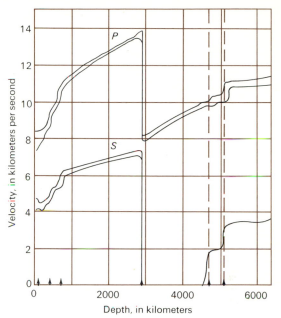

Figure 7-27
Precision bands for *P* and *S* velocities in the earth's interior below the crust. Arrows along the abscissa indicate radii near which discontinuities in one or both seismic velocities have been determined. [From B. A. Bolt, 1972, *Physics of the Earth and Planetary Interiors*.]

focuses light rays toward its thicker part because the speed of light is less in glass than in air. The earth's core thus acts as a huge converging lens for seismic waves.

This interpretation is the basis for the paths shown in Figure 7-28, which shows the earth with a central core of radius about 3475 km, surrounded by mantle that extends from the core boundary—the Gutenberg-Wiechert Discontinuity—to the Mohorovičić Discontinuity at the base of the crust. This shows how the lower velocities in the core explain both the shadow zone and the extraordinary strength of the *P* wave at 143°, for here are focused waves that impinged on a large segment of the core boundary.

The abruptness of the Gutenberg-Wiechert Discontinuity is strikingly shown by Canadian records of a South Pacific earthquake: at Toronto (epicentral distance 141°), only a very faint record was made, but at Ottawa, only 2° farther from the epicenter, the *P* wave was very strong. In Figure 7-28 the shadow zone includes the whole

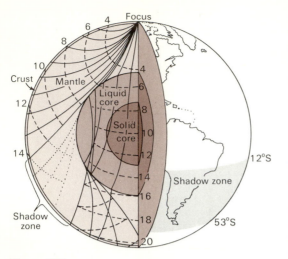

Figure 7-28
The shadow zone of an earthquake originating at the
North Pole. The cutout shows the effects of the
core on the pattern of wave paths and wave fronts.
The numbers are the travel times in minutes for each
of the paths shown. [Modified from Beno Gutenberg,
Internal Constitution of the Earth, 1951, Dover
Publications, New York.]

surface between 12°S and 53°S—that is, the belt
between 102° and 143° from the North Pole, the
assumed epicenter.

The *P* waves that penetrate the core to a total
depth of 5145 km have notably higher speeds
than those that merely penetrate the outer core,
indicating a sharp discontinuity at this level (Fig.
7-27). This is taken as the boundary of the **inner
core** with properties notably differing from those
of the **outer core**. What is the significance of the
three sharp discontinuities—the Moho, the
Gutenberg-Wiechert, and the Inner-Outer core
boundary—and the more diffuse but fairly distinct changes in mantle properties at about 360
to 400 and 650 to 700 km? Before discussing
these questions we must turn to some of the major
chemical features of the earth.

The Composition of the Mantle

The chemical composition of the outer crust
of the earth cannot be representative of the interior, for several reasons: (1) The mean density
of the earth (p. 116) is far higher than could be
reached by the self-compression of material of
the average composition of the crust. (2) As the

earth is a part of the solar system, it is reasonable
to expect its mean composition to be comparable
to that of the sun (taking account of the fact that
the earth's gravitational attraction is too little to
hold either H or He, now escaping into outer
space). The crustal composition is very different,
being especially deficient in iron (Chapter 19).
(3) The composition of the meteorites that have
been studied differs greatly from that of the crust.

Meteorites were generally thought to represent
fragments of a minor planet whose orbit was in
the asteroid belt (Chapter 19), and hence should
be more or less representative of the earth. They
are now thought never to have been assembled
into a major body, but are nevertheless samples
of planetary material. If then, we accept the
average meteorite as giving a clue to the composition of the earth as a whole, we find that the
earth must consist essentially of Fe, O, Si, and
Mg, with less than a percent each of Ni, S, Ca,
and Al (important constituents of sial; see Appendix IV, Table IV-2). If the *average* composition of the earth is the same as the average
meteorite, then almost all the Ca, Al, Na, Mn,
K, Ti, and P of the earth must be concentrated
in the crust; the mantle and core must contain
most of the Fe, Mg, Si, and O.

This inference is confirmed by the fact that
basaltic lava on Kilauea, whose journey from a
depth of about 60 km has been traced by seismographs as a series of shallowing foci, contains
inclusions of ultramafic rocks, virtually free from
any minerals except olivine and pyroxene. This
is also true of many other basalts, and experiments have shown that partial melting of such
rocks as garnet peridotite will yield basaltic melts
(Chapter 9). From this we conclude that the upper
mantle, below the M Discontinuity, is composed
of garnet peridotite, dunite, garnet pyroxenite,
or eclogite. As the upper mantle is heterogeneous, perhaps it is composed of all of these in
various mixtures.

Experimental work on many minerals has
shown that, at high pressures and temperatures,
many invert to new phases or break down into
others. Plagioclase, the commonest mineral of
the upper crust, is unstable at pressures greater
than 15 kilobars, equivalent to a depth of about
45 km, and inverts to garnet and pyroxene. The
germanium analogue of forsterite, Mg_2GeO_4,
inverts from an olivine-like space lattice to a
closer-packed structure like that of spinel under
high pressure. Thermodynamic calculations and

laboratory experiments suggest that the diffuse transition zone near 400 km, wherein both *P*-wave speeds and their rate of increase with depth increase, is caused by such inversions as that of olivine to the spinel structure or of olivine plus stishovite (the densest known form of silica) to a mineral of pyroxene composition having a close-packed space lattice like that of corundum.

The other diffuse zone of elastic change near 700 km may be caused by the breakdown of silicates to their constituent oxides: MgO (periclase), SiO_2 (stishovite), and FeO, as is suggested by high-pressure shock experiments. This material probably constitutes the entire lower mantle. Because of variations in temperature, mineral proportions, and minor constituents, the inversions take place over a considerable range of pressures, not at a single value.

A decrease in wave speed observed in a zone about 150 km thick just above the core is thought to be due to pockets of core material in the lower mantle. The Gutenberg-Wiechert Discontinuity is certainly a solid-liquid phase change, as shown by the loss of *S* waves, which indicates that the outer core is molten. The density, further discussed later in this chapter, also requires a change in chemical composition, a change to a molten mixture of metallic iron, nickel, silicon, and sulfur, a composition similar to that of many metallic meteorites. Although the differentiation of core and mantle is obviously very great, it is certain that the exclusion of the minor elements cannot have been absolute. Oxygen, however, so abundant in mantle and crust, is almost surely absent from the core. A further reason for considering the outer core molten is that the earth's magnetic field, known to be of internal origin (p. 122), seems to demand fluid motion of a molten, deeply buried conductor for its genesis.

The inner-outer core boundary at about 5145 km is sharp, and is surely a liquid-solid phase change, with the inner core a solid, for a shear wave with velocity of more than 3.5 km/sec has been revealed in it by the *P*-wave it generates on striking the boundary with the molten outer core.

WEIGHING THE EARTH

Strictly speaking, it is impossible to weigh the earth, for weight is defined as the gravitational pull of the earth on a mass. When we speak of weighing the earth, we are really speaking of determining the earth's mass. This problem has been attacked in many ingenious ways, but all have the same basis — comparing the attraction of the earth to that of a known mass. One of the most readily visualized experiments, though far from the most accurate, was that performed by von Jolly of Munich, in 1878.

Von Jolly mounted a balance at the top of a high tower with the usual scale pans attached (Fig. 7-29). Another pair of scale pans was suspended by wires about 20 m below. Two glass globes of equal weight and volume, A and B, were each filled with 5 kg of mercury; two other identical globes, C and D, were left empty. Globes A and B were put in the upper pans and C and D in the lower; they balanced. Then A and C, on the same balance arm, were interchanged, thus bringing the mercury in A a measured distance closer to the center of the earth. The increased attraction of the earth on the mass of mercury was measured by making a new balance; it turned out to be 31 mg.

Von Jolly then placed a lead sphere beneath one of the lower pans and repeated the measurements; this time A gained 0.59 mg more than it had before because of the attraction of the lead sphere. The centers of the lead sphere and

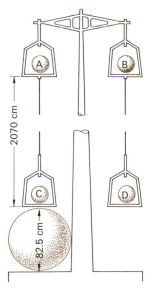

Figure 7-29
Von Jolly's method of "weighing the earth." [After J. H. Poynting, *The Earth*, G. P. Putnam's Sons, 1913.]

mercury globe were 57 cm apart. Since the lead sphere at a distance of 57 cm exerted a pull of 0.59 mg on the mercury, and the earth—at an effective distance equal to its radius, about 637 $\times 10^6$ cm—exerted a pull of 5×10^6 mg, the mass of the earth could be readily compared with that of the lead sphere.

Von Jolly's result was about 6.1×10^{21} metric tons—a figure so large as to be incomprehensible. When divided by the volume of the earth, this gave a mean specific gravity of the earth of 5.69. More accurate methods, too complex for discussion here, but using the same principles, give a figure of 5.516 as the best value.

The average density of the surface rocks is about 2.8, and very few minerals are as dense as 5.5. Obviously rock density is far greater at depth; in fact, the core must be denser than most natural substances we know.

Although we know the crust and mantle both to be inhomogeneous, the similarity of time-distance curves for most parts of the earth shows a general tendency for concentric shells to vary in elastic properties in accordance with their radial distance from the center. This means that assumptions about density at any particular level imply a definite contribution to the rotational inertia of the earth. Rotational inertia is a measure of the tendency of a rotating body to persist in its rotation against a braking action of any kind, such as, in the case of the earth, the tides (Chapter 16). Each particle of matter in a rotating body contributes to the rotational inertia in proportion to its mass and the square of its distance from the axis of rotation. (When a spinning skater pulls in his arms, he speeds up his spinning rate; when he throws them out, he slows down, for the product of mass times velocity, or momentum, must remain the same.) Therefore, when we assume a particular density at a particular distance from the axis of rotation, we can compute the rotational inertia contributed by that material. Estimates of the density distribution within the earth can thus be checked by their agreement or disagreement with the earth's rotational inertia as measured astronomically.

Professor Bolt (1972) has made a careful study of the density variations in the earth in conjunction with its elastic properties (Fig. 7-30). Although irregularities in rate of increase occur just below the Low-Velocity Zone, the density increases rather smoothly from about 600 km to the Gutenberg-Wiechert Discontinuity. Here the density jumps from less than 6 to about 10—notably higher than the density of either iron (7.9) or nickel (8.6) at the surface. The density of the inner core is quite uncertain, as it lies so near the axis of rotation as to contribute little to the earth's rotational inertia. The possibilities lie between slightly more than 12 and as much as 14.6, with the probability that the density at the earth's center is about 14.

In summary, then, the elastic properties of the mantle and core, together with the densities inferred for the various depths of the earth, suggest that the upper mantle, down to about 400 km, is composed of pyroxenite and peridotite, probably with some garnet. Between this depth and 600 km, the minerals gradually convert to denser phases. Below about 650 km, further phase changes convert these minerals into a mixture of oxides of iron, magnesium, and silicon, and as the core is approached, a few percent of admixed nickel. The outer core is almost certainly a molten mixture of iron, nickel, a little silicon, and sulfur, and the inner core is probably a solid of approximately the same composition, near that of metallic meteorites.

TEMPERATURE WITHIN THE EARTH

In every well and mine shaft, temperature increases with depth. The rate of increase varies geographically; in the hot spring area of Yellowstone Park, temperatures of several hundred

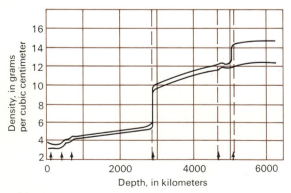

Figure 7-30
Precision bands for the density variations in the earth's interior. [From B. A. Bolt, 1972, *Physics of the Earth and Planetary Interiors.*]

degrees Centigrade are reached by the drill in a few tens of meters. Well temperatures in the plains of Hungary increase 1°C in as little as 7 m of depth — suggesting that the boiling point would be reached in as little as 600 m. But the temperature increase in the gold mines of the Transvaal, South Africa, is so small that mining goes on at depths of 3000 m, though not without some refrigeration of the ventilating air. The gradient is exceptionally low because of the low heat flow and the high thermal conductivity of the abundant quartzite nearby. The average rate of increase in the outer crust of the earth has been estimated as little more than 2°C in 100 meters. However variable the rate, the universal increase proves that the earth is losing heat to outer space, for heat flows only from bodies of higher temperature to those of lower. Rough estimates are that the flow from the interior amounts to about 0.03 percent of the solar energy received.

The loss of heat depends on two factors: the **temperature gradient** (the rate of temperature rise with depth) and the **thermal conductivity** (the amount of heat transmitted through a surface of 1 cm² in 1 second under a thermal gradient of 1°C/cm). Both gradient and conductivity depend on rock properties and structure, for heat can be refracted by differences in conductivity just as light is. The gradient is also disturbed in many places by seasonal and longer-term temperature changes, by ground-water movements (Chapter 14), rock structure, and topography. Accordingly, we have relatively few reliable measurements of heat flow from the continents, though more are being actively sought.

Strangely enough, it is much simpler to measure heat flow accurately on the sea floor than on the land. Except where turbidity flows have poured thick layers of new sediment onto the bottom (Chapter 16), the slow sedimentation does not seriously affect equilibrium, and climatic changes are negligible, for the temperature at the sea floor differs little today from what it was during the glacial times of the Pleistocene. The measurements are made by sinking a probe into the sediment to a depth of 2 to 6 or 8 m. The probe carries sensitive temperature sensors arranged along its length so that temperature differences can be read to a small fraction of a degree C, thus giving the thermal gradient. A core of sediment collected alongside at the same time is used to estimate the conductivity of the mud from the water content or to measure it more accurately by physical means.

The state of our knowledge of heat flow from various geologic provinces of Eurasia is tabulated in Table 7-3. The Sierra Nevada has low values, as does the intermountain plateau country, but in the California Coast Ranges, the Rocky Mountains, and the belt from southern Arizona north

Table 7-3
Mean heat flow from various geologic provinces of Eurasia.

Province	Number of measurements	Heat flow units* average	range
Precambrian folded regions	49	0.96	0.61–1.8
Precambrian shields	11	0.86	0.61–1.4
Precambrian platforms	38	0.99	0.70–1.4
Regions folded during Paleozoic	113	1.6	0.6–2.6
folded in Mid-Paleozoic	11	1.1	0.8–1.5
folded in Late Paleozoic	102	1.6	0.6–2.6
Regions folded in Cenozoic	82	1.7	0.65–3.3
Cenozoic volcanic areas	21	2.1	1.4–3.3
All Eurasia	255	1.49	

SOURCE: Data from E. A. Lubimova and B. G. Polyak, 1969. Data from North America have not been classified in similar tectonic provinces. The small part of the Canadian Shield within the United States has a heat flow comparable to the Eurasian Shields, about 0.8,* with but small departures. The northern Appalachians give a somewhat higher value, about 1.1 hfu. The mid-continent values average slightly higher. The available data for the Rocky Mountains to the Pacific Coast are shown in Figure 7-31.

*One heat flow 10^{-6} μcal/cm²/sec. One calorie is the amount of heat required to raise the temperature of 1 gram of water from 15° to 16°C.

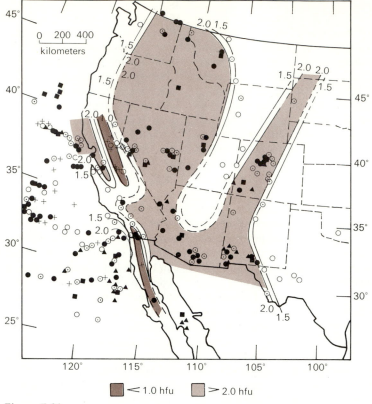

Figure 7-31
Measurements of heat flow in the western United States and adjacent parts of the Pacific Ocean. The contours define regions of high and low heat flux that would be measured in rocks with surface radioactivity appropriate to granodiorite. The plus signs represent values between 0 and 0.99; open circles, between 1.0 and 1.49; dotted circles, 1.5 to 1.99; solid circles, 2.0 to 2.49; solid triangles, 2.5 to 2.99; solid rectangles, > 3.0. A few measurements in the Pacific Coast province have not been plotted because of limitations of scale. [From R. F. Roy, D. D. Blackwell, and E. R. Decker, *in* Eugene Robertson (ed.), *The Nature of the Solid Earth*, McGraw-Hill Book Company, 1972.]

through Nevada, western Utah, Idaho, and probably eastern Oregon—all areas of abundant Cenozoic volcanism—the values are higher than 2 hfu. It is doubtless significant that much of the area of high heat flow corresponds with the area of thin crust shown in Figure 7-18 and of low *P* velocities shown in Figure 7-19. High heat flow along and on either side of the San Andreas fault in the California Coast Ranges may be in part due to friction along the fault.

The much more abundant heat-flow data from the oceanic areas has been summarized in Table 7-4 and in Figure 7-32.

It is surprising that, although the difference is probably not significant with so few data, the average heat flow through the ocean floor is slightly higher than that through the continental crust—very significant in view of the much greater concentration of radioactive elements in continental crust as compared with oceanic.

The data show that continental heat flow is generally lowest in the Precambrian shields, areas that have been tectonically inactive and exceptionally deeply eroded over large areas for more than 600 m.y. It is intermediate in the continental plates, wherein the basement has been long pro-

Table 7-4
Oceanic heat-flow data.

Area	Number of observations	Average heat flow (HFU)
Atlantic Ocean	406	1.43
Indian Ocean	331	1.44
Pacific Ocean	1322	1.71
Arctic Ocean	29	1.23
Mediterranean Sea	71	1.33
Marginal Seas	260	2.13
All oceans and seas	2320	1.65
All continental stations	255	1.49

SOURCE: Data from Herzen and Lee, 1969.

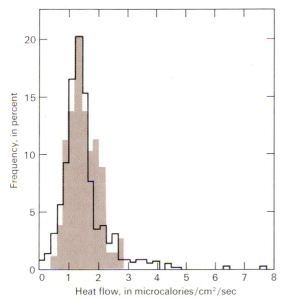

Figure 7-32
Histograms of heat-flow averages over areas of 9×10^4 square nautical miles. The solid-line histogram is for oceanic data, the broken line for continental. [From Von Herzen and Lee, 1969.]

the continental heat flows even from regions of Cenozoic volcanism; they seem restricted to narrow zones along the mid-ocean ridges and other topographic highs.

It was long thought that the loss of heat to outer space necessarily implies cooling of the earth; since the discovery that radioactivity generates much heat, it has become important to compare the heat being currently generated with the quantity being lost to space.

The elements whose radioactive disintegrations contribute most to the earth's heat budget are uranium, thorium, and potassium. Widespread sampling shows that these elements are all far more abundant in siliceous igneous rocks than elsewhere. Measurements on large composite samples show that the average uranium content of basalt is about 0.0001 percent, and that of granite four times as much — 0.0004 percent. Potassium content in these same composite samples behaves similarly: 1.07 percent in basalts; 3.70 percent in granites. Most rocks of whatever composition contain 3 or 4 times as much thorium as uranium.

Even the ultramafic dunites, peridotites, and pyroxenites, the least radioactive of surface rocks, contain enough heat sources so that if they made up the whole earth beneath the crust, it would be molten. The solidity of the upper mantle shows that radioactivity is concentrated toward the surface of the globe. The average granite is so radiogenic that a globe-encircling layer only about 13 km thick would supply all the heat being lost to outer space. But the granites, with trivial exceptions, are confined to the continents, and generally to their shallow crust. Yet the heat flow

tected from erosion, and is highest in areas of post-Paleozoic tectonism and volcanism. For example, near the Pleistocene rhyolitic intrusions of the Imperial Valley, California, a flow of 7 hfu has been measured. Though sampling in the trenches is inadequate, they do seem to have the lowest oceanic heat flow, perhaps because of rapid sedimentation or of subduction (p. 134). The highest oceanic heat flows exceed most of from the ocean basins is about the same as that from the Paleozoic mountain chains, and notably higher than from the shields. Clearly, the continental heat flow is largely from the crust itself; the thin and poorly radioactive basalts flooring the oceans cannot possibly supply the comparable heat flow. Most of the heat flowing through the ocean floor must come from the mantle, implying that the mantle beneath the oceans differs somewhat from that beneath the continents.

But, as we shall see in Chapter 8, much evidence points to the high probability that the continents have not been fixed in place throughout geologic time but have drifted from place to place over various parts of the mantle now overlain by the seas. Why, then, do the heat flow measurements on the continents not exceed those on the ocean floor? At first glance this seems to be a strong argument for fixed continents. It has been pointed out, however, that rock is a poor conductor of heat and that the sialic crust serves as a blanket impeding the flow of heat from the mantle to the surface. Instead, the heat from the subcontinental mantle goes to heating up the upper mantle, so that the LVZ beneath the continents is considerably thicker than that beneath the seas.

Many years ago the brilliant Norwegian geochemist V. M. Goldschmidt pointed out that the enrichment of the siliceous rocks in radioactive elements is due to the size of their ionic radii. Too large to be readily accommodated in the lattices of the early-crystallizing iron-rich and magnesium-rich minerals, they remain in the magma and rise along with the later fractions of the differentiation series (Chapter 9, pp. 194–197) to the higher parts of plutons and thus to shallow levels in the crust. Presumably they are virtually absent from all zones of the earth below the lithosphere.

Although the upward concentration of radioactive elements takes place chiefly via magmatic differentiation, hydrothermal solutions contribute a significant fraction. The average rock of the highest-grade metamorphic facies—the granulite facies, probably from the deepest crustal levels sampled—shows only 0.000015 percent thorium (15 parts in 10^8) and 0.000006 percent uranium (6 parts in 10^8), whereas the shallow granitic bodies and lower-grade metamorphic rocks contain 12 times as much thorium and 3 times as much uranium. Potassium, also with large ionic radius,

behaves similarly. The transfer of the radioactive elements from the metamorphic rocks must take place in hot water, for these rocks were not melted. Thus we can understand the lower heat flow from the shield areas, for although the shields are largely granite, they have surely been more deeply eroded than most of the crust. The shallow zones, presumably originally enriched in radioactivity, have been eroded away.

These facts support the conclusion that the continents result from magmatic differentiation that allowed the siliceous and more radioactive magmatic fraction to rise, leaving the more mafic and earlier crystallized fraction as a mildly radioactive residue at depth.

The extent of upward concentration of radiogenic heat is amazing, as is shown in New England by the rapid local variation in heat flow and by the remarkable agreement between the equivalent radioactivity of local rocks and the total heat flow at the same site (Fig. 7-33).

Both these features strongly suggest shallow sources for most of the heat. In fact, the Conway Granite (Triassic) in New Hampshire is so highly radioactive that if it were only 10 km thick it would supply the entire heat flow of the area. Obviously, some heat also comes from the mantle, so that the granite cannot be that thick; gravity measurements suggest about half that thickness.

Other evidence of strong upward concentration of radiogenic heat sources comes from the Sierra Nevada. The entire heat flow from the high eastern Sierra, underlain by a 50-km root identified seismically, could be supplied by a mere 15-km thickness of the exposed rock. Gravity measurements show that the granite does not extend as deep as the root, but it seems most unlikely that the rocks of the root are wholly free from radioactivity. On the western slope of the Sierra, somewhat older granitic rocks show a heat flow only half as high as that to the east, but the local rocks are also far less radioactive than the summit granites; it would require a thickness of 30 km to supply the local heat flow. These relations can be interpreted to mean that the older rocks of the western slope have been much more deeply eroded than those to the east; the originally more radioactive upper parts having been eroded away. This is the generally accepted reason for the low heat flows from shield areas.

Continental measurements of heat flow are still few and of such dubious accuracy (percent

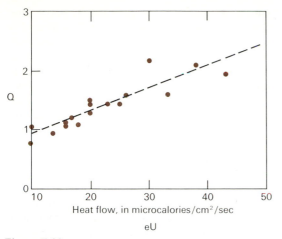

eU

Figure 7-33
Correlation between total heat flow, Q (in hfu) and local radioactivity of surface rocks in equivalents of uranium, ^rU, as indicated by gamma ray measurements in New England. [From F. Birch, R. F. Roy, and E. R. Decker, *in* E. Zen, W. White, J. B. Hadley, and J. Thompson (eds.), *Studies of Appalachian Geology, Northern and Maritime*, John Wiley & Sons, Inc., 1968.]

errors being as high as 20) that, as pointed out by A. H. Lachenbruch of the U.S. Geological Survey, the assumption of a mantle contribution of 0.4 hfu, together with an exponential decrease in radioactivity from the surface value to depths of 10 to 15 km, fits the Sierra Nevada province very well. In the Basin Range province to the east, the mantle contribution is much higher, 1.4 hfu; in the Precambrian shield, 0.8 hfu. The exponential decrease with depth is of course an artificial model that cannot be correct in detail, but its fit with the data is a very strong argument for marked upward concentration of radioactivity.

The upward concentration is obviously great but not complete. It is still uncertain whether the earth is cooling or heating. But any heat in excess of that conducted to the surface has not been enough to liquify any large part of the mantle, except in the plastic Low-Velocity Zone. Transmitted shear waves show that nearly all the mantle is solid and highly rigid under short-term stresses, even though it yields under long-term loads, as shown by isostasy. Nevertheless, seismograms of some Japanese earthquakes, recorded by the Russian volcanologist Gorshkov

in Kamchatka, show no *S* waves, indicating a molten reservoir in the Low-Velocity Zone beneath the intervening, actively volcanic Kurile chain. Magma bodies are also indicated below Hawaii, where seismic evidence has traced the rise of magma from a depth of at least 45 kilometers below the M discontinuity. Of course this is oceanic crust, which is not as differentiated as continental. But volcanoes exist on continental crust also, though few are found very far from existing or Cenozoic oceans.

Perhaps these magma pockets form because of localized radioactivity, thus explaining the relatively high radioactivity of surficial lava and high-level plutons. Or perhaps release of pressure along active faults lowers the melting point enough for fusion. Both factors may operate. Partial melting of a garnet peridotite—a rock thought to be common in the upper mantle—would selectively concentrate the radioactive elements upward; the magma so formed would rise through the more mafic materials with higher melting points.

That the relatively steep temperature gradient in the crust does not continue far down in the mantle is shown by the associations of diamonds in the few volcanic vents that contain them. The diamonds are invariably accompanied by garnet peridotite, eclogite, and other mafic rocks in fragments of variable size. Laboratory experiments indicate that these rocks could only be stable at considerable depths, estimated as at least 150 kilometers, but the temperature must not have exceeded 1100°C, for were it higher the diamonds would have been converted to graphite. Since many lavas come from shallower depths, yet have higher temperatures than this, such molten pockets must be of scattered and local occurrence in the upper mantle.

The molten outer core revealed by the extinction of *S* waves is almost certainly convecting, as is suggested by the dynamo theory for the origin of the earth's magnetism, discussed in the next section. The earth's field has been known to be of internal origin for several generations and thus far the only apparently adequate mechanism suggested is the circulation of a molten conductor—the metallic outer core.

Temperatures at the core boundary and at the earth's center are quite uncertain, as extrapolations from experiments are extreme. Estimates made at various times of the temperature at the core-mantle boundary have ranged from about 2500° to 7500°C. In a recent careful review, Professor Birch of Harvard University concluded that estimates of 4300°C for the core boundary and 5300°C for the earth's center are probably maximal, and that the actual temperatures should not be more than 1000° lower.

THE EARTH'S MAGNETIC FIELD

The compass has been used in navigation for centuries, but it was not until 1600 that Sir William Gilbert, Queen Elizabeth's physician, noted that the earth is a huge magnet whose poles are near the geographic poles. Magnetic lines of force emanate from the magnetic poles, so that the north-seeking pole of a freely suspended magnet is inclined downward in the northern hemisphere, is horizontal at the magnetic equator, is inclined upward in the southern hemisphere, and lies nearly in the plane of the magnetic meridian. Gilbert knew that this picture is oversimplified and that the needle does not behave so regularly.

The great German mathematician Gauss showed in 1835 that the main elements of the field are internal: atmospheric electricity modifies the field but slightly. But the phenomenon, long puzzling, deepened when Pierre Curie of France showed in 1895 that all magnetic substances lose their magnetism when heated above a certain temperature, the **Curie point**, which differs in different substances. No mineral has a Curie point higher than about 800°C, a temperature reached at depths of 30 to 35 km nearly everywhere the world over. The most important magnetic mineral, magnetite, has a Curie point of only 578°C, which seems to rule out a deep-seated source. But magnetism of the shallow zone above the Curie depth has long been known to be wholly inadequate to produce the existing field.

The puzzle went unsolved until, as has so often happened in the history of science, two men independently proposed identical answers—E. C. Bullard, a British geophysicist, in 1954, and W. M. Elsasser, an American, in 1955.

Their theory, that of the self-exciting dynamo, attributes the field to fluid motion in the outer core of the earth, assumed to be molten conducting metal. Motion of such a conductor generates a magnetic field, though of course a source of

energy is required to keep it going. For a time, this source was thought to be radioactivity, but more recent analyses of metallic meteorites supposedly similar in composition to the core, indicate too little radioactivity for the task. If, however, the inner core of solid metal has been slowly growing by crystallization from the overlying melt, the heat given off in freezing—the *latent heat* of crystallization—would suffice. Though this theory has so far met all tests, Sir Edward Bullard, one of its originators, has repeatedly said that it remains unproved. The theory has frequently been stated in reverse: that is, there must be a molten convecting core of conducting metal to explain the existence of the field! The evidence is indirect, but as Thoreau reminded us, some circumstantial evidence is beyond dispute, as when one finds a trout in the milk.

The rotation of the earth influences the convection currents, so that the field produced tends to be symmetrical about the earth's axis. The symmetry is, however, far from perfect: it is said to be approximated by the field expected from a short magnet at the earth's center, tilted 11.5° away from the axis along the meridian of 70°W. This is called the inclined **dipole field**. (Actually the present field is best modeled by an inclined dipole 442 kilometers from the earth's axis and 133 km north of the equator.)

This major, more or less fixed, field is modified by a transient field that migrates slowly westward, perhaps produced by eddies in the convecting core. The combined result is that the present magnetic poles are more than 800 km from the geographic poles: the north pole at 74°54' N, 101°W; the south at 70°S, 148°E, thus more than 15° and 20°, respectively, from the geographic poles. The northern pole is moving at the rate of about 8 km a year; the sourthern considerably faster, for it has moved 800 km between 1910 and 1960. A line connecting them misses the earth's center by several hundred kilometers.

The field is far from constant, either in orientation or strength. For example, the compass needle in London has swung through an arc of more than 36° on either side of the meridian in the past 400 years, and is now swinging back. At Cape Town the change in azimuth has been even greater. The inclination of local fields also fluctuates notably. Furthermore, the overall strength of the field has been diminishing about 0.05 percent a year ever since adequate measurements began in 1830. Studies of Japanese, Assyrian, and Roman pottery dated by C^{14} methods indicate that the magnetic field was 1.6 times its present strength 2000 years ago and only half the present value 5500 years ago.

The earth's field is not strong, but it suffices to magnetize many rocks, both igneous and sedimentary. When a magma or lava flow cools, its magnetic minerals crystallize and, as the temperature drops below the Curie point, become magnetized parallel to the local lines of force of the magnetic field. (Some minerals or mineral pairs are magnetized precisely opposite to the local field, but these can be recognized by their internal structure and magnetic behavior in the laboratory; they are so rare that we do not consider them further.) Metamorphic rocks behave the same way if their temperatures during metamorphism rise above the Curie points of their magnetic minerals; the rocks in a contact aureole of a pluton are magnetized parallel to those of the pluton. Detrital magnetic minerals carried into a sedimentary basin tend to align themselves parallel to the local field, and in most sandstones enough have done so to record it. By collecting carefully oriented specimens and determining the alignment of their magnetism in relation to azimuth and inclination to the horizon, students have been able to make rough determinations of the orientation and location of the magnetic poles at many times in the geologic past, thus throwing light on several fundamental geologic problems.

As seen in Figure 7-34, such observations require considerable care in interpretation. Finding the location of an ancient magnetic pole requires abundant statistics, and the attainable degree of confidence is often lamentably low. Yet even though accuracy leaves much to be desired, **paleomagnetism** has often been corroborative of conclusions from other lines of evidence. At present the magnetic field deviates markedly from symmetry about the magnetic poles. If we assume that the north-seeking pole of a compass needle points directly at the north pole and that the inclination of its needle is a measure of polar distance and then plot the inferred polar location from the many magnetic observatories in the international network, we would obtain the large scatter shown in Figure 7-34. Although the

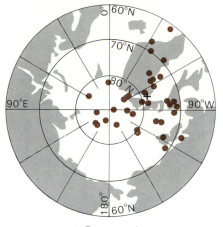

+ Present pole

Figure 7-34
Positions of the magnetic pole, assuming that the compass needle at each magnetic observatory of the world-wide network points directly toward the north magnetic pole and that the inclination of the needle is a measure of polar distance from the observatory. The wide scatter and deviations from the actual position of the north pole show the local deviations from an ideal dipolar field. [After R. R. Doell and Allan Cox, 1961.]

"center of gravity" of all the inferred poles would lie not far from the true pole, the scatter is over an angular distance of more than 30°. Inasmuch as the pole position deduced from the magnetic pattern of Holocene lava flows or sedimentary strata might well correspond with any one of the "virtual" poles in Figure 7-34, one might be reluctant to put much confidence in the method as a guide to the geographic pole.

But, instead of using an instantaneous picture of the field, as in this example, if we use a field averaged over a considerable time span, the picture is much more encouraging. Figure 7-35 shows the pattern of magnetic poles inferred by the same method, using the magnetic patterns in several sequences of lava flows and sedimentary rocks less than 0.5 m.y. old from widely scattered parts of the earth. This clearly shows that during the past half-million years the *average* position of the north magnetic pole has been close to the geographic pole. This is what would be expected if the dynamo theory of earth magnetism is correct, and is indeed a strong argument in its favor,

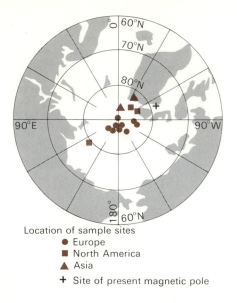

Location of sample sites
● Europe
■ North America
▲ Asia
+ Site of present magnetic pole

Figure 7-35
Holocene and late Pleistocene (present to 500,000 years ago) virtual magnetic poles. Equal area projection. [After R. R. Doell and Allan Cox, 1961.]

showing that the present displaced and asymmetric position of the poles is due to transient disburbances that would tend to average out into a symmetrical field over an uncertain time span, perhaps measured in tens or hundreds of years. Studies of magnetic directions in datable stratigraphic sequences representing considerable time spans — say 10^4 to 10^6 years — may thus give clues to the position of magnetic poles, and less precisely, to geographic poles at various times in the past.

REVERSALS OF THE EARTH'S MAGNETIC FIELD

The precisely reversed magnetism shown by certain unusual minerals and mineral pairs has been mentioned above. In several long series of lava flows or other strata, rocks that do not contain such minerals and are in all other respects identical to rocks above and below have been magnetized in precisely the opposite sense. Thus the north-seeking magnetic pole of the rock lies parallel to the south-seeking poles of the rocks above and below. The Japanese geophysicist Matuyama recognized in 1929 that many Pleistocene volcanic rocks are reversely magnetized. Since the development of the K-Ar method of

dating rocks, the polarity of relatively young rocks that can be closely dated has been intensively studied. Prolonged epochs have been recognized during which the field was as it now is, but these have alternated with comparable epochs when the field was reversed, as shown by like polarities of contemporary rocks from widely scattered localities.

The several long epochs of constant polarity have been named for geophysicists who have contributed to the study of magnetism. The youngest, *the Brunhes normal epoch*, extended from about 0.7 m.y. ago to the present; the preceding Matuyama reversed epoch extended from 2.5 m.y. to 0.7 m.y. but was interrupted briefly by the Jaramillo and Olduvai "events" of normal polarity. The Matuyama epoch was preceded by the Gauss normal epoch (3.36–2.5 m.y.), which was also interrupted by the Mammoth reversed event. Earlier than the Gauss was the Gilbert reversed epoch, which is of uncertain duration because the uncertainties of dating do not permit interruptions so short as the Olduvai and Jaramillo events to be established at such earlier times. From scattered data, the field seems to have been reversed for much of Triassic and Permian time, but for times before the Pliocene, short changes in polarity cannot be accurately fixed. The data are summarized in Figure 7-36.

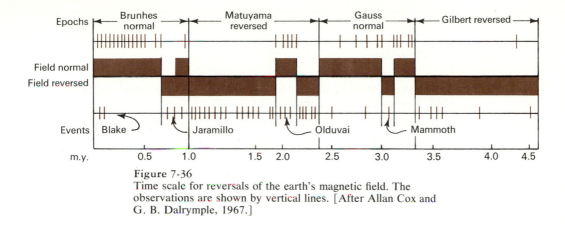

Figure 7-36
Time scale for reversals of the earth's magnetic field. The observations are shown by vertical lines. [After Allan Cox and G. B. Dalrymple, 1967.]

Variations in the strength of the earth's magnetic field have been noted above; perhaps it will diminish to zero at some time and then reverse in polarity as it has so often in the past. Such reversals do not conflict with the "dynamo theory" for the origin of the field; they may require only a differential motion between mantle and core opposite to that now prevailing. A study by American geophysicists Allan Cox and G. B. Dalrymple suggests that a reversal may take place in an interval between 1600 and 21,000 years long, perhaps in something less than 5000 years.

Questions

1. During many Central American earthquakes, well-built masonry structures have been destroyed while bamboo huts nearby were undamaged. Why?

2. The Bouguer anomalies for the Alps (Chapter 2) suggest a greater thickness of light rocks beneath the mountains than beneath the lowlands to the north. What does this imply as to the relative heat flows in mountains and lowlands?

3. During the San Francisco earthquake the porch was sheared off a frame house and moved several meters by the fault movement; the brick chimney, less than 10 m from the fault, was not injured. Well-built structures 7 or 8 km from the fault were demolished. How can you account for the differences in behavior?

4. Why is the M Discontinuity more significant than any higher ones?

5. What does the elastic rebound theory suggest concerning the distribution of changes in elevation near a fault while strain is building up for vertical displacement?

6. How do you account for the fact that magnitude and ground acceleration seem to be relatively independent of each other?

7. If molten material of the outer core does not transmit S waves, how could the existence of S waves in the inner core be demonstrated?

Suggested Readings

American Iron and Steel Institute, *The Agadir, Morocco, Earthquake of February 29, 1960*. New York: The Institute, 1961.

Iacopi, Robert, *Earthquake Country*. Menlo Park, Calif.: Lane Book Company, 1964.

Macelwane, J. B., *When the Earth Quakes*. Milwaukee: Bruce Publishing Company, 1947.

U. S. Geological Survey, *The Alaska Earthquake* (Professional Papers 541, 542, and 543-1). Washington, D. C.: Government Printing Office, 1965–1967, 1969.

U. S. Geological Survey, *The San Fernando, California, Earthquake of February 9, 1971* (Professional Paper 733). Washington, D. C.: Printing Office, 1971.

Scientific American Offprints

804. K. E. Bullen, "The Interior of the Earth" (September 1955).

825. W. M. Elsasser, "The Earth as a Dynamo" (May 1958).

827. Jack Oliver, "Long Earthquake Waves" (March 1959).

829. Joseph Bernstein, "Tsunamis" (August 1954).

855. Don L. Anderson, "The Plastic Layer of the Earth's Mantle" (July 1962).

CHAPTER 8

Renewing the Geologic Cycle—
Mountain-making

A quarter century ago, before the explosive increase in knowledge of the sea floor, of the history of the earth's magnetism, and of the minutiae of seismic processes made possible by newly devised instruments, it would have seemed absurd to begin a discussion of mountain-making with a description of submarine features. But the convergence of new data from these three seemingly disparate studies has produced a new theory—one that has fit so well with world-ranging and long-unexplained observations from many other fields that most geologists see new light on that great, long-standing enigma of geology, *the origin of mountains*. Although this newly popular theory leaves many data still in discord, it has met so many tests that were unmet by previous theories that we tend tentatively to accept it as the best available working hypothesis, subject, no doubt, to modification as new data are acquired.

In broad outline, this theory of plate tectonics considers the earth's crust to be divided into a few gigantic segments that are drifting over the Low-Velocity Zone in various directions with respect to each other. Mountains are raised along plate boundaries in collision. New oceanic crust forms along some oceanic ridges and rises, drifts

away on either side, and sinks into the mantle at oceanic deeps or other zones of postulated plate collision. This chapter discusses the evidence both for and against this hypothesis.

MOUNTAINS BENEATH THE SEA

The longest mountain ranges are mostly beneath the sea; the Mid-Atlantic Ridge, composed of many offset segments, lies almost precisely midway between the Old World and the New (Fig. 2-5). It differs from nearly all continental mountains in its great width and in its bilateral symmetry. It has a nearly consistent width of more than 1000 km, considerably wider than most continental ranges. The marked symmetry of the ridge (Fig. 8-1) is in sharp contrast with most continental ranges, such as the Alps (Fig. 8-24). For most of its length the ridge rises 2 or 3 km above the ocean floor to either side.

Longitudinal faults cut the ridge into a succession of terraces, and for much of its length it is split by a central graben. Iceland, which sits astride the ridge, is cut by gaping fissures that are widening at the rate of 3 or 4 cm per year, as measured by careful, repeated surveys. First

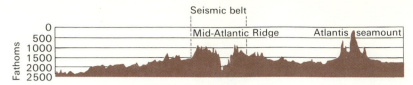

Figure 8-1
Cross section of the Mid-Atlantic Ridge at 30° N. Vertical exaggeration 40:1. Note the central fault trough and the rough symmetry. Atlantis seamount is an extinct volcano. The seismic activity along the ridge is confined to the central zone. [After C. L. Drake, *Philosophical Transactions of the Royal Society of London*, 1958.]

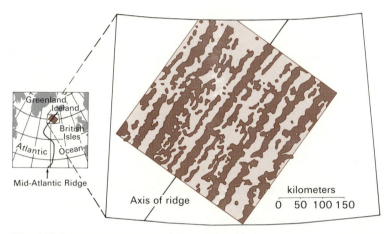

Figure 8-2
Magnetic anomalies in part of the North Atlantic (expanded view). Note the symmetrical arrangement of the linear anomalies on either side of the ridge axis. Positive anomalies, with field strength greater than the regional average, are dark brown; negative anomalies, with field strength less than the regional average, are light brown. [After Allan Cox, G. B. Dalrymple, and R. R. Doell, 1967.]

motions of earthquakes along the ridge are uniformly consistent with normal faulting, and thus with crustal extension.

The ridge is the locus of much volcanic activity from Iceland to Tristan da Cunha; heat flow along it is commonly three or four times the world average, and is much higher than to either side. Sonic sounding has failed to identify any M Discontinuity beneath the ridge. This is interpreted as indicating that rocks beneath are too hot to transmit elastic waves at speeds appropriate to most of the upper mantle and that the change in elastic properties with depth is gradual rather than abrupt.

The barren rocks of the central ridge (and of most other oceanic ridges) are mostly tholeiite, a variety of basalt poor in alkalies. Tholeiite, long known but not abundant on the continents, is the commonest lava on the ocean floor. Dredge hauls from the fault scarps include also somewhat metamorphosed ultramafic rocks, some showing sedimentary layering (Chapter 9).

Systematic arrangements of linear magnetic anomalies parallel the ridge on either hand (Fig. 8-2). These anomalies are interpreted as recording the earth's magnetic regime at the times that basalt, erupted at the core of the ridge (or dikes injected there), cooled through the Curie point. The possible mechanism of formation is diagrammatically shown in Figure 8-3.

The spacing of the stripes is in reasonable agreement with the duration of the various normal

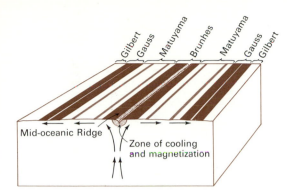

Figure 8-3
The possible mechanism of formation of the linear
magnetic anomalies. As the sea floor spreads, basalt
is continuously erupted at the ridge axis; upon
cooling, it records the successive magnetic episodes
of the past 3.35 m.y. since the close of the Gilbert
reversed epoch. [After Allan Cox, G. B. Dalrymple,
and R. R. Doell, 1967.]

and reversed magnetic regimes, on the assumption that the rate of spreading of the ocean floor
has been approximately constant during the past
3.5 m.y. Anomalies farther from the ridge represent still older reversals of the magnetic field but
cannot be so precisely dated; some occurred well
back in the Jurassic, as determined from the fossil
content of overlying beds. No strata older than
Jurassic have yet been found in the many deep
holes drilled in the ocean floor.

Ocean surveys and seismic studies show that
the Mid-Atlantic Ridge is not unique. As we saw
in Figure 2-5, and with slightly different emphasis
in Figure 8-4, comparable ridges accent the ocean
floor in many other places. Some, such as the
Carlsberg, Mid-Indian, and Pacific-Antarctic,
have medial fault troughs and parallel magnetic
stripes; a few, such as the Rio Grande Rise and
the Walvis Ridge in the Atlantic and the Ninety

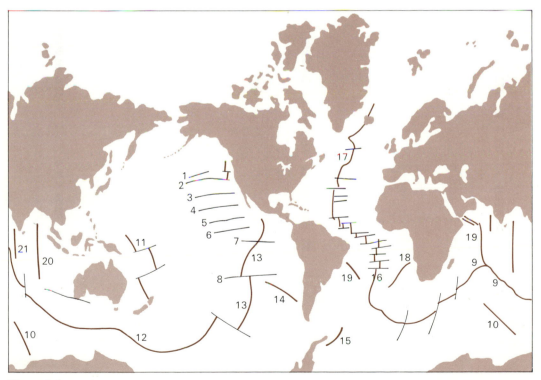

Figure 8-4
The oceanic ridges and rises and the fracture zones of large displacement: (1) Mendocino Fracture Zone,
(2) Pioneer Fracture Zone, (3) Murray Fracture Zone, (4) Molokai Fracture Zone, (5) Clarion Fracture Zone,
(6) Clipperton Fracture Zone, (7) Galapagos Fracture Zone, (8) Easter Island Fracture Zone, (9) Mid-Indian
Ridge, (10) Kerguelen-Gaussberg Ridge, (11) Melanesian Rise, (12) Pacific-Antarctic Ridge, (13) East Pacific
Rise, (14) Chile Rise, (15) Scotia Rise, (16) Mid-Atlantic Ridge, (17) Carlsberg Ridge, (18) Walvis Ridge,
(19) Rio Grande Rise, (20) Ninety East Ridge, (21) Chagos-Laccadive Ridge. [Modified from various sources.]

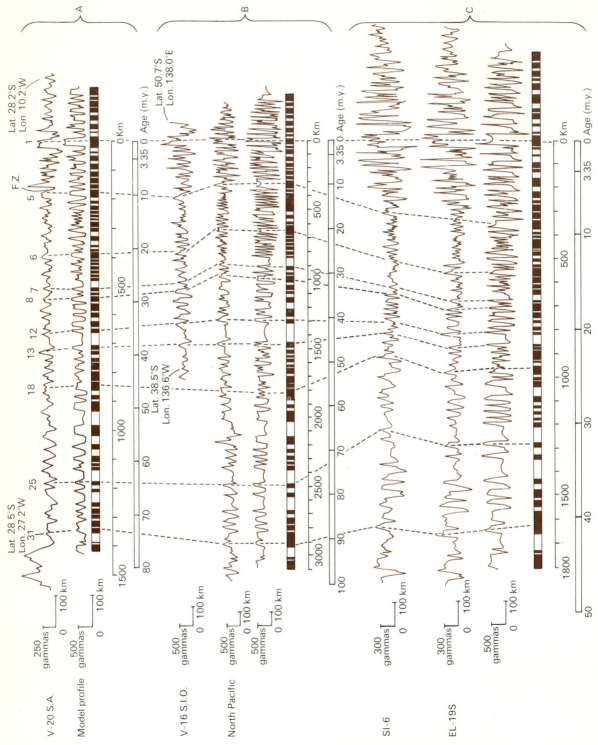

Figure 8-5 (*facing page*)
Sample magnetic profiles from various oceans. Beneath each profile is a theoretical profile computed on the assumption that each body, whether normally magnetized (black) or reversely magnetized (uncolored) is 2 km thick. With each is a time scale constructed by assuming an age of 3.35 m.y. for the end of the Gilbert reversed epoch (Chapter 7). Broken vertical lines connect similarly shaped anomalies thought to correlate (identified by numbers at the top). [From J. R. Heirtzler and others, *Journal of Geophysical Research*, 1968.]

East and Chagos-Laccadive ridges in the Indian Ocean, have no associated magnetic stripes; presumably these are not spreading, and have a very different origin. The East Pacific Rise stands much lower than the Mid-Atlantic Ridge, has no central fault trough, few other faults, and much less active volcanism. It does, however, have several narrow belts of high heat flow, each attributed to dike injections, but most of it has normal or even subnormal heat flow, even though the spacing of the magnetic stripes suggests that it is spreading considerably faster than the Mid-Atlantic Ridge.

Spreading Speeds

The spacing of the magnetic stripes differs from place to place, both along any single ridge and from one ridge to another. A comparison of the magnetic patterns of several ridges is given in Figure 8-5. (Note that very different horizontal scales have been used for the several samples because of the large differences in deduced rates of spreading.)

As Figure 8-6 shows, the spreading rates of the active ridges vary greatly, from as little as 1 cm/yr to as much as 12 cm/yr. (Because of the remarkable symmetry of the magnetic patterns on either side of the ridges, the rates are generally given as "half-rates".)

Modification of Sedimentation

The spreading of the ridges is recorded not only in the magnetic stripes, but in the differences in sea-floor sediments with distances from the active spreading lines. For example, the axial 20 km of the East Pacific Rise off Mexico is free of measurable sediment; the crust is too new to have received any. On either side of this and other ridges, both sonic sounding and drilling through the sea floor show that the sediment thickens systematically away from the ridges and that the

age of the oldest sediment resting on the basalt of the ocean floor increases correspondingly (Fig. 8-7). Of course the drill rarely penetrates the basalt for more than a few meters, and indeed metamorphism of the overlying sediment in some holes indicates that the igneous rock is intrusive and not basement. But it would be a remarkable coincidence, indeed, to produce so good a correlation of age with distance as is seen in Figure 8-7 were the agreement not significant. As mentioned before, the oldest sediment yet cored in any ocean is of Jurassic age and distant from any active ridge. It seems that all the ocean floors have been created by spreading since the Triassic. Some caveats to this inference are mentioned later (p. 137).

Subduction Zones

What happens to all the new oceanic crust formed by sea-floor spreading? One suggestion is that the new crust fills gaps created by expansion of the earth because the gravitational constant, G, is not really a constant and is diminishing with time. Thus the force of gravity acting to compress the earth may be becoming less and less effective. Two arguments oppose this assumption: (1) The electrostatic binding forces between the ions of a crystal lattice or even of a fluid, are so very much stronger than the gravitational attraction that even a considerable decrease in G would not seriously affect the volume of the earth. (2) The spreading goes on at tremendously different rates in various directions and has been doing so since at least Triassic time, yet if the earth were expanding, the spreading should be equal in all directions.

If, then, there are no grounds for thinking that the earth is expanding, we are faced with the problem of disposing of the new crust created at so many spreading ridges. The global map in Figure 8-6 gives us some clues.

A glance at Figure 7-17, a map of earthquake frequency, shows the record of heavy seismic

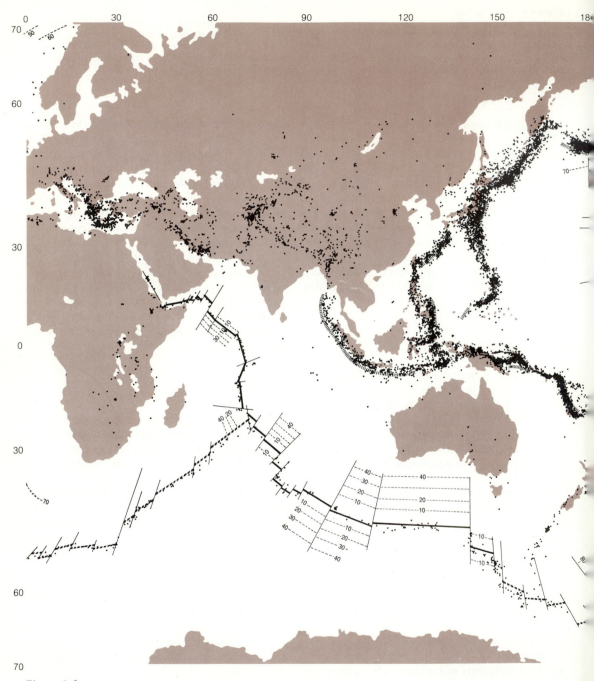

Figure 8-6
Positions of ridges, fracture zones, and trenches. Double lines are ridge axes; numbers near axes are half-spreading rates, in cm/yr. Small circles show poles of rotation (Fig. 8-14) of crustal plates: N. A. = North Atlantic; S. A. = South Atlantic; N. P., North Pacific; S. P., South Pacific; A, Arctic; I. O. Indian Ocean.

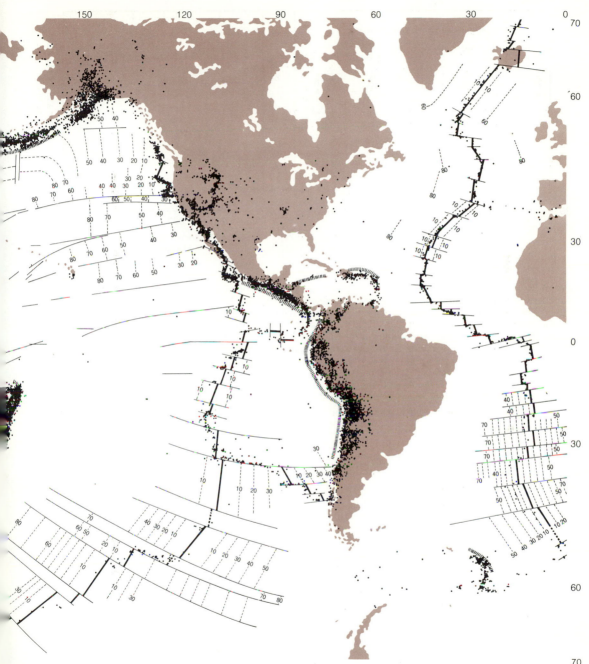

A plus within the circle indicates the pole was located from the intersection of great circles drawn normal to the fracture zones; an x in the circle means it was determined from spreading rates. [From J. R. Heirtzler and others, *Journal of Geophysical Research*, 1968.]

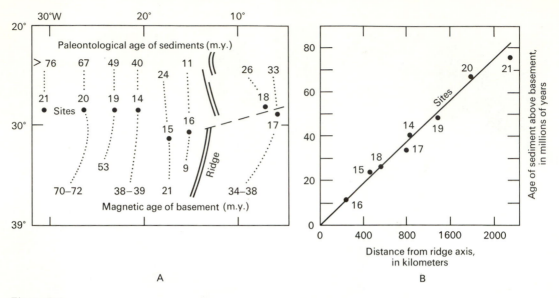

Figure 8-7
Comparison of basement age (inferred from the magnetic stripes) with the age of the oldest sediment resting directly on basalt (determined paleontologically, and referred to radiometric ages as in Table 6-1). (A) Location of drill holes. (B) Age of sediment plotted against distance from the Mid-Atlantic Ridge in the South Atlantic. [After Maxwell et al, "Deep Sea Drilling in the South Atlantic." *Science*, 168:1047–1059, 1970. Copyright 1970 by the American Association for the Advancement of Science.]

activity along the various oceanic trenches. We saw in Figure 7-15 that the earthquake foci on the continental side of the Kurile Trench lie at progressively deeper levels away from the trench. More recent work, with much improved seismographs, shows that the scatter is far less than appears in that figure, and, in fact, the foci define a tilted slab extending from the trench to depths as great as 700 km. This is typical of all trenches. The landward-descending seismic zone is named, from its discoverer, the American seismologist Hugo Benioff, the **Benioff zone**.

Inasmuch as the magnetic stripes of the sea floor record movement toward the trenches in many parts of the earth, the Benioff zones are interpreted as **zones of subduction** along which crust formed at the oceanic ridges dives down and is reincorporated in the mantle (Fig. 8-8).

Figure 8-8 shows a fault that surfaces in the Java Trench and dips northward at an angle of about 10°. The rocks in the hanging wall are greatly disturbed in a zone that thickens with depth. The fault is obviously a thrust along which the oceanic crust is descending beneath the island. The piling together of the sediments results in a marked negative gravity anomaly

that extends the full length of the Indonesian chain (Fig. 8-9). Such belts of anomalies have been found along the Chile Trench, Lesser Antilles Trench, Japanese Trench, and several others. As most trenches are close to land, most must be currently deepening. Otherwise they would soon be filled with sediment. In fact, sonic soundings show that the southern end of the Chilean Trench is indeed filled. By analogy with the Java Trench, and because of their distribution, all Benioff zones are considered zones of subduction along which surficial rocks are being swept down into the mantle. Early in the century the great Austrian geologist Ampferer noted that the complex geometry of the rocks of the Eastern Alps could only be explained in terms of subduction of great volumes of crustal rocks; the process is now thought to be operating in many areas of active mountain-building.

As exemplified in Figure 8-8, just inland from many trenches are coastal mountains largely made up of marine sedimentary rocks and andesitic volcanics. In many such ranges, complex masses made up of cherty sedimentary rocks, peridotite, serpentine, and tholeiitic pillow lavas —an assemblage precisely like the materials

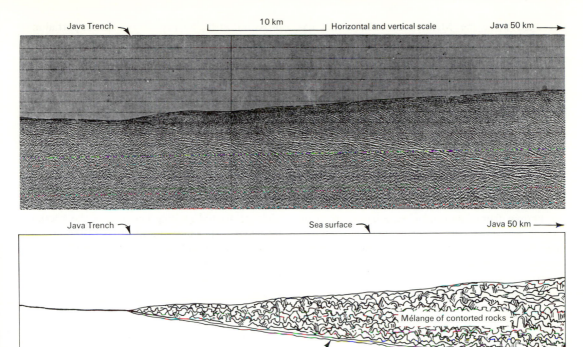

Figure 8-8
Seismic profile south of Java, showing the Benioff zone dipping northward from the trench, with much-disturbed beds in the hanging wall and the disturbed belt thickening northward toward the island. No vertical exaggeration. [Courtesy of Dr. R. H. Beck, The Hague.]

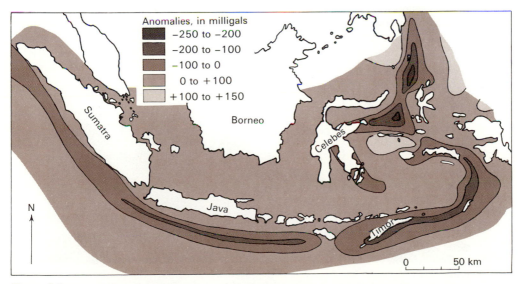

Figure 8-9
Map of Indonesia, showing regional gravity anomalies. [After F. A. Vening-Meinesz, 1934.]

dredged from the Mid-Atlantic Ridge and exposed on St. Pauls Rocks—are all jumbled together as though stirred in a giant mixing bowl, to form a **mélange**. This so-called **ophiolitic assemblage** is exposed only in mountain chains on the continents. It has long been recognized as of marine origin; we now have a clue as to the manner of its incorporation into the mountain structure: by being scraped off the suboceanic rocks as they descend a Benioff zone.

Included in many mélanges are bodies of "blueschist"—rocks containing the blue amphibole glaucophane and related minerals. Glaucophane is known from laboratory experiments to form under pressures equivalent to those expected at depths of 20 to 30 km but at temperatures much lower than would normally be found at such depths. Such conditions might well be expected in a subduction zone down which cold rocks were sinking into the mantle. Blueschists in coastal California are thus considered evidence of the former existence of a Benioff zone there, although no such zone exists there today.

With inland deepening of the Benioff zones at continental borders or island arcs, the volcanics become more siliceous and potassic. This was first noted in California, but is now recognized widely around the world. At first glance, this seems odd, for heat-flow data (summarized in Chapter 7) strongly indicate that K is largely concentrated in the shallow crust. Experiments in the presence of water vapor show, however, that rocks of the composition of tholeiite, when partly melted at low pressure, yield a liquid poor in K, but at higher pressures the first partial melt is both more siliceous and more potassic. Furthermore, mica-bearing sediment sinking along a Benioff zone would yield a more potassic magma as the increasing frictional heat melted the mica. The situation is diagrammed in Figure 8-10.

All these features are thus consistent with the interpretation of the Benioff zone as one of subduction. Friction of the descending rocks against the hanging wall locally distorts it so that chunks break off and are squeezed to the surface as ophiolites and blueschists; other masses are carried to successively greater depths, there to undergo partial melting because of the frictional heat and to produce the magmas of the volcanic belts inland of and paralleling the trenches. Such associations are found along the Japan, Chilean, Philippine, Indonesian, Kurile, New Britain, and Alaskan trenches. Some are associated with such trenches as the Mariana, Kermadec, Tonga, and lesser Antilles, where the rocks on both sides of the Benioff zones are oceanic.

As mentioned, many rock relations along the coast of California and Oregon suggest that a trench was once active there. Strengthening this conclusion is the fact that, although nearly comparable areas of the North American continent contributed sediment to the bordering oceans during Mesozoic and Cenozoic time, at least five times as much sediment of these ages remains off the Atlantic coast as off the Pacific: the sediment missing from the offshore Pacific may well have disappeared down a former subduction zone. The disturbed strata of the mountains in-

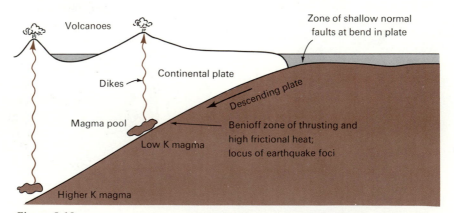

Figure 8-10
Professor Kuno's interpretation of the relations of the Japan Trench to earthquake foci (outlining the Benioff zone) and volcanism in Japan.

land from most trenches are expected results of the collision of crustal blocks driven by sea-floor spreading.

SEISMIC EVIDENCE

Shallow earthquakes near the trenches generally show first motions appropriate to normal faulting, perhaps because of tension in the descending block as it adjusts to the steepening trajectory. First motions associated with deeper foci are appropriate to thrusting, as exemplified by the Alaska earthquake of 1964. At depths of several hundred kilometers, faulting is again tensional, perhaps because partial melting of the downgoing slab so weakens it that the cooler, remaining fragments sink individually into the hotter mantle.

SUMMARY OF SEA-FLOOR SPREADING

The evidence cited suggests that along the spreading ridges magma is injected as dikes and effused as lavas. The heating caused by magma injection expands the central zone so that it rises high above the ocean floor. The higher temperatures of the axial zone must extend to considerable depths, perhaps to the Low-Velocity Zone. The crust cools as it drifts away from the ridge, the hot rocks contract, and the surface subsides until it reaches the normal level of the ocean floor, even though it receives increasing thicknesses of sediment while drifting.

The old crust, split at the ridges, plus the new crust formed there, drifts relatively away from the ridges, eventually to encounter a Benioff zone, commonly where continental and oceanic plates converge, but in the South Pacific, where two oceanic plates are colliding at the Tonga and Kermadec trenches, and, in the Himalayas, where two continents are colliding along the Indus-Brahmaputra zone.

The situation around Africa is noteworthy. On the west, south, and east are spreading ridges; between them and the continent, new sea floor is forming, but no subduction zone intervenes. Subduction may possibly be going on in the complex Mediterranean Basin, but certainly not elsewhere on the African periphery. The only conclusion possible is that *the ridges are pushing themselves away from the continent* — westward, southward, and eastward. The picture is similar in the northeast Atlantic; no subduction zone intervenes between the ridge and the European Coast, and although the whole Eurasian plate may be moving eastward toward the west Pacific subduction zones, it seems more likely that the ridge is moving westward. To the west of the Mid-Atlantic Ridge the first subduction zone is along the west coast of the Americas. The Chile Trench is apparently a zone of active subduction, but there is no such activity north of Mexico, though magnetic stripes and tectonic and magmatic features indicate a formerly active Benioff zone.

OBJECTIONS TO THE SUBDUCTION HYPOTHESIS

Sonic probing of the Chilean Trench and some others shows perfectly undisturbed stratification, with no trace of the tremendous deformation that is known from the Java Trench and that is necessary to form the mélange inland. A possible, though unconvincing, explanation is that the subduction zone may surface offshore of the trench and be deep enough beneath the trench so as not to disturb the overlying sediment. Or perhaps the zone surfaces on the inland trench wall, another ad-hoc argument.

Others have pointed out that tectonic activity in California, despite the absence of an offshore trench or zone of spreading, is probably as intense today as that along any Benioff zone, or as it was during the Cenozoic. This may mean that adjustment to the cessation of subduction is long delayed; the Cascade volcanoes may rise from a Benioff zone that only recently became inactive.

Despite the very suggestive association of many subduction zones with volcanic and mountain belts, it should be pointed out that many mountain belts bear no relation to subduction activity. The Rocky Mountains of North America formed in Cenozoic time far from any imaginable Benioff zone. Correlative older strata both east and west are so similar that they must have been deposited in associations little different from those of the present. Little crustal shortening has been involved, yet the mountains are as high as the Alps.

Fracture Zones (Transform Faults)

All the mid-ocean ridges are cut by fracture zones almost precisely at right angles to their trends. The map view of the offset ridges strongly suggests strike-slip faulting, and this is indeed the

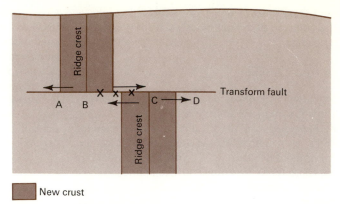

Figure 8-11
A transform fault. The apparent displacement of the ridge crests is left-lateral. The actual fault displacement is confined to the segment BC, where the relative displacement is right-lateral; on segments AB and CD, there may be little or no relative movement of the blocks. [Modified from J. Tuzo Wilson.]

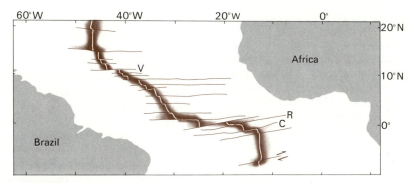

Figure 8-12
The offsets of the Mid-Atlantic Ridge between the bulge of Brazil and the Gulf of Guinea. R, Romanche Trench, displacement about 500 km. V, Vema Fracture Zone, offset about 300 km. C, Chain Fracture Zone, offset 300 km. Central graben of the Ridge left blank. [After B. C. Heezen and Marie Tharp, *Phil. Trans. Royal Society of London* v. 258, 1965.]

case. What appears, however, to be a left-lateral fault (Fig. 8-11) is shown by studies of first motion to have foci only between the ridges on opposite sides. These relations confirm the fact that the ridges are indeed loci of spreading sea floor, as first suggested by H. H. Hess of Princeton University, and that the offsets are along **transform faults**, as suggested by J. Tuzo Wilson of the University of Toronto.

Some transform faults offset the Mid-Atlantic Ridge in its great swerve between Brazil and the Gulf of Guinea by as much as 500 km and are still active (Fig. 8-12). Some in the Pacific, where no ridge is now active west of California, are even longer; the Mendocino Fracture Zone (Fig.

8-4) offsets the magnetic stripes by 1185 km, the largest fault offset known, and far greater than any continental fault thus far recognized.

PLATE TECTONICS

The relations just reviewed, suggesting creation of sea floor along spreading ridges and rises and its subduction and return to the mantle along Benioff zones, has given rise to the theory of plate tectonics, now the leading rationale of crustal deformation. The theory states that the earth's lithosphere is divided into a few large plates, each separated from the others by ridges, Benioff

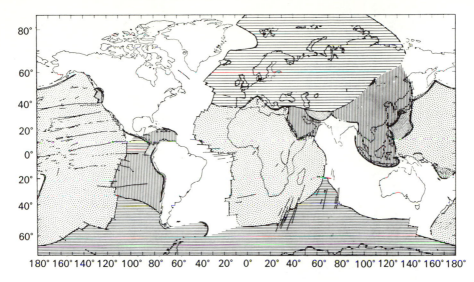

Figure 8-13
The major plates involved in global tectonics. The plates are considered to act as rigid blocks (an oversimplification); they are bounded by rises, transform faults, and trenches of young mountain ranges. [After W. J. Morgan, *Journal of Geophysical Research*, 1968.]

zones, and transform faults. The major deformations of the earth's crust have taken place along the Benioff zones; the older mountains, such as the Appalachians, the mid-Paleozoic mountains of Wales, Scotland, and Scandinavia, and the Urals (late Paleozoic) stand high today only because of isostatic response to erosional unloading, but their structure and the content of axial ophiolites suggest that they, too, mark the sites of older Benioff zones. (We have just pointed out that the Rocky Mountains are exceptions to such a generalization.)

Various authors recognize different numbers of plates, depending on the scale at which they consider the data; they all recognize the major plates shown in Figure 8-13, but many large plates include smaller ones, perhaps created by breakup of the larger ones during collision.

Poles of Plate Rotation

The Swiss mathematician Euler (1707–1783) proved that rigid elements of a spherical surface can be mutually displaced into any arbitrary orientation by a single rotation about a properly selected axis through the center of the sphere. Obviously the displacement near the poles of

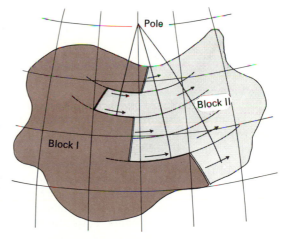

Figure 8-14
Transform faults on a sphere. The relative motion of blocks is about a pole whose position is fixed by the intersection of great circles drawn normal to the transform faults. [After W. Jason Morgan, *Journal of Geophysical Research*, 1968.]

rotation is least and, at the equator related to those poles, greatest (Fig. 8-14).

The mutual displacement at spreading ridges can be found from the magnetic stripes, and the various rates of spreading give clues to the poles

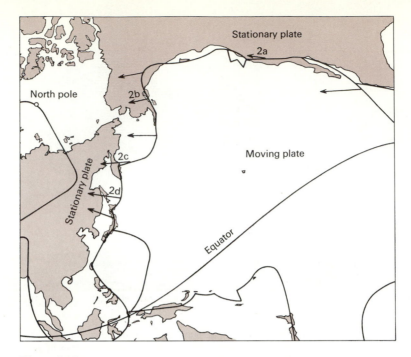

Figure 8-15
A Mercator projection of the Pacific Ocean with the pole of rotation at
50°N, 85°W. Arrows show direction of motion of the Pacific plate relative to
that containing North America and Kamchatka. If both plates are rigid, all
slip vectors should be parallel with each other and with the top and bottom
of the figure. Possible boundaries of other plates are sketched. [After Dan
McKenzie and R. L. Parker, *Nature,* 1967.]

of rotation of the bordering plates. Another clue
to the pole position is afforded by the transform
faults; if the plates are truly rigid, each transform
fault is a parallel of latitude on the sphere, and
its pole is the axis of Eulerian rotation. Pole
positions can thus be found by erecting great
circles normal to the transform faults and noting
their intersections. By reference to Figure 8-9 it
can be seen that pole positions determined by
both methods are reasonably close together,
another strong support for the hypothesis of sea-
floor spreading.

On the assumptions that the first motions on
the San Andreas strike-slip fault are those of a
transform fault separating the American plate
from the Pacific and that the first motions of the
thrust fault that produced the Alaskan earthquake
of 1964 indicate the direction of mutual displacement, Dan McKenzie and R. L. Parker,
then of the University of California at San Diego,
suggested that both could be accounted for by
rotation about a pole at 50°N, 85°W. They then

showed that about 80 percent of the shallow
earthquakes along the Pacific borders during the
preceding decade had first-motion vectors in
agreement with this (Fig. 8-15).

Before proceeding to further discussion of plate
tectonics, we turn to some of the structures characteristic of mountains and not found elsewhere
in the crust. If the plate-tectonics model is indeed
valid, these structures are its products.

Mountain Structure

As Figure 8-8 shows, rocks are highly distorted
along a subduction zone. Such disturbances have
affected crustal blocks many kilometers thick. In
many mountain ranges, erosion exposes rock
masses that have been crushed together by plate
collisions and later isostatically uplifted. Because
continental crust is less dense than oceanic,
wherever a Benioff zone separates continental
from oceanic crust, the continental crust tends to

ride over the oceanic, forming most of the exposed rocks. But the presence of some ophiolites and blueschists in many ranges shows that some oceanic crust has been squeezed into the sial of the continents. Conversely, much sediment, whose density is much less than that of mantle rocks, has disappeared off the Pacific Coast of the United States, very probably by subduction. At depth, of course, these sialic rocks must invert to denser phases, facilitating their sinking. Chunks of oceanic crust, such as the ophiolites, are common in Wales, the Alps, Pyrenees, Urals, Appalachians, Sierra Nevada, the Coast Range of California and Oregon, and many other ranges.

Unlike the largely submerged oceanic ridges, with their symmetric topography and magnetic stripes, all of these ranges are highly asymmetric. Folded rocks are consistently overturned toward one side of the range, and most thrusts travel in the same direction. The side toward which the thrusts have moved and the folds have been overturned is the **foreland**; the side from which the surface rocks have moved is the **hinterland**.

To illustrate the patterns of rock deformation in many mountain ranges, we turn to the most intensively studied range on earth, the Alps.

THE ALPS

A part of the great mountain system that stretches from the Pyrenees to Indonesia, the Alps are largely made up of Mesozoic and early Cenozoic rocks, resting unconformably on a basement of metamorphosed older rocks. In most of the range, the latest of a long series of deformational episodes dates from the Miocene. In the foreland to the north, relatively thin Mesozoic and southward-thickening early Cenozoic rocks unconformably blanket the highly deformed Paleozoic and older rocks. The Alpine Mesozoic strata differ greatly from those of the foreland and are vastly thicker; their source, as shown by sedimentary features, was in the southern hinterland, an area now drowned beneath the Mediterranean. A much simplified map of the Alps is shown in Figure 8-16.

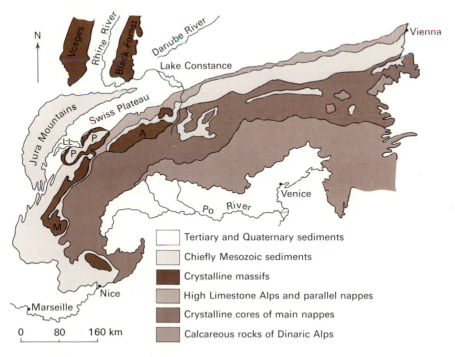

Tertiary and Quaternary sediments

Chiefly Mesozoic sediments

Crystalline massifs

High Limestone Alps and parallel nappes

Crystalline cores of main nappes

Calcareous rocks of Dinaric Alps

Figure 8-16
Generalized map of the western Alps. P, Prealps. M, Mont Blanc and Aiguilles Rouges massifs. A, Aar massif. LL, Lake Leman (Geneva). [After Rudolf Staub; redrawn from L. W. Collect, *Structure of the Alps*, Edward Arnold and Co., 1927.]

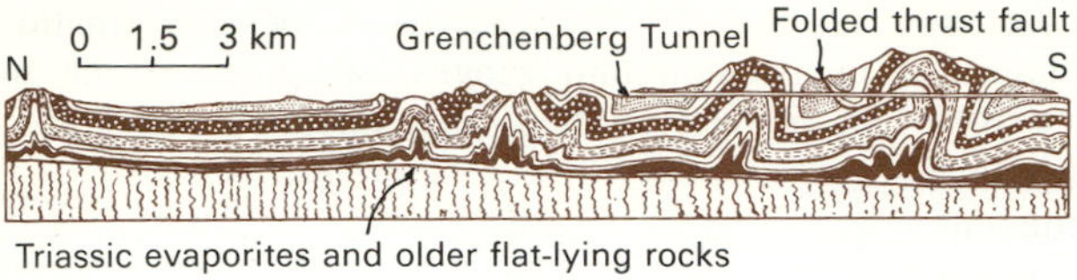

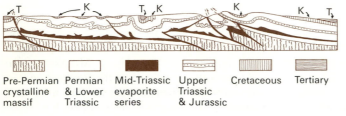

N 0 1.5 3 km   Grenchenberg Tunnel Folded thrust fault   S

Triassic evaporites and older flat-lying rocks

Figure 8-17
Cross section of the Swiss Jura, as interpreted by A. Buxtorf, 1908. Note the essentially undisturbed rocks beneath the Middle Triassic evaporites, from which the overlying beds are considered to be sheared off and independently crumpled—a décollement (French). [Redrawn from E. B. Bailey, *Tectonic Essays,* Clarendon Press, 1935.]

T   K   T, K   K   K   T

| Pre-Permian crystalline massif | Permian & Lower Triassic | Mid-Triassic evaporite series | Upper Triassic & Jurassic | Cretaceous | Tertiary |

Figure 8-18
Jura cross section as interpreted by D. Aubert, 1947. Note that the "ungluing" (décollement) of the folded rocks at the horizon of the Triassic evaporite has been brought about by thrusting within the basement—the pre-Permian metamorphic rocks and their passive cover of Permian and Lower Triassic strata.

The Jura Mountains, an outlying range that merges southwestward with the Alps, rise from the Swiss plateau and expose chiefly foreland Mesozoic strata in much broken anticlines, separated by almost flat synclines. The anticlines are larger and more broken toward the southeast (Fig. 8-17). Significantly, even the highest anticlines expose no rocks older than the evaporite series (salt and anhydrite) of Middle Triassic age. These and tunnel exposures prove that the strata above the evaporite series were torn loose and folded independently of their basement. Figures 8-17 and 8-18 show that the folds are *disharmonic* and do not involve the underlying rocks. The Jura lies farthest from the Alpine core and exhibits the least horizontal movement of any Alpine unit, but even here several kilometers of northwesterly movement was required to make the folds.

Buxtorf's 1908 interpretation seems much too simple: some Jura folds include Pliocene beds, and so are no older, yet Miocene strata of the Swiss Plateau lie almost undisturbed between the Juras and the Alps. Were the crumpling due to a push from the Alps, as the cross section implies, the plateau rocks should also have been folded, or, if they merely slid downhill, there should be a gap—a landslide scar—to mark the head of the slide. No scar exists, and, furthermore, the evaporites lie at a lower level beneath the Swiss plain than in the Jura: the slide would have to have been uphill.

Wells and gravity measurements have shown great local thickening of the evaporites beneath some of the folds. Some deep-lying faults have not cut through to the surface but have jammed the rigid basement rocks into the evaporites, injecting these plastic rocks upward. Décollement ("ungluing") has occurred, as Buxtorf thought, but the basement has also been involved in thrust faults (Fig. 8-18). As the thrust faults lie deep beneath the Swiss Plateau, its rocks were carried forward passively and remain undeformed. Recent studies and well records from the Appalachians have shown that there, too, the surficial structures are disharmonic with the more rigid rocks beneath—a relation that probably applies in all but the most simple folds.

The Swiss Plateau (Fig. 8-16) is underlain by Tertiary sandstones and conglomerates derived from the Alps, as known both from their southward coarsening and from pebbles of recognized source. Though broadly synclinal, the rocks are little disturbed up to the alpine border, where they are abruptly overridden by great thrust sheets of older rocks.

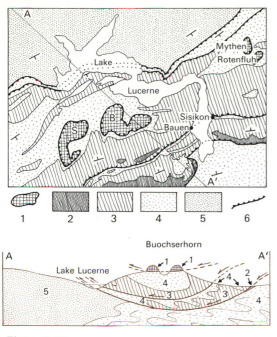

Figure 8-19
Geologic map and schematic section, showing
remnants of the great thrust sheets of the Prealps.
The symbols are: 1, klippes of far-traveled Mesozoic
rocks; 2, Jurassic; 3, Cretaceous; 4, chiefly Eocene;
5, chiefly Miocene; 6, thrust faults; S, Stanserhorn
peak; B, the Buochserhorn. [Simplified from E. B.
Bailey, *Tectonic Essays,* Clarendon Press, 1935.
The section is schematic and not from Bailey.]

THE PREALPS

Between the plateau and the main range of the
Alps lie the Prealps, composed of Mesozoic rocks
such as can only be matched on the south side
of the main chain. It is inferred that the Prealps
have been thrust over the entire range, a distance
of at least 100 km, and have come to rest as a
highly jumbled mass on the Tertiary strata of the
platcau and on some older, less far-traveled thrust
sheets (Figs. 8-19 and 8-20). The Prealps range
from about 1000 to 2500 m in height and are separated from the main range by a line of saddles in
the connecting spurs.

Deep erosion has cut away the former connection with the source of the thrusts, leaving the
Prealps as remnants isolated in a broad expanse
of younger rocks. Such remnants of thrust masses,
isolated by erosion, are *klippes* (German, *Klip-
pen,* "cliffs"). They were first recognized in the
Alps but abound in many other folded ranges.

Figure 8-19 includes a map and an interpre-
tative section. At the southeast end of Lake
Luzern, Eocene rocks dip beneath a thrust composed of Jurassic, Cretaceous, and Eocene strata;
these, in turn, at Sisikon and Bauen are overlain
by a higher sheet of Cretaceous rocks. On this
rests the highest and farthest-traveled sheet of all,
the Triassic rocks of the Stanserhorn, Buochserhorn, Mythen, and Rotenfluh. All the strata are

Figure 8-20
Sketch of the Mythen klippe. The steep peaks of white Mesozoic limestone rest on
Eocene shale. [After L. W. Collet, *The Structure of the Alps.* Edward Arnold & Co.,
1927.]

dated by their fossils; their present relations can only be explained by thrusting (Fig. 8-20).

The main chain of the western Alps includes four high-standing massifs of plutonic and meta-morphic rocks that locally retain unconformable covers of Permian and Triassic strata. Over these are draped scores of highly contorted thrust sheets of Mesozoic and Tertiary rocks, the so-called *nappes* (French), or Decken (German). Between the massifs, the nappes plunge toward an *axial depression*; over the massifs, they rise to *axial culminations*. Many nappes are so con-tinuous that, even though they are fantastically contorted, they can be followed for tens of kilo-meters both along and across the strike (Fig. 8-21).

The spatial arrangement of the nappes demands tremendous shortening of the outer crust. The shortening cannot be accurately measured, of course, as the rocks are in some places stretched and elsewhere thickened. Albert Heim and Joos Cadisch have independently estimated that a sedi-mentary basin originally at least 650 km wide, has been piled together in the range to a width of only about 150 km. The range records sub-duction of at least 500 km of crust. The crust beneath northern Italy must lie several hundred kilometers closer to the Swiss plateau than it did in Eocene time.

Figure 8-22 is a representative cross section of the northern Alps in eastern Switzerland. The northern extent of the thrust sheets can be ac-

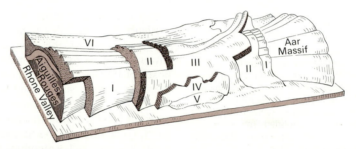

Figure 8-21
Block diagram modified from P. Arbenz, showing how thrust sheets may be preserved in axial depressions and eroded from culminations, allowing reconstruction of the structure to great heights over the culminations and great depths beneath the depressions. Diagrams such as Figure 8-22 are realistic rather than purely imaginative. The upwarped Aiguilles Rouges and Aar crystalline massifs are overlain by a succession of thrust sheets (I to V), eroded from the culminations but preserved in the depressions. It is seen that each thrust was derived from south of the massifs and driven northward over them. Some individual nappes have been traced for more than 50 km across the strike—a minimum measure of their distance of travel.

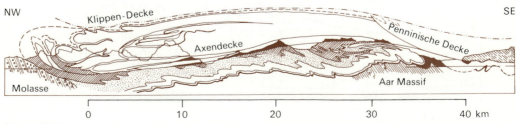

Figure 8-22
Cross section of the northern Alps in eastern Switzerland, just west of the Rhine. Permian and Triassic rocks are shown in solid black; all others are Mesozoic and Cenozoic. [After Arnold Heim, 1910.]

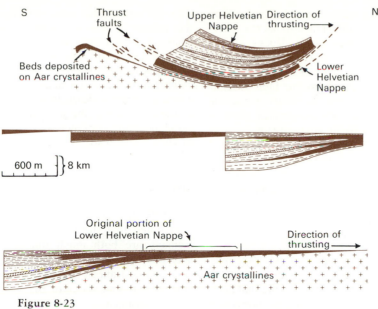

S   Thrust faults   Upper Helvetian Nappe   Direction of thrusting→   N

Beds deposited on Aar crystallines   Lower Helvetian Nappe

600 m }8 km

Original portion of Lower Helvetian Nappe   Direction of thrusting→

Aar crystallines

Figure 8-23
(Top) Present relations of nappes on the north flank of the Aar massif in eastern Switzerland. (Middle) The relations that would have existed had the lower thrust sheets always lain north of the massif and the higher sheet still farther north. (Bottom) The former relations had both sheets been thrust from the south, the higher from farther south than the lower. [After Arnold Heim.]

counted for by gravitational sliding, but the compression obvious at the southern end of the section was clearly not directly due to gravity. Still greater shortening is evident in areas farther south.

The evidence of shortening is not only geometric; facies contrasts (Chapter 17) of the nappes also demand great shortening. For example, Arnold Heim mapped the Cretaceous strata of eastern Switzerland and found the relations summarized in simplified form in Figure 8-22. Drag folds show that both nappes moved northward with respect to the rocks beneath, as the facies differences also demand. Had the nappes always remained north of the Aar massif, and the upper one north of the lower (their present relation), we would have had an original facies arrangement such as is shown in the middle of the figure. Obviously the arrangement in the bottom section is much more reasonable.

Flysch and Molasse The Alpine geologists long ago recognized that most of the youngest strata involved in the folding and nappes are inter-bedded shale and sandstone. Protuberances like those of Figure 17-10 on the base of many sandstone beds record sliding at the time of deposition on underlying shale. Some shales contain large isolated blocks of foreign rocks, as in Figure 17-23. Such associations of strata are called *flysch*— the bouldery variety, *wildflysch*. Flysch, considered premonitory of deformation, indicates the close approach of tectonic uplifts. Many ranges have apparently been formed by the subduction of island arcs like those of the northwestern Pacific. They contain oceanic rocks, such as ophiolites, but the youngest rocks incorporated in them, also shown by their sedimentary characters to be derived from the hinterland, hint of derivation from highland sources. Perhaps they came from islands moving landward toward the subduction zone during the closing of a sea, like the Sea of Japan.

During the Alpine folding, and during that of many other ranges, the drainage of the foreland was disrupted, forming many local basins. Into these poured sediments from the rising range— largely alluvial fans and lacustrine deposits now conglomerates and sandstones of the Swiss plain. This association of sedimentary rocks the Swiss have named the **molasse facies**. Rocks of this facies are associated with most mountain ranges. Appalachian representatives are land-laid Triassic rocks, locally preserved in fault troughs from Nova Scotia to the Carolinas. Along the cordilleras of North and South America, molasse fills many intermont basins and spreads far over the plains to the east. The Coast Mountains of British Columbia and the California Coast Ranges show excellent examples. Thus both flysch and molasse facies are valuable aids in reading geologic history.

SUMMARY OF ALPINE STRUCTURE

Although the few examples cited can only hint at the almost incredibly complex structure of the Alps, they suffice to demonstrate that here, as in many other ranges, prodigious horizontal forces, acting like a great vise, have shoved together parts of the upper crust (Fig. 8-24). During the process many rocks became dynamically metamorphosed: within a few meters a rock mass may display all transitions between shale and highly metamorphosed schist. Granitic rocks have been granulated and partly or wholly recrystallized into gneiss. These observations are significant, as they aid in identifying ancient mountain chains of the Precambrian Shields, now eroded to gentle relief. Plutons and metamorphic rocks are by no means invariably present in fold ranges, but they so commonly are that a related origin is strongly implied.

THE SUBSTRUCTURE OF FOLD MOUNTAINS

Some of the complex folding of the Alpine nappes, especially near the foreland, is readily explained by downslope sliding of a steeply uplifted crustal segment. But so simple a mechanism is wholly inadequate to account for many other features of the range. We saw in Chapter 2 that a great negative Bougeur anomaly marks the higher parts of the Alps. Seismic studies have shown that this is largely due to Airy isostasy— a sialic root projecting downward to a depth of fully 60 km; the deep crust has been deformed as well as the thin carapace. We see, in the metamorphic facies of the southern Alps, that some of the nappes have been under pressures equivalent to loads of 30 km of rock—a figure consonant

with the time lapse since the deformation and the measured rates of erosion now prevailing. These relations indicate that not merely the shallow crust but the deeper crust and upper mantle have been involved as well. Great lateral compression has been active far down into the earth, not merely downslope sliding. The nappes on the south side of the great massifs, even though standing at angles of more than 50°, show no signs of sliding backward, though there is some backfolding in that direction (Fig. 8-24).

GEOSYNCLINES

In 1859, the great American geologist James Hall noted that the folded strata of the Appalachians are both much thicker and richer in siliceous clastic rocks in every system than the correlative strata of the craton to the northwest (Fig. 8-25). The rocks of the Appalachians are largely sedimentary, though metamorphic, volcanic, and plutonic rocks abound in the core of the range. Fossils are nearly all of shallow-water organisms. Many rainprints, mudcracks, salt-crystal casts, and other features (Chapter 17) indicate deposition in shallow water, and some strata are even land-laid. Hall pointed out that shallow-water deposits recur many times throughout more than 10 km of Appalachian strata. Hall gave the name **geosyncline** to this exceptional thickness. Most, but not all the great mountain ranges are composed of similar, exceptionally thick sedimentary piles, and are **geosynclinal mountains.** Such are the Alps, the Caucasus, the Zagros, the American cordillera, and many other ranges.

So many ranges are geosynclinal that for many years it was thought that the formation of a geosyncline was a necessary preliminary to mountain-making. The recent recognition of the association of many young or growing mountains with subduction zones, however, seems to lessen the significance of geosynclinal accumulations in the mountain-making process. (Or perhaps a very considerable gap in our understanding is thus revealed?) At any rate, the earth's highest range, the still-growing Himalaya, is largely made up of crystalline rocks, though it does contain a geosynclinal thickness of sedimentary rocks at its junction with the Tibetan hinterland, more than 100 km from its frontal thrusts. The geosynclinal accumulation to the north obviously had nothing

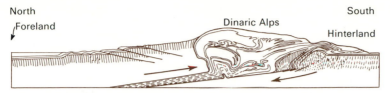

Figure 8-24
Idealized reconstructions of the structure of the Alps, neglecting erosion. [After Emile Argand, 1916.]

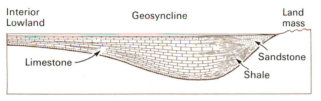

Figure 8-25
The thickening of strata from the Interior Lowland (the craton) into the Appalachian geosyncline. [After A. W. Grabau, 1924.]

to do with localizing the south-moving thrusts of the Siwalik foothills.

Thus not all fold mountains are geosynclinal, nor, as we have noted of the Rocky Mountains, are all of them associated with subduction zones. Nor does the accumulation of extremely thick sediments alone suffice to begin the mountain-building process. A great geosyncline in western Montana, northern Idaho, and southern British Columbia filled with more than 15 km of sedimentary rocks several hundred million years before the Cambrian. This great sedimentary pile, the Belt Supergroup, lay undisturbed except for gentle tilting until the onset of the Cordilleran mountain-building in Mesozoic time. The time gap between accumulation of a geosynclinal packet of strata and mountain deformation amounts to as much as all of Phanerozoic time. Many puzzles obviously remain as to the origin of fold mountains; we turn now to some of the other types of mountain structure.

FAULT-BLOCK MOUNTAINS

We have already discussed the greatest of fault-block ranges, the oceanic ridges. Many also accent the continents.

Ruwenzori, the highest mountain in Africa, is a huge block of Precambrian rock surrounded on all sides by high-angle normal faults, some of which, at least, are still active. Other mountains along the African Rift Valleys are fault blocks tilted like trap doors, such as the Aberdare Range of Kenya. Other examples are found in the eastern Sudan, Ethiopia, and Tanzania. In western North America, from central Mexico to Oregon, innumerable large and small isolated mountain masses rise above the desert plains and basins, in the so-called Basin-and-Range topographic province. All are separated from the intermont valleys by normal faults, some of great displacement.

These **fault-block mountains** have been formed from crustal segments of widely differing history. Some, indeed, involve geosynclinal rocks, but the facies trends of the strata are wholly independent of the fault trends, and in some an older range had been eroded to low relief before the faulting that produced the present high relief.

Some fault-block mountains of the western States, such as the Teton Range of Wyoming, are tilted blocks of metamorphic rocks capped by lava flows. The Sierra Nevada, the westernmost of these fault-block ranges, is composed largely of Mesozoic strata and plutons, with patches of Tertiary volcanics resting in great unconformity on its crest and gentle western slope. The great fault scarp forming its eastern slope is one of the most dramatic in the world.

Figure 8-26
Cross section of a relatively little-eroded fault-block range,
Steens Mountain, Oregon. Note that the right-hand fault is
somewhat older than the middle one, and is overlapped by
gravel; the middle fault cuts the gravel, so is younger. All three
faults, however, are relatively young, for their scarps have not
been much eroded. The marked bed indicates the amount of
displacement.

Steens Mountain, Oregon (Fig. 8-26), consists entirely of gently tilted Tertiary lava and tuff; the Wasatch Range, Utah, includes Precambrian and thick geosynclinal Paleozoic and Mesozoic strata, greatly deformed and injected by plutons in early Cenozoic time. The Little Hatchet Mountains of southern New Mexico are chiefly geosynclinal Lower Cretaceous strata with some Tertiary volcanics. Some ranges in Nevada expose more than 2500 m of volcanic rocks with no basement visible. Obviously such great variations in rocks and history set the fault-block mountains apart from the fold mountains, which, for long distances along strike, have undergone similar sedimentary and tectonic histories. The normal fault boundaries tell of crustal extension, while the fold mountains are products of crustal compression.

VOLCANIC MOUNTAINS

Although part of the elevation of the ocean ridges is clearly due to the active volcanism along them, they have been treated as fault-block mountains because of their association with crustal extension. The volcanoes of the island arcs and Hawaiian Islands are discussed in a later section. The Cascade Range, extending from northern California to British Columbia, and part, but not all of the Andes, are largely aligned volcanic cones built on variable basements of independent trends. Clearly the trends of the volcanic edifices are not controlled by the locally exposed structures of the basement rocks. They are more likely controlled by a currently active Benioff zone along which melting, due to frictional drag, is generating lava (Chile) or has until recently done so (Cascades).

Few such alignments of volcanoes exist on the continents; most such cones are isolated, like Kenya and Kilimanjaro in Africa, Büyük Agri Dagi (Ararat) in Soviet Armenia, and Demavend in Iran.

Volcanic structures are further discussed in Chapter 9.

UPWARPED MOUNTAINS

Some mountains are not readily classified in the above categories. The Adirondacks, the Labrador Highlands, the Black Hills, are a few of the many that could be cited. All are composed mainly of Precambrian metamorphic and intrusive rocks like those in the deeply eroded cores of fold ranges. But these mountains are not elongated parallel to their bedrock structures as are most fold ranges but trend independently of them. True, some faults border the Adirondacks locally, but they seem negligibly small and incapable of accounting for the relief of the range, as the Cambrian overlap onto the range is gently dipping and not faulted against the basement. All these mountains seem to be recently upwarped parts of the Precambrian shield, long ago deformed into typical fold mountains, but reduced by erosion to low relief before transgression of the Cambrian sea and deposition of a flat sedimentary blanket of nongeosynclinal thickness. Upwarping came long after the original folding, and is entirely independent of it.

Similar upwarping, long after folding and deep erosion, has gone on in the Appalachians, the Caledonian and Scandinavian ranges, the Urals, the Cape Mountains of Africa, and many other ranges. But these uplifts differ in that they trend with the bedrock structures. The mechanism of their uplift must differ greatly from that involved in upwarping of the mountains just discussed; it can hardly have been caused by renewed compression or by extensional processes; perhaps it is due to long-delayed isostatic response to unloading by deep erosion.

The Driving Force of Plate Tectonics

Perhaps the most disputed problem in all of geology is the operating mechanism of orogeny. Before the discovery of radioactivity, the crustal shortening in orogenic belts was thought by many to result from the accommodation of the crust to a shrinking of the globe owing to loss of heat—its secular cooling. The yielding was thought to be concentrated in the geosynclines because the great thicknesses of sediments made these places weaker than other parts of the crust. This hypothesis is still supported by a few, but it has been generally abandoned for several cogent reasons. First, the crust is shown by isostasy to be entirely too weak to transmit forces half way around the earth and thus account for the very uneven spatial distribution of the ranges. Second, we now know

from studies of first motion that areas of crustal expansion are as great as those of crustal shortening. Third, studies of heat flow and radioactivity make it unlikely that the earth is cooling appreciably, and in fact it may be heating up. Heat transmission through the mantle is so sluggish that even in four billion years no heat from depths of as much as 600 km can have reached the surface by conduction. Transfer of heat by radiation is less accurately estimated but is generally thought to be unimportant. Nearly all students now agree that the ultimate driving force must be the internal heat of the earth, of radioactive origin; still in dispute is the operative mechanism.

Some favor an idea first suggested by the distinguished British geologist Arthur Holmes (1931)—that convection currents in the mantle rise and diverge beneath the crust, dragging along the overlying material and thus creating new ocean. Where the currents descend, the crust is squeezed together and a mountain chain is formed. Changes of mineral phases in the belts of compression produce denser rocks, thus facilitating the sinking of the descending limb (Fig. 8-27).

This model would account well for the central fault trough on the spreading ridges, if the vis-

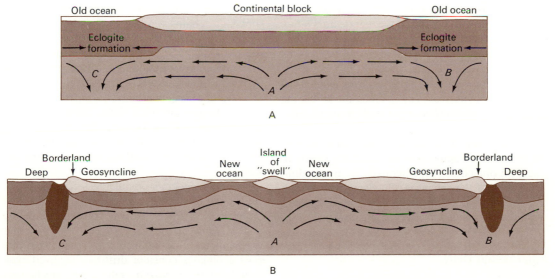

Figure 8-27
New hypothetical convection pattern in the mantle, as suggested by Arthur Holmes in 1931.

cosity of the mantle is low enough to permit convection—a point on which the opinion of geophysicists is strenuously divided. But it seems improbable that the transform faults that so commonly offset the spreading ridge would all be nearly at right angles to the ridge; one would expect considerably more irregularity, such as is seen in the patterns of the undoubtedly tensional Basin and Range faults and African Rift valleys.

Others have suggested that the uplift of the spreading ridges, and the gradient so formed, simply permits sliding without the help of dragging by a convection current. Here, again, it seems most improbable that a rising magma plume would be as straight as long reaches of the ridges are, and the pattern of transform faults remains just as puzzling as under the convection hypothesis. Further to question the sliding hypothesis is the fact that the East Pacific Rise, which stands only half as high above the general level of the ocean floor, is apparently spreading twice as fast as the Mid-Atlantic Ridge. Furthermore, a good deal of the relief of the spreading ridges is due to the fact that their rocks are hotter than those beneath the lower ocean floor and are thus less dense; this means that the relief is no advantage to the sliding, when account is taken of the whole lithosphere above the Low-Velocity Zone.

A third suggestion is that the subducted plate, even though less mafic than the mantle into which it plunges, is more dense because it is cold; thus it tends to pull the plate along as it dives. This must be a negligible factor in the operation, however, for rocks are very weak in tension, as shown by the normal faulting along the spreading ridges and subduction zones and in many other localities.

In summary, then, no mechanism thus far suggested as the engine of plate tectonics seems to meet all objections. As scientists, we must admit ignorance of the driving mechanism, even though the operation of the process seems consonant with a wide range of observed phenomena.

Continental Drift

Probably every child who has studied a globe has independently discovered that if South America were moved eastward it would almost fit snugly with the coast of Africa; North America would not fit so nicely with Africa and Europe, but even here the agreement is reasonably striking. The rough fit becomes even more persuasive when we consider the offset of the Mid-Atlantic Ridge in parallel with coastal trends near the Equator. About a century ago, Antonio Snider published a speculation that the Atlantic continents had once been joined and had since drifted apart. The idea seemed to his contemporaries so bizarre that it was ignored until about 60 years ago, when it was independently resurrected by the American geologist F. B. Taylor and the German meteorologist Alfred Wegener. If whole continents can move so far, the problem of mountain-building becomes incidental; the forces involved in raising the greatest of mountains are small compared to those needed to move continents for thousands of kilometers.

Wegener assumed that the continents float isostatically upon a dense substratum that is so weak it yields to very small forces, two of which he emphasized. The first is centrifugal: the continents stand higher than the oceans, and so are subject to a greater centrifugal force that tends to drive them toward the equator. The second force, the tidal attraction of the moon and sun, tends to drag the continents westward as the earth rotates. (It must not be forgotten that tidal forces act on rocks as well as on the seas and atmosphere.) Wegener thought the Alpine-Himalaya chain was formed by collision of Eurasia with Africa and India in response to the first force and the meridional Andes and Rockies were piled up by friction as the Americas drifted westward, pulled by the tides. He didn't explain why the first ranges are so far from the equator, nor did he explain why, if the continents could drift through the fluid beneath, that fluid could so oppose the drift that it lifted up the mountains 1500 km from the leading edge to form the Rockies.

Both of Wegener's postulated forces exist, but they are far too feeble to do the work he attributed to them. Nevertheless, so many features of the various land masses are suggestive of former arrangements differing from the present ones that most geologists consider drift has occurred even though we as yet have no satisfactory mechanism for bringing it about. For reasons subsequently discussed, most geologist postulate a post-Permian date for the beginning of traceable drift.

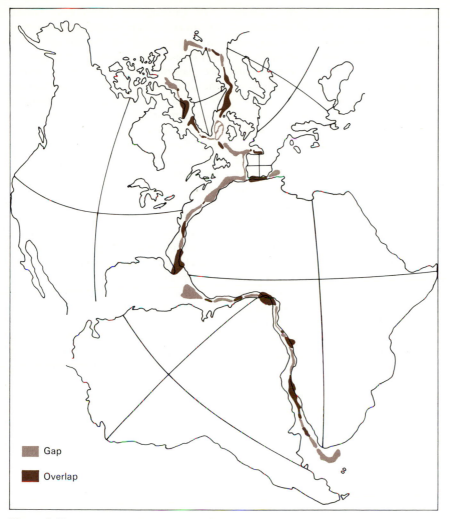

Figure 8-28
The possible fit of the continents before drift, on the basis of the 500-fathom subsea contour, which is everywhere on the continental slope. A present meridian and latitudinal circle is shown for each continent. [After E. C. Bullard, J. Everett, and A. G. Smith, *Phil. Trans. Royal Society of London* v. 238, 1965.]

The British team of Sir Edward Bullard, J. Everett, and A. G. Smith tried to fit the 500-fathom subsea contours of the continents together using a computer to apply Euler's theorem (Fig. 8-28). The fit of South America against Africa is indeed excellent (mean misfit about 130 km), the only large overlap, at the Niger delta, is an area of great post-Triassic sedimentation. The fit becomes even more convincing when radiometric age provinces are plotted (Fig. 8-29). P. M.

Hurley and colleagues have dated several tens of samples from both sides of the Atlantic and found that the boundary between plutons of two widely divergent age groups is virtually continuous from Brazil to Nigeria when plotted on the fit by Bullard and others.

The proposed fit of North America with the Old World is much less persuasive; in order to achieve the fit of the 500-fathom lines here, Bullard and his colleagues had to omit Iceland, which is of Tertiary age, and to ignore the 3 km of Mesozoic and younger rocks that cover the pre-drift basement at Cape Hatteras, where the pre-Jurassic 500-fathom contour may have been 200 km or more either east or west from the present one, as the magnetic stripes stop 250 km offshore. Also they retained the Strait of Gibraltar at its present width, despite the fact that the Betic Cordillera of southern Spain and the Atlas Range of Morocco, both of Cenozoic age, both underwent crustal shortening of well over 100 km. Before this happened, the Strait of Gibraltar must have been more than 300 km wide, as some paleomagnetic observations also suggest. The fit presented by Bullard's group also rotates the Iberian Peninsula to close the Bay of Biscay; some magnetic evidence suggests that this may well have happened. More seriously, their fit ignores the Paleozoic and older rocks of southern Mexico and Central America. Apparently Europe should have been placed at least 300 km to the north of the suggested position; this would give room for Central America, and is certainly not debarred by the geology or bathymetry of the eastern seaboard of the United States. In short, the fit of North America with Europe and Africa, suggested as support of their former junction, is highly questionable. A great deal of evidence for drift is much more persuasive than this map.

In Chapter 17 evidence is presented for a Devonian desert whose extensions into North America would very closely agree with its limit in Europe if the Atlantic were closed up. So, too, the Appalachian range would closely join with the mountains of Ireland and Scotland. Though the trend lines of the early Mesozoic ranges of the southern continents do not agree, closing of the Atlantic would bring them into close alignment. The diverse trends of such nearly contemporaneous ranges as the Alps and Carpathians shows that trends need not coincide for mountains to be closely related.

Stratigraphic similarities between South America and South Africa abound. In both, the oldest fossiliferous rocks are late Silurian or Devonian, containing fossils like those of the Falkland Islands but differing from those of the Northern hemisphere. In both regions, similar land-laid beds intervene between the marine Devonian and late Paleozoic glacial deposits—**tillites** (Fig. 8-31).

Tillite of roughly the same age is found also in Antarctica, Australia, and India (Fig. 8-31). The glaciers invaded South America from the east, the Falkland Islands from the north, Australia from the south—all impossible were the geography at the time of glaciation similar to that of the present. Moreover, an ice cap near 15°S left its northern moraines on the equator and its southern

Figure 8-29
Pre-drift fit of West Africa and eastern Brazil according to the reconstruction by Bullard and others. Solid circles indicate rocks about 2000 m.y. old; open circles, rocks about 550 m.y. old. The heavy broken line is the boundary between the provinces. [After P. M. Hurley et al., "Test of Continental Drift by Comparison of Radiometric Ages." *Science*, 157:495–500, 1967. Copyright 1967 by the American Association for the Advancement of Science.]

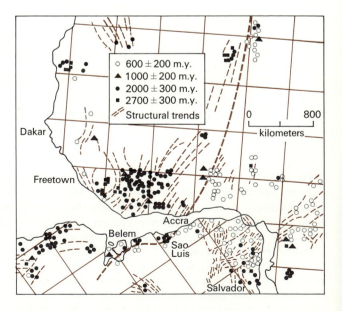

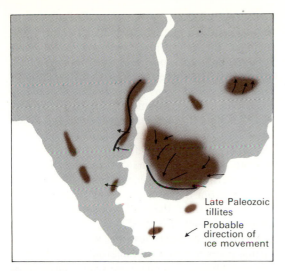

Figure 8-30
Early Mesozoic mountains of Africa and South
America as they would appear if the continents were
in the positions shown in Figure 8-28. Also shown
are the positions of late Paleozoic glacial deposits and
the direction of motion of the glaciers from which
they were derived, as discussed in Chapter 13.
[Modified from A. L. Du Toit.]

ones near the Cape of Good Hope. Inasmuch as
the area is virtually a plain, this implies the
improbable situation of an ice cap practically on
the equator, unless Africa was in a different
latitude at the time of glaciation. A roughly contemporaneous glacier in India centered on the
Aravalli Mountains north of Bombay in latitude
20°N. Rocks recognizably derived from here are
found in tillite hundreds of kilometers to the
southeast and northwest.

In all these glaciated regions, the tillite is
associated with fossils of two genera of fern-like
plants, *Gangamopteris* and *Glossopteris*, abundant in the southern continents and in India but
represented in the northern hemisphere only by
scant collections from central France, Russia, and
Siberia. These and other fossils show all the
tillites to be late Paleozoic but not quite contemporaneous. The terrestrial fossil reptile
Mesosaurus is found in Carboniferous strata of
Brazil and South Africa and nowhere else.

Du Toit, a distinguished South African geologist, pointed out that had the continents been
assembled in the pattern of Figure 8-31 they

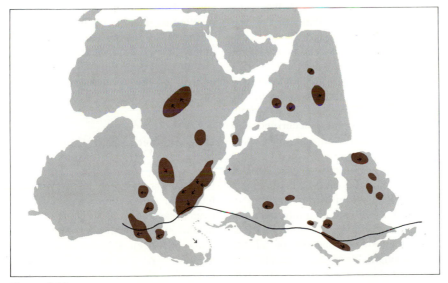

Figure 8-31
Du Toit's assemblage of the continents before early Mesozoic time. Tillites shown in
brown; arrows show direction of ice flow. Black line is the axis of the late Paleozoic
geosyncline found in all the southern continents. With the pole where the cross is, this
grouping places all the glaciated areas of late Paleozoic time in either polar or temperate
climatic zones. [Modified from Du Toit, incorporating results of recent Antarctic
research.]

could all have been in temperate or polar climates at the same time—impossible with their present relative positions no matter what polar wandering is assumed. Furthermore the late Paleozoic geosynclines of South America, South Africa, and Australia would form a continuous chain, marked in brown in the figure. Since Du Toit wrote, tillites of this age have been found in Antarctica, greatly strengthening his argument.

How, indeed, are we to account for continental ice sheets (Chapter 13) in the present tropics unless the continents have moved? Mere shifting of the earth's axis, which most astronomers consider physically impossible, would not solve the problem; the present relative positions of the ice sheets are such that no possible pole position would avoid leaving one or the other in the tropics.

The Glossopteris flora, abundant in the southern hemisphere and in India, and extremely rare elsewhere, has been considered strong evidence of a former closer grouping of these now scattered regions. Nearly a century ago, the name "Gondwana Land" was suggested by the great Austrian geologist Eduard Suess for a hypothetical grouping of the continents with the *Glossopteris* flora. He distinguished it from the northern lands of North American and Eurasia, "Laurentia," with their strikingly different floras and faunas. Du Toit thought that the Glossopteris coals of India, Australia, and South Africa represent cold peat bogs, like those of present-day Alaska and Ireland, and that the coals of the northern hemisphere represent tropical swamp deposits. The few Glossopteris occurrences in northern Asia and Europe were unknown when he wrote; they might still be explained by local conditions, as are the similarities in the floras of southeastern China and the south Atlantic States of North America. The known differences in flora and fauna of Laurentia and Gondwana Land were striking in Suess's day, and they still render unlikely some suggested groupings of all the continents into a single "Pangaea." The pre-drift arrangement must have been into two great supercontinents, separated by the Tethyan geosynclinal trough along which the Alps and Himalayas later rose.

Paleomagnetism has supplied further evidence of continental drift. We noted in Chapter 7 that the virtual magnetic poles for late Pleistocene and

Holocene time have a mean position very near the geographic pole, where it should be were the dynamo theory correct. Paleomagnetists thus believe that polar positions can be reasonably accurately determined if a fairly thick section of strata is sampled, so as to average the field over a time long enough to smooth out the nondipole disturbances. If the continents have drifted, each should show pole positions mutually consistent but different for times before and during the start of drift. As the continents drift through time, the pole determined from each should deviate less and less from those of the other plates until, in Quaternary time they all converge upon the geographic poles (Fig. 8-33).

Figure 8-32 illustrates the widely scattered virtual poles for the Carboniferous of all the continents. Although the data are scant, they suffice to show that none of the continents had the same relation to the magnetic poles that they have today. Either the geographic pole or the continents must have moved. The earth's angular momentum is great; like a gyroscope, it strongly resists change in its axis of rotation, and astronomers believe that the axis has never shifted more than a small amount with respect to the body of the earth. The apparent polar wandering must have resulted from shifts of the crust—drift.

Figure 8-33 shows the tentative trace of the poles through time, holding each continent in its present position, thus making the poles wander instead of the continents—a most unlikely assumption.

Deep Submergence of Continental Margins

Evidence shows that much of the sediment that filled the west Alpine geosyncline was derived from areas now sunk to depths of more than 2 km in the western Mediterranean. Obviously the crust has been thinned in some way. Seismic studies show a crust about 12 km thick—twice that of normal oceanic crust—and with elastic properties appropriate to a more than usually dense sima; before the Alpine orogeny this crust must have been both thicker and much more sialic. At the west coast of India north of Bombay, the flows of the great Deccan flood basalts of late Cretaceous and Paleocene age, nearly horizontal

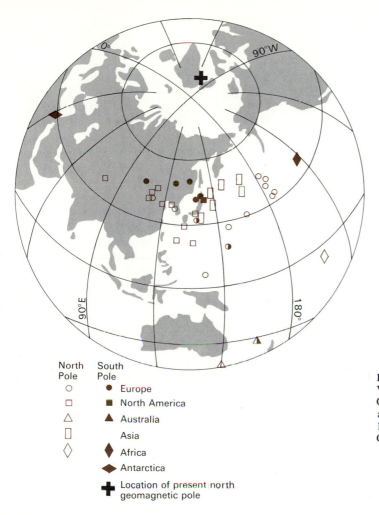
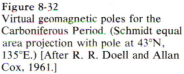

North South
Pole Pole

North Pole	South Pole	
○	●	Europe
□	■	North America
△	▲	Australia
▯		Asia
◇	◆	Africa
	◆	Antarctica
✚		Location of present north geomagnetic pole

Figure 8-32
Virtual geomagnetic poles for the Carboniferous Period. (Schmidt equal area projection with pole at 43°N, 135°E.) [After R. R. Doell and Allan Cox, 1961.]

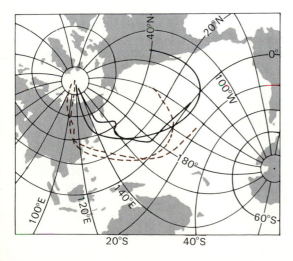

Figure 8-33
Polar wandering paths suggested for Europe and North America for different times. Brown lines represent the paths determined from North America. Black lines are those determined for Europe. [After S. K. Runcorn, *Phil. Trans. Royal Society* v. 258, 1963.]

for hundreds of kilometers to the east, abruptly bend into a steep monoclinal fold and descend to the deep ocean floor. Obviously the land formerly extended an unknown but considerable distance into what is now the Arabian Sea. A precisely similar situation is present in east Greenland, where the Tertiary flood basalts fold over a monocline and descend to the floor of the Atlantic. In all these places the crust must have been thinned in some manner.

On the Atlantic coast of the United States, explosion seismology has traced the nonconformity on which the Mesozoic rocks of the Coastal Plain overlie the ancient rocks of the Piedmont far out to sea, where it reaches depths of nearly 6 km before joining a normal oceanic crust (Fig. 8-34). As far out as the continental slope this nonconformity is continuous, and the rocks beneath have the same elastic properties as those of the Piedmont. Obviously this was formerly land, and its surface has subsided to depths far greater than can be accounted for by isostatic response to the sedimentary load.

Far from any plate margin, the former land surface in the Rocky Mountain region subsided beneath the mid-Cretaceous sea, locally to depths

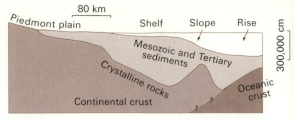

Figure 8-34
A cross section of the continental border off southern New Jersey. [After C. L. Drake, 1959, Pergamon Press.]

of 5 km, only now to stand at elevations of more than 4 km. Such excursions of height give reason to doubt that plate drifting is the only way to account for missing continental lands. Some of the mission fragments of Gondwana Land may not simply have drifted away, but may have sunk by abrasion of the lower crust as these crustal movements suggest. Much remains to be learned about our fascinating, puzzling earth—four billion years old but still endowed with energy literally to move continents!

Questions

1. In both the East and West Indies, volcanoes are aligned on an inner arc parallel to the arc of the ocean deeps and associated negative anomalies. The surface on which the deep-focus earthquakes occur slopes downward toward this inner arc. What does this suggest as to the source of the magmas?

2. What inferences can you make regarding the forces causing crustal deformation from such features as the San Andreas rift (Chapter 7)?

3. Many of the buildings of western England are roofed with slate, whereas those near London are chiefly tiled. Can you draw from this any inferences as to the regional geology of England?

4. Many deep wells drilled in the search for oil have shown the existence of a long, rather narrow mass of granite beneath the Carboniferous strata of eastern Kansas and southeastern Nebraska. What features of the rocks brought up by the drill would enable you to decide whether this mass is unconformably buried by the sedimentary rocks or whether it invaded the sediments after they were deposited?

5. The Triassic rocks of the Atlantic slope of North America are commonly thought to have been formed under both topographic and climatic conditions closely similar to those of the present Great Basin. What features would you expect them to have from which such an origin was inferred?

6. In western Nevada a fossil-rich Permian limestone lies nearly horizontally across upturned slates beneath. A few fossils identifiable as Ordovician in age have been found in the slate. What history is recorded by these relationships?

7. What features would enable you to distinguish a sill of granitic rock (whose lower contact only is exposed) from a block of granite thrust over flat-lying sedimentary rocks?

8. How would you distinguish a klippe (an erosional remnant of a thrust sheet) from the erosional remnant of a resistant bed in an undeformed sedimentary series? Assume in both cases that the remnant is of Carboniferous rocks resting upon flat-lying Devonian.

9. What do you infer from the absence of a root beneath the Northern Rocky Mountains about the applicability of Airy isostasy to the region? What do you infer from the presence of a root beneath the Alps?

Suggested Readings

Bailey, E. B., *Tectonic Essays, Mainly Alpine.* Oxford: Clarendon Press, 1935.

Cox, Allan (editor), *Plate Tectonics and Geomagnetic Reversals.* San Francisco: W. H. Freeman and Company, 1973.

Scientific American Offprints

814. Robert L. Fisher and Roger Revelle, "The Trenches of the Pacific" (November 1955).

855. Don L. Anderson, "The Plastic Layer of the Earth's Mantle" (July 1962).

CHAPTER 9

Constructive Part of the Geologic Cycle— Igneous Activity and Metamorphism

This chapter primarily concerns the emplacement in or on the earth's crust of rocks derived either from the mantle or from remelting of other rocks at shallower levels. Most obvious are the volcanic rocks, most of which are very probably derived from the mantle, though some of the more siliceous are probably melted from the crust.

VOLCANOES

Distribution

Most volcanoes lie on ocean ridges or above active or recently active subduction zones. Good evidence suggests that the Cascades lie above a subduction zone that has only recently ceased activity (Fig. 9-1). A few are not so situated— for example, the Hawaiian and Samoan volcanoes of the Pacific, the Cape Verde and Canary Islands of the Atlantic, Mount Tibesti in the eastern Sahara, and Mount Cameroon at the edge of the Niger delta. The volcanoes of east Africa, along the rift valleys, may be associated with an incipient spreading ridge.

With these few exceptions, most volcanoes are thus in orogenic environments (Chapter 8). This is surely significant, but may be overevaluated, for when the entire Cenozoic is considered, volcanoes were active at one time or another in the completely anorogenic Plateau Central of France,

in northern Ireland and Scotland, in the Rhine Valley, Hungary, northern Spain, Arabia, Palestine, northwestern India, all the western United States as far east as the Great Plains, and even in a minor way Virginia, Quebec, and New Hampshire.

Varieties of Volcanoes

Volcanoes are vents through which magma, pyroclastic material, and gases are extruded with widely varying energy. During human history, at least four-fifths—perhaps nine-tenths—of all visible eruptions have been explosive, producing far more pyroclastic material than lava, although most volcanoes produce some of both. During the past century about 1650 eruptions have been observed, of which 1526 have been explosive. The German volcanologist Sapper estimated in 1914 that since 1500, about 64 km^3 of lava had been erupted, and more than five times that much ash. These, of course, are considerable underestimates, for as we saw in Chapter 8, by far the most abundant lavas are being extruded along the spreading oceanic ridges, not only at the few visible island volcanoes but along their entire subsea lengths. Most of these eruptions are of lava, rather than ash, for the pressure of deep water inhibits explosive escape of volcanic gases. Accordingly, over the earth as a whole, emission of

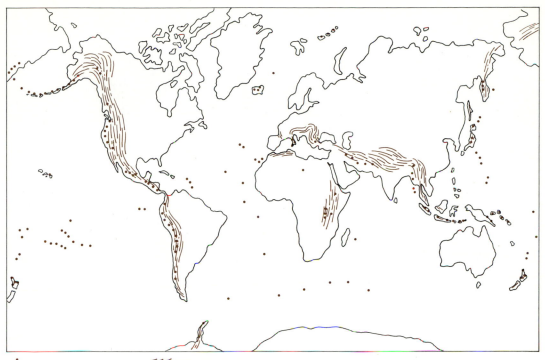

Active volcanoes   ⟋⟋⟋⟍ Mountains of Cenozoic age

Figure 9-1
The distribution of the active volcanoes of the earth. [Data from various sources.]

lava must far outweigh that of pyroclastic products, as, indeed, the geologic record definitely indicates.

Volcanoes vary widely in form and petrology because of differences in chemical composition of their eruptive products. Four principal classes are recognized:

1. Composite volcanoes are made up of both lava and pyroclastic rocks. Most are composed of andesitic or more siliceous magma. Some, like Vesuvius, are of subsiliceous rock, but all tend to be viscous and to flow only on fairly steep slopes.

2. Shield volcanoes are composed almost wholly of fluid basaltic (or, less commonly, andesitic) lavas; pyroclastics are trivial in amount. Their magmas are much more fluid than those of the composite volcanoes, and their slopes are rarely steeper than 5°, generally much less. The best examples are the volcanoes of Hawaii.

3. Flood basalts, as the name suggests, are nearly flat accumulations of basalt, emitted from fissures rather than from central vents—and so fluid that only a few cinder cones mark the sites of final activity along the conduits that fed the flows. Apparently the only example historically witnessed was the eruption of the Laki fissure in Iceland in 1783, but such eruptions have accumulated in many regions and in tremendous volume.

4. Pyroclastic flows may be emitted from fissures or from central vents in such gas-charged states as to flow almost like water and to accumulate, as they lose their content of gas and hence fluidity, in pond-like bodies of variable thickness. Almost all are derived from siliceous magmas, such as latite, rhyolite, and dacite, which when free of gas are extremely viscous, but when highly charged with gas are very fluid. But exceptions exist: the volcanoes Ulawun, on New Britain, and Manam, off New Guinea, both

160

Figure 9-2
Mount Rainier, Washington, a glaciated composite volcano. Three others rise above the Cascade Range: Mount Adams at left, Mount St. Helens at right, and Mount Hood, Oregon, just to the left of Rainier summit. [Photo by H. Miller Cowling, Spokane, Washington.]

emitted basaltic ash flows in precisely the same manner. Where thick enough to cool slowly the ash flow welds into a solid rock in which traces of former pumice fragments can be seen as flattened streaky masses (Fig. 9-11). Satellite vents of Mount Katmai, Alaska, have been historically active, as well as Pelée in the West Indies, and many others in New Zealand, the Andes, and Indonesia. Such flows have formerly been widespread in many other areas.

COMPOSITE VOLCANOES

Most terrestrial volcanoes are composite, erupting both pyroclastics and lava, either simultaneously or in turn. Examples are the graceful Fuji-san of Japan; the familiar Vesuvius, Stromboli, and Etna in Italy; Aconcagua in the Argentine; Büyük Agri Dagi (Ararat) in Turkey; and the many glacier-clad cones of the Cascades in the western United States. Mount Rainier is shown in Figure 9-2.

Hekla, 1947 The second largest eruption in Iceland's recorded history began on the morning of March 29, 1947, when a virtually continuous fissure 5 km long opened along the northeast-

trending summit ridge of Hekla volcano. Grayish-brown ash was ejected to the astounding height of 30 km, and within 20 minutes an ash jet extended downwind for 35 km or more, moving far faster than the wind, even in the stratosphere. Nearly continuous explosions hurled huge volumes of ash into the upper air, reaching the British Isles and Scandinavia the next day. During the early, violent phases, spindle bombs (Appendix III) of twisted lava as much as 15 m long and 4 m through were tossed several hundred meters, and one 50 cm in diameter was thrown 32 km. For more than an hour ash poured forth catastrophically, at a rate equal to the discharge of the Congo, and then began to diminish, the early continuous curtain of ash being limited to half a dozen craters and, after several days, to two, one on the summit, the other on the southwestern shoulder of the mountain, about 2½ km away.

Lava flows emerged within the first hour of ash eruption, and within a day had covered 15 km². The first flows, chiefly from the southwest crater, poured out at a rate estimated at 3500 m³ per second. In some areas a crust of congealed lava formed at the surface, but molten lava continued to flow through self-constructed tunnels for as much as a kilometer before emerging again. Such lava tubes abound in many volcanic fields. Once formed, they insulate the molten material within them, so that it retains its fluidity longer than that in open channels. Some tubes in the Hawaiian Islands are as much as 12 km long.

Lava continued to erupt for about 13 months, eventually covering 40 km² with an estimated volume of 0.8 km³. That first erupted was dacite (silica content 61.9 percent), followed by andesite (silica content 57.3 percent) and finally by basaltic andesite (silica content 54.25 percent), strongly suggesting that the magma beneath the volcano was stratified according to density, the more siliceous magma being the lightest and floating highest in the conduit. In the spring and summer following the eruption, great volumes of carbon dioxide, estimated at 24,000 tons, emerged from cracks in the lava and, being heavier than air, settled into depressions to form invisible gas ponds in which many birds and a few sheep were asphyxiated.

During this eruption the summit of Hekla gained 56 meters, reaching 1503 m, but at the end of the eruption it subsided about a dozen meters.

Usu, 1943–1945 The volcano Usu on the island of Hokkaido, Japan, behaved very differently. According to historic records the volcano had erupted five times before the eruption that began on the evening of December 28, 1943, with a series of severe earthquakes—more than a hundred a day for several days. Between March and mid-April of 1944, a block of ground about 4 km in diameter rose at the rate of as much as 30 cm a day. The ground, cracked and broken, was finally uplifted nearly 50 m before the first volcanic explosion broke through the corner of a cornfield.

On June 23 a column of steam and ash rose noiselessly from the field, soon changing into a series of mud eruptions that formed a crater about 35 by 50 m across. Mud overflowed from the crater, poured into a depression, and formed a hot mud puddle before the first broken fragments of cold lava were ejected, to heights as great as 800 m. Beginning just after midnight on July 2 and continuing for about 5 hours, explosions hurled powdered rock high into the air, producing a total of about 2×10^6 metric tons. Very little of this powdered rock was new magma; most was broken from the walls of the conduit that fed the volcano.

Eruptions continued for several months, by which time hotter blocks were being ejected, but none were incandescent. By the end of October the cornfield, which was originally 130 to 160 m above the sea, stood at nearly 300 m, making a plateau of half a km². Shortly after, solid but plastic dacite lava began to be extruded from a ring of craters, to form a symmetrical dome that eventually reached more than 400 m in height. Activity ceased in October, 1945. Evidently nearly all the magma emplacement was underground, very probably forming a bysmalith (p. 187); only a small fraction of the erupted material was new magma.

Many eruptions are far more powerful than either those of Hekla or of Usu.

Bezymianny, 1955–1956 Almost surely the most gigantic natural explosion of this century was that of the volcano Bezymianny, on the Kamchatka peninsula—fortunately in a very sparsely populated area, and under the perceptive observation of the distinguished Soviet volcanologist, G. S. Gorschkov.

After about three weeks of increasingly violent earthquakes, strong eruptions of ash began on the eastern flank of an old dissected volcano (which had been thought extinct) on the morning of October 22, 1955. The explosions grew in intensity until, by October 26, ash was falling heavily within a radius of 60 km.

In the summit crater of the old volcano a dome had been built at the close of the last prehistoric eruption; now a new crater formed on the east shoulder of the cone and encroached upon the walls of the old one, amid roaring explosions heard more than 100 km from the volcano. From the beginning of the eruption until the end of November, 25 mm (about 16 kg/m²) of ash had fallen on the village of Kliuchy, 40 km to the north. Late in November the shoulder crater had enlarged to a diameter of 800 m, engulfing part of the old crater rim and infringing on the central lava dome. The vigor of the eruptions fell off in late November and remained smaller for several months. A smooth plastic lava dome slowly grew to a height of about 100 m within the crater. The slopes of the volcano steepened from an average of about 30° to about 35°, and ash avalanched down them at speeds of 200 km/hour.

The eruption climaxed on March 30. Shortly after 5 P.M. a powerful earthquake wracked the whole region. A few moments later an atmospheric shock wave was widely felt, and, from Kliuchy, far to the north, a gigantic explosion cloud was seen flashing eastward on an oblique course, expanding upward from the crater. The base of the narrow explosion fan was between 6 and 8 km in the air and the top at least 35 km. Finely broken rock fragments that once made up most of the east side of the volcano—a volume of fully 0.5 km³—must have been blown skyward in a narrow vertical fan of ash that extended for at least 400 km out over the Bering Sea. Part rose into the jet stream, swept over the pole, and dropped fine dust over Great Britain the next day.

The whole east face of the mountain had disappeared, lowering the summit by 150 to 180 m. The ejection was at twice the speed of sound; boulders as much as 10 m in diameter were thrown more than 20 km.

Not all the eruptive products were in the upward-directed blast fan. Ash began to fall at Kliuchy about 5:30 P.M. and continued until 9:00 P.M., forming a layer 200 mm thick (24.5 kg/m²). For 18 km down the valley of the Sukhaia Haptiska, which rises at the east foot of the mountain, the stream was buried under a thick blanket of agglomerate, lava blocks, ash, river gravels, and sand, all mixed by the turbulent explosive jet. Through this jumbled cover innumerable fumaroles jetted steam into the air as the river boiled away beneath them. Part of the debris shot out from the crater, and the churning surface was still so hot that leveled trees were set afire at distances of nearly 30 km.

The hot ejecta melted the snow wherever it struck, forming a mudflow that reached thicknesses of 20 m in many places, and carried boulders scores of meters long in a thick chaotic boiling mass of broken trees, disrupted stream gravels, ash and mud of high consistency. The mudflow was so hot that it remained snowless throughout the following winter. Along axial parts of the mudflow, dense forest was completely swept away. Such mudflows are common in many pyroclastic eruptions; they are called by the Indonesian name, **lahars**, and are discussed later in more detail.

Gorschkov computed the energy involved in the great explosion of March 30 to have been about 4×10^{23} ergs, equivalent to the explosion of 10,000 tons of TNT and about half the energy of the Hiroshima atom bomb. The pressure required to hurl a half cubic kilometer of rock an average distance of 13 km was at least 3 kilobars—the pressure at a depth of about 9 km in the earth. Even so, the heat energy involved in the eruption was fifty times as great as the explosive energy. Comparable energy must have been expended in the 1929 eruption of Cotopaxi in Ecuador, for a 200-ton block was there hurled 15 km.

Mont Pelée, 1902   Despite the energy involved, Bezymianny caused no casualties, because of the sparse population of the region; Mont Pelée, on the crowded island of Martinique, French West Indies, though never so violent, has been one of the most lethal of all volcanoes. Although Mont Pelée had erupted nearly harmlessly in 1792 and 1851, it had been dormant for half a century when mild explosions began in the early spring of 1902. After several weeks, a dome of viscous lava began to grow in the crater, gradually displacing the water of the lake that still occupied it. Eventually, a boiling torrent broke through a notch in the crater wall, flooding the valley of the Rivière Blanche and wreaking much destruction on its way to the sea, several kilometers away. The notch, enlarged by the floodwaters, guided nearly all the following eruptions; unfortunately, it was in direct line with the largest city on the island, Saint Pierre.

On May 8 the hitherto mild explosions reached a crescendo: a tremendous blast of gas shot a torrent of incandescent ash horizontally through the gap in the crater wall at such velocity that much of it jumped the valley of the Rivière Blanche, crossed the low divide beyond, and swept down the western and southern slopes of the volcano at speeds approaching 3 km per min. (Ash clouds descending Mayon, in the Philippines, have been timed at 3.7 km/min.) Apparently the ash flowed almost without friction, doubtless because release of gas dissolved in the pumice and lava fragments acted both to buoy up the flow as a whole and to separate fragments of ash. As the gas contained such components as CO_2 and SO_2, both heavier than air, the gas emitted did not prevent the mass of ash from clinging to the ground.

Saint Pierre was overwhelmed by the flood of ash; within moments all but one person in a city of 30,000 people were dead—the sole survivor,

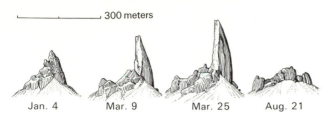

300 meters

Jan. 4   Mar. 9   Mar. 25   Aug. 21

Figure 9-3
Four stages in the rise of the spine of
Mont Pelée. [After A. Lacroix, 1904.]

a prisoner in a basement cell of the jail. The cloud passed out over the harbor, setting fire to ships and capsizing some by the momentum of its mass. Actually it was the heat and horizontal velocity of the ash that caused most of the damage; when it had settled to earth, the ash flow was only a few centimeters thick.

The distinguished French geologist La Croix was at a point of vantage to see most of the tragedy. He was much impressed with the boiling cloud of hot ash and gave it the name by which such eruptions are everywhere known: **nuée ardente** (glowing cloud).

For months after the catastrophe, minor nuées continued. In October a plug of extremely viscous lava began to emerge from the crater and to rise vertically high above the rim. As it rose, parts crumbled and fell off as stream jets blew from the base, finally leaving a **spine** that rose to a height of nearly 300 m (Fig. 9-3). Within a few months the spine crumbled to a heap of rubble.

About a quarter century later (1929), Pelée again became active and remained so for about three years. At first central eruptions blew away the remnants of the 1903 spine and excavated a crater in its place. Nuées began to pour forth in December. This time many people abandoned the city in panic, but fortunately no eruptions even approached the 1902 blast in violence, though many poured down the Rivière Blanche to its mouth at speeds of 1 km/min or faster.

Krakatau, 1883   Krakatau, an island in the Sunda Strait, between Sumatra and Java, was the site of a prehistoric volcano that collapsed, leaving a ring of islands surrounding a central depression beneath the sea. Within this ring several tuff cones grew to the surface and, like Surtsey (Chapter 4), were protected from destruction by a covering of lava; Krakatau, the largest, grew to a height of about 800 m. The volcano had lain dormant for two centuries when,

in May of 1883, it began to emit ash, which increased rapidly in volume; on August 27 the whole mountain exploded—probably the greatest natural explosion in human history. The entire structure collapsed, leaving a hole 300 meters beneath the sea. The ash was not made up of fragments of the rock composing the former island, but was virtually all glass, newly derived from depth. The collapse of the mountain came when the underlying magma chamber was emptied by the emission of a vast volume of ash. The energy involved has been computed as equivalent to that of 100 to 150 megatons of TNT, or five to seven times that of the Hiroshima bomb. Pressure in the magma chamber was at least 5 kilobars, equivalent to that at a depth of about 15 km in the crust.

The sound of the explosion was heard on the island of Rodriguez, 5000 km away, and in Australia. The barometric shock wave was recorded as it made three trips around the earth. Dust hurled into the atmosphere gave brilliant red sunsets the world over for more than two years. Fifty-ton chunks of rock were thrown for 35 km. Few people lived within direct killing range of the blast, but collapse of the volcano sent a gigantic sea wave fully 30 m high against the low-lying coasts of Sumatra and Java, overwhelming and drowning more than 36,000 people and razing the nearshore buildings for scores of kilometers.

SHIELD VOLCANOES

Excellent examples of shield volcanoes are those of Hawaii (Fig. 9-4). Material erupted from these volcanoes has been estimated to be 97 percent lava—so fluid that the lower slopes of the shields have gradients as low as 2°. Some pyroclastic material is erupted, but almost wholly at the beginning of an eruptive episode, soon giving way to the steady effusion of lava. As the pyroclastics

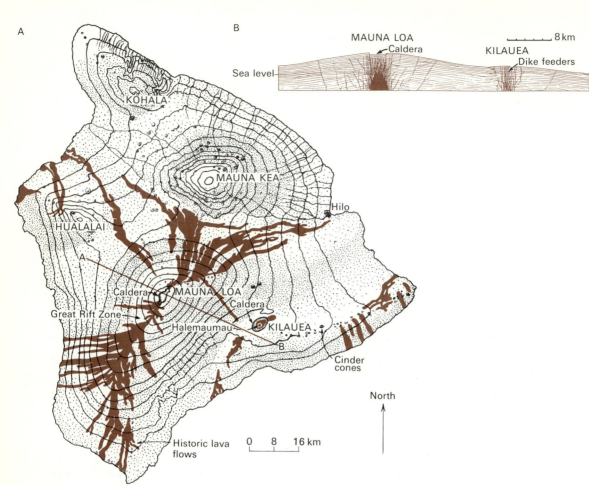

Figure 9-4
(A) Generalized map of Hawaii, an island built of five large shield volcanoes. Note the alignment of dike clusters, locally called "rift zones," across the Mauna Loa and Kilauea shields. (B) Cross section showing concentration of dikes along the rift zones. [After H. T. Stearns and G. A. Macdonald, 1946.]

are not thrown far, they accumulate on the upper slopes, which are a little steeper, as much as 5°.

Although more lava per linear meter obviously comes from the summit craters (to account for their greater height), an amazing amount is erupted from fissures in the rift zones far down the slopes of the shields (Fig. 9-5). Pyroclastics are far less abundant than in composite volcanoes, but their accumulations still suffice to make the rift zones stand out in the topography. Some lava fountains, spattering drops of lava, follow the rift zones for as much as a kilometer and rise skyward for a hundred meters or more as con-

tinuous curtains of incandescent lava. The driving pressure is so great that even where rifts cross valleys a hundred meters or more deep, lava fountains reach nearly uniform heights, ignoring the topography.

The rift zones of Hawaii are marked by lines of **spatter cones**—localized accumulations of bombs and lava feebly erupted from fissures formed late in an eruptive cycle. They are also characterized by **collapse craters** left by the withdrawal of yet uncongealed lava from a fissure by drainage to a lower outlet, either on land or below the sea. Similar but smaller features are called

pit craters; some have vertical or even over-hanging walls, and some are several hundred meters deep (Fig. 9-6).

Eruptions of the Hawaiian volcanoes are more frequent and less violent than those of composite volcanoes. (Etna may be an exception.) For many years Kilauean activity has been observed in greater and greater detail, by seismic studies (Chapter 7), tilt measurements, thermal measure-ments, and chemical analyses of gases, enabling geologists to foretell coming eruptions. Earth tremors from a deep source, say 60 km, gradually give way to those of shallower and shallower sources. Finally, just before an outbreak, they come from sources only 2 or 3 km below the surface.

While tremors are being monitored, precise measurements of surface tilt indicate that the volcanic edifice is swelling as the magma rises from depth, only to shrink again as activity dimin-ishes. Although Kilauean eruptions are episodic rather than cyclic or periodic, there have been so many that a monthly average discharge of about 9×10^6 m³ of lava has been maintained for the past 20 years.

Figure 9-5
Lava fountain a hundred meters high, along a rift-zone fissure early in a flank eruption of Kilauea volcano, Hawaii, Sept. 23, 1961. Although the dike crosses topography with relief of scores of meters, flow is not concentrated in the valleys, but extends unbroken the full length of the fissure, demonstrating the high pressure in the magma chamber. [Photo by J. G. Moore, U. S. Geological Survey.]

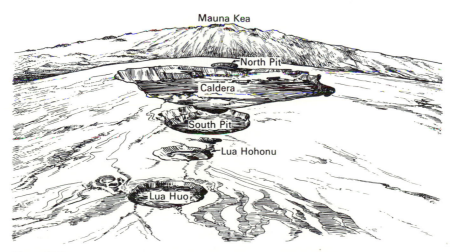

Figure 9-6
Sketch of the summit of Mauna Loa, showing the caldera and pit craters that follow the rift zone of fissures. [After H. T. Stearns and G. A. Macdonald, 1946.]

FISSURE ERUPTIONS

As Figure 9-4 shows, many historic eruptions on Mauna Loa and Kilauea have come from fissures that cross the shields along northeasterly trends. Some have issued from high on the shields, others from much lower; evidently the shields are built of many flows that emanated from a central crater and from fissures along these rift zones. The older, more deeply eroded shield volcanoes on the islands northwest of Hawaii expose hundreds of dikes, showing that they were similarly built. The fissure-fed flows from such shield volcanoes are transitional to the great flood basalts, next to be described. In fact, when dealing with an older pile of basalt, it is not possible to tell whether it is a shield volcano or a flood basalt unless the dike pattern is well exposed, a rare situation. Dikes of many shield volcanoes—but not those of Hawaii—tend to radiate from a central vent, whereas those of some flood basalts (Columbia River, next section) tend to be roughly parallel or *en echelon*.

FLOOD BASALTS

Probably the only historic flood eruption comparable in style but not in volume to many in earlier times was that along the Laki fissure, Iceland, in 1783. Great floods of incandescent, highly fluid basalt poured from the entire length of the 30-km fissure and spread over an area of about 600 km², reaching as far as 60 km from the fissure. Total volume was about 14 km³. The eruption left a nearly flat surface except where minor cinder cones formed in its fading phase. Great as this flow was, it was trivial compared to many earlier flood basalts, some of which accumulated over far greater areas and in greater aggregate thicknesses of successive flows. Some of the more voluminous accumulations are listed in Table 9-1.

Typical of the flood basalts is the Columbia River Basalt, of Miocene age, which covers more than 400,000 km² in eastern Washington, eastern Oregon, and northern Idaho (Fig. 9-7). Flow after flow accumulated there to thicknesses greater than 3 km, for a well that deep was entirely in basalt. Soil zones at the top of some flows record considerable interruptions in the volcanicity. Individual flows have been traced more than

Table 9-1
Some pre-Holocene flood basalts.

Area	Age
Canadian Shield	Precambrian (many)
Lake Superior district	Late Precambrian
Appalachian piedmont	Late Triassic
British Columbia	Triassic
Northwest India (Deccan plateau)	Cretaceous–Eocene
Coast Range, Washington	Eocene
Ethiopia	Eocene
Snake River plain	Pliocene
Virginia	Precambrian
Great Britain	Devonian
Siberia	Early Triassic
Paraguay and Brazil	Jurassic
South Africa	Triassic?
Northwest Scotland*	Eocene
Columbia Plateau	Miocene
Iceland	Miocene, Pleistocene

*Some geologists consider these to be flows from shield volcanoes.

200 km from canyon to canyon of the Columbia Plateau. Some flows are 35 m or more thick, and the average is about 25 m. One flow, the Roza Basalt member, has been mapped over an area of 55,000 km² and its volume computed at 2700 km³. The dike-filled fissures from which the flows were extruded lie in north-trending swarms, with the largest swarm along the east side. The basalt plateau sags in the middle; the margins rise to altitudes of 600 to more than 1500 m, but the highest flow in the middle of the basin is near sea level. Probably this sinking has been in response to the removal of the huge mass of magma from beneath the area. This mass of flood basalt is indeed huge, but the Jurassic flood basalt in the Parana Basin of Brazil, Uruguay, and Paraguay is probably half again as voluminous. The Eocene basalts of the Oregon and Washington coastal ranges are also more voluminous.

PYROCLASTIC FLOWS

We have noted that the ash emitted from Pelée as a nuée ardente was hot enough to set fires at distances of several kilometers, and that from Bezymianny at 30 km. Great as these explosions were, neither left more than a few decimeters of ash. More concentrated was the deposit of pyroclastic debris in the Mount Katmai area of Alaska, where a very different deposit formed.

Katmai and the Valley of Ten Thousand Smokes In June of 1912, Mount Katmai, situated near the end of the Alaska Peninsula and not hitherto considered active, was reported to have exploded with great violence. No witnesses were at hand, but from a distance a tower of black ash could be seen to rise high into the stratosphere, where some lingered to furnish brilliant sunsets the world around for several years. The Katmai cone, when visited four years later, was found to have an enlarged crater three km long, 2 km wide, and several hundred meters deep, a *caldera* (p. 171). About 11 km to the west and nearly 1300 meters lower, a small pumice cone, christened Novarupta, had formed, and below it for more than 20 km extended a body of glassy ash that filled the valley to a depth of more than 30 m in places (Figs. 9-8, 9-9). From this body rose a myriad of fumaroles emitting steam and sulfurous vapors—hence the name "Valley of Ten Thousand Smokes." Some of these fumaroles remained active for 40 years, many leaving their openings lined with sulfur, galena, mag-

netite, or other encrusting minerals. Only the throat of Novarupta continued to emit steam 50 years after the eruption.

The fact that the ash was practically confined to the valley suggests that it did not come from Katmai, as originally supposed; probably it was a gravity flow nonviolently spilled from Novarupta, at its head. The flow is considered transitional between the small nuées ardentes of Pelée and Bezymianny and the great welded tuff deposits of Yellowstone Park, the Great Basin of western America, the Andes, Indonesia, Turkey, and New Zealand.

Representative of these partially welded pyroclastic flows (*ignimbrites*) are many in northern New Mexico, described by R. L. Smith and Roy Bailey, U. S. Geological Survey. Here many pumice deposits, each of almost uniform thickness, were spread one on top of another in such quick succession that erosion did little to alter the surface of one flow before the next was deposited. Some individual flows cover as much as 300,000 km² and are about 100 m thick. The flows are noticeably layered. The basal layer, a few meters thick, is incoherent (probably because it cooled quickly on contact with the ground); in most flows this layer is overlain by a layer of annealed glass, formed by the compaction and

Figure 9-7
Flows of flood basalt near Vantage, Washington. The cliff is 330 m high.
[Photo by Washington Department of Conservation and Development.]

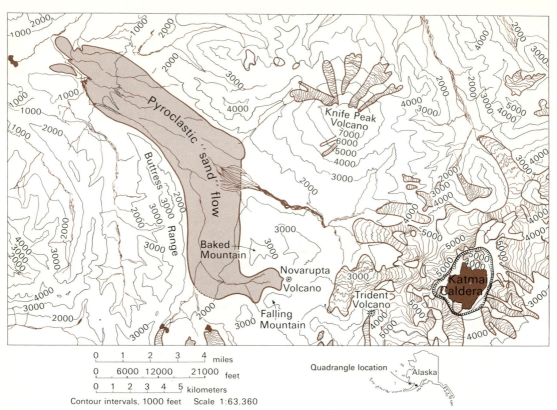

Figure 9-8
Sketch map showing the ash flow in the Valley of Ten Thousand Smokes, Katmai Volcano, and Novarupta.

Figure 9-9
The downstream end of the ash flow in the Valley of Ten Thousand Smokes.
[Photo by A. C. Waters.]

Figure 9-10
Welded ash flow in Valles Caldera, New Mexico, showing even top and layering.
[Photo by Roy Bailey, U. S. Geological Survey.]

complete welding of pumice fragments into a nonporous obsidian of variable thickness, in places as much as 10 meters but generally a meter or two thick. Above this glassy layer is a layer of variable thickness, from a few to several scores of meters thick, in which the ash is slightly more porous and its components are welded together and stretched out parallel to the flow surface, which, when originally formed, was virtually horizontal because of the fluidity of the pyroclastic mass (Fig. 9-10). Evidently the flow was still so hot when it came to rest that the pumice fragments were able to flatten out and coalesce into larger units—in the glass layer, even to the exclusion of all pores. Above the glass layer, where pressure was less, a little porosity remains (Fig. 9-11).

As some ash flows have been welded to within a few meters of the top and others only 2 or 3 m thick have been welded throughout, their temperatures must have remained nearly magmatic to permit welding under loads so light. Experiments suggest that temperatures of nearly 800°C might be needed, though with greater overburden, welding may proceed at temperatures as low as 500°C.

Because some ash flows are so very thick and cover such vast areas, it seems impossible that they could have come from central vents. They

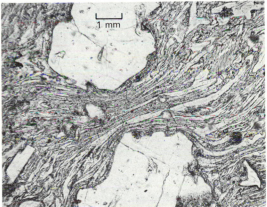

Figure 9-11
Thin section of welded tuff, showing deformation by compaction of the broken pumice fragments. [Photo by R. L. Smith, U. S. Geological Survey.]

were more probably erupted from fissures, even though few have been identified; A. C. Waters has traced an ash flow in central Oregon to an unbroken junction with an inclined dike several kilometers long whose upper hundred meters consists of pyroclastic material identical with that of the surface flow, thus demonstrating that the fragmentation of the magma began below the

Table 9-2
Some Cenozoic ash flows.

Locality	Area covered, in square kilometers	Volume, in cubic kilometers
Toba Tuffs*, northern Sumatra	20,000 to 30,000	1500–3000
Rotorua area*, North Island, New Zealand		750
Bishop Tuff, southeastern California	1100	165
Quichapa Formation, southwestern Utah	23,000	7000
Rattlesnake Tuff, central Oregon	14,000	2000
Superior Dacite Tuff, central Arizona	1100	500
Potosi Volcanic Group*, Colorado–New Mexico	640	10,000
Tambora, Sumbawa, Indonesia, 1815		100
Mazama Ash flow, Crater Lake, Oregon		40–55

*Composed of lithic pyroclastic flows.

surface. Similar relations have been reported from the Sudan. Comparable fissure erruptions probably formed the vast bodies listed in Table 9-2.

PILLOW LAVAS

In Samoa and Hawaii basaltic flows entering the sea have been seen to break up into irregularly rounded masses, from a few centimeters to 2 m

Figure 9-12
Underwater photo of pillow lava off Hawaii. Note starfish at bottom center of photo. [Photo by James G. Moore, U. S. Geological Survey.]

across, which resemble a stack of pillows. Scuba divers have photographed such forms where historic flows have entered the sea off Hawaii (Fig. 9-12). In many parts of the world, such **pillow lavas** are interstratified with marine sedimentary rocks from early Precambrian to Holocene age. They are common on the mid-ocean ridges and rises and are by far the commonest igneous rocks of the oceans—perhaps, indeed, of the whole earth. Most are basaltic or andesitic; a few are dacitic, but none of rhyolite have been reported.

Pillow lavas also form where flows enter freshwater lakes and ponds. The pillows commonly have glassy borders. In fact, glassy breccia commonly fills the interstices between the pillows, and in time alters to a yellowish clayey mineraloid, *palagonite* (Fig. 9-13). Apparently slight differences in temperature, viscosity, and rate of flow determine whether the lava forms pillows or breaks up into breccia fragments of glass. Palagonite alters further to a clay that has often been mistaken for mudstone, which it strongly resembles on casual examination.

EXPLOSIVE ERUPTIONS ON CONTACT WITH WATER

Most pillow lavas apparently form quietly, but locally lavas explode on entering water bodies and break up into fragments of frothy glass, to form piles of debris at the shore. **Phreatic explosions** (from the Greek word for "well") commonly take place when magma or merely hot rock

Figure 9-13
Altered pillow breccia of the Siletz Basalt (Eocene), western Oregon. Isolated large pillows are enclosed in a matrix of broken pillow fragments or pillows and palagonite breccia. The cavity fillings of white carbonate and zeolite minerals emphasize the details of the vesicular parts of the pillows. [Photo by Parke D. Snavely, Jr., U. S. Geological Survey.]

comes in contact with water. For example, in 1924, the lava pool in the summit of Kilauea was drained to a considerable depth by a fissure eruption far down the side of the shield. The ground water within the cone was able to drain down into the heart of the volcano, where it was vaporized and exploded into steam, throwing fragments of the vent walls high into the air. No fresh lava appeared, merely rocks broken from the fissure walls. The eruption illustrated in Figure 9-14 was triggered largely by the steam formed by the under-water exposure of lava. Such steam eruptions tend to form flat-bottomed pits (**Maars**), surrounded by low accumulations of the debris blown out of them by steam.

Such eruptions must often have taken place, for many sedimentary sequences contain thick layers of tuff breccia with dense lava fragments toward the base and more pumiceous and vesicular fragments toward the top. Clearly such breccias are formed by the underwater explosion of lava and the distribution of fragments by turbidity flow.

Myojin Reef volcano   At Myojin Reef on the Mariana arc, about 400 km south of Tokyo, submarine eruptions of the kind that produce such deposits were observed from many ships during several months in 1952. Some turbulent eruption clouds burst the surface and spread ash over the sea. One such cloud, on September 14, 1952, engulfed the Japanese research ship Kaiyo Maru with the loss of all on board, 31 scientists and crew. Very likely the 1883 eruption of Krakatau sent turbidity currents of volcanic debris coursing from its underwater vent.

CALDERAS AND VOLCANO–TECTONIC
DEPRESSIONS

Most craters are symmetrical hollows shaped like inverted cones in the top of conical volcanoes, and are kept open by explosive activity. Few such craters are more than 2 or 3 km across, and most are less than 1 km in diameter and 300 m deep. Craters on most shields and many composite volcanoes occupy roughly central positions within a caldera—a larger flat-bottomed depression that tends to be circular or elliptical but may be angular in part. Many calderas are as much as 20 or 30 km in diameter, but most are much smaller, perhaps 3 to 15 km in maximum diameter and 300 m deep.

The Mauna Loa caldera is about 3 km across and 300 m deep (Fig. 9-6). Within the caldera a fluctuating lava pool occasionally spills lava over the caldera floor and at other times is crusted over. Kilauea, a second and entirely independent volcano, as shown by the differing eruptive cycles of the two, forms a slight rise in Mauna Loa's southeastern flank, about 30 km from the summit, and has its own summit caldera about 2 km across. Halemaumau is a deep pit crater lying within this caldera (Fig. 9-15).

The clue to the formation of these calderas seems to be given by the frequent drainage of lava lakes formed early in an eruptive cycle. The withdrawal of lava to lower outlets on the cone removes support for the roof rocks of the intravolcanic magma chamber and brings about collapse of larger or smaller rock masses.

Magma chambers may also be drained with similar collapse of the overlying edifice by pyroclastic eruption, as at Krakatau. Another example is the formation of the caldera containing Crater Lake, Oregon. About 6600 radiocarbon years

Figure 9-14
Underwater eruption of Capelinhos volcano, the Azores. [Photo by U. S. Air Force.]

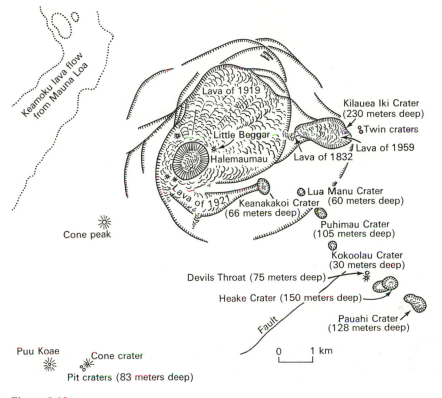

Figure 9-15
Kilauea Caldera, Halemaumau, and nearby pit craters. [After H. T. Stearns and
G. A. Macdonald, 1946.]

ago (as dated from a charred Indian moccasin found at the base of a pyroclastic flow), a caldera was formed at what is now Crater Lake, Oregon, high in the Cascade Mountains (Figs. 9-16, and 9-17). Glassy ash and pumice in tremendous volume was erupted from a former volcano—to which the name Mount Mazama has been given and whose truncated cone now holds the lake. The ash from this eruption has been recognized far and wide over much of the western United States. It formed a blanket many meters thick around the depression in which Crater Lake lies. This blanket is wholly of glassy lava, not at all like the largely crystalline lava that makes up the frustum of the remaining cone. The huge volume that must once have formed the summit of Mount Mazama must have collapsed into the eviscerated magma chamber from which the glassy ash was discharged (Fig. 9-16).

Lake Taupo, a water body 45 by 25 km in area

on the North Island of New Zealand, was evidently formed in much the same way, by the subsidence of a block of land from beneath which magma had been drained by eruption. Of comparable size is the Lake Toba depression, about 100 km long and 10 km wide, in northern Sumatra; its subsidence can clearly be related to the effusion of great volumes of siliceous pyroclastics. These downwarped and only partially downfaulted areas are similar in origin to many calderas but have shapes so different from them that they are classed as **volcano-tectonic depressions**.

CINDER CONES

Cinder cones are accumulations of pyroclastic material, chiefly bombs and scoria (Appendix III), surrounding a central vent. They form on all kinds of volcanoes, though more commonly on

174

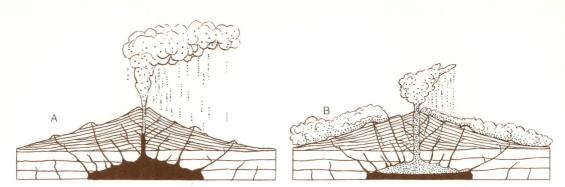

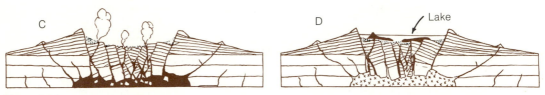

Figure 9-16
The evolution of the caldera of Crater Lake, Oregon. (A) Early eruption from the summit crater of prehistoric Mount Mazama. (B) Great eruption of pumice and ash flows, with the mountain top beginning to collapse into the magma chamber. (C) Caldera formed as discharge of magma was nearly complete. (D) Formation of the lake and crystallization of remaining magma after minor eruptions into the lake. [Modified from Howel Williams, 1942.]

Figure 9-17
Crater Lake, Oregon, occupies a crater about 10 km in diameter, formed as depicted in Figure 9-16. Mount Scott, a parasitic cone, can be seen at left center. The top of Wizard Island, a minor cone formed after the collapse, can be seen at the right center above the rim of the caldera. [Photo by Oregon State Highway Commission.]

Figure 9-18
Cinder cones aligned along a fissure, Cascade Mountains, Oregon. Mount Jefferson, a large composite volcano, appears in the distant right. [Photo by Oregon Department of Geology and Mineral Industries.]

the less siliceous ones, and many are aligned along fissures radiating from the central crater (Fig. 9-18), although many others show no such association. Most cinder cones are less than a kilometer across and between 30 and 200 m high, but Paricutín, an andesitic cone that grew in a Mexican corn field between 1943 and 1952, attained a height of about 400 m (Fig. 9-19). Lava that poured from its base eventually covered an area of about 8.5 km², burying the villages of Paricutín and San Juan de Parangaricutiro. The ash flow killed a large area of forest.

Figure 9-19
Paricutin volcano, a large cinder cone in eruption, July 1945. Lava emerging from the base has covered the village of San Juan de Parangaricutiro until only the church steeple projects above the flow. [Photo by Tad Nichols, Tucson, Arizona.]

Columnar Jointing

Rocks, like most substances, expand on heating and contract on cooling. A lava or pyroclastic flow may come to rest while still red hot; its contraction on cooling decreases its volume, causing cracks to form. Because the surface is first to cool, the cracks propagate into the mass in directions normal to the surface. The common result of the cooling is to divide the rock, whether lava or pyroclastic, into columnar segments at about right angles to the cooling surface. As the flows are more or less homogeneous laterally, the columns tend to be six-sided—the pattern that forms ideally when a solid mass cools away from a plane surface. But in nature there are enough modifying factors so that the pattern is never perfect. The columns generally have from four to seven sides, though most have six (Fig. 9-20).

Surface Patterns of Lava

Many lavas emerge from vents as viscous streams whose surface tends to be ropy, with surface striae parallel to the current flow or to the stretching of the lava (see Fig. 9-21). Such flows are called **pahoehoe** (Hawaiian for "ropy"). The same flow may suddenly change its surface to a jagged clinkery mass, reminiscent of furnace clinkers (see Fig. 9-22) and roll along with a tremendous roar as chunks of solid crust fall off the front of the flow only to be overrun by the steep creeping wall of lava. This kind of flow is called **aa**; it is thought to develop as the lava loses volatiles and thus becomes less fluid.

Lahars

Among the most dangerous of volcanic phenomena are the volcanic mudflows, or **lahars**, like those mentioned in connection with the eruptions of Bezymianny and Pelée. Ash columns rising into the upper atmosphere commonly trigger lightning storms and furnish nuclei for the condensation of rain, which falls in torrents and produces mudflows of sufficient density to sweep even large boulders along at tremendous speeds. Such mudflows have been catastrophic on the lower slopes of Vesuvius and especially so in Indonesia, whence the name is derived.

Figure 9-20
Columnar jointing in basaltic lava. The upright columns formed by the shrinkage of the thick flow as it cooled from the base upward. Nearby exposures show that the flow was three times as thick as this remnant; the upper part, which cooled from the surface down, has here been eroded away. Where the two joint sets interfere near the middle of the flow, jointing is highly confused. This zone is here missing because of erosion. Columbia River Basalt, Maury Mountains, Oregon. [Photo by A. C. Waters.]

Some lahars are formed by lava melting great volumes of glacial ice, a common happening in Iceland. A notable prehistoric lahar apparently caused by this mechanism originated on the northern flank of Mount Rainier in the Cascades of Washington about 5700 radiocarbon years ago. It formed a gigantic wall of mud as much as 150 m high that rushed down the valley of the White River, sweeping forest-clad hills bare and leaving its "high-mud" mark high on the valley walls. It flowed for more than 100 km down the valley to the waters of Puget Sound, where it spread over a dozen km^2, leaving a deposit more than 80 m thick. The total volume was about 1/2 km^3. Composed mainly of fine ash in a thick slurry, the mudflow carried with it boulders as

Figure 9-21
Pahoehoe flow of basalt with ropy surface, Hawaii. [Photo by Whitman Cross, U. S. Geological Survey.]

Figure 9-22
Aa flow of basalt, Kilauea. Note the jagged surface. [Photo by G. A. Macdonald, U. S. Geological Survey.]

much as 20 meters across, huge trees, and probably masses of ice. It was cold enough so that it started no fires but swept all before it.

The most destructive lahars have been caused by volcanic eruptions displacing the water of crater lakes. Such overflow of the crater lake of Kelut volcano, Java, in 1919, caused a lahar that killed 5000 persons. Afterward tunnels were driven to drain the crater and prevent another such water mass from accumulating; this was successful for more than 30 years, but lava filled the crater again in 1951, displaced the smaller lake that had formed, and killed several hundred more.

On Christmas eve, 1953, the crater lake of Ruapeho volcano on the North Island of New Zealand was displaced by volcanics. The resulting flood swept away a railroad bridge near Waiouru and caused a wreck in which 151 passengers died.

Lateral Eruptions of Pyroclastics

We have noted that both Pelée and Bezymianny emitted laterally directed jets of pyroclastic materials. This is a common phenomenon. But even pyroclastics that rise vertically from the feeding orifice send out lateral streams of ash. One mechanism, called "base surge," operates when an explosion at shallow depth in the volcano throat is able to peel back the lip of the crater, at the same time spreading material in all directions upward and outward. Such eruptions in 1965 from the Taal volcano in Luzon, which erupted from a sublacustrine vent, sent showers of muddy ash in every direction, causing tremendous destruction and the deaths of 189 persons. The horizontally directed ash showers coated the trunks of trees facing the vent with layer after layer of mud (Fig. 9-23).

Even when an explosion takes place well down in the crater throat, so that the ash is directed vertically, a considerable horizontal movement soon develops because the pressure of the pyroclastic material falling back deflects the rise of later ejections.

Identification of Ash Falls from Distant Sources

Glassy ash, such as that of Mount Mazama, contains so few crystals and thus so high a proportion of glass that it can be identified by its chemical composition and optical features over very wide areas. If, however, the ash contains already crystallized minerals of varying density, the composition of the fall may change markedly along the path of travel. The denser minerals, which are also the least siliceous, settle out earlier than the less dense minerals or glass, so

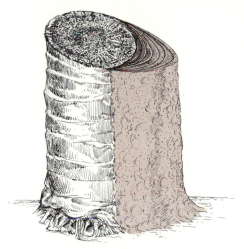

Figure 9-23
Section of a palm tree plastered by mud eruptions during the 1965 eruption of Taal volcano, Philippine Islands. [Drawn from a photo by James G. Moore, U. S. Geological Survey.]

that the chemical composition becomes increasingly siliceous with distance of travel. Figure 9-24 shows the change in composition of observed ash falls with distance in the Argentine and in Indonesia.

Volcanic Gases

For many years geochemists have tried to determine the composition of volcanic gases, but until recently results have been disappointing because of the rapid reactions that proceed at magmatic temperatures and during cooling. Recently developed analytical techniques permit analysis within a few seconds of the time of collection, before the gases have a chance to react. It has long been thought that water vapor is an invariable component of volcanic gases, but recent work indicates that it is absent from some. Studies of the isotopic composition (p. 420) suggest that most water in volcanic gases is meteoric—that is, recycled rainwater, and not **juvenile** (appearing at the surface for the first time). Of 26 analyses (by the new techniques) of gases from one vent at Stromboli, all taken within 2 hours, one showed 51 percent water, nine ranged from 30 to 47 percent, thirteen from 3 to 25 percent, and three showed no water at all. Of

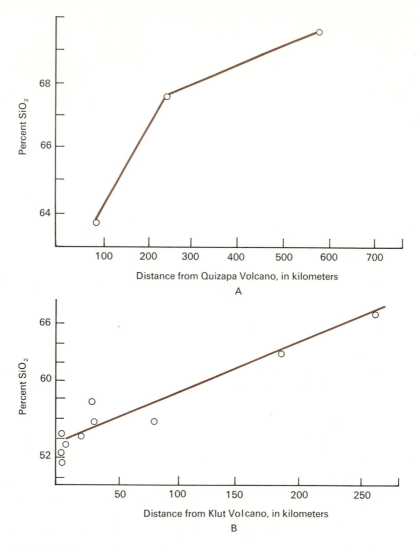

Percent SiO₂

Distance from Quizapa Volcano, in kilometers

A

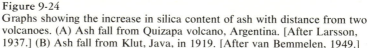

Distance from Klut Volcano, in kilometers

B

Figure 9-24
Graphs showing the increase in silica content of ash with distance from two volcanoes. (A) Ash fall from Quizapa volcano, Argentina. [After Larsson, 1937.] (B) Ash fall from Klut, Java, in 1919. [After van Bemmelen, 1949.]

10 samples taken within 11 minutes from a vent on Etna five had more than 50 percent water, three had less than 50 percent water, and two had none. In most of both sets of analyses carbon dioxide was at least as abundant as water and generally much more so. This preponderance of CO_2 over H_2O is found in analysis of the contents of minute bubbles enclosed in most igneous rocks, and is probably important in the genesis of magmas.

Some of the best older analyses of volcanic gases are included in Table 9-3.

Although recent work has tended to minimize the importance of water as a volcanic product, it is nevertheless a dominant component of many volcanic gases. Perret, an American volcanologist, noted that during the Vesuvius eruption of 1906, gas—almost wholly steam—was emitted for many hours from the whole crater, 700 meters across, with such tremendous velocity that even at a height of 13 km the steam column was only about 5 km in diameter. Even though water may be largely adventitious in volcanic rocks, it may be very important in the volcanic mechanism, and

Table 9-3
Representative analyses of volcanic gases.

Analysis	1	2	3	4	5	6	7
CO_2	21.4	40.9	4.6	10.1	2.1	10.4	25.9
CO	0.8	2.4	0.3	2.0	0.6	8.3	*
H_2	0.9	0.8	2.8	0.2	0.4	1.1	—
SO_2	11.5	4.4	4.1	—	0.01	—	0.0
S_2	0.7	—	—	0.5	0.9	1.3	—
SO_3	1.8	—	—	—	—	—	—
Cl_2	0.1	—	—	0.4	0.3	0.4	—
F_2	0.0	—	—	3.3	1.5	0.0	—
HCl	—	—	0.6	—	—	—	—
N, Ar. etc.	10.1	8.3	4.5	0.9	0.6	7.2	11.1
H_2O	52.7	43.2	83.1	82.5	93.7	71.3	63.0

*Included with N.

1. Kilauea, Hawaii; average of 10 collections in 1917–1919. (Jaggar, 1940).

2. Nyiragongo, Congo, 1959 (Chaigneau, Tazieff, and Fabre, 1960).

3. Surtsey, Iceland; average of 11 samples, Oct. 1964–March 1967 (Sigvaldsen and Elisson, 1968). Some samples contained 4.7 percent H_2.

4. Mont Pelée, West Indies; gas extracted from spine of 1902 (Shepherd and Merwin, 1927).

5. Lassen Peak, California; gas from dacite of 1915 (Shepherd, 1925).

6. Niuafo'ou, Oceania, between Samoa and Fiji; gas from basalt of 1929 (Shepherd, 1938).

7. Kozu-shima, Japan; gas from rhyolite (Iwasaki, Katsura, and Sakato, 1955).

certainly some must be newly discharged from the depths of the earth in order to account for the oceans (Chapter 19).

Fumaroles from the ash flow in the Valley of Ten Thousand Smokes, Alaska, were still emitting HCl at the rate of 1.25 million and HF at the rate of 20,000 tons a year fully 10 years after the eruption. HCl was also abundant at Kilauea during the eruption of 1959–1960.

INTRUSIVE IGNEOUS ROCKS

Volcanic Substructures

Lavas and pyroclastic rocks, we know, come from depth; we now turn to the substructures of some volcanoes that have been eroded enough so that their feeding channels are exposed. As noted in discussing certain pyroclastic flows of Oregon and the Sudan, extrusive materials can sometimes be traced without break into dikes that fill the fissures from which they were derived. Such exposures are few, for their preservation requires that erosion not completely remove the flow, but cut deeply enough to expose the dike. In some places one can identify a feeding conduit even though the flows from it have been completely eroded away. For example, high on the east flank of the Oquirrh Range, Utah, a round tower-like body of latite about 300 m in diameter rises 30 or 40 m above the surrounding country. The center of this **volcanic neck** is massive, but entirely surrounding it is a wall of such regularly jointed columns as to seem the work of a mason; nearly normal to the wall are horizontal columns that converge toward the center (Fig. 9-25).

Larger volcanic necks may cool to form vertical columns near the surface, but at depth, where the greatest heat loss is to the walls, the joints curve and approach the horizontal (Fig. 9-26).

Other volcanic plugs, or necks, can be identified by dikes radiating from them. Shown in Figure 9-27 is one of the many that mark the sites of extinct volcanoes in northeastern Arizona and northwestern New Mexico.

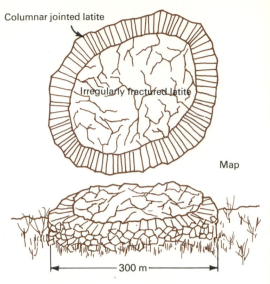

Columnar jointed latite

Irregularly fractured latite

Map

300 m

Figure 9-25
Step Mountain, Utah. A diagrammatic sketch, showing arrangement of columns circumferential to the central pipe. The rising magma was chilled along its border and the columns extended inward as the mass cooled.

Figure 9-26
Devils Tower, Wyoming, a Tertiary volcanic plug in the Black Hills. [Photo by N. H. Darton, U. S. Geological Survey.]

Figure 9-27
Volcanic neck, northeastern Arizona. The man is standing on one of several dikes that radiate from the plug. The former volcano has been completely eroded away except for the feeding vent in the underlying sandstone. [Photo by Warren Hamilton, U. S. Geological Survey.]

DIKE SWARMS

Dikes abound near volcanic centers; the rifts of both Kilauea and Mauna Loa (Figs. 9-4, 9-6) must clearly be continuous dikes at depth. The dike swarms of the British Isles (Fig. 9-28) are most abundant near the old volcanic centers; in general they do not radiate but form subparallel bundles, trending northwesterly. Some are so long—the Cleveland dike more than 180 km— that they can hardly have issued from a central vent but must have been fed from below, as with the fissures of flood basalts. The map is highly generalized; for each dike shown, there are scores of others.

SILL SWARMS

The commonest sills are dolerite, just as the commonest volcanics are basaltic. They abound in Antarctica (Fig. 4-7) and South Africa, where the Karoo dolerites comprise hundreds of sills

ranging from a few meters to more than 300 m thick and cover several thousand km². Their total bulk compares with that of the Columbia River Basalt. The great Whin sill of northern England is known from outcrops and drilled wells to cover an area of more than 4000 km². Its average thickness is about 30 m, but in places it is twice that. Only a few other sills are associated with it.

CAULDRON SUBSIDENCE

We have noted the formation of calderas caused by the emptying of subjacent magma chambers; deep erosion of volcanic structures exposes relations that reveal much about the emplacement of plutonic rocks. A classic example is from the region of Glencoe, in the Scottish Highlands (Fig. 9-29). Here an oval patch of Devonian lava about 15 by 8 km in area is surrounded on three sides by Precambrian schist and gneiss from which it is separated by a shear zone of finely ground

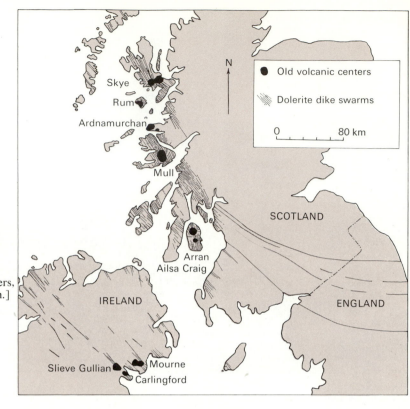

Figure 9-28
Dolerite dike swarms and
eroded volcanic centers of
the northern British Isles.
[After J. E. Richey and others,
Geological Survey of Britain.]

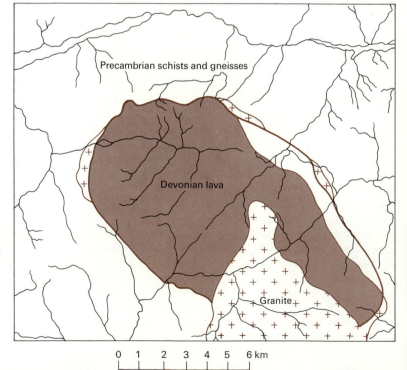

Figure 9-29
Simplified map of the cauldron
subsidence of Glencoe,
Scotland. [After C. T. Clough,
H. B. Maufe, and E. B.
Bailey, 1909.]

breccia along which patches of granite have been injected. Granite also invades the southern side of the oval, and is obviously occupying space formerly occupied by lava. All along the shear zone the lavas have been sharply turned up and locally overturned. It seems obvious that the lava is preserved in the midst of the schist and gneiss only because it had subsided into an elliptical caldera. The underlying magma was injected into the fault zones and engulfed the southern end of the down-dropped block. The lava was colder than the granite it dropped into, and the granite is chilled against it, with no evidence of any thermal metamorphism; on the other hand, the gneiss and schist show growth of new minerals in a zone near the main granite mass and also along the outer contacts of the wedges injected along the fault zone. Quite evidently the subsidence took place while the granite magma was still molten.

The mechanism of the cauldron subsidence suggested to its discoverers, the distinguished trio of C. T. Clough, H. B. Maufe, and E. B. Bailey, that collapse of volcanic structures into an underlying chamber is a common mechanism of underground magma emplacement. The Glencoe example would be like the left-hand scheme or perhaps the third-from-left scheme in Figure 9-30. They did not discuss the very important question as to how the magma chamber itself came to form (see p. 188).

RING DIKES AND CONE SHEETS

At Kilauea the volcanic cone dilates at the onset of an eruption and settles back at its close. Similar expansion and contraction are recorded in the dike patterns of several Tertiary volcanic centers in northern Ireland and Scotland (Fig. 9-31). One series, the **cone sheets**, dip inward toward a focus

Figure 9-30
Diagrams illustrating various kinds of cauldron subsidence with and without extrusion of lava to the surface. Magma, black; crust, hatched. [After C. T. Clough, H. B. Maufe, and E. B. Bailey, 1909.]

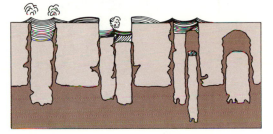

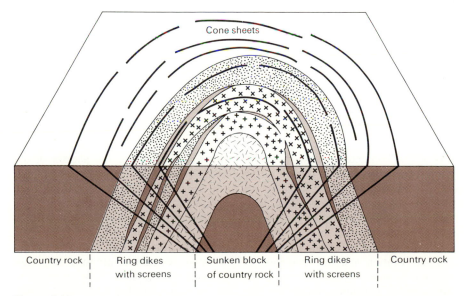

Figure 9-31
Cone sheets and ring dikes, as interpreted in several British Tertiary volcanic centers (Mull, Ardnamurchan, Slieve Gullion) by E. M. Anderson, E. B. Bailey, and J. E. Richey. [After the Mull Memoir, 1924.]

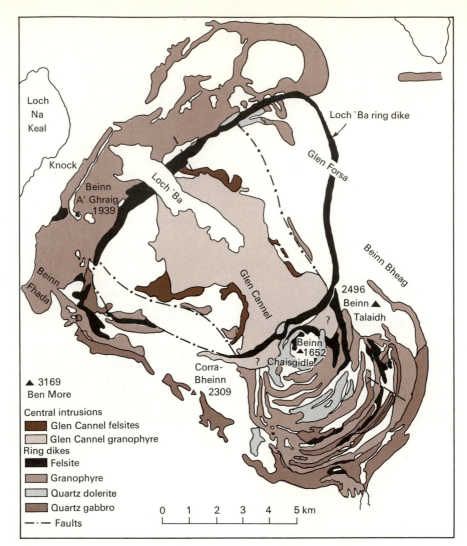

Figure 9-32
Ring dikes on Mull (the cone sheets are too small to show on this scale). [After E. B. Bailey and others; Mull Memoir, 1924.]

at a depth of a few kilometers; these dikes are commonly thin and interrupted, though some circumscribe a considerable part of a circle; the other set, the **ring dikes**, usually much less abundant and composed of generally thicker dikes, dips outward. The two sets have been referred respectively to the upward magmatic pressure at the onset of an eruption and the relaxation of pressure as the magma retreats, allowing the structure to subside. The cone sheets are like the cracks made in a window pane by a BB shot, conical, with the apex toward the point of impact;

the ring dikes have the pattern of a collapsing arch. The magma emplacement made possible by the ring-dike mechanism is illustrated in the right-hand example of Figure 9-30.

The actual maps from which the idealized drawing in Figure 9-31 was inferred are represented in Figures 9-32 and 9-33. On the island of Mull two centers of pressure and release can be recognized: the earlier one, lying just southeast of the prominent Loch Ba ring dike, shows many incomplete ring dikes of classical regularity; the Loch Ba ring dike marks the fracture that allowed

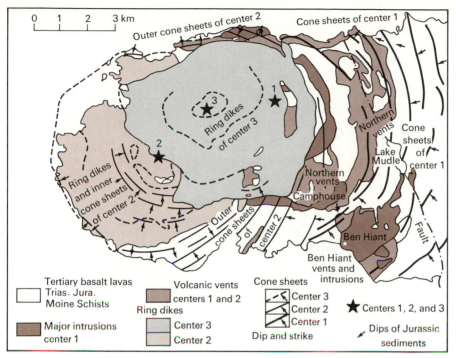

Figure 9-33
Simplified map of Ardnamurchan, showing ring dikes and cone sheets about three centers.
[After J. E. Richey, Ardnamurchan Memoir, 1930.]

subsidence of the block enclosed by it and the emplacement of the intrusions within it.

The nearby Ardnamurchan Peninsula has a more complex history; three centers can be recognized, as shown by the curvatures of the several ring dikes and cone sheets, even though two have been practically covered by the latest ring dikes of Center 3.

Both round and angular masses of granite are so associated with rhyolite and ring-shaped dikes in Nigeria as to indicate emplacement by combinations of cauldron subsidence and cone-sheet injection (Fig. 9-34).

Laccoliths

In many places magma has been injected along a bedding surface as in a sill, but instead of lifting the strata uniformly, has bulged the roof upward into a dome. Such a congealed mass is a *laccolith* (Fig. 9-35).

Some laccoliths give no evidence of the feeding conduit; presumably they were fed from below. Others, such as those of the Henry and La Sal mountains of Utah, were fed laterally from larger central masses, stocks, as shown by their exposed connections (Fig. 9-36).

Most laccoliths whose depth of formation can be established have formed 3 km or less below the surface; some that evidently began to form as laccoliths have broken through to the surface along marginal faults. These are bysmaliths, with examples known from Arizona, Iceland, the Yellowstone Park region, and elsewhere. Probably the 1943–1945 eruption of Usu (p. 161) was started by the emplacement of a bysmalith.

Batholiths and Stocks

The intrusive bodies so far discussed have systematic relations to their wall rocks. Most plutons are not so readily classified, however, and are

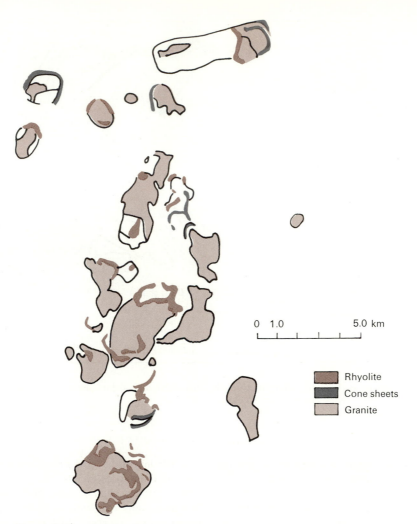

0 1.0   5.0 km

■ Rhyolite
■ Cone sheets
■ Granite

Figure 9-34
Greatly simplified sketch map of part of the Nigerian tin-bearing granite field.
[After Jacobson, McLeod, and Black, 1958, much generalized.]

called by the noncommittal name of *stocks* if they cover an area of less than about 100 km²; if larger, they are *batholiths*. Some can be shown to have congealed at shallow depths of a few hundred meters; others are inferred, from their relations to their wall rocks, their crystallinity, and their metamorphic envelopes, to have formed at far greater depths, approaching 30 km.

Batholiths were named by the great Austrian geologist Eduard Suess, and defined by him as "bottomless," which he of course did not intend literally but merely that their bases are neither

visible nor inferable in detail. Many are seen to be as much as 4 km thick, and gravity and seismic measurements (Chapter 8) indicate that many are far thicker, but of course all do end downward.

The most abundant plutons are granitic or granodioritic in composition and are exposed along existing mountain chains or in areas whose geology indicates a former mountain chain (Chapter 8). Some are found in anorogenic areas, however. They are especially numerous in areas of early Precambrian rocks. One of the greatest batholiths forms a virtually continuous mass, 30

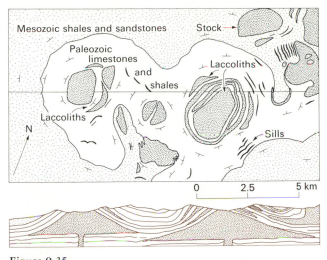

Figure 9-35
Map and cross section of a group of laccoliths and stocks in the Judith Mountains, Montana. [After W. H. Weed and L. V. Pirsson, U. S. Geological Survey.]

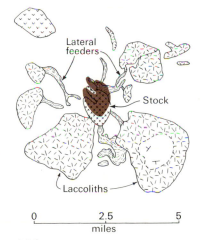

Figure 9-36
The Middle Mountain stock, La Sal Mountains, Utah, showing conduits leading from the central stock to nearby laccoliths. [After C. B. Hunt, U. S. Geological Survey, 1958.]

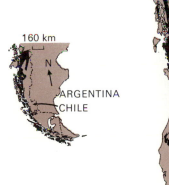

to 250 km wide, extending from southeastern Alaska, through the Coast Mountains of British Columbia to the northern Cascades of Washington (Fig. 9-37). Other huge bodies are in Idaho and Montana, in the Klamath Mountains and Sierra Nevada of California, in southern Cali-

Figure 9-37
Large granite batholiths (black) of western North America (right) and of southern South America (left). [After geologic maps of North and South America. Geological Society of America.]

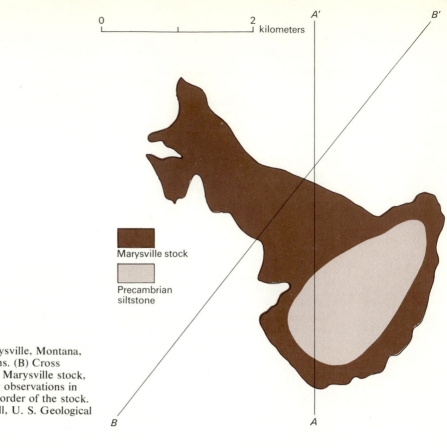

0   2
kilometers

Marysville stock

Precambrian
siltstone

Figure 9-38
(A) Map of the Marysville, Montana,
stock and its environs. (B) Cross
sections through the Marysville stock,
largely controlled by observations in
the mines near the border of the stock.
[After Joseph Barrell, U. S. Geological
Survey.]

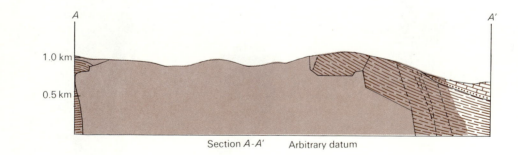

Section *A-A'*   Arbitrary datum

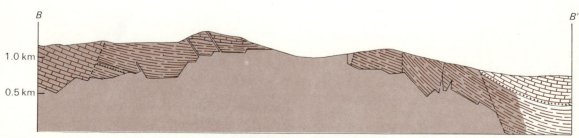

Section *B-B'*   Arbitrary datum

fornia and Baja California, and in the Andes from Colombia to Patagonia. Stocks and smaller batholiths crop out in nearly but not all mountain ranges. Africa south of the Sahara has huge areas of granite and related rocks, many not associated with present or former mountains; the same is true in Finland.

CONTACT RELATIONS AND EMPLACEMENT

The widely varying relations of plutons with their wall rocks give important clues to their mode of emplacement, and in some cases to their genesis.

STOPING

A most instructive set of contact relations is seen near Marysville, Montana. Here a Mesozoic pluton has invaded a moderately disturbed shale and limestone terrane of late Precambrian age. Gold and silver ores, thought to have been derived by volatile transfer from the stock (Chapter 18), occupy narrow veins in the intrusive near its contacts and extend into the wall rocks (Fig. 9-38,A). The tunnels and shafts dug in mining thus expose the contact relations in three dimensions (Fig. 9-38,B). The workings reveal that the roof of the stock has hardly been disturbed by the intrusive, but that the contacts are angular and planar, as though formed along joints and bedding partings already in the shale and limestone before the stock was emplaced. In fact, thin stringers of the intrusive penetrate joints and, in places, have virtually isolated blocks of the country rock within the pluton. The wall rock has been highly altered near the contacts (as shown by the light brown shading in Fig. 9-38,B) by heat and gases from the intrusive; new minerals have formed, such as pyroxene, which is of greater density than either the unaltered rock or the solidified quartz diorite. The density difference between wall rock and quartz diorite magma must have been somewhat greater. These facts seem to indicate clearly that the magma made its way upward by following the joints, prying off blocks of the country rock, and causing them to founder in the molten mass. This process, so similar to that used in extracting ore by overhead stoping, has been called "emplacement by stoping." It has been the operative mechanism in the final emplacement of many plutons, both large

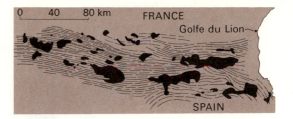

Figure 9-39
Map of cross-cutting batholiths in the Pyrenees. The lines are fold trends. [After R. A. Daly, *Igneous Rocks and the Depths of the Earth,* McGraw-Hill Book Company, 1933.]

and small. Another example is seen in Figure 9-39, for the fold axes are sharply cross-cut by the intrusives without disturbance of their trends.

In southwestern England, seismological studies of the kind described in Chapter 7, together with gravity measurements over a closely spaced network (granite is less dense than its wall rocks), have shown that the five exposed granite plutons between Dartmoor and Lands End and the offshore pluton of the Scilly Islands all merge downward into a single body about 200 km long and 30 to 40 km wide. The granite of this mass has a remarkably flat bottom at a depth of about 12 km below sea level and overlies somewhat denser material, perhaps residual from stoped blocks of country rocks or earlier-crystallized parts of the granitic magma. The M Discontinuity is not depressed beneath the Cornish batholiths as it is beneath the Sierra Nevada batholiths (Fig. 9-40).

FORCEFUL INTRUSION

Even at greater depth (indicated by textures of the intrusives and by both textures and mineral content of the wall rocks, p. 192), stoping is still a common mode of emplacement (Fig. 9-40). But here many batholiths have obviously shoved the wall rocks aside, as shown by the sweeping deflections of the wall-rock schistosity away from its regional trend and into parallelism with the walls of the pluton (Fig. 9-41). Such parallelism of contact with wall-rock schistosity, produced as the slowly crystallizing magma was dragged past the walls, characterizes many plutons. So deeply were their magmas emplaced that the wall rocks were not very much cooler than the magmas (see Chapter 8). Accordingly there is no appreciable

192

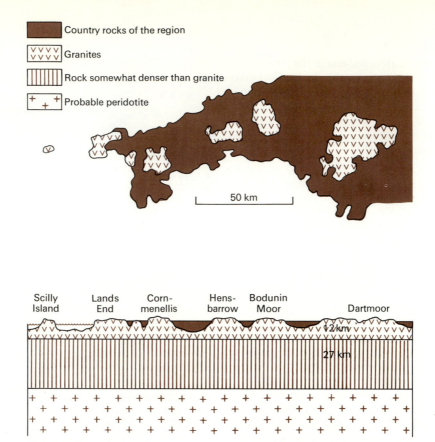

Figure 9-40
The granite mass of southwestern England, as inferred from geophysical studies. [Modified from description by Bott and others, in *The Mechanics of Igneous Intrusion,* Memoir of the Liverpool Geological Society, 1970.]

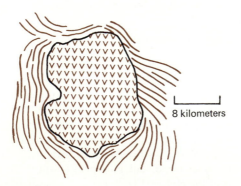

Figure 9-41
Bidwell Bar pluton, California, showing deflection of the schistosity of the country rocks into conformity with the pluton walls, except on the eastern side. It seems clear that much of the granite rock made room for itself by shoving the walls aside, though probably stoping took place along the eastern border. [Simplified after R. R. Compton, 1955.]

quenching of magma nor lessening of grain size along contacts, as is common in quickly chilled borders of shallow intrusives.

MIGMATITES

Many deep-seated plutons so intimately inject their walls and become interfingered with slabs of host rocks as to form **injection gneiss**, or **migmatites** (mixed rocks), hundreds of meters thick (Fig. 9-42), roughly parallel to the structures of the wall rocks. Such mixtures have occasioned much debate: some geologists think that injection gneisses are formed by forceful injection of magma along surfaces of schistosity, splitting the schist; others think the granite lenses formed by partial melting of the wall rocks and were in process of being squeezed out of the rock when they crystallized; still others think that volatiles intro-

Figure 9-42
Migmatite, Clear Creek Canyon, Colorado. Note the complex folds and mixing of light-colored granitic lenses and dark, schistose wall rocks. [Photo by Warren Hamilton, U. S. Geological Survey.]

duced ions of such elements as Na, K, and Si that reacted with the wall rocks to form new granite. Many bodies lack diagnostic features that would indicate whether all or any of these processes participated. In some contact aureoles fossil shells made of calcium carbonate have been converted completely to garnet with such faithful preservation of minute sculptured markings as to allow identification of the fossil species. Clearly Si, Mg, and Fe have been introduced, and C and Ca removed, ion by ion, to bring about this change. Such atom-by-atom exchange of materials, preserving the form of the original object (as in the preservation of wood structures in the opal of petrified wood), is called **metasomatism**. It obviously accounts for the growth of such minerals as K-feldspar in sandstone near deep-seated

plutons, and some geologists think that some entire granite plutons may be so formed.

For example, some plutons fade into their wall rocks so gradually that no distinct contact can be found over a width of many hundreds of meters. Some sedimentary rocks may be thoroughly made over chemically but still retain unmistakable sedimentary features such as cross-bedding. Layers of marble or quartzite may extend for hundreds of meters into and even across entire plutons, bounded on either side by granite. Metasomatic granites unmistakeably border these septa.

But not all plutons that show some "ghost stratigraphy" are so formed. Many bands of country rock that have essentially the same attitude both within (surrounded by the intrusive rock) and without a pluton are pendants of country rocks hanging from a former roof. Some are due to wedging fingers of magma separating the wall rocks along surfaces of ready parting, such as schistosity, while the whole mass was being forcibly injected. Each pluton must be separately studied if its emplacement is to be understood.

Over large areas, particularly in the Precambrian Shields, plutons and wall rocks have flowed more or less coherently as viscous masses of only slightly varying plasticity (Fig. 9-43). Clearly the wall rocks and the plutons were at the same temperature and both were largely crystalline, but with enough volatile fluxes to heal the fractures in crystals broken during the motion. In places motion continued after cooling, so that the crystals remained broken and great belts of breccia mark the borders of the moving mass.

Some granite undoubtedly forms by metasomatism. Some geologists think that all granite does—that it forms virtually solid bodies at depth, which, being less dense than their surroundings, rise slowly to shallower depths as a balloon rises in the air or a piece of wood pushed into tar rises to the surface. If the mass rises fast enough to retain heat at sufficiently shallow depths, where both pressures and melting points are lower, it melts and forms the magma whose liquid nature is so obvious in volcanic and shallow plutonic phenomena. Most geologists agree that this may indeed happen, but most also think it exceptional; we have ample evidence that mafic magmas, whose chemical and isotopic character imply birth deep within the earth, can evolve to produce siliceous compositions by various processes soon

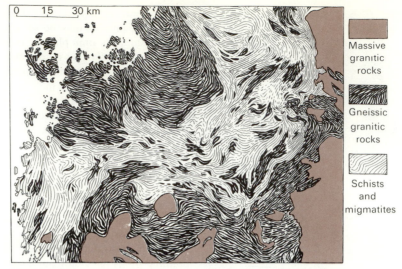

Massive
granitic
rocks

Gneissic
granitic
rocks

Schists
and
migmatites

Figure 9-43
Concordant batholiths and migmatites in Finland. The dashes indicate the
general trends of foliation. [After Martti Saksela, 1935.]

to be discussed. A mechanism exists for forming granitic magma by means other than replacement of sedimentary rock, even though that has indeed happened on a gigantic scale.

MAGMATIC DIFFERENTIATION

We noted that at Hekla, lavas became increasingly mafic in composition as eruption proceeded. Many other volcanoes have erupted lavas ranging widely in composition—for example, the Kamchatkan volcano Kliuchevsky and its satellite craters erupted lavas ranging in silica content from 54.5 percent (May 1937) to 51 percent (February 1938). Lavas from the volcanoes of the Krakatau Islands ranged in silica content from 50 percent to 70 percent over a cycle of centuries. Many granitic plutons are cut by successions of dikes: a salic group including porphyritic and equigranular rocks of different compositions, and commonly an associated group of mafic rocks with phenocrysts of pyroxene, hornblende, and biotite in a fine-grained groundmass. How do such contrasting magmas arise?

A leading student of this question, the late Norman L. Bowen of the Carnegie Institution of Washington, presented in 1928 a theory of **crystallization differentiation**. Professor R. A. Daly of Harvard had noted that basalt is far the most widespread igneous rock; it is found in rocks of all ages on continents and ocean floors, oceanic islands, mountains, plateaus, and plains. He hypothesized that basaltic magma is the parent of all other igneous rocks. Bowen followed Daly's suggestion and experimented with various mechanisms of crystallization differentiation, the most obvious of which is exemplified by the differentiation of many large sills and other floored intrusives.

Differentiated Sills

The Palisades of the Hudson, facing New York from the New Jersey shore, are the eroded edge of a dolerite sill nearly 300 m thick. The sill is not homogeneous, but is layered (Fig. 9-44). Along the base and locally along the roof, selvages of fine-grained basaltic rock contain sparse crystals of olivine. Above the basal zone rests a 3-to-6-meter layer of olivine-rich dolerite, averaging about 25 percent olivine. The bulk of the sill overlying this layer contains no olivine, and its pyroxene content gradually diminished upward. Were the minerals of the olivine-rich and overlying variable zones uniformly distributed throughout

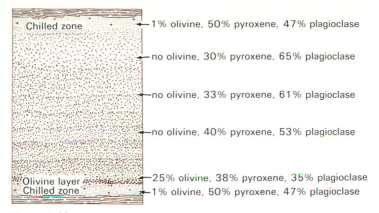

1% olivine, 50% pyroxene, 47% plagioclase

no olivine, 30% pyroxene, 65% plagioclase

no olivine, 33% pyroxene, 61% plagioclase

no olivine, 40% pyroxene, 53% plagioclase

25% olivine, 38% pyroxene, 35% plagioclase

1% olivine, 50% pyroxene, 47% plagioclase

Figure 9-44
Diagrammatic section of the Palisades sill, New Jersey, showing the vertical variations in texture and mineral composition. [After Frederick Walker, 1940.]

the sill, the main body would have had the same composition as the fine-grained selvages along the border. Geologists infer that the fine-grained border zone is a chill zone, quickly crystallized when the magma came in contact with the cold country rock. The olivine phenocrysts in the chilled zone show that this mineral formed early in the cooling history of the magma; its accumulation in the olivine-rich layer indicates that, being denser than the magma, it settled out before any considerable amount of other minerals had formed. The remaining magma was thus more salic than the original magma frozen in the chilled zone. Similarly, the pyroxene tended to settle but could not do so as cleanly as the olivine because plagioclase was crystallizing at the same time, interfering with the separation, and the magma was becoming more and more viscous as temperature fell. The Palisades sill is an excellent example of crystallization differentiation. Large changes were brought about by the gravitative settling of early-crystallizing minerals.

The gravity mechanism illustrated in the Palisades sill could obviously produce dikes of varying composition if, after the olivine layer accumulated, a fissure broke through the roof of the sill and allowed the residual magma to rise into it.

Another way in which liquids may be separated from already crystallized minerals is by **filter-pressing**, a process in which a mass of mostly crystallized magma is thought to be squeezed in

Figure 9-45
Schistose, gneissic, amphibolitic, and granitic rocks in complex interrelations. Near Riggins, Idaho. [Photo by Warren Hamilton, U. S. Geological Survey.]

some way so that the still liquid magma is pressed out, just as water is squeezed from a sponge. This must surely go on at depth, and perhaps Figure 9-45 illustrates a product of such action.

Before Bowen presented his theory, he and his colleagues had determined the order in which minerals crystallized from simple magmatic melts as they cool. In some experiments, the early-formed minerals were separated from the remaining magma, thus preventing their reaction with it during further cooling; in others, such reactions were allowed to go on. Bowen worked out the following sequence for the evolution of a basaltic magma on the assumption that early-formed crystals are removed as they crystallize:

Bowen Reaction Series.

(Early crystals, largely removed as formed)

Olivine
      ↘
  Pyroxene   Anorthite-rich plagioclase
      ↘   ↙
    Amphibole   Intermediate plagioclase
         ↘   ↙
      Biotite   Albite-rich plagioclase
          ↘   ↙
         Potassium feldspar
               ↓
           Quartz

(Late crystals formed from residual magma)

Olivine and anorthite-rich plagioclase are the principal early minerals to crystallize as a basaltic magma cools. Unless removed in some way, much or all of the olivine reacts with the remaining liquid at lower temperatures to form pyroxene, and the anorthitic plagioclase reacts to form plagioclase continually enriched in albite. No Ca ions are removed from the crystals; instead, higher proportions of sodium ions are added, either throughout the crystals if the cooling is extremely slow, or in an outer zone of the crystals if the cooling is a bit faster. If none of the magma is removed from contact with the crystals, the mass ultimately crystallizes to produce basalt—an aggregate of pyroxene and intermediate plagioclase, with or without olivine.

If the earlier-formed crystals are removed, however, the modifying reactions cannot proceed. Olivine contains a higher percentage of magnesia and relatively less silica than the magma, and anorthitic plagioclase more calcium and less soda and silica. Their removal leaves the magma relatively depleted in magnesia and calcium and relatively enriched in silicon, sodium, potassium, and iron. The magma composition thus becomes andesitic rather than basaltic. Without further crystal removal, the residue of the magma congeals as a diorite (if plutonic) or as andesite (if erupted to the surface).

If the pyroxene and intermediate plagioclase are removed, and hence unavailable for further reaction, the residual magma continues to become more siliceous and alkaline, eventually forming a rhyolitic melt. Only about 10 percent or less of the originally basaltic magma would be available as rhyolite, however, as the great bulk would have crystallized earlier to form peridotite, gabbro, diorite, and similar rocks.

In support of this scheme, Bowen pointed to the olivine cumulates near the floors of many sills, like that of the Palisades, and to the quartzose rocks often found near their roofs. The small rhyolitic flows from dominantly basaltic and andesitic volcanoes he explained as residua of fractional crystallization within the volcanic edifice. In later work, with his colleague O. F. Tuttle, he pointed to the fact that the great bulk of granitic rocks are within a very small range of compositions near to one-third orthoclase, one-third albite, and one-third quartz (Fig. 9-46).

Experiments show that mixtures of these three components in these proportions have the lowest melting temperature of any magma-like mixture. This is indeed a strong argument that most granites are magmatic rather than metasomatic products, for it is difficult to see why replacement should result in such a concentration near the composition of the lowest melting mixture. Though thus probably magmatic, the bulk of the granite may not be derived from basaltic liquid, but from the melting or partial melting of other rocks, as will be discussed.

OBJECTIONS TO BOWEN'S THEORY

Despite these arguments, Bowen's theory is widely questioned. The principal objections are quantitative: granite seems to be far too abundant relative to other plutonic rocks to fit this theory. The huge batholiths of the American Cordillera and of the Precambrian Shields are not accompanied by plutons of compositions intermediate between basalt and rhyolite in anything like the quantities implied by this theory. It must be remembered that even with the most efficient separation of early crystals, nine-tenths of an original basaltic magma would be congealed by the time its residue reached a granitic composition. But the gigantic granitic plutons are not accompanied by vastly greater masses of diorite, gabbro, and peridotite, nor does geophysical work

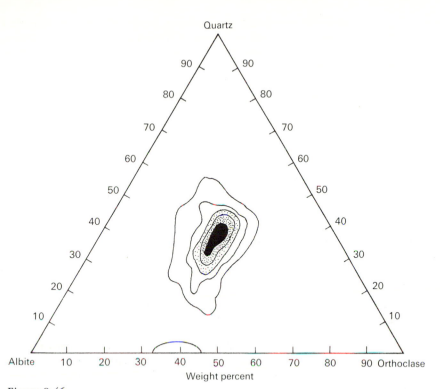

Figure 9-46
The concentration of granitic compositions about the lowest melting temperatures in mixtures of granitic components. A plot of 571 analyses. [After Tuttle and Bowen, 1958.]

(Chapter 7) suggest that such masses are buried deeper in the crust.

Granite, granodiorite, and quartz diorite are almost restricted to the continents; the few granitic oceanic islands, such as the Seychelles (a "micro-continental fragment" of Gondwana Land) and Iceland, are very exceptional, as noted in Chapter 8. The basaltic lavas of Hawaii begin their upward journey from depths of 60 km, as shown by earthquake records (Chapter 7). The islands rise from oceanic depths of 5 or 6 km, and the peak of Mauna Loa is another 4 km above the sea. In this 10-km-thick pile of lava, insulation should have been ample to permit full evolution of a granite by fractional crystallization of the abundant basaltic magma available if this were the dominant process involved—yet there are no granites. Though there are many exceptions, most Phanerozoic granitic batholiths lie in the axial regions of mountain ranges. Although most Precambrian batholiths are not associated with present-day mountains, the structure of their country rocks suggests that they, too, were formed chiefly in connection with mountain-making (Chapter 8). Why, if basaltic magma is the direct ancestor of granite, do we find such a tectonic restriction, while basalt itself is almost ubiquitous? None of the great accumulations of flood basalt show any relation to contemporary mountain-making.

THE ORIGIN OF MAGMA

Magmas at Plate Divergences

The basalts that arise at ocean ridges result from the partial melting of the mantle rocks. As noted in Chapter 8, the upper mantle is made up of garnet peridotite or some very similar rock. At depth, though very hot, it remains solid because of the high pressure, but on rising on the limb of a convection cell, it reaches a level where the pressure permits partial melting and the generation of a tholeiitic basalt magma. As we noted in mention of the Hekla eruptions, the magma may differentiate by gravity high in the volcanic edifice, so that the first eruptions are more salic than the later ones. Such differentiation does not seem to go on in the elongate eruptive columns that feed the submerged part of the ridges, as the basalts of the ocean floor seem to be remarkably uniform in composition wherever dredged.

Convergence-zone Magmas

The association of much of the Pacific "Ring of Fire" with subduction zones has been pointed out in Chapter 8. The association is such as to suggest that the frictional heat developed along the boundaries of the descending crustal slab suffices to melt part or all of the less refractory minerals within it. The downgoing slab contains some, and in places a great deal of, sedimentary rock, muddy sediment, and graywackes saturated with water. Laboratory experiments show that water is a powerful flux; many minerals melt in the presence of water at temperatures hundreds of degrees lower than their melting points when dry.

Where subduction zones separate oceanic plates, as along the Tonga, Kermadec, and Aleutian deeps, the magma rising from the subduction zone is basaltic or andesitic. But where the collision is between ocean plate and continent the magma generated is more siliceous and increasingly so with increase in depth to the zone. This is well illustrated by the contrast between the magmas of the Aleutian chain and those of the Alaska Peninsula, its eastern continuation. The Aleutian volcanoes are nowhere more siliceous than andesite. But where the subduction zone passes on strike under the sialic Alaskan Peninsula, the lavas are rhyolite, quartz latite, and similar quartzose rocks, and the plutons are granodiorite and quartz monzonite. Distances from the Aleutian Trench differ only moderately. This contrast in chemical composition of the magmas may be due to contamination of a magma that on formation at the subduction zone was basaltic but absorbed stoped sialic rocks in sufficient quantity to become quartz monzonitic. Or perhaps the abundant graywacke that makes up so much of the Alaska Peninsula was itself abraded along the subduction zone and melted there. It may be significant that careful laboratory work has produced melts of granodioritic com-

position merely by melting graywacke. In any event, it seems unlikely that the Peninsular rocks and the huge batholiths of Alaska, British Columbia, and the western United States are derived directly from the mantle. Were they derived from a basaltic magma, there should be vast accumulations of mafic rocks if the Bowen scheme of crystallization differentiation were operative. It must be remembered that a granitic residual magma formed by crystallization differentiation cannot be more than 10 percent of the bulk of a basaltic parental magma. But the few mafic bodies associated with the great batholiths are small and widely scattered. It seems almost certain that the batholiths have been derived largely by melting of sialic crust; their differentiation from a mantle composition was principally by sedimentary rather than magmatic processes.

Intraplate Magmas

Although most of the earth's magmas are associated with either divergent or convergent plate boundaries, many notable volcanic fields are active far from any plate boundary. Kilauean basalt has been traced, by the seismic activity caused by its motion, from depths of 60 km to the surface. Experiments indicate that melting of 15 or 20 percent of a garnet peridotite at pressures corresponding to this depth would produce magma of Kilauean chemical composition. But why is the magma localized beneath Hawaii? The Hawaiian chain, all volcanic, is consistently younger from northwest to southeast—60 m.y. old at Midway and active today at Hawaii, intervening islands being progressively older in northwestward succession. This has inspired the suggestion that the whole Pacific plate, carrying the Hawaiian Islands passively along, is drifting over a "hot spot" in the mantle. This may indeed be true, but other linear chains, such as the Line Islands in the South Pacific, seem all to be of about the same age. Certainly the alignments of continental volcanoes is quite inconsistent with drifting over fixed "hot spots"; the volcanic chains of the Andes, Indonesia, the Cascades, and the African Rift Valleys are not oriented by drifting over fixed spots, nor are the isolated volcanic districts in the heart of the American Plate, in Montana, Wyoming, Colorado, New Mexico and west Texas, to be explained by hot spots. Nevertheless, of the many hundreds of known volcanic centers, two sets can be selected to fit the "hot-spot" idea.

Some of the continental plates have undergone great changes in elevation during their drift. For example, during the Mesozoic a great submergence of the North American continent permitted the accumulation of geosynclinal thicknesses of sedimentary rock along the site of the eastern cordillera, which now stands as much as 3 or 4 km above the sea. Much igneous activity has been associated with this uplift. During the drifting of the continent, the thickness of the lithosphere presumably varied notably, either because parts of the lithosphere were worn away by frictional drag or because it accumulated portions of the asthenosphere. A highly speculative suggestion is that the late Tertiary activity in the San Juan Mountains and in Yellowstone Park has been due to exposure of magma bodies in the Low-Velocity Zone—the asthenosphere.

ISOTOPIC CLUES TO MAGMA SOURCES

The considerably rarer element rubidium has the same valence and nearly the same ionic radius as the much more abundant element potassium; accordingly, like potassium, it is concentrated in the crust rather than the upper mantle. As we noted in Chapter 6, one of its isotopes, ^{87}Rb, is radiogenic, converting by beta decay to ^{87}Sr. Because of the relative concentration of Rb in the crust, the ratio of radiogenic ^{87}Sr to the non-radiogenic stable isotope ^{86}Sr is notably higher in crustal rocks (averaging 0.7090 or higher) than in rocks derived directly from the mantle (averaging 0.7035). Thus, when allowance is made for the age of the rock, and the $^{87}Sr/^{86}Sr$ ratio at the time of its formation can be computed, we have a clue to the source of the magma. If the ratio is low, say 0.704 or less, the magma is likely to have been formed in the mantle and to have been little contaminated on its way through the crust; if the ratio is relatively high, say greater than 0.705, the magma has either been contaminated by assimilating crustal materials during its ascent, or it has been formed by direct melting of crustal rocks.

These conclusions are not absolute determinations. The wide variations in velocity of P_n seismic waves, mentioned in chapter 7, indicate that the mantle is far from homogeneous, and the very existence of subduction zones shows that the upper mantle is continually being contaminated by downgoing crustal rocks. But it is surely highly

Figure 9-47
Coarse-grained feldspathic gneiss with interlayered amphibolite (dark gray). Note local contortion and faults. Llano County, Texas. [Photo by Sidney Paige, U. S. Geological Survey.]

suggestive. A careful study, by Carl Hedge of the U. S. Geological Survey, of the isotopic ratio of many rocks has shown that siliceous volcanics tend to have notably higher ratios than mafic ones, strongly suggesting that their higher silica content is due to assimilation of crustal material.

METAMORPHIC ROCKS

Not all metamorphic rocks are visibly associated with plutons, but many are clearly related to them and show gradational changes that vary with the distance from igneous contacts. It has thus long been recognized that the mineralogical and textural alterations produced when such supracrustal rocks as sandstone, shale, and limestone are changed into quartzite, schist, and marble, respectively, are responses to high temperatures and pressures, which are characteristic of active plutons. Near many plutons the country rocks have been recrystallized into coarse-grained feldspathic gneiss with strong banding (Fig. 9-47). Although at first glance such rocks seem to be stratified and thus of sedimentary origin, in many places they can be traced into massive granite cut by basaltic dikes. Either plutonic or supracrustal sedimentary rocks can be squeezed and forced to flow to produce such patterns and have done so in many places where no batholiths are present.

Some metamorphic rocks have flowed to produce patterns as complex as any seen in a mixing bowl (Fig. 9-48). The fact that these rocks were originally stratified is known only from their chemical composition, for the bedding has been completely obliterated. Although early in the development of the science all metamorphic rocks were considered Precambrian, purely on the basis of their complex banding and unusual mineral composition, it is now known from tracing metamorphic rocks into less and less metamorphosed terrains that such characteristics are not criteria of age, even though the older the rock the more likely its metamorphism.

Metamorphic Zoning

It was long ago realized that metamorphic effects of deep-seated processes differ from those of shallower levels. A giant step forward in their understanding was made by the meticulous field studies of the British geologist George Barrow more than sixty years ago. Barrow found a systematic change in the mineral content of metamorphic rocks surrounding a group of plutons in the Grampian Highlands of Scotland. He studied the zoning of minerals. The first zone encountered away from the granite is characterized by coarse-grained feldspathic gneisses and schists containing wisps of *sillimanite* (Al_2SiO_5). Farther away the schists are finer grained and contain tabular crystals of *kyanite*, a mineral identical in composition with sillimanite, but of denser space lattice. Still farther out are schists with little crystals of *staurolite* ($2FeO \cdot 5Al_2O_3 \cdot 4SiO_2 \cdot H_2O$). This succession of **zones** is consistent about the pluton even where a single belt of schist can be traced from one zone to another; since the zoning cannot be attributed to differing chemical compositions, it must be caused by differences of temperature and pressure at differing distances from the pluton.

The zoning farther from the central heat source was studied by later workers who found the staurolite zone to be succeeded outward by a zone containing the iron-alumina garnet *alman-*

Figure 9-48
Flowage structure in marble and quartzite. The original beds were limestone and chert of Permian age, Big Maria Mountains, southeastern California. The unmetamorphosed, fossiliferous equivalents of these rocks are exposed in the Grand Canyon of the Colorado, 200 km to the northeast. [Photo by Warren Hamilton, U. S. Geological Survey.]

dine, then by one wherein the schists grade into *biotite* phyllites and, furthest outward, a *chlorite* zone, evidently nearly the lowest zone of metamorphism. Approaching centers of most intense metamorphism, then, in pelitic rocks we commonly reach in succession the zones of *chlorite*, *biotite*, *almandine*, *staurolite*, *kyanite*, and *sillimanite*. It must not be thought that such zoning is universal, for obviously the chemical composition of the original rock may be such that the characteristic minerals do not form in every rock. In pelites, the commonest sedimentary rocks, most of these zones develop. In basaltic lavas, zoning is also found, but the guide minerals are different because of the different chemical composition. The British petrologist C. E. Tilley has suggested that lines connecting points of first occurrence of these guide minerals are **isograds**, that is, lines of equal metamorphic grade.

Barrow's zones have been supplemented by the concept of **metamorphic facies**, developed by the great Finnish petrologist Pentti Eskola. A meta-

Table 9-4
Mineral facies of rocks of basaltic composition.

← Rising temperature			

Rising pressure ↓

Diabase facies	Amphibolite facies	Epidote amphibolite	Greenschist facies
Pyroxene hornfels			
	Hornblende gabbro facies		
Gabbro facies			
Granulite facies			
		Glaucophane schist facies	
Eclogite facies			

SOURCE: After Pentti Eskola, 1921.

morphic facies is a group of rocks whose minerals are the stable phases under the prevailing temperature, chemical composition, and pressure. Obviously the minerals we see are mostly metastable at room temperature and atmospheric pressure, but their existence is testimony to the general sluggishness of silicate reactions at low temperatures; most rocks retain at least remnants of the minerals formed under the highest temperature and pressure to which they have been exposed. Eskola pointed out that rocks with virtually identical chemical compositions—that of basalt—can be found with half a dozen different mineral compositions, depending on the conditions under which it crystallized (see Table 9-4).

In like manner, rocks of other compositions form their own characteristic facies under changing environmental conditions.

Eskola's inferences as to temperatures and pressures under which the various facies form are wholly qualitative, based on geological inferences. With the development since about 1945 of presses and furnaces capable of giving high pressures and temperatures simultaneously, great progress has been made by experimental petrologists in simulating the actual conditions under which the various facies form.

Figure 9-49 illustrates the results of some experiments under high pressure and temperature with specimens of composition Al_2SiO_5, corresponding to that of three minerals: sillimanite, kyanite, and andalusite. Experimental difficulties are great, and considerable uncertainty continues as to the boundary between the stability fields of andalusite and sillimanite. A charge of this composition, held under a pressure of 5000

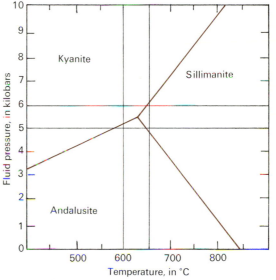

Figure 9-49
Phase boundaries for aluminosilicates of the composition Al_2SiO_5. The position of the triple point varies somewhat in the presence of impurities, and it is not uncommon to find more than one of the phases in a metamorphosed rock, though one can generally be seen to be in the process of altering to another. [From S. W. Richardson, M. C. Gilbert, and P. M. Bell, *American Journal of Science* v. 267, 1969.]

bars (about equal to that at a depth of 15 km in the earth) and heated to a temperature of 600°C will form, at equilibrium, crystals of kyanite. If the temperature is raised to about 700°C, the kyanite usually inverts to sillimanite. It is thus easy to see why, in the Barrow scheme of metamorphic zoning, the innermost zone, next the granite, contains sillimanite rather than kyanite. At the higher temperatures of the sillimanite zone, the micas become unstable, their water is driven off, and they recrystallize as feldspar and garnet.

It should be emphasized that not every rock subjected to a given temperature and pressure will adjust its mineralogy promptly to that appropriate to those conditions. Active shearing, for example, continually exposes new surfaces to contact with interstitial fluids; fine-grained rocks, in which surface-to-mass ratios of the minerals are large, react much more readily than do coarse-grained rocks. Another factor is the availability of water, an essential constituent of the amphiboles and micas. For these reasons, some adjacent metamorphic rocks are identical in composition but differ in mineral content because one has failed to reach equilibrium with the prevailing environment.

Questions

1. Why are cinder cones more in evidence along a dike after the dike has fed lava flows and is in process of congealing than when the dike first breaks through to the surface?

2. How can you distinguish a deposit of welded tuff from a flow or domical protrusion of flow-banded rhyolite?

3. Obsidian from a Japanese volcano has a specific gravity of 2.60; pumice lapilli from the same volcano floats two-thirds submerged in water. Approximately what volume of ejected pumice would be required to account for the subsidence of the top of the volcano, making a caldera 4 km in diameter with average depth of 300 m? (Assume that the caldera was the result of piecemeal subsidence by lack of support after the pumice lapilli had been ejected from a magma chamber directly beneath the site of the caldera.)

4. Draw cross-section diagrams of the following igneous bodies: (a) a basalt flow, (b) a ring dike, (c) a differentiated sill, (d) a laccolith fed from a stock, (e) a concordant batholith.

5. Draw a sketch or a geologic map showing a portion of the edge of a discordant batholith. Show and label the following typical features: (a) a discordant contact, (b) inclusions, (c) a series of dolerite sills that are older than the batholith, (d) a rhyolite dike that is younger than the batholith, (e) a series of sedimentary rocks that are older than the batholith, (f) a welded tuff younger than the batholith.

6. Explain how differentiation by crystal settling in an igneous mass works.

7. What is the evidence that some granites are of metamorphic origin?

8. How could you tell that a granite body had been injected into cold rocks rather than into a zone of high metamorphic intensity where the wall rocks were nearly as hot as the granite?

9. Why is it more difficult to make synthetic metamorphic minerals in the laboratory than it is to make evaporite minerals?

10. Many greenschists, composed of chlorite, epidote, and albite show well-developed pillow structures. Would you expect pillows to be associated with chlorite-bearing mica schists typical of the Barrow chlorite zone or with highly feldspathic sillimanite-bearing gneisses? Why?

11. Assuming the chlorite schist to be in contact with the sillimanite gneiss, what inference can you draw regarding the nature of the contact?

12. Why is an isograd not a true measure of metamorphic temperature?

Suggested Readings

Buddington, A. F., "Granite Emplacement with Special References to North America." *Geological Society of America Bulletin*, v. 70, 1959, pp. 671–748.

Bullard, F. M., *Volcanoes*. Austin: University of Texas Press, 1962.

Eaton, J. P., and K. J. Murata, "How Volcanoes Grow." *Science* v. 132, no. 3432, 1960, pp. 925–938.

Hunt, C. B., "Structural and Igneous Geology of the La Sal Mountains, Utah" (U. S. Geological Survey, Professional Paper 294-I). Washington, D. C.: G.P.O., 1958. [pp. 305–364].

Macdonald, G. A., and D. K. Hubbard, *Volcanoes of Hawaii National Park*. Hawaii Nature Notes v. 4, 1951.

Read, H. H., *The Granite Controversy*. London and New York: Interscience Publishers, 1957.

Shelton, J. S., *Geology Illustrated*. San Francisco: W. H. Freeman and Company, 1966.

Tuttle, O. F., and N. L. Bowen, *Origin of Granite in the Light of Experimental Studies*. Geological Society of America Memoir 74, 1958.

Scientific American Offprints

819. O. Frank Tuttle, "The Origin of Granite" (April 1955).

822. Howel Williams, "Volcanoes" (November 1951).

854. Edwin Roedder, "Ancient Fluids in Crystals" (October 1962).

CHAPTER 10

Climate, Weathering, and Soils– Conditioners of Erosion

The constructional aspects of the geologic cycle have been outlined in the preceding chapters; here we begin to treat of the wearing away of the rocks exposed to erosion through tectonic and igneous processes.

Volcanic, plutonic, and metamorphic rocks are all formed under temperatures and pressures that do not prevail generally over the earth's surface. Most of their minerals are thus unstable and tend to change into others more nearly in equilibrium with their new environment. Most such changes are very sluggish, but the tendency is universal. Even sedimentary rocks whose minerals were stable as to temperature and pressure are out of equilibrium under a different chemical environment, and thus tend to decompose or disintegrate. This chapter treats of the factors that produce such changes.

SOLAR RADIATION

The sun radiates energy in a wide range of wavelengths, from at least as long as 30 km (long radio waves) to as short as 0.1 Å (shorter than X-rays). Part of this radiation is absorbed in the air, as has long been known, but the amount could not be measured until rockets became available to carry instruments above the atmosphere. The effective top of the atmosphere is at about 90 km, though it extends hundreds of kilometers farther. At this distance a surface at right angles to the sun's rays receives, on the average, between 1.90 and 2.00 langleys of energy per minute, with the average about 1.97. (A langley is 1 gram calorie per cm^2.) Energy absorbed at this rate is enough to melt a gram of ice in about 40 minutes and heat it to the boiling point in less than another hour. Though the radiation certainly fluctuates, as shown by the changing sun spots and solar flares, this so-called "solar constant" is not thought to vary by as much as 2 percent.

The wavelengths of radiant energy depend on the temperature of the source. Because the sun is so hot—over 6000°C—more than 95 percent of the energy is carried by waves between 2900 and 25,000 Å long. As much as 42 percent is in the visible spectrum (Fig. 10-1); the human eye has so evolved as to be most sensitive to the most abundant radiation. The maximum is in the green wavelengths vital to photosynthesis and thus to virtually all life, even though only a small fraction of 1 percent of the solar energy received is so utilized. A little more than 50 percent is in the infrared wavelengths, and only about 7 percent in the ultraviolet. Nearly all the ultraviolet is absorbed in the atmosphere, where it dissociates molecular oxygen (O_2) to form ozone (O_3) and atomic oxygen.

from clouds, and another sixth from the atmosphere. Thus only about 27 percent of the energy impacting the earth from the sun reaches the surface directly. Thirty-four percent is absorbed by the atmosphere and clouds, about equal to the amount they radiate to the surface.

Figure 10-2 shows 100 percent of the incoming radiation as shortwave (an exaggeration). Of this, we have seen that only about 60 percent reaches the earth's surface, directly or by refraction; 15 percent goes to heat the atmosphere, 33 percent is radiated to space from the clouds, and 9 percent from the air. The long waves radiated from the earth carry 8 percent directly to space, 16 percent to heat the atmosphere (of which 4 percent is returned by turbulent air); 25 percent is transferred from the surface to the air by evaporation, and 50 percent radiates directly from the air to outer space. On the average the earth is getting neither hotter nor colder; it must therefore radiate into space as much heat as it receives. But being thousands of degrees cooler than the sun, it radiates at far longer wavelengths—about 99 percent at wavelengths between 3000 and 90,000 Å (Fig. 10-3). The wavelengths of the earth's radiation are much longer than most of the sun's, and they greatly affect the climate. Both water vapor and CO_2 selectively absorb and are heated by these wavelengths. The atmosphere is thus much warmer than it would be without them. This so-called "greenhouse effect" may, indeed, eventually become even more significant climatically as the CO_2 content of the atmosphere continues to increase.

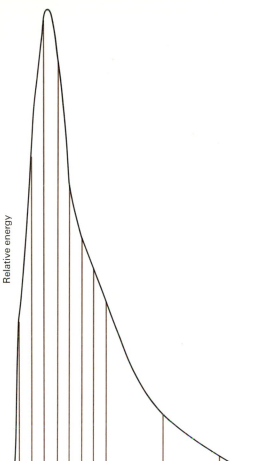

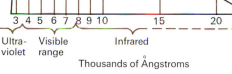

Thousands of Ångstroms

Figure 10-1
The relation between wavelengths and energy of solar radiation. [Data from Moon, 1963; modified in presentation.]

This absorption is fortunate for life, as even small amounts of ultraviolet may cause painful sunburns, and long exposure is lethal.

The atmosphere also absorbs much other solar radiation; only about a quarter of it penetrates directly to the surface (Fig. 10-2). Sunlight striking molecules of air is refracted and reflected into diffuse sky radiation. About a sixth of the incoming solar energy reaches the surface through diffusion

COMPOSITION OF THE ATMOSPHERE

Hundreds of chemical analyses of the atmosphere at points widely distributed over the earth indicate that the three major constituents—oxygen, 20.946 percent; nitrogen, 78.084 percent; and argon, 0.934 percent—are in virtually identical proportions everywhere. So, too, with the other inert gases: He, Ne, Kr, and Xe, all present in very minor amounts.

But the extremely important minor constituent CO_2—essential to plants and thus to all life—varies widely in amount from place to place. Over the Arctic and Antarctic seas there is less of it than

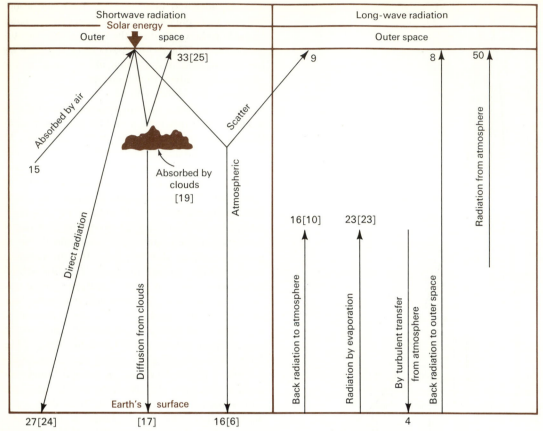

Figure 10-2
Mean heat balance in the northern hemisphere (that in the southern would not differ greatly, though the average temperature there is about 2°C lower). The figures represent percentages of the incoming radiation. [Modified from Baur and Phillips, *Beitrag zur Geophysik,* 1934. The figures in parentheses are independent estimates by H. G. Houghton.]

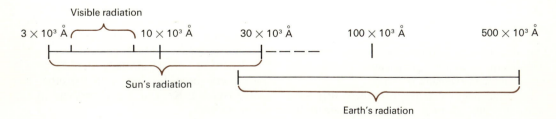

Figure 10-3
Wavelengths of the sun's and earth's radiation (logarithmic scale). Though an overlap exists, most solar energy is in much shorter wavelengths, and all terrestrial radiation is in the far infrared.

on the average because of its ready solubility in very cold water. It is also constantly increasing over the earth as a whole because of the burning of organic fuels. In 1900 the average atmospheric content of CO_2 was close to 0.030 percent, but the burning of more than 200 billion (2×10^{11}) tons of carbon-bearing fuel raised the content to at least 0.034 percent by 1950. It is of course still increasing at the rate of about 0.2 percent a year, with possibly significant climatic effects because of the greenhouse effect. It is estimated that by A.D. 2000, global temperatures might rise by half a degree C, but this would be partly counteracted by the increasing reflection of incoming radiation due to the increasing opacity of the more and more polluted air.

Another vitally important minor constituent of the atmosphere is water vapor, which makes up as much as 4 percent of the total mass in warm tropical regions or as little as 0.010 percent in polar air. Like CO_2, it is important in the greenhouse effect.

ATMOSPHERIC CIRCULATION

We are normally insensible of the fact that air has weight, but the destruction wrought by hurricanes makes evident that a vast weight of air is in motion. The weight of air overlying any particular spot varies, as shown by barometric fluctuations. (A barometer is simply a device for weighing the column of air above it.) Atmospheric pressure (weight per unit of area) is usually recorded in **millibars**. (A millibar, one-thousandth of a **bar**, is the pressure of 1 g/cm²; a bar is 1 kg/cm₂.)

As the earth is nearly spherical, the zonal distribution of heat received at the surface varies with latitude; for example, no heat reaches the region above the Arctic Circle at the winter solstice, except for that brought by the winds. Another obviously important factor is cloud cover, which varies greatly with latitude and altitude. It has been estimated that, on the average, the earth receives about 40,000 cal/cm²/year, but in clear cloudless deserts perhaps almost 20 times as much.

Almost half the earth's surface lies between the latitudes 30°N and 30°S, and as this area receives solar heat more nearly at right angles to the surface than do other zones, it receives considerably more than half of the earth's annual heat budget—the

driving energy of the atmospheric circulation. On the average, about 1 m of water is evaporated from the oceans annually, largely from the low latitudes. Although the various regions differ greatly in detailed circulation because of the uneven distribution of land and sea, the gross scheme is as outlined in Figure 10-4.

In the equatorial doldrums the air temperature averages about 26° to 28°C. This light, warm air rises to a height of as much as 18 km—the tropopause. In rising, it cools, drops much rain, and drifts both north and south, where it sinks to lower and lower altitudes. As each higher parallel of latitude has a smaller circumference than lower ones, the cool air drifting north and south becomes constricted and compressed to a greater density. It therefore sinks in the Horse Latitudes. As it descends it is heated by compression, just as a bicycle pump warms as it compresses air. The result is that the skies in the Horse Latitudes are generally clear and rain is rare. Many of the earth's deserts are found here.

Being a gas, air tends to flow from areas of high to those of low pressure. But the rotation of the earth causes a complication, known from its discoverer, a French physicist, as the **Coriolis Force**. Though the force operates regardless of direction of motion, it is perhaps most readily visualized by thinking of the course of a rocket fired due southward from the North Pole along the meridian of Greenwich. Because the pole has little motion in longitude, the rocket also has little. But the meridian beneath the rocket is moving eastward at the rate of 15° per hour; at the latitude of London, this is about 150 m/sec. If the flight from the Pole to London requires an hour, the rocket will strike, not on the meridian of Greenwich, but 540 km to the west. *With respect to the earth's surface,* it will have followed a path curved to the right, even though its actual flight path was straight. This is true in general: all moving masses tend to be deflected into paths curved to the right in the northern hemisphere and to the left in the southern. The Coriolis force is tremendously important in influencing fluid motion, either of air or water. Winds and ocean currents on the earth do not flow from areas of high pressure directly toward areas of low pressure but instead flow more nearly parallel to lines of equal pressure rather than at high angles to them.

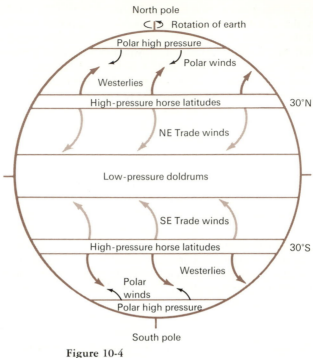

Figure 10-4
General circulation of the atmosphere.

As seen in the circulation pattern of Figure 10-4, part of the air descending in the Horse Latitudes moves toward the equatorial doldrums along paths that curve southwestward in the northern hemisphere—the Northeast Trade Winds—and northwestward in the southern hemisphere—the Southeast Trade Winds. Where these warm winds blow over broad expanses of sea, they pick up much water by evaporation. The Trade Winds rising over the island of Hawaii, for example, dump more than 12 m of rain per year on the windward slopes at altitudes above 1300 m.

Other parts of the descending air flow poleward and curve into the prevailing Westerlies, characteristic of high latitudes in both hemispheres. These are much stronger in the "Howling Fifties" of the southern hemisphere than in the northern, as virtually no land masses obstruct them there.

Because the air in polar regions receives little radiation at any time, and especially little in winter, it is cold—averaging −48°C at the South Pole and −17°C at the North Pole. It is therefore relatively dense and tends to flow equatorward beneath the warmer air of the prevailing westerlies. This forces the lighter air upward, where it cools and drops its moisture at the Polar Front, important in modifying the weather, especially in winter. The interaction of air masses along the Polar Front causes cyclonic storms.

Water vapor is important in transferring heat from low to high latitudes. It takes heat to evaporate water—the "latent heat of vaporization"—and air currents carrying water vapor carry this heat with them. When the air rises and cools, the vapor condenses either as rain or snow, and the latent heat is given off to the surroundings. The heat transferred by this mechanism is much greater than might be thought from the relatively low content of water vapor in even the most humid air, because air itself can carry but little heat—its **specific heat** is very low while that of water is very high. The excess heat of the tropics is transferred poleward in both hemispheres, for the tropics are not getting warmer nor the polar regions colder over long time spans. For the earth as a whole, about 90 percent of the temperature redistribution is carried out by the winds, about 10 percent by ocean currents.

Were the earth's surface uniform and at uniform altitude, the atmospheric circulation would closely accord with the scheme just outlined. But the real surface is covered by irregularly distributed lands and seas, high mountains, and sheltered valleys. The lands warm and cool much more quickly than the seas; the actual circulation pattern is far from simple, though a fairly systematic distribution of climates is discernable in the generalized map in Figure 10-5.

CLIMATE AND VEGETATION

The main factors controlling climate are temperature, precipitation, and evaporation. The relation between precipitation and evaporation—the **precipitation effectiveness**—is critical to vegetation. Where summers are short and temperatures generally low, as in high latitudes and elevated regions, temperature is more critical than precipitation effectiveness in controlling vegetation. Here three climatic and vegetational provinces are distinguished: taiga (coniferous forests), tundra (mosses, lichens, and stunted trees), and permanently frozen ground.

Wet climates with tropical rain forests are typical near the equator, especially near east coasts, and wet climates with coniferous forests flourish in high middle latitudes on west coasts. Arid climates are found chiefly (1) in the Horse Latitudes, where the descending air is cloudless and warming; (2) in the Trade Wind belts, where the winds blow for long distances across warm lands; and (3) in the rain shadows of high mountains. Figure 10-5 is of course greatly generalized; in fact, the boundaries shift from year to year.

The soils of the earth vary greatly in physical and chemical properties, and a major factor controlling their development is climate and the kind of vegetation that the climate favors.

WEATHERING

Deep road cuts commonly penetrate firm rock, which grades upward into a zone of broken rock with softened and discolored rock particles and finally into loose soil near the surface. Many rocks, though black or gray in deep quarries, are yellow or brown in surface exposures. In some the yellow color merely stains the surface and cracks, but in most it is pervasive and accompanied by drastic changes in mineral composition and firmness. We infer that exposure to air and moisture and to the action of organisms that live at shallow depths has brought about the changes. We say the altered rock is **weathered** and that much soil has formed by the weathering of underlying rocks.

In Chapter 3 we noted that flint (quartz) enclosed in limestone is unstable at very high temperatures and reacts with the calcite to form a new mineral, wollastonite, stable under the changed conditions. Conversely, most minerals of volcanic, plutonic, and metamorphic rocks are unstable under conditions at the earth's surface; they tend to react with the oxygen, carbon dioxide, and water either to dissolve or to form new minerals, most of which are **hydrous** (chemically combined with water). Even sedimentary rocks that formed at surface temperatures and pressures may come to be subjected to chemical environments vastly different from those under which they formed. Their decomposition may be exceedingly slow, but it can commonly be inferred with confidence.

SOIL

Soil is weathered, unconsolidated rock debris mixed with at least a little organic matter. The mixture of these two kinds of material is what distinguishes a soil from any other unconsolidated material, such as beach sand, also readily separable into individual grains. Without organic matter or evidence of weathering, they are not soils. (Foundation engineers are less stringent and call any uncemented, incoherent but completely nonorganic material "soil"; applying the term to the material on the surface of the moon is also a loose usage.)

Five factors principally influence soil character: climate, living organisms, parental rocks, topographic situation, and time. Of these, climate is dominant; soils in areas with similar climates tend to be similar even though parental rocks differ, whereas similar rocks in different climates yield

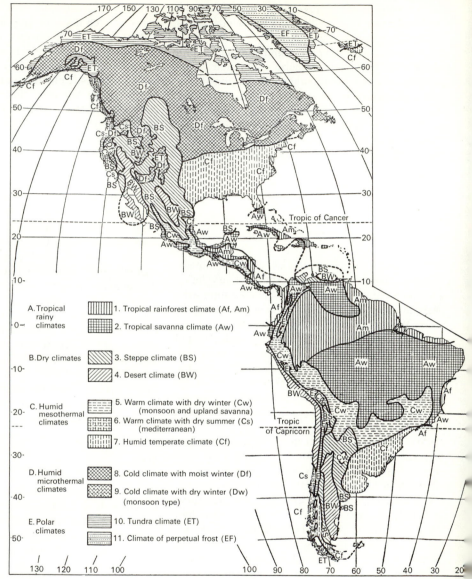

Figure 10-5
The principal climates of the earth. [From Trewartha, *An Introduction to Climate*, New York, McGraw-Hill, 1954.]

differing soils. Thus there is considerable similarity between the climatic map in Figure 10-5 and the map of the great soil groups in Figure 10-6. Climate controls the kinds and amounts of vegetation and soil-inhabiting organisms such as bacteria, algae, amoebas, mites, springtails, nematodes, and earthworms. But in a single climatic environ-

ment the local microclimates differ significantly, and thus the soils on well-drained hillslopes differ from those on valley bottoms; those on north-facing slopes from south-facing. The type and thickness of a soil also depend on the length of time it has been forming. For example, soils forming in the Midwest of the United States on

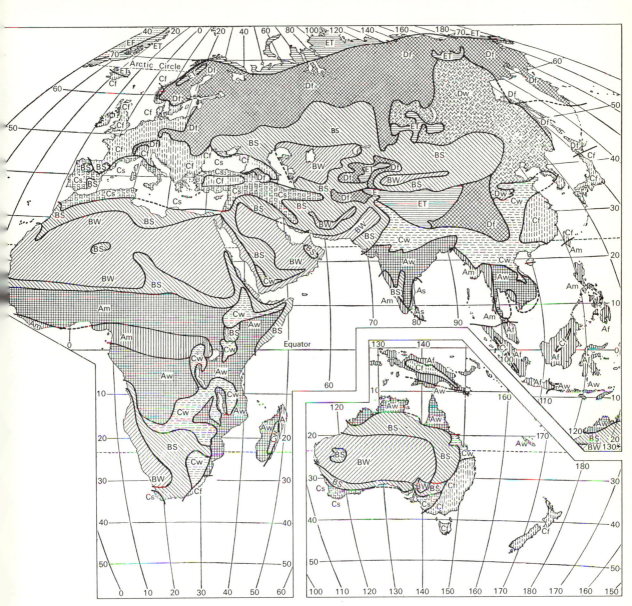

ANALYSIS OF WEATHERING

glacial deposits a few thousands of years old (Chapter 13) differ notably from those formed on similar material after many thousands of years.

Even casual inspection shows that weathering involves both *disintegration* and *chemical decomposition* of rocks. Disintegration, the mechanical loosening of the rock fabric, affects the rock composition only slightly. But it does expose more surfaces of the component mineral grains to the chemical agents of decomposition. The complex silicate minerals formed under high temperatures and pressures tend to alter (chemically decompose) into hydrous silicates, hydrous oxides, and other minerals stable under surface conditions; some elements are dissolved and carried away.

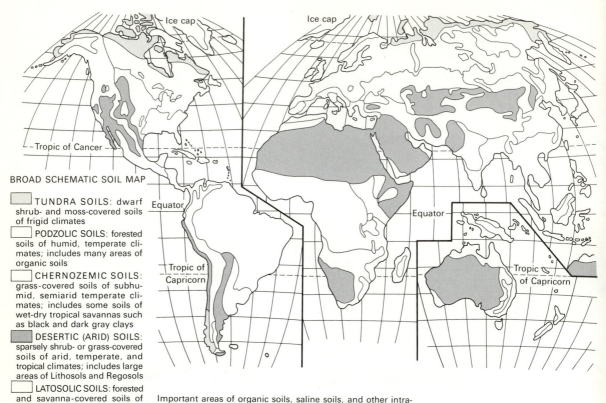

BROAD SCHEMATIC SOIL MAP

TUNDRA SOILS: dwarf shrub- and moss-covered soils of frigid climates

PODZOLIC SOILS: forested soils of humid, temperate climates; includes many areas of organic soils

CHERNOZEMIC SOILS: grass-covered soils of subhumid, semiarid temperate climates; includes some soils of wet-dry tropical savannas such as black and dark gray clays

DESERTIC (ARID) SOILS: sparsely shrub- or grass-covered soils of arid, temperate, and tropical climates; includes large areas of Lithosols and Regosols

LATOSOLIC SOILS: forested and savanna-covered soils of humid and wet-dry tropical and subtropical climates

SOILS OF MOUNTAINS: stony soils (Lithosols) with inclusions of one or more of the soils listed above, depending on climate and vegetation, which vary with elevation and latitude

Important areas of organic soils, saline soils, and other intrazonals are omitted as well as very important bodies of alluvial soils along such great rivers as the Mississippi, Amazon, Nile, Niger, Ganges, Yangtze, and Yellow.

Figure 10-6
Map of the great soil groups of the world. In each group a characteristic soil profile forms, though there is great variation within each. Many large areas of alluvial soils, such as the valleys of the Mississippi, Ganges, and Amazon rivers, are neglected in this chart. [After the Department of Agriculture Yearbook for 1957.]

Frost Action

On freezing, water expands 9 percent—a fortunate property, for if it shrank, as do most liquids on solidifying, most of the earth's water would be frozen at the bottom of the sea. The force of crystallization is impressive; at the not extreme temperature of −22°C, it is more than 22 bars (kg/cm²)—enough to burst steel pipes. Freezing of wet soils expands them upward; thawing leaves them open and spongy.

Solid rock conducts heat much better than broken fragments; accordingly, the bottom of a pebble 10 cm thick cools much more rapidly than sand at the same depth when exposed to subfreezing surface temperatures. Thus freezing of water-soaked gravel tends to sort the components according to size: ice forms most quickly beneath the coarser particles, pushing them toward the surface faster than the finer. Both stones and fence posts tend to be pushed out of the ground during deep frosts. Figure 10-7 shows the vertical sorting of gravel on a flat area in Greenland, brought about by this process.

In tundra at high latitudes this sorting is guided only by random size differences, but the process eventually forms mounds of larger upwelling fragments that slowly glide toward lower ground, there to accumulate in crude polygonal patterns (Fig. 10-8). Such areas of "patterned ground" are now active only under rigorous climates. The existence of similarly patterned ground in parts of Oregon, Pennsylvania, Ireland, and New Zealand where freezing and thawing are not active suggests that climates were once much colder there.

Breaking of rock by freezing requires that the water be confined. Inasmuch as freezing proceeds from the surface downward, water in the deeper crevices may be sufficiently confined as to produce extensive frost shattering. Many high mountain peaks are so mantled by frost-wedged rubble that a rock in place is hard to find. Of course, the colder the climate, the deeper the freeze, but frost wedging is undoubtedly more effective in climates with many freeze-thaw cycles than in those with fewer, though deeper, freezes. For example, Montreal has five times as many freeze-thaw cycles as Ellesmere Land in Arctic Canada.

Figure 10-7
Vertical sorting of gravel by size on a flat area near Thule, Greenland. [Photo by Dr. Arturo Corte, National Cold Regions Laboratory, Hanover, New Hampshire.]

Exfoliation

The parting of scale-like layers from an exposed or soil-covered massive rock (not a platy rock, such as shale or schist) is **exfoliation**, a common mode of rock disruption. The separated sheets may be flat or curved, paper thin, or meters thick, a centimeter or hundreds of meters long. Two kinds are distinguished: one caused mainly by mineral decomposition is described later; the other, which goes on in virtually unaltered plutonic masses, is the splitting off of huge curved

Figure 10-8
Patterned ground on a gentle slope near Thule, Greenland. [Photo by Dr. Arturo Corte, National Cold Regions Laboratory, Hanover, New Hampshire.]

Figure 10-9
Granite dome showing coarse exfoliation, Sierra Nevada. The sheet-like slabs are several meters thick. [Photo by G. K. Gilbert, U.S. Geological Survey.]

Figure 10-10
Thin exfoliation slabs in granite of the high Sierra, California. Scale given by figure in right lower third. [Photo by N. King Huber, U.S. Geological Survey.]

plates (Fig. 10-9) to form the giant domes and stairways of Yosemite in California (Fig 10-10), Stone Mountain in Georgia (Fig.10-11), and the famous Sugarloaf in the harbor of Rio de Janeiro. These domes in plutonic rocks that were formed at depth may be due to the release of weight of overburden as it is eroded away, just as cracks form parallel to quarry faces in some granite quarries as adjoining blocks are removed.

Other Agents of Disintegration

Minor agents of disintegration include the crystallization of soluble salts washed into rock pores in deserts, the intense heat of forest fires, the shock and heat of lightning, and the impact of rocks falling or rolling downhill.

It was once thought that day-to-night temperature changes brought about rock disintegration by cyclic expansion and contraction. But many thousand cycles of heating and cooling in the laboratory have been ineffective in disruption. Surfaces of polished granite at Aswan, Egypt, have been fully exposed to the tropical sun for 3000 years, yet still retain their polish. Slow as

215

Figure 10-11
Stone Mountain, Georgia, a granite dome shaped by exfoliation. [Photo by Warren Hamilton, U.S. Geological Survey.]

chemical action is in the deserts, it is clearly more effective than temperature changes alone in disrupting rocks.

Chemical Weathering

A major factor in rock disintegration is chemical decomposition. Feldspars, micas, amphiboles, pyroxenes, and other minerals generally form under conditions very different from those prevailing near the earth's surface; they are here unstable and tend to react to form minerals at equilibrium under surface conditions. Feldspars swell as they weather into clay, and many other minerals decompose similarly. Such swelling in time causes even massive granite or gneiss to disintegrate into loose sand. Alternate swelling and shrinking of clay as it is wet by the rain and dried in the sun aids the disruption.

Most minerals react very slowly; they remain metastable for long times. Most chemical reactions double in rate for each 10° rise in temperature, but reactions proceed even in glacial environments, where ground temperatures of 40°C have been measured. Most reactions go on more readily in the presence of liquid water because of the *polarity* of the water molecule. In water molecules, the hydrogen atoms are on the same side of the large oxygen atom, so that each molecule is, in effect, a minute magnet that acts to loosen the bonds of the ions at the surface of minerals in contact with water. Rocks therefore decompose in perhaps a few years in the humid tropics but may remain almost unaltered for centuries or even millennia under extreme desert conditions.

Microscopic and chemical studies show that a principal agent of mineral alteration is carbon dioxide, which dissolves in water to form the weak but abundant carbonic acid (H_2CO_3). Rain condensing from a cold cloud dissolves CO_2 from the air in much higher proportion than its low ratio in the air (0.03 percent); instead of forming 0.03 percent of the falling rain, it may be as high as 1 percent. On dissolving in water, the carbon dioxide combines to form ions of hydrogen and bicarbonate:

(1) $\quad H_2O + CO_2 \;\rightleftharpoons\; H^+ + HCO_3^-$
$\qquad$ water $\quad$ carbon $\qquad$ hydrogen $\quad$ bicarbonate
$\qquad\qquad\quad$ dioxide $\qquad$ ion $\qquad\quad$ ion

In the soil the carbon dioxide content of the water increases still more. Plant tissues are mainly carbohydrates — compounds of carbon, hydrogen, and oxygen. Where oxygen is available, bacteria feed upon decaying organic matter and produce additional CO_2; the content may rise in the soil pores to as high as 10 percent. So high a concentration yields a strongly acid solution that is extremely effective in weathering.

Another solvent brought down by the rain, though much less important, is ammonium nitrate, formed by lightning discharges. Analysis of rain in France suggests that more than 3 kg/hectare is brought to earth in solution yearly.

Organisms in the soil are amazingly abundant — in temperate humid regions, as much as 30 or 40 metric tons/hectare. Damp soil may contain between 2 and 14 million bacteria per cm^3, or one every 50 microns! Amoebas feed on the bacteria, and other microorganisms feed on the amoebas. The tremendously complex interaction of these plants and animals with the minute nematode worms, the larger earthworms, and the mites, ants, and termites that crowd the soil develops an organic complex known as *humus*, which dissolves in water to form a weak acid that attacks the silicate minerals of the rock and soils.

Weathering of Limestone

The temperatures and pressures at the ground surface differ little from those in shallow marine waters in which limestone was formed, except in the great excess of CO_2 in the soil environment. In damp climates this suffices to reverse the stability of calcite, the main mineral of limestone; instead of accumulating, it dissolves. Only a little dissolves in pure water, to produce a few ions of Ca and CO_3:

$$(2) \qquad \underset{\text{calcite}}{CaCO_3} \;\rightleftharpoons\; \underset{\text{Ca ion}}{Ca^{++}} \;+\; \underset{\text{carbonate ion}}{CO_3^{--}}$$

But if the water in contact with the calcite already contains some CO_2 in solution, the carbonate ion produced by the reaction of Equation 2 reacts with the H^+ ion of Equation 1 to form more bicarbonate ions:

$$(3) \qquad \underset{\substack{\text{hydrogen}\\\text{ion}}}{H^+} \;+\; \underset{\substack{\text{carbonate}\\\text{ion}}}{CO_3^{--}} \;\rightleftharpoons\; \underset{\substack{\text{bicarbonate}\\\text{ion}}}{HCO_3^-}$$

Thus water containing excess CO_2 can dissolve much more calcite than pure water because its hydrogen ions can combine with carbonate ions to form bicarbonate ions. This lessens the tendency for the carbonate ions to reunite with calcium ions to again form calcite, a possibility indicated by the left-pointing arrow in Equation 2.

Combination and dissociation of ions is continually going on in solutions at rates governed by the temperature and by the abundance of the several ions present. When the rates are such that there is no change in the abundance of any ion, the system is said to have reached **equilibrium**. Thus, at equilibrium, in the reaction of Equation 2, as much calcite is being precipitated from solution as is being dissolved. But if the ions of carbonate on the right are combined with H^+ to form bicarbonate ions, there are fewer to react with Ca^{--} to form calcite; more calcite will dissolve. At equilibrium, the reversing reactions equal the progressing ones; the solution is **saturated** with calcite. But in a well-drained soil the water slowly percolates downward, leaving room for unsaturated water to take its place near the surface; calcite is dissolved along every crack and pore of the rock. Where the underground circulation is vigorous (Chapter 14), large caves may be dissolved. Clay impurities in the original rock do not dissolve, as clay is a stable phase under surface conditions. It accumulates at the surface as the surrounding calcite is leached away. A few millimeters of clay may be all that remain from many meters of dissolved limestone.

The leaching of calcite from the limestone does not produce any new minerals; it merely changes the proportions of those originally present. The original limestone was of course formed of minerals relatively stable under surface conditions in sea water that was constantly exchanging CO_2 with the air. It is the excess of CO_2 and humus in the soil that renders the rock unstable. Many sedimentary rocks, such as shale, are also nearly stable chemically under surface conditions, for their minerals were formed and are stable there. But the calcite-cemented sandstone we described in Chapter 4 would be reduced to loose sand by solution of the cement.

But igneous and metamorphic rocks were formed under conditions vastly different from the temperatures and pressures of the earth's surface. Most of their minerals are unstable under these conditions, and soils formed on them contain many new minerals formed at the expense of the original ones. The weathering of granodiorite furnishes an example.

Weathering of Granodiorite

The minerals composing a granodiorite differ in such physical properties as compressibility and thermal expansion, as well as in chemical stability. Their grain boundaries are always differentially stressed as temperature and pressure change and thus become favored sites for attack by solutions; the rock tends to disaggregate into a mass of non-coherent mineral grains.

In general, minerals that crystallize early in magma consolidation (Chapter 9) are more unstable under surface conditions than those that crystallize later at lower temperatures. Deep cuts show biotite and amphibole to be altered at greater depths than potassium feldspar, showing their greater susceptibility. Though amphibole alters a little more readily than biotite, the products are so similar that the alteration of both may be illustrated by that of biotite. The chemical reactions list only the final products, not the complex ionic reactions of intermediate steps:

Quartz is generally stable chemically, though it becomes stained and mechanically broken. It does dissolve, but very slowly.

Advanced weathering of granodiorite in a humid temperate climate thus yields stained quartz grains embedded in newly formed clay, commonly stained yellow by iron rust ("limonite"). The alkalies K and Na and the alkaline earths Ca and Mg are largely removed in solution, though some Mg and K remain, held to the clay by feeble electrical forces. Leaching of the soluble ions goes on much faster in humid than in arid regions. As potassium is an important plant food, its retention is agriculturally important. The slow leaching of arid soils is partly responsible for the exceptional fertility of much newly irrigated land in the arid West of the United States, in the Murray Valley of Australia, and in the valleys of the Nile and Euphrates.

Thus the final soils derived from the sedimentary rock limestone and the plutonic rock granodiorite are both rich in clay; in the limestone soil it is merely concentrated by solution of the calcite, but in the granodiorite soil the clay is a new mineral. In both soils the clay is stable in temperate

$$2KMg_2Fe(OH)_2AlSi_3O_{10} + \tfrac{1}{2}O_2 + 10H_2CO_3 + nH_2O$$

biotite   oxygen carbonic acid   water

$$\rightarrow 2KHCO_3 + 4Mg(HCO_3)_2 + Fe_2O_3 \cdot H_2O + Al_2(OH)_2Si_4O_{10} \cdot nH_2O + SiO_2 + 5H_2O$$

potassium   magnesium   limonite   clay mineral   quartz   water
bicarbonate   bicarbonate   (iron rust)   or soluble
(soluble)   (soluble)   silica

Plagioclase, a solid solution of anorthite and albite, is next most susceptible of the grandiorite minerals, the more so the higher the anorthite content.

$$CaAl_2Si_2O_8 \cdot 2NaAlSi_3O_8 + 4H_2CO_3 + 2(nH_2O)$$

anorthite   albite   carbonic   water
acid

$$\rightarrow Ca(HCO_3)_2 + 2NaHCO_3 + 2Al_2(OH)_2Si_4O_{10} \cdot nH_2O$$

calcium   sodium   clay mineral
bicarbonate   bicarbonate
(soluble)   (soluble)

(The groups indicated as soluble do not actually form, but their components remain as ions.)

Potassium feldspar, next to quartz the most resistant mineral of the granodiorite, slowly weathers somewhat similarly:

$$2KAlSi_3O_8 + H_2CO_3 + nH_2O \rightarrow K_2CO_3 + Al_2(OH)_2Si_4O_{10} \cdot nH_2O + 2SiO_2$$

potassium   carbonic   water   potassium   clay mineral   soluble hydrated
feldspar   acid   carbonate   silica or quartz
(soluble)

or cold conditions. We shall see, however, that in monsoon climates even clay is unstable and decomposes still further.

Exfoliation by Chemical Weathering

Large-scale exfoliation, which forms granitic domes and similar features, has been attributed in large part to relief of pressure by unloading. Another kind of exfoliation, clearly a product of chemical weathering, produces exfoliation on a smaller scale.

Most rocks are cut by cracks or joints that divide the mass into smooth-sided blocks a few milli-

meters to a few meters across. Blocks of medium-grained rocks, such as dolerite and graywacke, and of coarser-grained granite and gneiss commonly encase less-altered cores in thin concentric layers of crumbly weathered rinds like onion peels (Fig. 10-12). Swelling due to formation of clay detaches each shell in turn from the less-weathered core of the joint block. Although the original block of fresh rock might be nearly cubical, weathering attacking from two block edges and at corners from three surfaces ultimately yields a nearly spherical core.

THE SOIL PROFILE

Weathering begins at the surface and proceeds downward, perhaps ultimately reaching depths of many meters in a systematic progression. Soil scientists have described the succession of zones as the **soil profile** (Fig. 10-13). Most mature soils — those that have been forming long enough to be fully developed and whose rate of formation is in rough equilibrium with the rate of removal of the surface layer — are divisible into three zones. We take, as an example, the soil profile of the western Sierra Nevada intercanyon uplands, developed on granodiorite seen in the nearby canyon walls. The surface layer — the **A-horizon** — is a red-brown sandy loam (a mixture of quartz sand, silt, clay, and partly decomposed plants). In size and shape, the quartz grains are like those of the fresh granodiorite. At a depth of about 30 cm, the clay content increases notably, marking the subsoil, or **B-horizon**. Evidently some of the clay of the A-horizon has been washed downward, filling pores in

Figure 10-12
Exfoliation in granodiorite. The spherical forms have developed by rounding of originally angular blocks. [Photo by Eliot Blackwelder, Stanford University.]

Figure 10-13
Soil profiles on Sierra Nevada granodiorite (left) and Kentucky limestone (right).

Sandy loam (A-horizon)

Sandy clay (B-horizon)

Transition zone to crumbly granodiorite (C-horizon)

Fresh granodiorite

Humus-rich clay (A-horizon)

Clay with limestone fragments (B-horizon)

Fresh limestone

the B-horizon. Sixty to 120 cm from the surface, the B-horizon becomes paler and sandier, with feldspar as well as quartz making up the sand. Some of the feldspar is partly altered, but some is virtually fresh. A little farther down, pearly yellow flakes of mica can be seen to have been derived from the original flashing black "books" of biotite in the fresh rock. The transitional interval between the zone of high clay content and the virtually unaltered granodiorite is the **C-horizon**. This whole soil is obviously *residual* from the alteration of the granodiorite.

Most soils have profiles of two or more horizons that differ from place to place because of differing parental rocks, microclimates, vegetative cover, and maturity—that is, the length of time the soil has been forming. The residual soil on limestone in Kentucky differs notably from the Sierra upland soil. The Kentucky climate, though also temperate, is more humid, and the rock parental to the soil is mineralogically simpler. The A-horizon is black humus-rich clay that passes downward into a mixture of lighter-colored clay and corroded fragments of limestone—a B-horizon. But there is no recognizable C-horizon. Clearly, the difference in parental rocks had an influence on the differences in these two soil profiles, but the effect of climate is dominant, for the A-horizons of both are chiefly clay; only the quartz in the Sierra soil betrays its source. Thus mature profiles of ancient buried soils may give strong clues to the climates under which they formed.

Soils on Unconsolidated Transported Material

Many soils—in fact the great majority of those of agricultural importance—have not developed directly from bedrock but from alluvial material that had been more or less weathered before it was transported. Such loose alluvium allows ready access to air and water; typical A- and B-horizons form readily in favorable climates. The soils of the Mississippi, Ganges, Nile, Euphrates, and thousands of lesser floodplains are examples. Those of the Great Valley of California are also typical. Some residual soil of the Sierra, described above, has been carried by streams and spread over thousands of square kilometers, admixed with fragments of unaltered granodiorite. Where undisturbed long enough these materials have de-

veloped a mature soil profile. The brown silt A-horizon, 60 to 90 cm deep, contains sporadic quartz pebbles. Faintly bounded clumps of closely spaced angular quartz grains in clay are relicts of thoroughly weathered granodiorite pebbles. The B-horizon, 30 to 60 cm thick, is a "hardpan" of massive clay whose original pores have been filled by particles washed down from the A-horizon. The C-horizon is relatively unaltered gravel.

Other soils on transported material differ greatly. In eastern Massachusetts, for example, a soil formed on a mixture of finely ground fresh rock, clay, and boulders transported by glaciers that once covered the area (Chapter 13), has only an A-horizon—a dark, humus-rich clayey silt, 7 to 10 cm thick, directly overlying unweathered material. This soil is obviously immature; it has been forming in a relatively cool region only since the glaciers retreated about 12,000 years ago (Chapter 13); the California gravels were deposited in a warm climate beyond the glacial limit, and their soil has been forming a much longer time.

Effect of Slope

Because slope affects the rate of runoff and hence both infiltration of water and removal of surface material, soils are generally much thinner on slopes than on hilltops. A study in Manitoba showed the A-horizon to be twice as thick on flats as on slopes of 10°.

Soil Color

The color of a soil is a clue to the climatic conditions under which it formed. Iron-containing minerals give soils red, brown, or yellow hues; organic matter turns them black or gray. Weathering conditions (climate) determine which color dominates.

Iron forms two kinds of ions: ferrous, with two, and ferric, with three positive charges. The loss of one electron changes a ferrous to a ferric ion. The presence of oxygen facilitates this loss; the change is called **oxidation**. FeO is a ferrous compound, Fe_2O_3 a ferric. When hydrous ferric oxide ("limonite") is formed during the decomposition of a ferrous silicate, such as biotite, it incorporates additional oxygen, either from the air or from solution in water. Finely divided limonite stains soil yellow. But if the soil is repeatedly wet and dried, as in a climate with a warm, dry season, some limonite may lose its water, changing to hematite; the soil then is red. At intermediate stages, or if manganese oxides are abundant, the soil is brown.

Reduction of iron from the ferric to the ferrous state goes on almost equally easily. Pale greenish, dark gray, or black soils generally form where the iron has been reduced by reactions with plant carbon or by sulfur bacteria. This commonly happens in water-soaked soils from which oxygen is excluded. A black soil may thus indicate swampy conditions under which the iron has been reduced and largely combined with S, forming black hydrous iron sulfide. Such soil is rich in humus because oxygen is unavailable. Soils rich in organic matter are generally dark, but many well-drained forest soils of the temperate zone contain so much humus that acid formed from it dissolves and washes nearly all the iron out of the A-horizon, making it very light gray.

WEATHERING IN DIFFERENT CLIMATES

We saw that limestone and granodiorite both yield clay-rich soils in moist temperate climates. In either temperate or tropical humid climates, rocks rich in ferromagnesian minerals and carbonates weather more rapidly than quartz-rich rocks, as these minerals are far more susceptible to decomposition. Thus basalt and limestone weather rapidly, granodiorite slowly; and quartzite is almost immune.

In the driest deserts, such as those of central Australia or southwest Africa, the chief weathering process seems to be slight chemical attack. Soils in such areas contain little organic matter and few bacteria, so that CO_2 is less abundant than in humid soils. Obvious chemical changes are few, but the surface layer does disintegrate to yield coarse, only partly weathered soil. The thin A-horizon is slightly leached of carbonates and iron and thus is paler than the clay-enriched B-horizon.

In semiarid regions such as the High Plains of the United States, the soils are rich in calcium carbonate and clay minerals. Here crusts, veinlets, and nodules of calcite precipitate in the pores of the B-horizon, locally cementing it into a nodular rock, **caliche**. The carbonate accumulates be-

cause rainfall is light during most of the year, and water sinks only a little way into the soil before it is held by capillarity and slowly evaporates, leaving its dissolved load to precipitate. In more humid climates the rainfall is abundant enough to carry the dissolved load out of the soil to permanent streams.

In more arid regions, in playas, or where irrigation has been heavy and subsoil drainage is poor, the soil water rises almost to the surface (Chapter 14) and precipitates salts even more soluble than calcium carbonate, forming "alkali-soils." Such soils, encrusted with sodium carbonate and sodium sulfate, are poisonous to most plants and are useless to agriculture unless they can be drained and the soluble salts washed out.

In Arctic regions and in alpine heights elsewhere, frost action is dominant, but in subarctic Finland, Canada, and Siberia chemical weathering produces very siliceous clay soils. The acid formed from the tundra vegetation suffices, even at the low prevailing temperatures, to leach most calcium, magnesium, and iron from the soils.

Dramatic testimony to the importance of climate in weathering is the contrast between the valleys formed in humid limestone belts, such as the Shenandoah Valley in Virginia, and the bold limestone ridges rising above the desert of arid Arizona. Only quartzite forms similarly prominent ridges in the humid Appalachians.

Laterite

The soils of the rain forests of the Congo and Amazon basins have not been adequately studied, but appear similar to those of moist temperate regions. Most of the open savanna belts near the rain forests, however, have tough to extremely hard yellow and red-brown soils whose upper layers dry to become brick-like and are locally used for building, hence the name, **laterite** (from *later*, Latin for "brick"). Parts of India, Southeast Asia, Nigeria, Brazil, and the Caribbean region have such soils.

Because laterites differ so from more familiar soils and also illustrate one way a chemical element may become so concentrated as to form a valuable ore, we give additional details.

Laterites differ widely in composition but most contain aluminum hydroxides and iron hydroxides

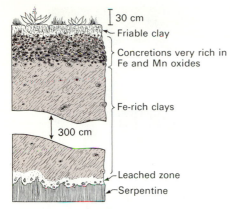

30 cm
Friable clay
Concretions very rich in
Fe and Mn oxides

Fe-rich clays

300 cm

Leached zone
Serpentine

Figure 10-14
Profile of Cuban laterite. The concentration of iron is
sufficient to make the concretionary zone a marginal
iron ore. [Data from H. H. Bennett and R. V. Allison,
1928.]

and oxides, either separately or together, along
with a little residual quartz. A rare variety, **bauxite**,
almost pure hydrous aluminum oxide (Al_2O_3
$\cdot nH_2O$), is the principal ore of aluminum. Another, diagrammed in Figure 10-14, is derived
from serpentine, a rock that contains virtually no
aluminum but much magnesium and iron. The
laterite has concentrated enough hydrous iron
oxides to be a rather poor ore of iron.

In most laterites, nealy all silica of original
silicates has been leached away by rainwater,
along with the easily soluble alkalies Na and K
and the acid-soluble Ca and Mg. The simultaneous leaching of all these is puzzling. Silica is only
slightly soluble, but it does dissolve fairly readily
in alkaline solutions—those richer in OH^- (hy-
droxyl) ions than in H^+ ions. Calcium and magnesium are not very soluble in such solutions, but
dissolve readily in acid solutions (H^+ ions abundant). Yet all three have been leached from the
laterites.

Most laterites form in regions with marked wet
and dry seasons—monsoon climates. During the
dry season the organic acids are probably so
completely oxidized to carbon dioxide, which
escapes into the air, that the dry soil contains no
acid-producing materials but some Na and other
ions capable of forming an alkaline solution. At
the onset of the first rains, silica might be carried
off in the temporarily alkaline solutions before new

vegetation and its decay renew the supply of
carbonic acid.

Some complex hydroxides of Fe, Mn, Mg, and
Al in the laterites are formed indirectly from compounds that are at first easily soluble but gradually
stabilize and become highly insoluble before the
onset of the next wet season.

Residual laterites (many are transported, not
residual) are characterized by a pale zone of
leaching just above the parent rock, and a dark-
brown *concretionary zone* at or near the surface
(Fig. 10-14). Each is usually only a few meters
thick but in places may thicken to more than a
hundred meters—a fact difficult to understand in
view of the lack of porosity of the rock. The concretionary zone is a concrete-like mass composed
mainly of either dark brown "limonite" or of
many limonite nodules (**concretions**) of pea or
marble size tighly cemented into a solid mass.

Leaching is so complete in many laterites that
some elements essential to plant life are entirely
removed, and others, such as P and Fe, are precipitated as anhydrous iron oxides, phosphates,
and other compounds so nearly insoluble as to be
practically unavailable. The lateritic soil of the
Hawaiian Islands contains abundant iron, but
pineapple plants must be fed soluble salts of both
Fe and P in order to thrive there.

The fate of the silica leached out during later-
ization is unknown. Extensive siliceous crusts
have been reported from Angola, just southwest
of the Congo Basin, and from elsewhere in tropical
Africa, but laterite is not known nearby. Possibly
nodules of chert form beneath the laterites. It
appears that silica is more soluble in tropical than
in temperate climates, for the silica content of the
streams in tropical British Guiana is about twice
the world average of streams draining similar
rocks.

RATES OF WEATHERING

Weathering rates vary greatly with climate and
rock type. Igneous rocks polished by glaciers
10,000 years or more ago still glisten (Fig. 10-15)
in the cool temperate climate of northern New
York; even the rock flour finely ground by the
same glaciers is little decomposed. But roughly
contemporaneous glacial deposits in Iowa have
fairly well-developed profiles.

Figure 10-15
Glacial polish on igneous rock, Adirondacks, New York, unscarred by 10,000 years of weathering. [Photo by V. C. Kelley, University of New Mexico.]

Chemical weathering proceeds most rapidly in the moist tropics. The A-horizon formed on ash from the Krakatau eruption of 1883 in Indonesia was, in only 60 years, as much as 5 percent poorer in silica and 2 percent richer in alumina than the C-horizon. These differences show extremely rapid decomposition of the glassy pumice—the parent rock; a crystalline rock would undoubtedly weather much more slowly. But even in the tropics, after weathering has produced a deep blanket of decomposed material, the product reaches virtual equilibrium with its environment. The arched roofs of Angkor Wat, Cambodia, built of laterite, stand nearly intact after seven centuries of neglect. The jungle crowds close, and plants spring from every cranny, but the delicately sculptured walls are only slightly marred; the laterite, already nearly in equilibrium under the climate, has been slow to change further. In contrast, a well-developed soil profile has formed on burial mounds left by the Huns in the Volga valley eight centuries ago. The stream sediments of the Volga have weathered far more, even in a cool climate, than the Cambodian laterite, itself a product of weathering.

Many other data appear conflicting: The nearly negligible weathering of the glacial debris of New York and New England took place under a climate not unlike that of western Europe. But in Nor-

mandy, soil with an A-horizon 9 cm thick and a B-horizon 31 cm thick has formed on a refuse pile of oyster shells at a castle abandoned in 1066. Perhaps becuase of the porosity and solubility of the shells, the weathering in Normandy has been many times as great in 900 years as in ten times that span in New England. Soil reclaimed from the sea in the Netherlands originally contained 10 percent calcium carbonate. In 300 years the surface layer has been completely leached of Ca. Measurements on ancient structures in Northumberland, England, suggest that limestone has been dissolved at the rate of about 2.5 cm in 300 years, but the heights of nearby pedestal rocks, which support glacial boulders, suggest a much slower rate.

Biotite scattered throughout sand in Wisconsin has been converted to a hydrous clay mineral (vermiculite) simply by growing four successive crops of wheat. After only 50 years, abandoned plowlands in eastern North Carolina developed new A- and B-horizons whose combined thickness is 12 cm.

Old gravestones also give interesting data. In less than 80 years, the inscription on marble in memory of Joseph Black of Edinburgh, the discoverer of carbon dioxide, was rendered illegible, chiefly by the action of the gas he discovered. Acid solutions gave access to water, which, on freezing, disrupted the rock. Faced limestone in Edinburgh has lost an average of about 1 cm per century; in small nearby towns with less coal smoke; the loss is far less. Slate, little attacked by acid solutions, has been barely roughened in the same time.

Weathering is slowest in a hot, dry climate. Colossal statues carved in granitic rock from near Aswan, Upper Egypt, were set up at various dates between 2850 and 313 B.C. at Luxor, in Middle Egypt, where the average rainfall is 1 to 4 mm/yr. Similar blocks were used to face some of the pyramids near Cairo, in Lower Egypt. These were studied in 1916 by D. C. Barton, an American geologist. He estimated that the average rate of exfoliation of this granite in Middle Egypt was about half that in the slightly moister climate of Cairo, where it was about 1 or 2 mm per thousand years. Yet the blocks of porous limestone of the pyramids have weathered many times as fast, as much as 1 cm in 50 years.

Two obelisks of Aswan granite, each bearing many deep-cut hieroglyphs and each now called "Cleopatra's Needle" (Fig. 10-16), stood for about 3500 years in Egypt with only slight weathering. One, removed to London, has weathered notably but not disastrously; the other, set up in Central Park, New York, about 1880, has been so attacked by frost, water, and air rich in carbon dioxide, that, despite application of shellac-like preservatives in the past fifty years, much of the pictured story is completely illegible. Frost wedging alone has produced more weathering in ninety New York winters than all processes of weathering in twenty times ninety years in Egypt.

SUMMARY

Soil profiles develop through the rotting of rock. The minerals of even the sedimentary rocks, formed under surface temperatures and pressures, are vulnerable to changes brought about by differing chemical environments; those of igneous and metamorphic rocks are nearly all outside their

Figure 10-16
Cleopatra's Needle, Egypt (left), and in Central Park, New York City (right). [Photo courtesy of The Metropolitan Museum of Art, New York.]

fields of stability at the earth's surface. Accordingly, most of the earth's surface is mantled by a variable thickness of decayed and broken rock.

Climate is the dominant influence in weathering, although the composition of the parental rocks is also important. Although the variety of soils is great, most fall into one of three major groups:

a) Soils of humid regions, in which calcium, magnesium and sodium are leached and in which aluminous and iron-bearing minerals are washed downward into the subsoil. These are acid soils, and silicon is concentrated near the surface. In the evergreen forests of high latitudes, the A-horizon below the plant litter can be almost white.

b) Soils of the savannah tropics, in which iron- and aluminum-bearing minerals are concentrated near the surface and other elements are selectively removed. These are lateritic soils.

c) Soils of arid and semiarid regions, where the supply of water is not sufficient to leach the soil and carry away the solutions to streams. Instead, the soil system is essentially closed; all the solutes remain in it and are merely redistributed and precipitated to form caliche, alkali-soils, hardpan, and subsurface accumulations of compounds of calcium, sodium, and even potassium.

While the soil profile is immature the source rock is important in governing the soil composition, but given a long time it becomes less so. In mature soils the effects of differing source rocks become less and less significant, so that ultimately soils of a mature region derived from such different rocks as slate, granite, and basalt become closely similar.

Topographic factors also influence soil formation. Drainage, vegetation, humus accumulation, soil bacteria and doubtless other factors of microclimate all depend to some extent on topographic situation. So, too, does the rate at which the surface layer washes away or is buried under washed-in material.

The rate of soil formation varies under different circumstances from decimeters per century to centimeters per thousand years. The mean rate for the earth as a whole has been estimated at 1 cm per century. Since it normally takes thousands of years for a soil to adjust to any considerable climatic change, the soil character may long record the conditions under which it formed; in this way the prehistoric climates of many regions can be shown to have differed greatly from those of the present.

The range in surface temperature on the earth is at least 45°C, with marked latitudinal differences that greatly modify the weathering from place to place. The geologic record indicates that the earth has normally been without ice caps comparable to those of the present, though there have been times of far larger ones. Generally, though, the temperature range was smaller than now, with correspondingly more uniform rates of weathering and soil formation the world over.

Questions

1. In Oregon, a laterite developed on basalt is now covered by a second basalt flow, which has a soil composed chiefly of clay minerals. Explain.

2. Western Nevada is semiarid, with dry summers; eastern Iowa is moist, with considerable summer rainfall. Assuming that the soils are derived from similar parental rocks, how should the characteristic soil profiles of the two areas differ, if at all?

3. Why does calcium carbonate accumulate in the A-horizon of some soils formed on basalt in Nevada, whereas it is practically absent from this horizon in soils developed on limestone, predominantly composed of calcium carbonate, in Kentucky?

4. Few of the soils of extreme western Texas are mature, whereas most of those of eastern Texas are mature. What hypotheses occur to you as possible explanations of this fact?

5. The Hagerstown Valley of Western Maryland is underlain by limestone; the Piedmont, to the east, by gneiss and schist. How would you expect the soils of the two areas to differ, if at all?

6. Name the chief minerals in granite, and tell what happens to each of them when granite weathers in a moist temperate climate.

7. Volcanic tuff containing abundant fragments of pumice is successfully used as a building stone in southern Arizona, but not in Alaska, where similar volcanic rocks are widespread. Why?

8. In the great Chicago fire, pillars of granite in burned-out buildings were greatly spalled and cracked, but pillars cut from limestone withstood the flames with much less damage. Can you suggest why?

Suggested Readings

Bridges, E. M., *World Soils*. Cambridge University Press, 1970.

Carroll, Dorothy, "Rainwater as a Chemical Agent of Geological Processes." U.S. Geol. Survey Water Supply Paper 1135G, 1962.

Carroll, Dorothy, *Rock Weathering*. New York: Plenum Press, 1970.

Coughnan, F. G., *Chemical Weathering of the Silicate Minerals*. New York: Elsevier, 1969.

Geikie, Archibald, "Rock Weathering as Illustrated in Edinburgh Churchyards," *Proceedings of the Royal Society of Edinburgh*, V. 10, 1880, pp. 518–532.

Goldich, S. S., "A Study in Rock Weathering." *Jour. Geology*, v. 46, pp. 17-58, 1938.

Hunt, C. B., *The Geology of Soils*. San Francisco: W. H. Freeman and Company, 1972.

Keller, W. D., *The Principles of Chemical Weathering*. Columbia, Missouri: Lucas Brothers, 1955.

Persons, B. S., *Laterite—Genesis, Location, Use*. New York: Plenum Press, 1970.

Reiche, Parry, *A Survey of Weathering Processes and Products*, (rev. ed.). Albuquerque: University of New Mexico, 1950.

Scientific American Offprints

821. Charles E. Kellogg, "Soil" (July 1950).
823. Gilbert N. Plass, "Carbon Dioxide and Climate" (July 1959).

835. Ernst J. Öpik, "Climate and the Changing Sun" (June 1958).

841. Victor P. Starr, "The General Circulation of the Atmosphere" (December 1956).

847. Joanne Starr Malkus, "The Origin of Hurricanes" (August 1957).

849. Walter Orr Roberts, "Sun Clouds and Rain Clouds" (April 1957).

870. Mary McNeil, "Lateritic Soils" (November 1964).

CHAPTER 11

Downslope Movement of Soil and Rock

Every object resting on a slope is acted on by a force tending to pull it downhill. From the principle of the parallelogram of forces, we know that a kilogram weight resting on a 30° slope, this force is 0.5 kg (Fig. 11-1). Any factor that overcomes an object's cohesion with its base causes downslope motion. Such factors may range from earthquake shocks to the relatively mild forces exerted by the freezing of water, the expansion of wet clay, the buoyancy of water in a saturated mass of soil and rock, a heavy snowfall, the tread of a sheep, or even the pelting of raindrops and hail. Though the processes governed by most of these forces act relatively slowly, geologic time is long, and they act on every slope on the landscape; unquestionably these processes are among the mose effective of erosional agents. We begin with the seemingly weakest and pass ultimately to the more dramatic.

RAINWASH

Rainwash includes the pelting of raindrops and grades indistinguishably into rill formation and stream flow, but think first of raindrops impacting the ground. Their effect is most marked on slightly permeable materials, whereas creep is more effective on permeable ground. Large rain drops may attain an air speed of 25 km/hr; at higher air speeds the drops break into smaller ones that fall more slowly. But in violent storms, downdrafts in the turbulent air may hurl the drops against the ground at speeds of 100 km/hr; they are not negligible missiles, for they knock loose grains about at considerable speed. Hailstones are even more effective. Of course their effectiveness depends greatly on the vegetational covering and the grain size of their targets.

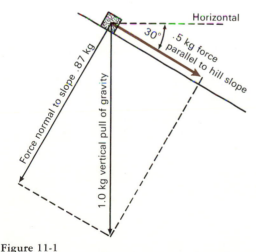

Figure 11-1
The resolution of the forces of gravity, directed vertically downward, into forces parallel and normal to the slope on which the object rests.

SLOW DOWNSLOPE MOVEMENTS

Creep

The most widespread of downslope movements is **creep**, the slow movement of soil and weakly consolidated particles downhill. Even on grass-covered and forested slopes in most regions, the

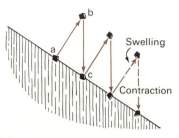

Figure 11-2
One mechanism of creep. The arrows show the gradual downslope movement of a particle of swelling clay, alternately wetting and drying.

Figure 11-3
Paths formed by grazing animals on a steep slope in central Oregon. After each rain, hooves push the softened soil a little farther downslope. [Photo by A. C. Waters.]

surface material moves downhill at rates of as much as a millimeter a year; on barren slopes of fine-grained material, the rate may be more than 100 times as great. Several mechanisms are involved. The lattices of most clay minerals expand on wetting. A similar result comes from the freezing of water, increasing 9 percent in volume. By either process, as shown in Figure 11-2, a particle of soil is pushed up at right angles to the surface to some point such as *b*. When the clay dries or thaws, the soil contracts, but not along the same path as its expansion; it tends to slump down toward *c*. If the soil is incoherent, the particle may slide even farther.

Other processes that contribute to creep include wedging by plant roots, the moving of soil by earthworms and rodents, and the tramping of muddy soil downslope by grazing animals (Fig. 11-3). Any random movement of loose material gives gravity a chance to displace it downslope.

Creep is of course most active in the surface layer, and becomes less so at depth. In a road cut on an Idaho hillside, weathered and softened boulders, which must originally have been nearly round, like the less weathered ones beneath, are stretched out parallel to the slope; the surficial ones, into thin ribbons of clay, the deeper ones less so (Fig. 11-4). Even gently slopes are enough to produce detectable creep (Fig. 11-5). The effect is locally very striking (Fig. 11-6). At some points

Figure 11-4
Decayed boulders on an Idaho hillside, stretched into spindles by creep. [Photo by S. R. Capps, U. S. Geological Survey.]

Figure 11-5
Alternating sandstone and shale beds in the Haymond Formation (Pennsylvanian) near Marathon, Texas, a typical flysch deposit (Chapter 17). The sandstones show graded bedding. Note creep to the right at the right side of the picture and to the left at left, despite the very low hillslopes. [Photo by Earle F. McBride, University of Texas.]

Figure 11-6
Bending of thin vertical strata by creep on a steep hillside in Washington County, Maryland. [Photo by George W. Stose, U. S. Geological Survey.]

in Arkansas, field geologists working without the advantages of exposures in deep road cuts, were so deceived by creep that they interpreted every hillslope as parallel to bedding; when deep cuts became available, the strata turned out to be almost vertical across several kilometers of country previously thought to be folded into shallow anticlines and synclines. Very commonly creep tilts trees, telephone poles, and fences (Fig. 11-7); as it is nearly ubiquitous in some degree, its aggregate effect in sculpturing the landscape and reducing it in altitude is very great, though the rate of creep varies widely with climate and slope.

Measurements in the semiarid West of the United States and in Arctic Greenland, two very different climates, yielded creep rates of about 6 cm/yr on slopes of about 25°; on very steep slopes of 39° in both Sweden and Colorado, the rate was

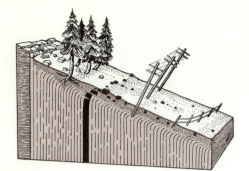

Figure 11-7
Common effects of creep. [After C. F. S. Sharpe, *Landslides and Related Phenomena,* Columbia Univ. Press, 1938.]

Figure 11-8
Solifluction lobe on Victoria Island, Canada. The row of stakes, originally straight, was used to measure rates of movement of the lobe. [Photo by A. L. Washburn, Univ. of Washington.]

mate mixing of A- and B-horizons farther up slope, is sharply separated from the underlying fresh rock by a smoothly scraped surface. The movement of the surface layer is recorded by tilted fence posts and telephone poles, but the lack of scars or irregularities on the surface indicates general movements, not concentrated into land-slide tongues.

Solifluction

In arctic and subarctic regions, and above timber-line in temperate zones, downslope movements called **solifluction** (literally: "soil flow") are inter-mediate between creep and debris flows. Frost and other weathering agents produce abundant fine rock fragments that spread downslope in arcuate lobes when saturated with water that sporadically freezes and melts. Most movement takes place immediately after the spring thaw. Rows of stakes driven into a lobe on Victoria Island, northwest of Hudson Bay (Fig. 11-8) showed movement of a little more than 4 cm in about a month.

Rock Glaciers

In many mountainous areas, even including Ruwenzori on the equator, large lobate piles of angular rock fragments are festooned in or at the mouths of open valleys (Fig. 11-9). These are **rock glaciers**, so called because they resemble the debris masking the ends of most glaciers. Indeed, excavations expose cores of ice at shallow depths. The external ridges are strewn with talus piles, suggesting that the fragments are released individ-ually from the ice core.

Gros Ventre Debris Flow

The Gros Ventre River is a tributary of the Snake River south of Jackson Hole, Wyoming. In the spring of 1908 rock debris began to flow slowly down the Gros Ventre Mountains on slopes of 10° to 20°. These slopes nearly parallel the bed-ding of underlying Jurassic shale with thin inter-beds of sandstone, all thoroughly soaked by a series of heavy rains and melting snow. The sliding was at first imperceptible to the eye, but

about 10 cm/yr for the surficial layer. These are very high rates, far higher than on the grassy slopes of northern England, where even a 33° slope crept only 0.18 mm/yr and another 22° to 30° slope only 0.25 mm/yr. But even the slowest of these rates is not negligible in geologic time.

Creep passes imperceptibly into downslope sliding of definitely bounded masses. On many slopes in the California Coast Ranges, a homo-geneous surface layer, doubtless formed by inti-

Figure 11-9
Rock glacier on Cerro del Plomo, Chilean Andes. Note the smooth talus at the left side and front of the glacier. [Photo by Kenneth Segerstrom, U. S. Geological Survey.]

became obvious within a few weeks. Telephone poles tilted slowly downhill, snapping the wires. A wagon road paralleling the river was twisted and broken beyond repair; eventually it was so churned up that even traces of it were hard to find. The slide moved faster in the rainy season and slowed down in the dry months; in the wet spring of 1909 it overwhelmed the river by a dam several tens of meters high that the river was unable to overrun for nearly two years (Fig. 11-10).

FASTER DOWNSLOPE MOVEMENTS

Mudflows

Long-continued heavy rainstorms may so saturate surface material that instead of individual particles being driven downslope, the whole of a considerable thickness becomes a mass of mud and moves like a lava flow. In 1965 continued torrential rain

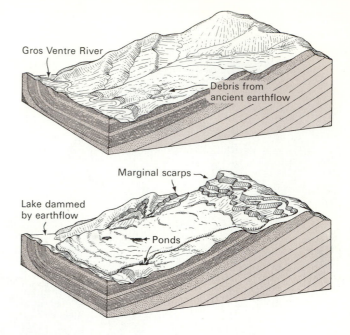

Figure 11-10
The south slope of the Gros Ventre River valley before (top) and after (bottom) the 1909 debris flow. [After Eliot Blackwelder, Stanford University, 1912.]

that fell upon the small coal-mining town of Aberfan, in Wales, so saturated a huge dump of waste rock from the mines that it became liquefied, flowed downvalley, and buried a school, killing more than 100 children.

In northwestern Europe and eastern Canada, many valleys are floored by unconsolidated clays and finely ground rock flour deposited in lakes dammed by the Pleistocene glaciers described in Chapter 13. Some of these clay plains are very extensive. During spring thaws the weakly consolidated sediments become so saturated with water that the rock flour acts as a lubricant; masses several meters deep and covering several square kilometers may break loose and roll rapidly downvalley as a mudflow. The front of one such flow moved at 10 km/hour. In 1893 such a mudflow traveled down a Norwegian valley so fast that 111 persons were overrun by it; others escaped only because their log houses floated atop the flow.

Arid regions show still another kind of mudflow. Desert cloudbursts falling on nearly barren slopes may lubricate rock waste of all sizes up to and including boulders several meters in diameter, so that sodden masses move downcanyon with nearly vertical fronts several meters high, sweeping all loose material before them, only to slow and stop where the channel widens and water can escape.

Underwater Mudflows

Mudflows can be started in several ways. At Zug, in Switzerland, a retaining wall that was built along the lakeshore proved to be an effective dam: unexpectedly, ground water began to rise into previously dry cellars behind it, buoying up the sand and clay on which the city was built. In the spring of 1887 the ground began to move, and a small section of the retaining wall and three houses sank beneath the lake. Three hours later a whole section of the city—several streets and houses—suddenly settled beneath the lake. The average drop was about 8 m, and some buildings moved 10 to 20 m toward the lake.

The city was built on a small delta of incoherent silt and fine sand that mobilized when saturated and gave way under the weight of the buildings. The flow excavated a trench 60 m wide and as much as 6 m deep in the sand of the lake bottom, and deposited the material as a debris tongue at the end of the trench. The end of the tongue was 40 to 45 m lower than the land at the head; the slope was less than 3°. About half the deposit was derived from the trench excavated in the lake bottom (Fig. 11-11).

Many other slides have originated by the same mechanism as the one at Zug. Water impounded

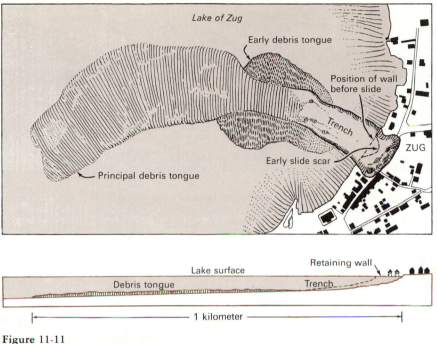

Lake of Zug

Early debris tongue

Position of wall
before slide

Trench

ZUG

Early slide scar

Principal debris tongue

Lake surface

Retaining wall

Debris tongue

Trench

1 kilometer

Figure 11-11
Map and cross section of the underwater mudflow at Zug, Switzerland.
[After Albert Heim, 1888.]

behind dams can loosen incoherent rocks of reservoir walls, so that they give way and slide into the reservoir. This has happened on a large scale at Roosevelt Lake on the Columbia River above Grand Coulee Dam; the reservoir on the Eel River in northern California has lost much of its capacity because of such slides.

Slides Triggered by Earthquakes

Similarly saturated sediments may remain stable for many years, only to be mobilized by the shaking of an earthquake. During the Alaska earthquake of 1964, the docks and part of the waterfronts of both Valdez and Seward (Fig. 11-12) slid out into the deep fjords on which the towns were situated. Loss of life was considerable at Valdez.

The 1964 Alaska earthquake also caused sliding at Turnagain Heights, Anchorage. Anchorage is built on glacial outwash gravels a few tens of meters thick toward the north but thinning to a very few meters toward the south. The gravels rest

on a thick, weakly consolidated, flat-lying deposit of water-saturated clay, the Bootlegger Cove Clay (Tertiary). Turnagain Heights, in the southwestern part of the city, commands a magnificent view of Cook Inlet and the snowcapped peaks of the Alaska Peninsula beyond. It forms a flat a few tens of meters above the sea. The darkness of Arctic evening had just settled over the town when the earthquake struck. The heights began to heave, shake, and slide toward the sea; power lines snapped. In total darkness and falling snow, scores of houses and their terrified residents rode the slide down the bluff. An area of many thousand square meters, gliding on a mass of clay rendered semiliquid by the shaking, broke into crazily tilted blocks and moved on a seaward-flattening gradient toward the sea (Fig. 11-13). Houses were canted at high angles, many were broken into fragments; porches drifted away from entrances; chasms several meters deep and wide opened and closed. With water from the ruptured mains adding to the liquefaction of the clay, nearly all of this pleasant residential area was destroyed. Fortunately, casualties were few, but property destruc-

Figure 11-12
Waterfront at Seward, Alaska, before (left) and after (right) the Alaska Earthquake of 1964. The small boat harbor, railroad yards, the large docks, and other waterfront facilities were swept away by underwater landslides. Windrow-like heaps of overturned railroad cars and other debris were later rolled inland by a tsunami. [Photos by U. S. Geological Survey.]

Figure 11-13
Tilted blocks in the Turnagain Heights landslide, Anchorage, Alaska. [Photo by U. S. Geological Survey.]

tion was virtually complete for several hundred meters inward from the bluff (Fig. 11-13). Surveys and drilling carried on after the slide indicated the mechanism of motion illustrated in Figure 11-14.

The hazards of building on Turnagain Heights had been specifically pointed out in a geologic report more than a decade earlier, but, as commonly happens, the potential hazard, though recognized, was discounted because of the superb scenic values. Comparable hazards are commonly ignored in many earthquake-threatened areas of California, Yugoslavia, Italy, and other places.

RAPID SLIDES, FLOWS, AND FALLS

Talus

The most abundant, and in aggregate most voluminous rock falls are the countless small fragments, in the millimeter to meter range, that, loosened by frost, rain, or wind, drop from cliffs or steep canyon walls, bound downslope, and accumulate in **talus piles** (Figs. 11-15, 11-16, 11-17). The slope angle of talus is nearly uniform as it grows; in dry climates nearly 40°, but in more humid ones only about 30°. This angle is called the **angle of repose** because it is the steepest slope the talus material can maintain without rolling. Steep-walled valleys

in arid regions are widened chiefly by the fall and continued downward rolling and sliding of large and small talus fragments.

RATE OF TALUS FORMATION

Climbers on steep slopes of closely jointed rock in lofty, snow-clad mountains know well the danger of rock fragments tumbling from the peak above and bounding, whirring, and crashing down long talus slopes. Where frost works on well-jointed rocks (Figure 11-15), talus accumulates rapidly. But the mere presence of talus is no proof of concurrent rapid erosion of the cliffs above. In southern Arizona, great talus piles below many granite cliffs are composed of huge, thoroughly weathered blocks that could not, in their present state, have withstood the impact of their fall. They must have weathered since falling—a slow process in that arid land. They are one of many indications that the Arizona climate was formerly wetter and cooler than it is today.

Small Rapid Slides

Most rapid gravity movements involve only small volumes of soil or rock, but their aggregate effects are large. In humid temperate regions small rapid

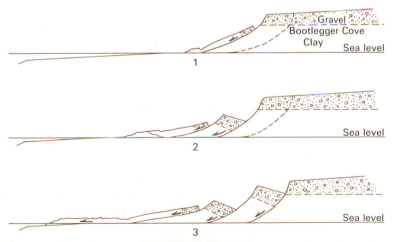

Figure 11-14
The pattern of movement of the Turnagain Heights landslide, as determined from surveys and drilling data. [After U. S. Geological Survey Circular 491, 1964.]

236

Figure 11-15
Talus slopes masking most of the walls of Salmon River Canyon at Riggins, Idaho. [Photo by Warren Hamilton, U. S. Geological Survey.]

slides abound in soil or weak sedimentary rocks after unusually heavy rains. The depression or scar left at the head may be several meters wide. At the base of the slide the material piles up in crumpled and disordered masses as seen in Figure 11-18, an unusually large but typically shaped example. Such avalanching of water-soaked soil down the walls of Hawaiian canyons was estimated by C. K. Wentworth to be lowering the general land surface at the rate of 6 cm/century.

Large Rockfalls; Alpine Examples

A tremendous rockfall took place at the little Swiss village of Elm in 1881. A steep crag, 600 m high, on a ridge, had been undercut halfway up by a slate quarry. In about 18 months a curving fissure slowly grew across the ridge about 350 m above the quarry, roughly normal to the bedding and

foliation of the slate. In late summer, runoff from heavy rains poured into the fissure, saturating the shattered rocks. Late one September afternoon, two small earth slides started just above and on either side of the quarry; suddenly the whole mass outlined by the fracture slid down, filling the quarry and shooting forward into the valley in free fall (Fig. 11-19). On striking the valley floor the churning mass ran up the opposite slope to a height of 100 m, turned and shot down the valley in a debris stream that swept all before it, killing 115 people. Ten million cubic meters of rock fell an average of about 450 m and spread as rubble over the 3 km² to depths of 10 to 20 m.

All observers agreed that the slide was in free fall below the quarry floor. The debris at the front of the slide traveled 5.5 km in less than a minute; the average velocity was 155 km/hour. To attain such speeds the mass had to be in free fall for much of its descent, and was thus in part buoyed up by the air compressed beneath it. That such

Figure 11-16
Sheep Rock, John Day State Park, Oregon, showing sheet talus at the upper left, ravine trains in center and right. Nearly all the fragments are from a basalt flow capping the peak. [Photo by Oregon State Highway Commission.]

Figure 11-17
Talus slope at the base of a basalt butte, Grand Coulee, Washington. [Photo by Washington Department of Conservation and Development.]

Figure 11-18
Slide near Orinda, California. Four-lane highway and large dump trucks give the scale. [Photo by Bill Young; courtesy of the *San Francisco Chronicle*.]

Figure 11-19
Cross section of the rock fall and slide at Elm, Switzerland, showing the original position of the slide block. [After Albert Heim, 1882.]

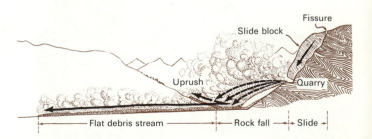

compression can be very great has been shown by many snow avalanches in free fall, for the blasts of trapped air frequently knock down well-built masonry structures. The supporting effect of such an air cushion must be important in the distant travel of large rockfalls such as Elm. Air cushions are not essential, though, as great rockfalls are obvious on the moon, which has no atmosphere.

THE VAIONT RESERVOIR DISASTER

The worst dam disaster in history took place on October 9, 1963, at the Vaiont Reservoir in the Italian Alps, when a mass of rock, more than 240 million m³, slid down the valley wall from heights as great as 600 m and filled the whole reservoir from the dam to a point 2 km upstream to depths as great as 175 m above the flow line (Fig. 11-20).

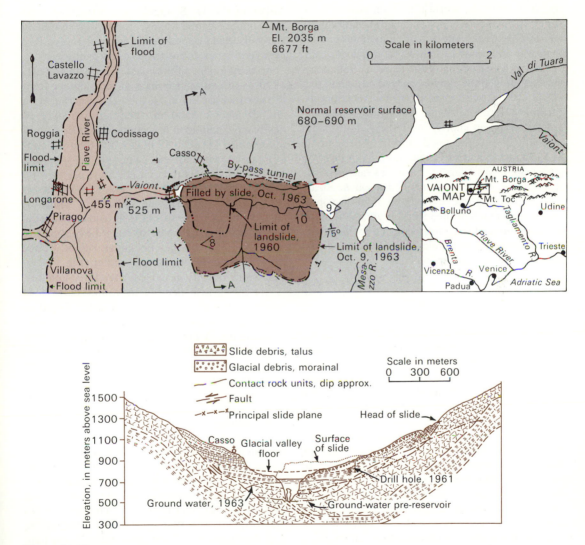

Figure 11-20
(A) Map of Vaiont Dam area and Piave Valley, showing area of slide and flood caused by it. (B) North-south cross section of Vaiont slide and reservoir canyon, showing slide surface and water levels in 1960 (*pre-Reservoir*) and 1963, when the slide took place. Position of section marked on A. [After George A. Kiersch, *Trans. Amer. Soc. Civil Engineers* v. 34, 1964.]

The shock of the slide was recorded on seismographs as far away as Brussels. Rocks and water were hurled up the opposite wall of the valley to heights of 260 m, and the water displaced from the reservoir went over the dam as a deluge 100 m high. The water was still 70 m high when it burst upon the heavily populated valley of the Piave River 1.5 km below the dam. Here it surged both up and downstream over a belt more than a kilometer wide and many kilometers long, causing the loss of nearly 3000 lives. The dam itself, the second highest in the world (265.5 m) proved to be superbly engineered and remained structurally intact after withstanding a load many times greater than was anticipated when it was designed.

During construction of the dam, in 1960, excavations showed the canyon walls to be under great elastic strain, with tendencies for rock bursts. George Kiersch, an American geologist who examined the area shortly after the slide, thought this was due to the rapid cutting of the canyon in Pleistocene and Holocene time. Although the catastrophe was triggered by man's raising of the water table, the condition of high elastic strain revealed during the excavation doubtless prevailed over much of the reservoir area, and in the normal course of stream erosion and spalling from the valley walls the slide would have occurred anyway, but not catastrophically.

Earthquake-triggered Rockfalls

Another phenomenon triggered by the Alaska earthquake of 1964 was the rockfall on Sherman glacier. The glacier is in the Chugach Range, near the head of the Copper River delta and close to the epicenter of the quake. The shocks broke loose a gigantic mass of rock, estimated at more than 23 million m^3, which crashed down one wall of the glacier and raced 150 meters up the other, finally settling down to cover more than 6 km^2 of the glacier to depths averaging 1 to 3 m but locally up to 30 m. Much of the slide did not scrape off the wet snow that lay on the glacier at the time of fall; the slide appears to have ridden on a cushion of air compressed beneath it (Fig. 11-21).

Comparable rockfalls have clearly taken place many times in the earthquake-prone area of southern Alaska. At Lituya Bay, farther southeast, a minor earthquake in 1958 dislodged a mass of rock estimated at 90 million tons, which fell into the bay from a maximum height of 1000 m. The disturbance of the bay was equivalent to what might be caused by dropping 2000 battleships from a height of a kilometer. The gigantic wave formed was so powerful that it swept the dense mature evergreen forest—containing many trees half a meter in diameter—cleanly off the promontories and islands to heights of as much as 500 m, leaving naked rock where once a rain forest had flourished (Fig. 11-22).

Studies of tree distribution and ages by Don Miller of the U. S. Geological Survey showed that similar gigantic waves, probably generated by rockfalls, had swept different parts of Lituya Bay at several times during the past few centuries.

An earthquake off the coast of Peru on May 31, 1970, loosened a huge mass of rock and ice from the northwest peak of Nevados Huascarán, at an altitude above 5500 m. The mass ricocheted downslope, adding to its volume on the way, until it contained between 50 million to 100 million m^3 of material. Much of the 3-km total fall was free, and the velocity by the time the mass reached the foothills was between 280 and 335 km/hr. It traveled west down a ravine, split into two tongues, leaped over a ridge 140 m high, and overran the city of Yungay and part of Ranrahirca, burying more than 18,000 people under rubble a few meters to several tens of meters thick (Fig. 11-23). The whole tragedy lasted less than three minutes. As the rubble avalanche swung round sharp bends in the guiding ravines, it cast aside boulders weighing many tons, throwing them as far as 1600 m in the air. A much smaller rockfall only eight years before had struck some of the same towns and killed 3500 persons; it was not triggered by an earthquake.

Other Slides

Nevados Huascarán, gigantic as it was, is not the largest historic rockfall. Early in 1911 an enormous mass—estimated at 6 billion metric tons—of broken rock plunged down the Pamir Mountains of central Asia to form a giant dam across the Murgab River. Water accumulated behind the dam to form Lake Sarezkoye, now 75 km long and 500 m deep. The lake has never overtopped the slide, but discharges by seepage through it, making a new source for the Murgab, 150 m below the lake level.

Figure 11-21
Air photo of rock fall on Sherman Glacier, Alaska, brought about by the earthquake of 1964. Note stream-lines radiating from head of slide. [Photo by Austin Post, U. S. Geological Survey.]

Figure 11-22
Promontory in Lituya Bay, Alaska, swept free of trees to heights of more than 500 m by the gigantic wave of 1958, caused by a rock fall. [Photo by Don Miller, U. S. Geological Survey.]

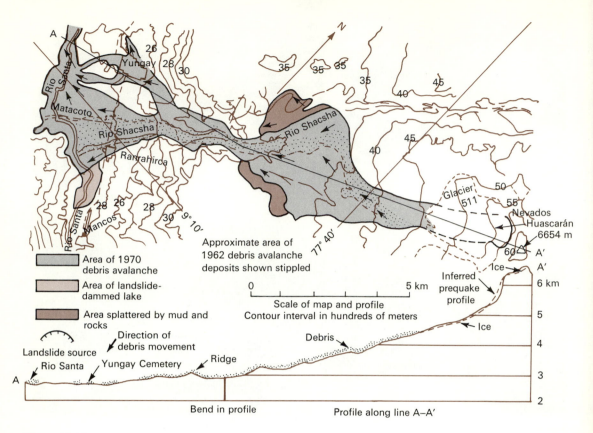

Figure 11-23
Generalized map and section of the Huascarán rock avalanche. [After G. Plafker and G. Erickson, U. S. Geological Survey.]

The great Indus River has several times been dammed by rockfalls in its canyon through the Himalayas. In December 1840, a rockfall dam 300 m high backed up a lake more than 60 km long. When the river overtopped the dam, rapid downcutting released a huge flood which overwhelmed a Sikh army encampment thoughtlessly placed on a floodplain near Attock.

PREHISTORIC SLIDES AND DEBRIS FLOWS

Thousands of ancient slides have been recognized in all parts of the world. One, just over the hill from Elm is the enormous Flims landslide, at least 11,000 million m³, a thousand times the volume of the Elm slide and twice that of the Pamirs. Long ago, probably in Pleistocene time, it slid down the surface of an old fault and blocked the valley of the upper Rhine to form a lake. After the lake overflowed, the Rhine slowly cut a gorge 600 m deep and 15 km long through the great slide, a gorge that connects today's broad and open valley segments above and below.

A complex lobate mass of debris extends north from the San Bernardino Mountains into the southern Mojave Desert in southern California (Fig. 11-24). The breccia lies in front of the outcrop of two thrust faults. On the lower one, granite overlies Late Tertiary sedimentary rocks; on the upper, crystalline limestone overlies the granite. The breccia rests on Tertiary sedimentary rocks; it may have slid in Pleistocene time. A broad lobe of poorly cemented limestone breccia extends 7 or 8 km out onto the desert floor. It is somewhat

Figure 11-24
Dissected breccia lobe on the Mojave Desert. Note how the breccia laps up against the range of hills on the left. The lobe is 4 km across at its near end. [Photo by R. C. Frampton, Claremont, California.]

eroded — one stream has cut entirely through it — but its abrupt margins are still between 8 and 30 m high. This is apparently the remains of a giant rock fall, perhaps a hundred times the size of that at Elm. Several features suggest this: first, the predominance of limestone, unmixed with other rocks; second, the contact of the breccia with low hills shows deflections like that at Elm; and finally, erosion exposes an underlying breccia of Tertiary strata, probably plowed up as the limestone breccia poured over it.

Slides or Flows Preserved in Sedimentary Rocks

The form and textures of some breccias, now parts of sedimentary formations, suggest that they are ancient slides or debris flows. Most convincing are distortions in thin-bedded marine sediments and lenses of heterogeneous debris that lie in troughs resembling that formed by the Zug slide. Such features have been recognized in many parts of the world.

SIGNIFICANCE OF DOWNSLOPE MOVEMENTS

Grand Canyon Example

The Grand Canyon of the Colorado River not only gives evidence of great erosion, but allows a rough estimate of the parts played by different processes in forming it. Figure 11-25 is a structure section of the Grand Canyon and its tributary, Phantom Creek, approximately south to north.

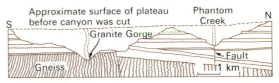

Figure 11-25
Section across Grand Canyon, Arizona. The vertically trending broken lines above Granite Gorge and Phantom Creek indicate the narrow trenches that would have been made by stream erosion alone.

The section shows the wide upper canyon cut in the nearly horizontal strata, and Granite Gorge, cut in the gneiss that noncomformably underlies the strata. Phantom Creek canyon is cut entirely in the bedded rocks. The total depth of the canyon is about 1½ km and its breadth from rim to rim about 12½ km. The inner Granite Gorge is about 400 m deep at the line of section. The walls of Granite Gorge slope uniformly; those of the upper canyon and Phantom Creek are in steps, the best-consolidated strata forming cliffs and the less-consolidated gentler slopes. Phantom Creek follows a normal fault of about 45 m vertical displacement; presumably it is fixed in position because of the ready erosion of the fault breccia. The broken vertical lines on the right in Figure 11-25 indicate a trench as wide as the creek bed, and represent the downcutting by the stream. But if the stream directly eroded only this narrow vertical slot, what processes account for the much wider valley? Judging the past by the present, the other processes include rainwash coursing down the canyon walls in minor rills and deep gullies and many kinds of downslope movements, among which the fall and continual streamward movement of talus is notably important.

The Grand Canyon itself has been cut by the Colorado River and widened by like processes. Since the course of the river was not determined by a fault, the river may have wandered somewhat, as suggested by the wavy broken lines. Similar relations seem probable for streams and stream systems in other regions.

Downslope movements of rock and soil are of course not limited to the matched sides of valleys; they operate on any slope and to its very summit. They have reduced the summit ridge between Phantom Creek and the Grand Canyon well below the level of the plateaus to north and south, which have themselves been lowered also.

EFFECT OF PLANT COVER

Vegetative cover, which is largely governed by climate, greatly affects downslope motion. Where vegetation flourishes, roots bind the soil, so that downslope movements are largely limited to creep; minor differences in rock resistance are commonly masked, and topographic profiles are smooth; in arid regions largely bare of vegetation, little soil accumulates, the resistant rocks stand out boldly, and rock contrasts are emphasized. The differential etching of rocks by erosive processes — **differential erosion** — is emphatic.

SUMMARY

In terms of the areas affected, downslope movements and rainwash are the most important of all erosional processes. For every square meter directly subject to stream erosion, there are hundreds or even thousands of square meters over which soil is slowly creeping — flowing a centimeter or two downhill after frost heaving or clay hydration, being pushed downslope by burrowing rodents or the feet of grazing animals — and gliding and falling in debris flows and slides or as talus. Of course if there were no streams to remove the debris, erosion would be almost infinitely slower than it is.

ENGINEERING APPLICATIONS

Gravity movements of soil and rock affect many structures, such as roads, dams, and buildings. Clearing highways of the debris that slides or rolls from the slopes above and repairing sections of roadbed that have slumped downslope are costly necessities. Many newly built railroads and highways have had to be rerouted within a few years to avoid slide areas or cliffs that shed many talus fragments.

Careful geologic inspection can often reduce the necessity for such changes. Areas of active sliding are readily recognized; hillslopes are generally hummocky with undrained depressions, scars at slide heads are common, and trees are tilted downhill. A new highway or railroad cut obviously increases danger of sliding because the excavation removes support from the slope above. Many an ancient slide that had been almost inactive for decades has been reactivated by excavating material from its toe. Excavation of the Gaillard Cut of the Panama Canal removed support for a hillside composed of weakly consolidated volcanic ash. The slide that ensued has gone on for decades and extended headward for several kilometers. Even though the surface slopes only a few degrees, heavy rainfall keeps the clayey ash thoroughly saturated.

The cost of slide damage cannot always be avoided by changing the proposed site of a structure or relocating one already built. An oilfield near Ventura, California, is an example. Wells drilled through a slide were slowly bent downslope, and some well casings were sheared off at the basal slip surface. Movement was most rapid in winter when the ground was saturated by seasonal rains. To save the valuable oil field, the entire hillside was paved with asphalt, and galleries were driven to drain the slide surface of such water as penetrated the pavement. These measures were effective.

Construction of the huge Grand Coulee Dam on the Columbia River was threatened by the creeping of a huge mass of water-soaked silt toward the excavation for the north abutment. To halt this slide, the engineers drove pipes into it and circulated a refrigerant that froze the pore water, binding the clay together until construction was complete.

Similar problems must be faced when such structures as large bridges and dams must be sited on weak clay or water-soaked silt. Many soils behave plastically and flow radially from beneath the load. One pier of the San Francisco Bay Bridge has much enlarged footings to spread the load of the heavy structure over a larger area.

Many laboratory tests have been devised to determine the load various kinds of clay, sand, and other loose foundation materials will bear. These tests are the basis of the engineering science **soil mechanics**. It includes field determination of such conditions as the attitude of stratification or other slip surfaces, amount of contained water, and slopes of the surface, plus the laboratory determination of properties of the materials. Such information often makes it possible to forestall or control gravity movements, even where weak materials will have to support massive structures.

Questions

1. On steep slopes in snowy country even trees rooted in rock crevices have trunks bent downhill near the ground before becoming vertical a few feet above. Why?

2. In warm humid regions compact clay-rich soils are more subject to creep than sandy or gravelly open soils, but the latter move readily in arctic regions. Why?

3. What structures in a marine sedimentary rock might record an ancient submarine flow?

4. Building sites on a hill with a fine view are restricted to two locations. Both are underlain by weak shale. At one site the beds dip steeply into the hill; at the other they dip parallel to the slope. Which do you prefer?

5. Basalt cliffs on the Columbia Plateau have large piles of coarse talus at the base; on the Colorado Plateau, with similar climate, equally high cliffs of sandstone have little or no talus. Why?

Suggested Readings

Eckel, E. B., ed., *Landslides in Engineering Practice* (Highway Research Board, Special Report 29). Washington, D.C.: National Research Council, 1958. [Of special interest because of applications to highway and construction problems, but also contains an excellent chapter by D. J. Varnes, *Landslide Types and Processes.*]

Heim, Albert, *Bergsturz and Menschenleben* (Beiblatt zur Vierteljahresschrift der Naturforschenden Gesellschaft in Zürich, No. 20). 1932. [Data concerning Swiss and other landslides, rock falls, and debris flows, including Elm and Flims.]

Howe, Ernest. *Landslides in the San Juan Mountains, Colorado, Including a Consideration of Their Causes and Their Classification* (U.S. Geological Survey, Professional Paper 67, 65 p.). Washington, D.C.: G.P.O., 1909. [Many photographic illustrations.]

Kiersch, George A., "*Vaiont Reservoir Disaster.*" *Civil Engineering*, March, 1964.

Sharpe, C. F. S., *Landslides and Related Phenomena*. New York: Columbia University Press, 1938.

U.S. Geological Survey, *Alaska's Good Friday Earthquake* (Circular 491). 1964.

CHAPTER 12

The Hydrologic Cycle and the Work of Streams

The moisture precipitated on the land as rain and snow follows divergent paths (Fig. 12-1): (a) Some promptly evaporates from the ground surface and from the vegetation on which it falls. (b) Some is absorbed by the roots of growing plants and is soon transpired back into the air through the leaves (though a little is locked up in the plant tissues until the plant dies and rots). (c) Some seeps into the soil and rocks where it is temporarily stored as **ground water** (Chapter 14) in the pores, cracks, and larger openings of the soil and underlying rocks. Some ground water returns to the surface by capillary action and evaporates; some is delivered to plant roots; and much reappears at lower elevations in springs or seepages. (d) Much water from rain and snow runs off in surface rills, brooks, and rivers.

Runoff is defined as the total discharge of water by surface streams—not merely that in rills during rains, but also the increments from springs and seepages that go to erode the lands.

The part of the precipitation that evaporates or is transpired is called the **evapotranspiration factor**. The equation

Precipitation = Runoff + Evapotranspiration

is approximately correct over a long time for in-land areas but not for the porous lavas of the Hawaiian Islands or the cavernous dolomites of the Yugoslavian coast, where much of the ground water flows directly into the sea. A minor correction is also needed for the water that chemically combines during weathering.

An estimate of the distribution of the earth's waters is given in Table 12-1.

FACTORS AFFECTING RUNOFF

Measurements made in many drainage basins reveal a great range in the ratio of runoff to precipitation. Among the many factors affecting it are the following:

Amount and Duration of Rainfall

The distribution of rainfall throughout the year greatly affects runoff. Rain uniformly distributed in many small showers may be largely evaporated or absorbed by the ground before it gathers into streams, but during violent rainstorms infiltration is too slow to capture much water, and evaporation is negligible. Rapid melting of winter snow is another common cause of floods.

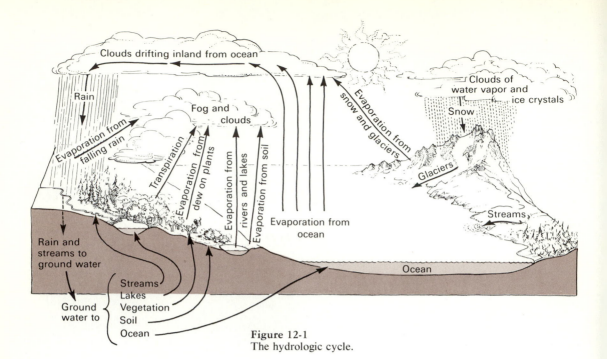

Figure 12-1
The hydrologic cycle.

Table 12-1
Estimated distribution of the world's water.

	Volume, in 10^6 km³	Percent
Surface water on the continents		
Polar ice caps and glaciers	7300	2.24
Fresh-water lakes	30	0.009
Saline lakes and inland seas	25	0.008
Average in stream channels	0.28	0.0001
Subsurface water on the continents	2000	0.61
Total water on the lands	9360	2.87
Atmospheric water	3.1	0.001
Oceanic water	317,000	97.1
Total world supply of water	326,000	100

SOURCE: After R. L. Nace, U. S. Geological Survey, 1960.

If annual rainfall exceeds 200 cm, the ground is generally waterlogged, and most rain runs off even though it is uniformly distributed throughout the year. In deserts, on the other hand, the scant rainfall is usually quickly absorbed by the parched ground or evaporated; little runs off except after cloudbursts.

Permeability

Soils and rock differ greatly in **permeability** (ability to transmit water). Most rain that falls on the porous ash of a fresh cinder cone disappears to join the ground water, but nearly all falling on a similar slope of shale runs off because the pores in the shale are so minute that the rock is impermeable, even though fully as porous as the ash. A sandy loam absorbs more than ten times as much water in a given time than does clay. Ground saturated with water or filled with ice favors runoff. Hydrologists measure "infiltration capacity" as the maximum rate at which soil under given conditions can absorb rainfall. Measurements made by playing sprinklers on test plots and collecting the runoff show the rate of infiltration to be high at the beginning but to diminish quickly to a fairly constant rate differing with each soil.

Vegetation

All vegetation hinders runoff; matted sod or the leaf carpet of a forest absorb rain like a blotter. Earthworms that live in plant-rich soil aid infiltration by their tunnels. Coniferous forests may hold much snow on their branches, increasing evaporation.

Temperature

Temperature profoundly affects runoff. Loss by evapotranspiration is much greater in warm regions than in cold; for a given rainfall, the higher the temperature, the less the runoff. Data for the United States are shown in Figure 12-2.

Slope

Slope is obviously a major factor; flat ground holds much water in shallow puddles, favoring both infiltration and evaporation.

All these factors result in great variations in runoff: in the western United States, the range is from 6 mm (6 percent of the total) on the parched lowlands of southwestern Arizona to more than 200 cm (75 percent of the total) on the rain-drenched western slopes of the Olympic and Cascade mountains of Washington. East of the Mississippi, a smaller range holds—only locally less than 25 cm or more than 75 mm.

Water Available for Erosion

Meteorologists estimate that every year the atmosphere carries enough water over the United States to form a layer nearly 4 m thick; all but about a fifth, 75 cm, passes by. The average runoff of the country as a whole is less than a third of this, about 22 cm. The difference, 53 cm, is lost by evapotranspiration. Stream gaging shows that the rivers carry an average of 50,400 m³/sec to the sea; about 1500 km³/yr. Roughly a third of this is carried by the Mississippi.

For the world as a whole, data are less complete. Table 12-2 gives two sets of estimates.

We can summarize the data as follows. The average annual precipitation falling on the lands

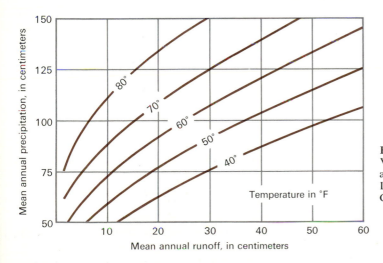

Figure 12-2
Variation of runoff with rainfall and temperature. [After W. B. Langbein, Circular 52, U. S. Geological Survey, 1949.]

Table 12-2
World distribution of runoff.

Region	Area, in 10^3 km²	Runoff, in centimeters	
Europe, including Iceland	10,372	26.0*	(21.0)
Asia, including Japan and Philippines	45,336	17.0	(21,7)
Africa, including Madagascar	31,972	20.3	(9.1)
Australia, including New Zealand	8,541	7.6	(17.2)
South America	19,280	45.0	(37.3)
North America, including West Indies and Central America	21,925	29.5	(20.3)
Greenland and Arctic Archipelago	4,164	18.0	
Malayan Archipelago	2,811	160.0	
Mean	144,401	27.0*	(20.4)

SOURCE: From W. B. Langbein; after L'vovich.
*Other estimates differ somewhat; one set is shown in parentheses.

is a little more than 1 m. Each year about 160,000 km³ falls as rain or snow. The rivers return to the ocean between 19 and 27 cm, totalling 37,000 km³ of water. This is the amount available for erosion of the land.

Energy Available from Runoff

On the average, the land stands about 800 m above sea level; the runoff thus falls an average of about 800 m as it flows from its source to the sea. This is only a rough average: from a coastal plain, the fall may be only a few meters; from Mount Everest, more than 9 km. The energy developed by the world's streams is truly stupendous. Imagine 37,000 km³ of water tumbling over a waterfall 800 m high! Such a waterfall could continually supply about 72 horsepower for each km² of land. Although no stream works at full capacity, it is easy to see why running water is the great leveler.

RATE OF DENUDATION

Streams carry dissolved and clastic (broken) materials, wearing away the mountains, hills, and plains; here we look into the rate at which this is done. Clearly the rate varies tremendously from place to place. In humid areas vegetation retards erosion; measurements have shown that brushlands are more readily eroded than forests, virgin

grassland much less so. Other factors being the same, steep slopes are more vulnerable than gentle ones, and impermeable soils more vulnerable than permeable ones, though the latter allow water to go underground, there to dissolve much rock. Erodibility of soils is closely related to dispersal into separate granular particles. Studies in California show that clays with adsorbed calcium and magnesium ions are sticky and thus more resistant to erosion than soils depleted in these ions. Well-cemented or massive rocks are obviously far less readily eroded than fine-grained unconsolidated silts. These are well-established differences; unfortunately, quantitative data are very sparse.

Streams carry rock material in three ways: in **solution**, in **suspension**, and as a **traction** (or **bed**) **load**, dragged and bounced along the bottom. The first two are easily measured by sampling during measurements of stream discharge. Samplers that reach the stream bed obviously disturb the sediment transport, so that data on bed loads are not very accurate. The mechanics of each method of transport are discussed later; here we consider only general results.

The dissolved load is only a small fraction — as little as 5 percent — of the total load in the streams of arid regions. It increases with rainfall, and in such humid areas of low relief as the South Atlantic and eastern Gulf States, the dissolved load is more than half the total. This is true also of the St. Lawrence drainage, most of whose clastic load is trapped in the Great Lakes. Estimates for the world as a whole suggest that about 30 percent of stream load is in solution.

Clastic loads obviously vary widely also, both from one stream to another and from time to time in the same stream. The Niobrara River in Nebraska carries half its load as bed load; the Mississippi only 7 to 10 percent. The Colorado River sediments trapped in the delta of Lake Mead are all so fine that the bed load upstream must be negligible. Perhaps 10 percent of the clastic load of most streams is bed load.

Not only is accurate measurement of stream load difficult at any time, the long-term average is still harder to arrive at. The load of the Delaware was five times as great one year as it had been the year before. Most hydrologists think it likely that the exceptional "fifty-year flood" carries off as much sediment as is normal for an entire year. The common large floods—about two every three years—probably transport most of the sediment. Because recording times rarely extend beyond a few decades, our estimates of denudation are only rough, and all are perhaps somewhat below a true figure.

Each year the Mississippi pours almost 500 million tons of dissolved and clastic material into the Gulf of Mexico. Though generally thought a muddy stream, this is only about 0.5 percent of the load. Even the Missouri—the "Big Muddy"—rarely carries more than 2 percent. This is high for a big river, though exceeded by the Colorado, Hoang Ho, and a few others.

Many smaller streams in arid or semiarid regions carry enormously greater loads per unit volume than the large rivers. The San Juan, in southwestern Colorado, carried more than 75 percent by weight of silt and sand during one flood; such highly loaded streams grade into mudflows. A small drainage basin in the Loess Hills area of Nebraska yielded in one year more than 30,000 kg of sediment per km². The hills are made of weakly consolidated silt. In humid New York or Connecticut, comparable streams carry less than 1 percent as much from a terrain of much more consolidated rocks. The variation of sediment yield with precipitation in the United States is shown in Figure 12-3. The maximum rate is in areas of only 38 cm of rainfall, where the runoff is only 12 mm! Here the vegetative cover is poor and root-binding of soil negligible.

Professor Sheldon Judson of Princeton University has made careful estimates of erosion

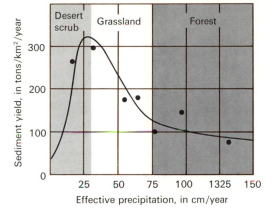

Figure 12-3
Relation of precipitation to annual sediment yield, in tons per square kilometer. [Data from W. B. Langbein and S. A. Schumm, U. S. Geological Survey, 1958.]

rates (the mean rate of lowering the land) in the United States, taking account of both chemical and detrital fractions. He found the rate to range from a high of 166 cm per thousand years in the Colorado drainage to a low of 38 cm per thousand years in the Columbia River watershed, with an average for the whole country of 61.4 cm per thousand years. Even the slowest of these rates is notable, implying, as it does, a reduction of the Columbia Basin by 38 meters in a million years; the Colorado rate implies a lowering by 166 m in 1 m.y. Yet some geologists believe they can recognize landscapes carved in Oligocene time—40 m.y. ago—in the area of Pikes Peak!

Measurements of pegs driven into weakly coherent shale on the Navajo Indian Reservation in Arizona showed erosion of 5.7 mm/yr on slopes between 20° and 40° during a period of 14 years, but less than half this rate on slopes below 10°, though still higher than in more humid regions. Studies of root systems and tree-ring ages of the long-lived bristlecone pines in the White Mountains of California show that these steep mountain slopes are being cut down at rates between 240 and 360 mm per thousand years—say 1000 m since the beginning of the Pleistocene. Such rates of mountain erosion are to be contrasted with the rate of 50 mm per thousand years for the Mississippi Basin. They are scores of times greater than the rate for the Hudson Bay lowlands.

Although such estimates and measurements can be multiplied many times, all lack precision when extrapolated over geologic times; they certainly do, however, give pause to suggestions that any landform we see today dates from far in the geologic past. Were the present rate of erosion for the United States maintained without further uplift or isostatic rebound, the country would be reduced to sea level in about 14 m.y., a brief time geologically. Of course this would not happen, because isostatic uplift would go on as the rock load on the asthenosphere diminished, and erosion rates would diminish with relief.

Changes in Erosion Rates

For many reasons the loads of present-day streams are higher than the average for the geologic past. Rivers in populous areas carry industrial and municipal wastes in addition to their normal loads. Enormous areas are roofed and paved, preventing water from entering the ground and very appreciably adding to flood heights, though of course protecting the area they occupy. Cultivated fields furnish vast tonnages of sediment not naturally available. Although the changes in erosion rates brought about by man cannot be precisely evaluated, it must be remembered that grasses did not evolve until Miocene time, and average rates of the past may not have been so much less than those of today as might at first appear. Studies of the sediment volume brought to the Gulf of Mexico from the north since the Jurassic and of the volume accumulated off the eastern coast of North America since the Triassic suggest that the present rate is much less than twice the average of pre-human times.

FLOW IN NATURAL STREAMS

A stream flows because of the downslope component of the pull of gravity. Strictly speaking, this component is proportional to the sine of the angle of slope (Fig. 11-1), but since most streams slope at very low angles, the tangent, called "s," is nearly equal to the sine and is easier to measure, so it is used in all hydraulic engineering. If a stream slopes only 12 m/km—a steep but not unusual gradient—the water should accelerate at a rate of almost 12 cm/sec/sec if the accelera-

tion is not restrained in some way. In an hour, such unrestrained flow would attain a velocity of more than 1500 km/hr!

No natural stream even remotely approaches such a velocity. Most flow at speeds below 8 km/hr; the swiftest large river ever gaged in the United States is the Potomac: 25 km/hr in the flood of March 1936. The greatest velocity ever recorded is 33 km/hr in the Lyn River, England, in August 1952. Clearly, the flow is restrained.

The only restraining agent, aside from the trivial resistance of the air, is the friction of particles of water against one another, against a stream's beds and banks, and against particles of sediment suspended in it.

If streams of dye are injected into most natural streams, they do not travel in parallel lines as they do in water flowing very slowly through a glass tube (**laminar flow**, Fig. 12-4, Upper); instead, they swirl about and are quickly mixed, showing that the flow is **turbulent** (Fig. 12-4, Bottom).

Experiments have shown that the kind of flow in open channels depends on Reynolds' number, R:

$$R = \frac{\text{velocity} \times \text{depth} \times \text{density}}{\text{viscosity}}$$

The numerical value of R obviously depends on the units used in measuring velocity, depth, and viscosity; whatever the units, though, at lower values the flow is laminar, at higher values, turbulent. Thus any increase in velocity, depth, or density of a fluid (increasing the numerator in Reynolds' number) favors turbulent flow; any increase in viscosity (increasing the denominator) favors laminar. In laminar flow, one layer of water slides past another; the frictional loss is proportional to the difference in velocity of the fluid layers involved and thus to the mean velocity. In turbulent flow, however, mixing takes place

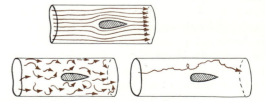

Figure 12-4
(Top) Laminar flow. (Bottom) Turbulent flow.

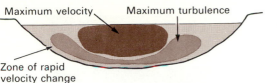

Figure 12-5
Distribution of velocity and turbulence in a straight, symmetrical channel. [After John Leighly, 1934.]

by swirls and eddies; the frictional drag is proportional to the *square of the velocity*, thus greater. Since virtually all natural streams are turbulent, this explains why streams even in flood rarely attain speeds of more than 10 km/hr. A narrow film of water at the stream boundaries is in laminar flow, but the turbulence increases abruptly inward. Farther from the banks, changes are less abrupt (Fig. 12-5).

In straight channels the velocity may be nearly constant across most of the central part of the channel, with the thread of greatest velocity near the middle and 0.4 of the depth above the bottom. But in curving channels the fastest flow is shifted by inertia far toward the outside of the curve—an important factor in molding stream channels, as discussed later.

If velocity is to remain uniform, friction must precisely balance the acceleration due to gravity. That the friction develops heat is shown by the fact that a meltwater stream on a glacier, even at freezing temperatures, is able to melt its channel down into the ice. The downvalley force of a stream is the product of the *weight of the water*, *W*, times the *slope*, *s*, the weight being given by multiplying the area of the stream channel, *A*, by the length (arbitrary), *L*, and the unit weight of the water, *W* (Fig. 12-6). The frictional drag is given by multiplying *drag per unit area*, *T*, by the *surface area of the wetted channel*, *A*, the product of the *wetted perimeter*, *p*, and the length, *L*. Thus

$$WALs = TpL \qquad (1)$$

or

$$T = \frac{WAs}{p} \qquad (2)$$

Since *A/p* (the cross-sectional area divided by the wetted perimeter) is the average depth, *d*, this may be written $T = Wds$; that is, the drag is proportional to the product of slope and depth.

As the flow is turbulent, the drag—the shear stress along the channel boundary—is proportional to the square of the velocity.

Natural streams obviously differ greatly in average velocity, and in any one, velocity fluctuates from hour to hour. The variations result from many factors, but chiefly from those in **slope**, *s*; **channel shape** *A/p*; **discharge** (which affects both *A* and *p*); and the **roughness**, *r*, of the channel. Increase in either volume or slope increases velocity; the first reduces the ratio of frictional drag to water volume, and the second increases the downstream component of gravity. Increases in either wetted perimeter or roughness cause greater friction and lower velocity. Hydraulic engineers have found experimentally that:

$$V^2 \propto \frac{As}{rp} = \frac{ds}{r} \qquad (3)$$

Velocity also varies with cross-sectional shape of the channel. Figure 12-7 shows three channels with equal areas but different shapes. The semicircular channel has the smallest wetted perimeter and is thus most efficient. Because channel

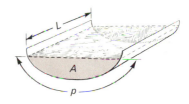

Figure 12-6
Cross section of flow in a stream channel.

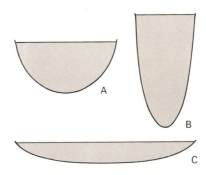

Figure 12-7
Three channel cross sections of equal area. Channel A contains this area within the smallest possible wetted perimeter and offers the least frictional resistance to running water. [After W. W. Rubey, 1952.]

B offers more friction than A and its steeper walls are less stable, it will tend to widen into a more "efficient" channel. Such a channel can only form in highly coherent material.

Mountain torrents flowing at several meters per second over resistant rocks tend to have semicircular channels (Fig. 12-7, A). As tributaries join and flow increases downstream, the channel generally widens more than it deepens, tending toward a channel like C, chiefly because the lower valley banks are generally alluvial and less coherent than those upstream. Cross section A is favored by engineers constructing irrigation canals.

Roughness caused by sediment and other irregularities in a channel greatly affect velocity and channel shape. The effect on velocity is shown in Figure 12-8. Pebbles on the bed of Pole Creek, near Pinedale, Wyoming, average 1.28 cm in diameter. The creek's slope is 0.003. The Hoback

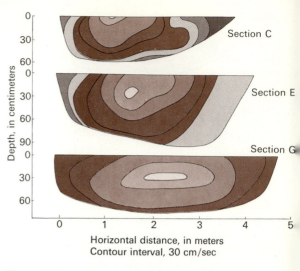

Figure 12-9
Velocity distribution in the channel of Baldwin Creek, near Lander, Wyoming, at three sections along a straight reach a few scores of meters long. Flow near bankfull. [After *Fluvial Processes in Geomorphology* by L. B. Leopold, M. G. Wolman, and J. P. Miller. W. H. Freeman and Company. Copyright © 1964.]

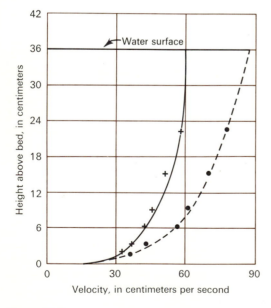

Figure 12-8
Current-meter measurements of velocities as a function of depth; the examples are chosen because their depths are identical, but they differ in slope and in mean size of their bed materials (channel roughness). [After *Fluvial Processes in Geomorphology* by L. B. Leopold, M. G. Wolman, and J. P. Miller. W. H. Freeman and Company. Copyright © 1964.]

River, near Bondurant, Wyoming, with the same depth and a considerably steeper slope of 0.0051 does flow faster, but not proportionately so, because gravel composing its bed averages ten times as large, 12.8 cm. The ratio of the square roots of their velocities is 1:1.13 (Equation 3); the ratio of their slopes is 1:1.7. Many measurements show that the mean velocity is close to the average of the velocities at 0.2 and 0.8 of channel depth.

Most streams do not flow fastest at their surface. Figure 12-9 shows three cross sections along a straight stretch of Baldwin Creek, Wyoming. The pale central areas indicate the swiftest current; in two sections they are below the surface, as is common in small rivers and in curves of larger rivers.

The **discharge**, Q, is the quantity of water passing a given point during a unit of time. Clearly Q is a function of gradient, velocity, and channel size.

$$Q = AV_m \qquad (4)$$

in which V_m = mean velocity. But A equals surface width times mean depth:

$$Q = AV_m = WD_mV_m \qquad (5)$$

where W = width.

We saw in Figure 12-8 that the grain size of bed material affects the roughness of the channel. Many stream variables are interrrelated. In general, where a stream is deepening its channel, it is lowering its slope; where it is depositing sediment, it is steepening it, but there are local exceptions to these generalizations.

Some factors fluctuate greatly even within the hour. Floods rage after heavy seasonal rainfall, rapid snowmelt, or even after brief cloudbursts; during a drought, a sizeable river may dwindle to a chain of stagnant pools. All changes in discharge cause adjustments in velocity and load transport.

STREAM LOADS

The Suspended Load

A stream's suspended load consists of two classes of material. One, the **wash load**, consists of clay particles so fine — even colloidal (10^{-5} cm or less in diameter) — as to remain in suspension almost indefinitely, though of course they would eventually settle out of stagnant water. Even sluggish streams carry such fine sediment with about the same velocity as the water. Only in stagnant backwaters is much wash load deposited; it is virtually absent in channel sediments. The amount of wash load depends not so much on stream characteristics as on such factors as rain intensity, grain size of surface material, and grass cover.

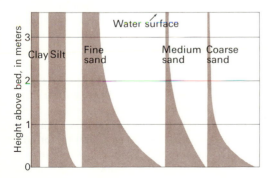

Figure 12-10
Graph showing variation in sediment size and volume with depth. Missouri River near Kansas City, Missouri, Jan. 3, 1930. Note nearly uniform distribution of clay and silt through all depths and the concentration of coarser particles near the bottom. [After L. G. Straub, in *Hydrology*. Courtesy of Dover Publications.]

Turbulence enables streams to carry much coarser material in suspension. Sampling of many streams shows that only unusually swift rivers are turbulent enough to lift particles larger than medium-sized sand from their beds (Fig. 12-10). Upward currents pick up and transport fine material; it is only kept in suspension by other upward threads. As a stream flows down a slope, downward currents are of course even more abundant than upward. But since both are distributed at random, an individual particle may be suspended for a long time before settling to the bottom or being carried there by downward swirls. While in suspension it moves downstream at about the average velocity of the current.

Bed Load

Some material too coarse to be lifted is pushed or rolled along the bottom; this is the **bed load**, supported in part by water but chiefly by the other grains on the channel floor. The bed load may be boulders, gravel, sand, or even coarse silt, depending on the stream flow and terrain. Observations through windows in the walls of experimental flumes show that most bed-load grains roll, some slide, others bounce or momentarily vault into suspension. As velocity over a sandy bed increases, particle motion progresses from (1) short leaps of a few grains through (2) spasmodic movement and deposition of groups of grains to (3) smooth general transport of many grains. At still higher velocities the bed load blends almost continuously with the suspended load. In general the ratio of bed load to suspended load is greater in smaller than in larger streams, but exceptions abound.

Dissolved Load

Streams also carry much material in ionic solution; the ions are part of the fluid and move with it. They are derived chiefly from ground water that has slowly filtered through weathering soil and rocks. Little is dissolved from the channel walls except by streams flowing on limestone or gypsum.

Few rivers carry more than a thousand parts per million (0.1 percent) of dissolved load; a general average might be closer to 200 parts per million (ppm). Yet this is not negligible. Even the lofty Wind River Range of Wyoming is being lowered by solution alone at the rate of one meter in about 140,000 years (northeast flank) to one meter in 275,000 years (southwest flank). (The northeast flank retains snow longer and has more limestone.) These figures have been corrected for the fact that precipitation in this area contains about 6 ppm of dissolved matter. Clearly, erosion by solution is important, even in mountainous country; in areas of low relief and slower runoff, such as the southeastern United States, solution is lowering the land at the rate of a meter in 25,000 years. Here the dissolved loads considerably exceed the solid loads of streams.

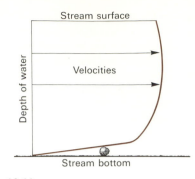

Figure 12-11
Hydraulic lift caused by the steep velocity gradient close to the stream bed. [After W. W. Rubey, 1938.]

COMPETENCE AND CAPACITY

The force exerted by a stream on particles of its bed acts in three ways: (1) as the impact of a moving water mass against the upstream surface of the particle, (2) as the frictional drag of the current across its surface, and (3), since fluid pressure varies inversely with current velocity, by the pressure difference tending to lift the grain from the bottom into the speedier current. For a grain to move, the sum of these forces must exceed the grain's inertia, measured by its submerged weight.

(1) The impact of the mass of water that strikes a grain is proportional to the square of the velocity of the water ($\frac{1}{2}mV^2$). Theoretically, then, the diameter of the largest grain a current can move varies as the square of the velocity. Since the volume and weight of a spherical grain vary as the cube of the diameter, the weight of the movable grain varies as the 6th power of the velocity. This theoretical relation, the so-called "6th power law," was suggested before 1830. Unfortunately, it is of little value: most grains on the bottom are partly shielded by their neighbors, and it is impossible to measure stream velocity at the very bottom.

(2) The second mechanism, frictional drag of the column of water overlying the particle, is effective in proportion to its downstream component, thus depending on depth times slope. As with the impact mechanism, few grains are fully exposed to this drag.

(3) The third mechanism, hydraulic lift, is illustrated in Figure 12-11. Because the pressure in a moving fluid is inversely proportional to the velocity—a fact utilized in aspirating pumps and water injectors for steam engines—a large pressure difference exists across a thin layer of water at the stream bottom. The pressure in the stagnant water beneath a sand grain lying in this layer is greater than that at the top of the grain, thus tending to lift it into the stream. Once raised into the main body of the stream, where velocities are more nearly uniform, the lift diminishes, and the grain may settle again to the bottom. This mechanism tends to select certain-sized grains because the thickness of the layer of highest velocity-contrast obviously affects the favored grain size.

These three mechanisms together determine the maximum size of particles a stream can transport. As the velocity increases, all three mechanisms are strengthened. The bed load depends on the velocity of the water within a very few grain diameters of the bottom. This is always less than the mean velocity, and in big rivers very much less. Accordingly, bed load is, in general, proportionately much greater in small than in large streams.

The maximum particle size a stream can carry is its **competence**. The competence of many natural streams is enormous; many have moved boulders more than 3 m in diameter. When the St. Francis Dam in southern California broke in 1928, the flood tumbled blocks of concrete (20 × 18 × 10 m) weighing 10,000 tons for a kilometer downstream. A flood in the Wasatch Mountains in 1923 carried boulders weighing 80 or 90 tons for nearly 2 km. The Lyn flood in England in 1952 moved boulders weighing 15 tons.

In well-sorted material, particles of fine sand

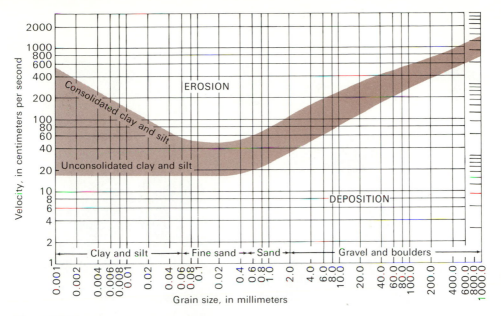

Figure 12-12
Curve showing least velocity of flow at which sedimentary particles (quartz in the sand sizes, other materials in coarser and finer sizes) begin to erode. The width of the brown band indicates approximate variations, depending on stream depth and cohesiveness of material. Note that fine sand erodes most readily; both finer and coarser materials require greater velocities. These curves are based on experiments with materials of carefully chosen sizes, not upon mixtures of a wide size range. Such mixtures act as though one-third were coarser and two-thirds finer than the actual mean grain size. [After A. Sundborg, *Geografiska Annaler* v. 38, 1956.]

(0.06 to 0.5 mm in diameter) are most readily moved (Fig. 12-12). The same or a greater velocity is needed to move smaller grains, perhaps because they do not project above the film of laminar flow lining the channel. An emphatically higher velocity is needed to erode fine silt or clay if they are closely packed or slightly consolidated.

Strangely, in poorly sorted sediments, grains of sizes between 1 and 6 mm diameter—and especially between 2 and 4 mm—are very much less abundant than those either larger or smaller. Apparently grains in this size range are selectively washed out of coarser material and selectively kept in motion when finer particles settle out. This is an empirical fact, well-supported by much sampling in many streams, yet no really adequate explanation has been advanced.

Stream competence should not be confused with **capacity**—the total load a stream can carry. Capacity depends not only on velocity (which alone governs competence), but also on the total volume—the discharge of the stream.

MECHANICS OF STREAM EROSION

Abrasion

Stream sediments in motion constantly jostle each other and dash against the channel boundaries. Sharp corners of joint blocks are broken or rubbed off. A few hundred meters of travel produces subrounded gravel; the longer the travel and the coarser the particles, the better the rounding. Grains of silt and fine sand are little rounded because their small inertia lessens their impacts. Part of the rounding, though, is caused by solution, as suggested by experiments with material in rotating barrels; frequent changes of water speeded rounding.

Sediments also become finer downstream because, with fluctuations of stream flow, travel is intermittent, and they may lie exposed on gravel bars for months or years, subject to weathering like any other rock. Thus abrasion, solution, and

Figure 12-13
Potholes in the bed of Susquehanna River, Conowingo Falls, exposed during the drought of 1947. [Photo by *Lancaster Intelligence Journal*; courtesy of H. H. Beck, Franklin and Marshall College.]

weathering all act to lessen the grain size of sediments downstream.

Below Great Falls, Maryland, where the Potomac River rages in flood through a narrow gorge, 40-cm boulders have been thrown as high as 20 m above the channel bottom. The impact of such missiles shatters both boulder and bed. Eddies in flooded streams round the projecting rocks and drill cylindrical pits (potholes) into the stream bed (Fig. 12-13).

The extreme turbulence along steep cascades and at waterfalls gives water tremendous erosive power. At Niagara, water and entrained rock debris strike the base of the 50-meter high fall at about 80 km/hr and scour deep plunge pools that undermine the lip of the fall, making it susceptible to caving (Fig. 12-14). Thus undercutting at the plunge pool and caving of the overhanging lip contribute greatly to the erosion of some stream valleys. Below Niagara Falls, for instance, the 7-mile gorge between Kingston and the Falls was mainly cut by headward migration of the waterfall, not by slow downward erosion of the channel.

Importance of Flood Erosion

The flood discharge of most rivers is many times the low-water flow: of the Mississippi, 30 times; of the Vistula, 120 times; of the Ohio, 313 to 1 for annual high and low stages. As the discharge increases, so does the velocity. The Columbia low water flow in March 1945 was 2200 m³ at a speed of 57 cm per second. Following the early June snowmelt, the discharge was thirteen times as great and the velocity nearly six times that at low water.

The stream regime at flood is vastly different from that at low water: In 1951 the U.S. Geological Survey closely observed that of the San Juan River at Bluff, Utah. On September 9 the flow was only 18 m³/sec over deposits of sand, silt, and gravel that nearly filled the deep bedrock channel (Fig. 12-15). By mid-September the rising water mobilized part of the bed; by the time the flood crested in October all sediment was in motion and the channel bottomed in bedrock. With the falling waters, the sediment again began to fill the channel. Velocity at low water was quite inadequate to handle it all.

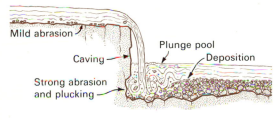

Figure 12-14
Erosion in the plunge pool of a waterfall.

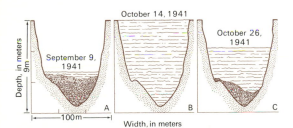

Figure 12-15
Changes in the channel of the San Juan River near Bluff, Utah, during the flood of October 1951. (A) September 9, before the flood, discharge only 18 m³/sec. (B) At flood height, 1700 m³/sec, when the river had mobilized all the sediment in its channel. (C) An early stage in recession of the flood, October 26. [After L. B. Leopold and T. Maddox, 1953.]

Such fluctuations in channel depth with river stage are remarkable. During excavation of the foundations for Hoover Dam, a railroad tie was unearthed beneath 20 m of coarse gravel. This depth of scour and fill is notable, but a flood in Kanab Canyon, Utah in 1883, cut a channel 15 m deep and 80 m wide in less than 8 hours.

Bed Form

Changes in discharge of streams flowing in sandy channels produce changes in shape of the stream bed. Few beds are really smooth, but if they are at low stream velocity they soon become rippled as the velocity rises. Ripples in sand are usually less than 30 cm long and 0.5 to 6 cm high, their amplitude independent of both stream depth and sand size. Though the water flowing over ripple troughs is less speedy than that over the crests, the difference is at first not enough to influence the water surface, which may remain smooth or be rippled independently of the bed (Fig. 12-16). Sand grains climb in linear trains up the gentle upstream faces of the ripples and roll down the steep downstream faces; the ripples migrate slowly downstream. Increasing velocity changes the ripples into larger forms called dunes (Fig. 12-16, B). A **dune regime** may form in either fine or coarse sand; the coarser the material, the shorter and steeper the dune. Dunes may range from 60 cm to many hundred meters in length, and from 6 cm to many meters in height. At lower stream velocities, the dunes may carry ripples but these disappear at higher speeds. In large streams channel dunes may reach great size: in the Mississippi, at Memphis, heights reach 10 m; in the Amazon, more than twice this.

Dunes disturb stream flow much more than ripples; stream lines do not follow the dune form, but shoot over the crest to form a vertical eddy in its lee. The lower countercurrent may run at half to two-thirds the speed of the main current. Flow over troughs is slower than that over crests, so that the water surface bulges over them (Fig. 12-16, C). Streams in the dune regime carry much more sediment than those with rippled beds.

Ripple and dune bottoms typify streams in the **lower flow regime**. With increasing velocity the dunes are washed out, so that a smooth stream

260

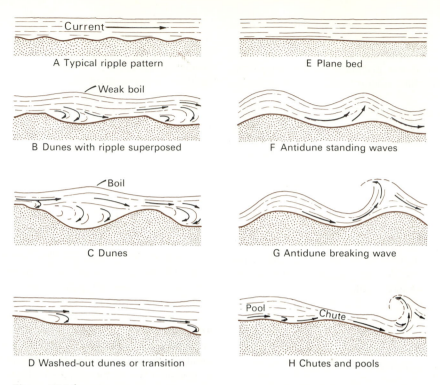

Figure 12-16
Idealized diagram of the forms of bed roughness in an alluvial channel. [After R. B. Simons, E. V. Richardson, and Carl Norden, *in* Special Publication No. 12, Society of Economic Paleontologists and Mineralogists, 1965.]

bed evolves—the **transitional regime**. The entrained sediment—often as much as 6000 ppm—mostly remains near the bottom, and the stream bottom and surface are both smooth.

At still higher velocities, the smooth bed is transformed into **antidunes**, in which the wave form of the dunes travels upstream, as do the overlying water waves. The downstream sides of the dunes are eroded while the upstream sides are rapidly added to. This is the **upper flow regime**. At still higher velocities the antidunes break up, all available bed material is thrown violently into suspension, and sediment concentrations may reach 40,000 ppm even when little fine material is available and as much as 600,000 ppm (60 percent solids!) if it is.

At the highest velocities—short of those in waterfalls—the surface becomes a series of alternating troughs and bulges that migrate upstream. This regime is only reached in rock-walled or artificial cemented channels, for if the walls are readily erodible the stream widens its channel and thus reduces its own velocity.

We noted that the formula for Reynolds' number has viscosity in the denominator. Thus any decrease in viscosity increases the Reynold's number and the tendency toward turbulence. A study of the flow of the Rio Grande below Las Cruces, New Mexico, found that in winter, when the water temperature was about 5°C, the river bottom was virtually flat, but in summer, with the same discharge, the bottom was dune-covered. The water temperature was 21°C and the viscosity enough lower to notably increase velocity and turbulence. The winter temperature of Mississippi water at St. Louis is 0.5°C; in summer it reaches 31°C. At comparable discharge the flow regime must vary considerably with season, but no systematic observations have yet been recorded.

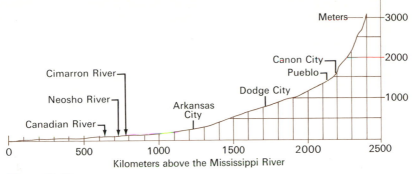

Figure 12-17
Long profile of the Arkansas River from its source at Tennessee Pass, Colorado, to the Mississippi River. Note that vertical exaggeration is almost 5000 to 1. [Modified from Henry Gannett, U. S. Geological Survey.]

Source of a Stream's Load

As we noted in Chapter 11, although a stream erodes its channel, by far the greater part of its load is contributed by downslope movements, solution by tributary ground water, and rainwash.

THE LONG PROFILE AND THE CONCEPT OF GRADE

The **long profile** of a stream is a plot of the elevations of points on the water surface against distances along it (Fig. 12-17). The vertical scale must always be greatly exaggerated, for almost all streams are hundreds of times longer than their vertical falls.

Long Profile of the Arkansas River

No two streams have identical long profiles, but that of the Arkansas is representative of many. The Arkansas rises in the southern Rocky Mountains, flows across the Great Plains, and joins the Mississippi about 730 km above its mouth. Above Canyon City, Colorado, the gradient is steep and irregular; the river plunges swiftly through deep canyons cut into the resistant granites and gneisses of the Front Range. In contrast, the lower 300 km has a regular gradient, nowhere steeper than 0.00018, or 18 cm/km. It winds in serpentine bends (as shown in Fig.

12-25) on a gently sloping plain of sand and silt like that now being carried by the stream.

Between these extremes the gradient of the middle course varies regularly and gradually, so that the profile is smoothly concave upward, resembling half an hyperbola, but in detail far from a simple mathematical curve.

Graded Profile in Relation to Stream Gradient

The profiles of most large streams resemble that of the Arkansas, though details vary greatly. What does this similarity mean? We have noted that an underloaded stream may cut into its bed, thus flattening its gradient and slowing down. But it may also flatten its gradient by becoming more sinuous without downcutting. Similarly, an overloaded stream may steepen its gradient by depositing some of its sediment or by straightening its course. By these processes a stream over time tends to adjust its course to the average discharge and sediment load. As these parameters are constantly changing, the river never attains an absolutely stable profile; it is constantly modifying its channel, but over a considerable time span it settles down to fairly stable pattern. The factors tending to make the profile flatter in its lower course are many but important. Among them are: limitations in downcutting imposed by base level, increase in discharge downstream, and downstream changes in grain size and load.

BASE LEVEL

No stream can deepen its channel far below sea level; even the Amazon is less than 100 m deep at its mouth. This limiting depth is **base level**. A stream cannot lower its bed below this; it can deposit sediment to steepen its gradient, but it cannot cut deeper to lower it.

Although the ocean is the ultimate base level for streams, a lake or a particularly resistant mass of rock through which a stream can deepen its channel only slowly (as with the gneiss in the Royal Gorge of the Arkansas) may form a **temporary base level**—temporary, because in the long course of geologic time even the largest lakes are either drained by downcutting at the outlet or filled by sediment poured into them; the most resistant rock ledge will eventually be sawed through just as the channel of the Royal Gorge has cut down at least 500 m since encountering the gneiss. The flattening of streams above such obstructions is thus geologically temporary.

INCREASE IN DISCHARGE DOWNSTREAM

At first glance it seems odd that the Arkansas profile is everywhere graded when the lower reaches slope only at 18 cm/km while between Pueblo and Dodge City the slope is fully 1 m/km. And why are the gradients of different segments transitional, so that the headward steepening is along a smooth curve?

A partial answer is that the stream is not of constant size. The Arkansas, like most rivers, grows by the confluence of tributaries and by seepage of underground waters. It is a gaining stream; only in deserts do streams diminish downstream (Chapter 14). As a stream acquires tributaries, the ratio of frictional surface to channel cross section diminishes. Figure 12-18 shows

Channel cross sections

6 banks

2 banks

Figure 12-18
Decrease in channel friction due to confluence of tributaries.

three channels, each a meter deep with vertical sides. Their confluence eliminates 4 m of wetted perimeter (this is an exaggeration, of course, for the stream would either have to widen or deepen). The energy hitherto dissipated in friction against the four eliminated channel sides is available below the junction to speed up the water, eroding the channel and lowering the gradient.

CHANGE IN GRAIN SIZE AND LOAD DOWNSTREAM

Confluence of two streams affects amount and kind of load as well as discharge. A steeper, and thus more competent, tributary may load the main stream with material too coarse to move on its normal grade. The steep tributaries of the Colorado in the Grand Canyon, though dry for most of the year, dump huge fans of coarse debris into the river when in flood, forming **debris dams** (Fig. 12-19).

The debris dams of the Colorado are conspicuous; less obvious changes in grain size cause more subtle changes in stream regimen. Thus the Platte dumps so much sand into the Missouri channel as to cause a hump in its profile. Sampling of the channel of the Mississippi shows a marked decrease in grain size downstream (Table 12-3).

The average composition of delta sediments, 70 percent silt and clay, confirms this general trend. The Rhine shows a marked decrease in pebble size downstream between Basel and Bingen, along with a flattening of the slope, although discharge increases only slightly. Attrition and weathering on gravel bars, which decreases pebble size and therefore channel roughness, is one factor allowing most streams to flow on ever declining slopes. Indeed, the velocity of many large graded rivers increases near their mouths even though gradients lessen. Increased discharge and depth, diminishing grain size of load, lessened friction against smoother banks of sticky clay instead of sand and silt—all these together outweigh the effect of lessened gradient. We thus see that gradient is far from being the sole control of channel action or even velocity. Stream regimen depends on the interaction of many factors: slope, discharge, grain size and amount of load, channel form, wall characteristics, water temperature, vegetation, and doubtless still others.

Figure 12-19
Rapids over a debris dam at the junction of Tapeats Creek (left) with the Colorado River.
[Photo by B. M. Freeman, U. S. Geological Survey.]

Table 12-3
Percentages of different grain size fractions of lower Mississippi sediments below Cairo, Illinois (Approximate only).

Grain size	Kilometers below Cairo					
	160	480	800	1120	1440	1600
Gravel	29	8	14	5	trace	none
Coarse sand	30	22	9	8	1	none
Medium sand	32	50	46	44	26	9
Fine sand	8	19	28	41	70	69
Silt	trace	trace	2	1	2	10
Clay	trace	trace	1	trace	1	10

SOURCE: Data from C. M. Nevin; derived from U. S. Waterways Experiment Station, Vicksburg, Mississippi.

CHANNEL PATTERNS

We have already noted that the lower Arkansas winds in serpentine bends called, from a river in Turkey, **meanders**. Its course is several times as long as a direct route. Between Pueblo and Dodge City, however, before irrigation controlled its regimen, the river was **braided**; that is, it was subdivided into a plexus of interconnected small channels between many shifting gravel bars and sandbars. Elsewhere the channel is relatively straight, though only for short distances. The Mississippi River has a typical meandering course south of Cairo (Fig. 12-20). The lower Amazon, despite its relatively low clastic load, braids instead. But meanders are also common in small creeks, as are braids in sediment-loaded rills that form after every rainstorm.

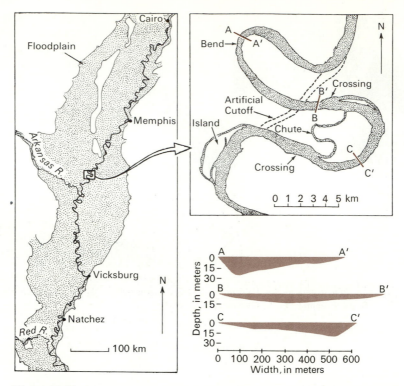

Figure 12-20
(Left) The meandering course of the Mississippi below Cairo. (Right) Map and cross sections of three bends. The artificial cutoffs were dug in 1941 and 1942. [After H. N. Fisk, Mississippi River Commission, 1947.]

Origin of Meanders

Why does a stream meander instead of flowing in a straight course? Some streams follow meandering course carved in bedrock (Fig. 12-21), but this is very uncommon; most meanders are on floodplains, and those in bedrock are generally considered to have formed on floodplains and been incised as the stream lowered its channel. The problem is to account for floodplain meanders.

Some stream curves have obvious causes: deflection by landslides, by tributaries whose loads are too coarse for ready removal, by fallen trees and similar random obstacles. But most meanders are regular, much more so than can be accounted for by these causes; a general cause must be operating.

Observation shows that meandering streams have relatively light bed loads. Their banks have a high proportion of sticky clay and are much

steeper than those of braided streams. Most are near base level, and the only way they can lower their gradients is to become more sinuous. Other meandering streams not close to base level become sinuous because of the difficulty in entraining fine sediments in absence of a bed load (Fig. 12-12).

A stream in random turbulence must attack now one bank and then the other. As the water rounds a bend, however formed, its inertia carries it toward the concave bank, bringing about an almost-or-quite-imperceptible uptilting of the water surface, thereby increasing its slope there. This tilt tends to make the stream run toward the opposite bank as it flows downstream; thus the curvature of the bank tends to reverse. The tilting of the water surface at curves and the reversal of tilt in the intermediate reaches produces an almost uniform water slope along the middle of the stream course. This uniform slope means that the steam is expending its energy almost uni-

Figure 12-21
Entrenched meanders of the San Juan River, Utah. [Photo by Tad Nichols, Tucson, Arizona.]

formly in accord with the relatively uniform rate of change of direction.

At the crossover between bends, the channel of the Mississippi may be only 3 m deep while at the concave bank of a bend it is 30 m deep and the bank nearly vertical, a most unstable situation. Caving off of such banks and deposition on the convex bank moves the meander laterally (Figs. 12-22, 12-23). Most material so entrained is deposited on the next crossover or on the inside of the next bend.

By this process of lateral migration, the stream in time reworks deposits of old meander channels and forms new deposits, generally at a slightly lower level. This is the **floodplain**. In a century and a half of meandering, the Klaralven River in southern Sweden had destroyed an average of 13,000 m² of arable land yearly along a 100-km length of the valley. At the same time it rebuilt about the same area of floodplain at elevations about 4 m lower. Thus about 50,000 metric tons of sediment was removed yearly from this stretch of valley.

Even meandering streams that are deeply incised in bedrock attack their concave banks and thereby undercut them. The remarkable natural bridges of southeastern Utah were formed in this way (Fig. 12-24).

Figure 12-22
Caving bank of the Belle Fourche River at a meander near Shoma, South Dakota. [Photo by N. H. Darton, U. S. Geological Survey.]

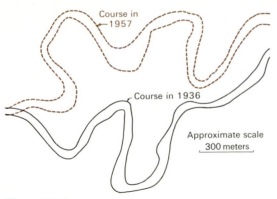

Course in 1957

Course in 1936

Approximate scale
300 meters

Figure 12-23
Migration of meanders of the Arkansas River in
eastern Colorado between 1936 and 1957, from aerial
photographs. At the left (upstream) end of the figure,
the channel was fixed by concrete bridge abutments;
elsewhere it was not artificially restrained.

Origin of Braided Channels

In braided streams the channel constantly subdivides around low alluvial islands that grow from
bars on the stream bed. The bars develop by
accumulation of material too coarse for the
stream to move except in flood. This coarser
material increases channel roughness, still further
reducing the stream's competence; the bar grows
in both height and length and eventually becomes
an island, perhaps temporarily anchored by
vegetation. Other things being equal, a coarse
bed load favors braiding; a fine bed load, meandering.

Braided channels characterize heavily loaded
streams with easily erodible banks, such as those
that carry glacial outwash, but they are by no
means confined to these. Indeed, one of the

Figure 12-24
Double Arch natural bridges, Arches National Monument, Utah. [Photo by Tad Nichols, Tucson, Arizona.]

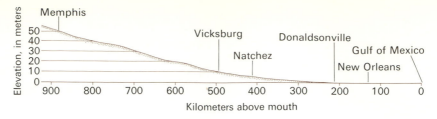

Figure 12-25
Long profile of the lower Mississippi River at low water. [After H. N. Fisk, Mississippi River Commission.]

finest examples of a braided stream was formerly the Platte River in Nebraska, far from any modern glacial influence. It is now controlled by dams and is modifying itself toward a single channel.

Yet it seems that the braided course of the Amazon can hardly be thus explained, for its clastic load is not large. Perhaps the ready erodibility of its sediment permits overloading, but this is mere speculation. Other factors not yet recognized may be involved.

SUMMARY: CHANNEL PATTERNS IN
RELATION TO GRADE

We have seen that discharge, amount and character of sediment load, velocity, gradient, width and depth of channel, roughness, channel patterns, acquisition and deposition of sediment are all closely interrelated: the changes in any one are reflected in corresponding changes in all the others, always in a direction of compensating for the first change. It is for this reason that attempts at flood control have nowhere been wholly successful, for the quantitative changes to be expected from any engineering projects cannot be accurately foretold.

Departures from the Ideal Graded Profile

The smoothly concave curve of the ideal profile is closely approached but probably never attained by any stream. Even the profile of the lower Mississippi, probably as well graded a stream as could be found, shows many slight but abrupt changes in slope (Fig. 12-25). This is normal; we have seen that inflow of most tributaries requires change in gradient. Dynamic equilibrium may be closely approached in each segment, but the result is not an ideally smooth profile.

A principal value of the concept of the graded stream is that notable departures from it point up modifications of stream regimen. Lava flows, landslides, drifting sand dunes, and faults may displace a channel, destroying its graded condition and making the profile highly irregular. Glaciation in the not distant past has greatly modified the courses and slopes of nearly all the major streams in Canada and the northern United States as well as much of northern Europe (Chapter 13).

Earth movements are also disturbing factors. Some anticlines in south-central Washington State have been folded rapidly and recently enough to introduce convexities into the stream profiles (Fig. 12-26).

LAKES

All lakes are major deviations from the graded profile. Many are formed by landslides or blocked by lava flows (Chapter 11), some by glacial erosion (Chapter 13), and some by earth move-

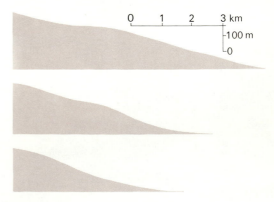

Figure 12-26
Profiles of streams flowing down flanks of anitclines in south-central Washington. [From maps of the U. S. Geological Survey.]

ments — the rise of anticlines or fault blocks across a stream channel. Each lake is obviously an interruption of the graded profile.

STREAM DEPOSITS

Large trunk rivers commonly meander; much more rarely they braid within broad, smooth valleys sloping so gently as to seem flat. The floodplains are so called because, on an average of about twice in three years, a flood stage of the river escapes its channel and overflows the valley floor. Some floodplains are little wider than the stream channel; the channels impinge on bedrock at almost every bend. Most, though, are far wider, and their streams everywhere meander through alluvium.

Low indistinct ridges — **natural levees** — border most river channels on floodplains. They are highest next to the stream bank and slope gently away from it. When a stream overflows, the velocity and turbulence quickly drop as it overtops its banks; the coarsest part of the suspended load drops first, thus building up the levees. Finer sediment is carried farther and distributed widely over lower ground. Since natural levees interfere with tributary drainage, the floodplain behind them is generally partly occupied by shallow lakes and swamps. Bartholomew Bayou and Macon Bayou parallel the southern Mississippi for many kilometers, prevented from joining it by the natural levee.

The silt forming natural levees seldom accumulates to great thicknesses, even over long periods. The Nile floods that renewed the soils of Lower Egypt for millennia before the high dam at Aswan was completed, have buried some Egyptian structures 10 m deep — as much as 2.5 m since the Arab conquest of A.D. 1300; this amounts to only about 30 cm per century. In much of the Nile valley, the accumulation is only a third as much. The muddy Hoang Ho of China, when in flood, is 40 percent silt and 60 percent water; it is raising its floodplain three times as fast, at 90 cm per century. But the contribution of overbank deposits to the formation of the floodplain is far less than that of channel deposits left by meanderings of the main stream. Were the overbank deposits predominant, floods would become increasingly rare as the levees built higher; this is surely not true. Most floodplains contain crescentic lakes and other relics of abandoned meanders, not masked and healed by overbank silts. These **oxbow lakes** (Fig. 12-27) are abandoned river bends left wherever a stream

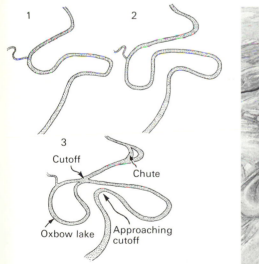

Figure 12-27
(Right) A river meandering on a broad floodplain in northern Australia. Note the low ridge deposits on the insides of bends and the abandoned meander channels, including oxbow lakes. (Left) Possible stages in the evolution of the meander cutoff that has recently taken place in the center of the view.
[Photo by U. S. Air Force.]

crossed the neck between bends and thus shortened its course. The drawing in Figure 12-27 illustrates a hypothetical sequence that might have produced the pattern shown in the photograph.

Nevertheless, most floodplains are deeply alluviated, especially near the sea. The Mississippi River Commission used several hundred drill holes and water wells to study the deposits of the floodplain. Stream deposits range from 30 to more than 125 m in thickness. Near Natchez the alluvium is about 80 m thick, with its base 65 m below sea level. The base of the alluvium is not a smooth plain but a shallow, steepsided valley with many tributaries. In California, the Santa Ana River also has buried its earlier deposits by as much as 40 m at the river mouth. Most large rivers on all continents show similarly filled channels near the sea. These fills doubtless arise from a considerable world-wide rise of sea level that followed the glacial epoch that ended about 10,000 years ago (Chapter 13).

Deltas

On entering quiet water, such as a lake or the sea, a stream drops at least part of its load because of lessened velocity, though if it is heavily silt-laden, it may sink and proceed along the bottom as a turbidity current (Chapter 16), spreading sediment widely over the basin floor. Measurements in northern Sweden showed that a density difference of only 0.0003 suffices to permit a density current (which may or may not be sediment laden) to form. Along an exposed coast, however, a greater contrast seems needed.

Where the Nile emerges from its valley near Cairo, it splits into distributary channels, which subdivide further, eventually to enter the Mediterranean along a front more than 200 km long. From its roughly triangular shape, Herodotus called the distributary plain a **delta**, from the Greek letter Δ.

Deltas may be triangular, generally with a convex sea margin, or irregular with lobe-like extensions, such as the "bird-foot" delta of the Mississippi (Fig. 12-28). The bird-foot distributaries are flanked by low natural levees, and the whole delta surface is gradually sinking, partly because of crustal warping under load, but mainly because of compaction of the delta clays. Delta shapes and sizes depend on the local waves and tides. The Mississippi and Colorado empty into relatively tideless gulfs and have prominent deltas. The muddy Tiber carries 4 million m³ of sand and mud to the tideless Mediterranean and is extending its delta at the rate of 9 m/yr. The Po, advancing into the Mediterranean, has left the Roman naval base of Ravenna 10 km from the shore. But the mighty Columbia and the Congo have no deltas at all. Tides and waves scatter the Columbia's load for scores of miles along the coast, and much of it goes down a submarine canyon to the deep sea. Most of the Congo's sediment is trapped in Stanley Pool a few kilometers upstream from its mouth; what

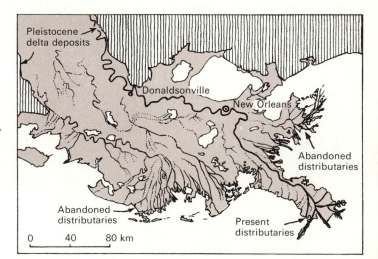

Figure 12-28
The delta (brown) of the Mississippi River. Note the old meander courses, the many lakes, the abandoned distributaries, and the bird-foot pattern of the active distributaries. [After H. N. Fisk, Mississippi River Commission.]

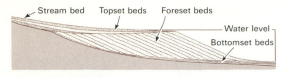

Figure 12-29
Diagrammatic section of a simple delta.

Figure 12-30
Small alluvial fans formed when leaks in a high-level canal caused rapid gullying near Leadville, Colorado. [Photo by M. R. Campbell, U. S. Geological Survey.]

little remains flows down a deep submarine canyon into the depths of the Atlantic. The heavily laden Niger, 1400 km to the north, supplies a huge, rapidly growing delta.

Sand-laden streams flowing into a deep still body of water deposit their sediment in characteristic patterns (Fig. 12-29). The thin overbank sediments topping the delta are flat-lying—**topset** beds. Thicker beds at the front of the delta slope at the angle of repose as **foreset beds**. The finest sediments remain longer in suspension and finally settle as **bottomset** beds in strata parallel to the sea floor, eventually to be covered by the foreset strata of the advancing delta pile.

Most large deltas in stormswept seas are much more complex. The Mississippi delta, for example, shows little discordance between topset, foreset, and bottomset beds except beneath rapidly advancing lobes; it is further described in Chapter 17.

Alluvial Fans

In arid and semiarid regions, a stream emerging from a steep valley onto a more gently sloping lowland forms an **alluvial cone** of intermediate slope whose apex gradually builds back into the steep valley. On emerging onto the flatter slope, both depth and gradient of the stream diminish abruptly, and the stream drops much of its load. In dry regions much water is promptly lost by infiltration in the loose sediments of the cone, hastening deposition.

The slopes of alluvial fans vary with the size of the stream and grain size of its load. Small streams with coarse loads may build slopes as steep as 15° (Fig. 12-30). The slopes of many larger fans (Fig. 15-4) decrease from 3° to 5° at their apexes to less than 1° near their bases. A decrease in average grain size parallels the decrease in slope. Unlike deltas, fans have no long foreset beds, though, as they build up, shorter and more steeply inclined beds come to overlie the gently sloping distal beds. Many deltas include some mudflow deposits. Bedding in such fans is very irregular.

STREAMS AND LANDSCAPE SCULPTURE

Landforms are products of the interplay of constructional and degradational processes. Earth movements and volcanism build up parts of the crust; erosion constantly works to level all to the sea. Most hillslopes owe their shapes to downslope movements and stream erosion.

The form of hillslopes varies greatly from region to region, depending on differences in rocks, soils, vegetation, climate, and the stage of development of the drainage. But within an area of similar rocks, structure, and climate, a characteristic hillslope angle tends to develop.

The Soil Conservation Service of the U. S. Department of Agriculture has extensively studied the influence of various vegetational covers on soil erodibility: for a given soil, both volume of runoff and soil erosion increase with slope up to an angle of about 12°, but the longer the slope, the less soil is removed per unit of length. Some sediment on long slopes is deposited farther down, reducing the slope, because the greater the distance the more infiltration of the soil. On relatively impermeable soils this is ineffective, and such soils tend to wear back at a constant angle. This effect was demonstrated by S. A. Schumm of the U. S. Geological Survey in his study of slopes in the badlands of western Ne-

Figure 12-31
Book Cliffs, northeast of Grand Junction, Colorado, showing contrasting slopes carved in shale and sandstone strata. [Photo by G. B. Richardson, U. S. Geological Survey.]

Figure 12-32
Monument Valley, Utah. Mesas and pillars of sandstone rising above shale slopes. Two structural terraces sustained by resistant sandstone beds are visible in the distance. [Photo by Tad Nichols, Tucson, Arizona.]

braska. He found that slopes on siltstone retreat at a constant angle, whereas those on shale tend to flatten because of creep. For parallel retreat to continue, it is of course necessary that the sediment be continually removed from the foot of the slope.

Hillslopes thus vary greatly in steepness, depending on the cohesion of the rocks. In Figures 12-31 and 12-32, the coherent sandstones stand in nearly vertical slopes, whereas the less coherent, interbedded shales yield much gentler ones, scored by gullies. The contrasts are more conspicuous in arid than in humid country, where they are blurred by soil creep.

A tributary stream cannot deepen its channel below the level of its junction with the main stream—a strong argument that streams in general cut their own valleys. But a stream can lengthen by **headward erosion** until halted by competing streams, similarly extending headward on the opposite side of the divide. If, as is usual, one of the competing streams has a lower base level or is eroding less-resistant rocks, it may be able to extend headward at the expense of the less-effective stream, robbing it of tributaries.

This may result in beheading the stream, an act of **stream piracy** (Fig. 12-33). The result may be that a formerly large stream is diverted to a new course, leaving its old course occupied by an **underfit** stream, much too small to have carved the valley it inherited.

A stream whose upper course has been captured is of course less competent and may become incapable of keeping its channel clear of alluvial fans from the sides, thus permitting lakes to form along it as in Figure 12-33. Because streams from one side of a divide generally have an advantage over those of the other, either in rock weakness or lower base level, it is very common to find lakes near mountain passes.

Stream Patterns

Stream patterns are generally affected by rock structure. Where rocks are flat and a single stratum may cover a considerable area, tributaries generally subdivide headward like the limbs of a tree, forming **dendritic** patterns (Fig. 12-34, Left). Streams extend headward by following

Figure 12-33
Unaweep Canyon, on the Uncompahgre Plateau, Colorado, the former course of the Gunnison-Uncompahgre River, which was beheaded by the capture of its upper course by the headward growth of a tributary of the Colorado River along the weakly resistant Mancos Shale (Cretaceous). Downcutting of the Gunnison-Uncompahgre was slowed by the resistant Precambrian granite. [Photo by C. B. Hunt, U. S. Geological Survey.]

274

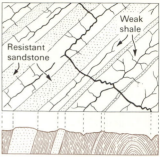

Figure 12-34
Structural control of stream patterns.
(Left) Dendritic drainage. (Right)
Trellis pattern.

Figure 12-35
A broad structural plain along the Colorado River, Arizona, formed where weak shale overlies a more resistant
rock. The shale is in turn overlain by more resistant massive sandstone, remnants of which form cliffed buttes
in the center of the view. [Photo by John S. Shelton and Robert C. Frampton.]

rock weaknesses; where strata are tilted so that
rocks of different resistance occur at the same
elevation, rivers selectively follow the less-
resistant beds, crossing the more-resistant at high
angles to their strike (Fig. 12-34, Right) either by
piracy or by downward erosion from a higher
plain. In steeply dipping beds a **trellis** pattern
develops. Other examples of **structural control**
of stream patterns are the concentric patterns on
eroded domes and anticlines (Figs. 5-11, 5-12)
and the strikingly linear courses of streams along
some faults (Fig. 11-25). Though a stream system

is in constant change because of climatic variations, earth movements, and its own regimen,
it accentuates differences in rock resistance
wherever it cuts downward.

Structural Terraces

Etching out of the less–resistant rocks by rainwash, creep, and stream erosion produces striking
effects even in areas of flat-lying strata. Here
streams tend to be uniformly spaced, cutting

Figure 12-36
Newly entrenched channel formed after relative uplift of the Tobin Mountain block, Nevada, at the time of an earthquake in 1915. The former valley floor now forms paired terraces. [Photo by Ben M. Page, Stanford University.]

Some terraces are normal products of stream erosion and downcutting; many are modified by tilting. Several terraces along the Mississippi approximately parallel the river slope between Cairo and Natchez, and range from about 20 to a few scores of meters above it. South of the Mississippi–Louisiana line, the terraces converge and lie on Pleistocene alluvium instead of on older rock. Near Baton Rouge they disappear beneath the present floodplain. This pattern of warped terraces is striking evidence of slow earth movements: the delta has subsided while the area upstream has progressively risen.

Most terraces, however, are neither caused nor modified by earth movements but result merely from normal downcutting of the stream or an increase in its competence because of climatic change, decrease in load, or some other factor independent of tectonics.

readily through the less-resistant rocks but perhaps arrested in downcutting by a resistant stratum that makes a widespread temporary base level. Above this bed headward erosion may extend the drainage over a wide area, stripping off the weak strata to form a **structural terrace** or plain at the level of the resistant bed. (Figs. 12-32, 12-35). After deep subsequent erosion to a lower base level, in an arid or semiarid climate, the eroded edge of a resistant layer that formerly supported a structural terrace may form the flat top of **mesas** and **buttes** whose tops are remnants of the plain.

Stream Terraces

Relatively few terraces are structural. Far the most are wholly stream deposits; many others are cut in bedrock but have veneers of river gravel on their flat surfaces. These are obviously remnants of older floodplains, now incised either by uplift (Fig. 12-36), lowering of sea level, climatic change affecting the stream regimen, or mere downcutting toward base level.

Other terraces caused by faulting lie on the west side of the Panamint Range in California (Fig. 12-37). After faults formed a series of small cliffs across the huge alluvial fans, the streams cut deep trenches in the uplifted parts of the fans and dumped steep new fans on the downdropped blocks.

PENEPLAINS

Deductions from stream behavior suggest that long–continued stream erosion should reduce even the most resistant rocks nearly to base level. In the absence of earth movements, broad plains should ultimately develop, with the interstream divides reduced to gentle slopes. Such a hypothetical surface is a **peneplain** (from the Latin, paene, "almost"). Great areas of more or less accordant summits—in Colorado, 55 peaks stand between 4100 and 4260 m in altitude—have been thought to be remnants of peneplains formed by erosion near base level. Similarly, in the Appalachians of Pennsylvania, many ridges stand at approximately the same elevation for long distances, and have been widely considered remnants of peneplains formed at much lower elevations. Indeed, such accordant summits may be in part residual from peneplains, but in view of the rapid erosion of steep mountains—at least 0.5 m per thousand years—the Colorado Mountains have been lowered several thousand meters since Miocene time, when the major uplift of the range took place.

Surfaces of low relief are of course lowered less rapidly, but even in the valleys of Pennsylvania the surface must have been lowered by several hundred meters since the early Tertiary. It seems that such rough accordance of summit levels as appears in high and steep mountains are

Figure 12-37
Faulted and entrenched fans at the mouth of Tuber Canyon, Panamint Range, California. Note the new fans growing on the downthrown block and their relation to the channels on the upthrown block. [Photo by John S. Shelton.]

generally better explained by downcutting of sub-equally spaced drainage channels dissecting the surface at roughly equivalent rates. If the surfaces are residual from a former plain, that plain was at a considerably higher elevation than the present surfaces unless it is very young indeed.

Consider, for example, the Ozark Plateau, in southern Missouri and northern Arkansas. Stream gaging and analysis of the runoff prove that the plateau is being lowered at the average rate of 20 m per million years by solution alone. A peneplain formed in the Oligocene, say 30 m.y. ago, would have been lowered 600 m by this process,

and still more by removal of suspended and bed loads. Few landforms older than Pleistocene remain except those that have been protected by unconformably overlying rocks.

THE UPPER DOMINANT LEVEL OF THE EARTH

The analysis of the graded streams and base level give insight into the control of the upper of the two dominant levels of the earth's surface that were pointed out in Chapter 2 and illustrated in

Figure 2-4. Base level—the surface of the sea—puts a limit on downcutting of streams; their upward concave profiles assure that more land is near sea level than at any other elevation, and the deposition of sediment off the shore assures that the nearshore subsea profile will also slope gently in general. We have already seen that the second prevalent level is accounted for by the differing densities of continental and sea floor rocks.

Questions

1. Would you expect the dissolved load per cubic meter of water to be higher in the Columbia (high rainfall) or in the Colorado River (low rainfall)? Why?

2. At most stream junctions the surface of the tributary and of the main channel are identical in elevation. Why?

3. Engineers have made many artificial cutoffs (Fig. 12-20) in the lower Mississippi and other meandering rivers. Considering the nature of meandering streams, can you suggest reasons for these projects?

4. List several criteria for distinguishing between floodplain, delta, and alluvial-fan deposits in ancient sedimentary rocks.

5. Suggest how a change in climate might produce stream terraces in areas with which you are familiar.

6. The St. Lawrence, one of the great rivers of the continent, has no delta, even though it runs into a landlocked estuary. Can you suggest why?

7. The longitudinal profile of most large rivers resembles the land portion of the graph in Figure 10-4. Can you offer any explanation of this?

8. At what point of an alluvial fan is the sediment coarsest? Why?

9. The Yazoo River, on a common floodplain with the Mississippi, parallels that stream for scores of kilometers before joining it. Why doesn't it join sooner?

10. Assuming flat-lying strata, would you expect stream spacing to be closer on a thick sandstone bed or on a thick shale formation? Why?

Suggested Readings

Fisk, H. N., *Fine Grained Alluvial Deposits and Their Effects on Mississippi River Activity*. Vicksburg, Mississippi: Waterways Experiment Station, 1947.

Gilbert, G. K., *Geology of the Henry Mountains*, (U.S. Geographical and Geological Survey of the Rocky Mountain Region, 1877). [Pages 99–150, *Land Sculpture*—a classic paper, outlining the principles of stream erosion and applying them to the origin of the landforms of central Utah. A milestone in geology.]

Leopold, L. B., M. G. Wolman, and J. P. Miller, *Fluvial Processes in Geomorphology*. San Francisco: W. H. Freeman and Company, 1964.

Rubey, W. W., *Geology and Mineral Resources of the Hardin and Brussels Quadrangles, Illinois* (U.S. Geological Survey, Professional Paper 218). Washington, D.C.: G.P.O., 1952. [Pages 101–137, *Physiography*. Pages 129–136 give a clear, concise account of the modifications a stream undergoes in adjusting its grade.]

Schumm, S. A., and M. P. Moseley, Slope Morphology. Benchmark Paper v. 6.

Sundborg, Äke, *The River Klarälven, a Study of Fluvial Processes, Geografiska Annaler* v. 38, 1956, pp. 127–316. [A good description of the hydraulics of river channels and their relation to the morphology of a particular river.]

Scientific American Offprints

817. W. D. Ellison, "Erosion by Raindrop" (November 1948).

826. Raymond E. Janssen, "The History of a River" (June 1952).

836. Gerard H. Matthes, "Paradoxes of the Mississippi" (April 1951).

869. Luna B. Leopold and W. B. Langbein, "River Meanders" (June 1966).

CHAPTER 13

Glaciers and Glaciation

Glaciers are slow-moving, thick masses of ice; **snowfields** are thinner, almost motionless masses of permanent snow (Fig. 13-1).

Since the seventeenth century most glaciers have been shrinking; on the average, snowfall has not balanced the melting. Exceptions are common, though; of two glaciers heading in the same New Zealand icefield, one is advancing, the other retreating. The Taku glacier in southeastern Alaska advanced 5 km in the past fifty years while ten others nearby have retreated. While its neighbors were retreating, the Bruggen glacier in Patagonia advanced 7 km between 1830 and 1962.

Tidal records from all oceans prove that sea level has been rising about 1 mm a year on the average for the past century and a half while a general glacial retreat has gone on, emphasizing that the glaciers are storage places in the water economy of the earth. We saw in Table 12-1 that they store 2.24 percent of the earth's water — 77 percent of all not in the oceans.

THE SNOWLINE

The lower limit of elevation of permanent snow is the **snowline**; its altitude varies with latitude, snowfall, temperature, wind direction (which controls drifting), and topography (which controls both snowsliding and shading from the sun). Mean annual temperature decreases with both altitude and latitude; precipitation also varies with latitude, being less in the Horse Latitudes than at those either higher or lower. Accordingly, the snowline is at 4100 m on Mount Kenya and at 4400 m on Mount Kilimanjaro in equatorial Africa, at 5900 m in Tibet in the rain shadow of the Himalayas, and at more than 6100 m in the Andes of northern Chile (Fig. 13-34). On the dry eastern slope of the St. Elias Mountains on the Alaska-Yukon border, the snowline is at 2300 to 2600 m, but on the western slope nearly 1500 m lower. It is lower in well-watered Norway than in far colder but dry Taimyr Peninsula of Siberia. Permanent glaciers are lacking in most of Siberia, in Canada east of the Cordillera, except in the Arctic Islands, and in Alaska north of the Brooks Range. The mean annual temperature is low enough to preserve ice, but snowfall is too low to supply it. In these areas, beneath a few decimeters of summer-thawed, winter-frozen tundra is a great thickness of permanently frozen ground, **permafrost**, in parts of Siberia as much as 1600 m thick and in much of Alaska 400 m or more — 600 m in the Prudhoe Bay oil field. Permafrost underlies 20 percent of the land area of the earth. Apparently a mean annual temperature of $-1°$ to $-8°$ C is needed to produce it.

Figure 13-1
Snowfields above the head of a valley glacier in Alaska. Clearest relations are at right foreground and left background; much of the flat in mid-distance is a snowfield, but it merges into moving ice and becomes a glacier below the crevasses that mark the head of the steep slope in the center of the photo. [Photo by U. S. Air Force.]

SNOWFIELDS

Permanent snowfields cover all but the steepest and windiest slopes above the snowline (Figs. 13-1, 13-13). Excavations in snowfields, where downslope movement is slight, show that annual additions of snow are distinguishable by slight textural variations. The beautiful geometric patterns of new-fallen flakes (Fig. 13-2) do not persist to depth. Instead, only a few meters below the surface, snow has recrystallized into **firn** — small granules about a millimeter or so in diameter that grow steadily larger downward. At depths of a few tens of meters the firn granules themselves recrystallize into large interlocking grains that include air bubbles surviving from the intergranu-

lar air pockets of the firn. In places the ice is added to by the freezing of summer meltwater. Inasmuch as all precipitation contains a little salt (from sea spray) in solution and salt water freezes only at lower temperatures than pure

Figure 13-2
Forms of fresh snowflakes. [After A. E. H. Tutton, 1927.]

water, a concentrated salt solution forms a sub-microscopic film along grain boundaries, lubricating intergrain movements. Samples from a depth of 500 m indicate that such films may make up as much as 3 percent of the mass of an Alpine glacier.

The melting and freezing is not entirely controlled by the temperature. Water expands 9 percent on freezing; pressure therefore lowers the melting point. (This is why a snowball sticks together when squeezed and released: the water melted at grain contacts by the pressure has refrozen to lock the grains together.)

For these reasons most mountain glaciers, even as far north as central Greenland and Spitzbergen are always, except near the surface where winter cold may prevail, at a temperature very near the pressure-melting point appropriate to the depth of the ice. These are **temperate glaciers**. **Polar glaciers**, such as those of Antarctica and

northern Greenland, are colder than the pressure-melting point, and their salty intergranular films are also frozen.

Crevasses even in temperate glaciers are deep enough to reveal all stages in the recrystallization of snow through firn to solid ice. In stagnant ice fields single crystals of ice as much as 24 cm across have been found. Strangely enough they enclose air bubbles at roughly the same spacing as in the firn; the bubbles have not been excluded as the crystal grew. We can thus trace the metamorphism of a sediment, snow, into a metamorphic rock, ice.

KINDS OF GLACIERS

The topography over which a glacier flows largely controls its form. **Valley glaciers** (Fig. 13-3) are ice streams flowing down steep-walled mountain

Figure 13-3
Steele glacier, Yukon Territory, a surging glacier. [Photo by Austin Post, U. S. Geological Survey.]

Figure 13-4
Hanging glacier on the north side of Mount Athabaska, Jasper National Park, Alberta. [Photo by Warren Hamilton, U. S. Geological Survey.]

valleys. Fed by snowfields above, such glaciers may extend far below the snowline. The glacier ends where the front melts as fast as it is replenished from above.

Glaciers occupy lofty mountain valleys the world over—even in the tropics, as in the Carstenz Range in New Guinea, Ruwenzori in Uganda, and Cotopaxi in the Ecuadorian Andes. The valley glaciers of the United States, except those of Mount Rainier, where as much as 23 m of snow fall in a single year, are short ice streams only a few scores of meters thick. Many are hardly distinguishable from snowfields, into which all gradations exist. The Rocky Mountains, Cascades, and Sierra Nevada hold hundreds of small irregular ice masses—**cliff glaciers**, or **hanging glaciers**, in well-shaded clefts opening out over steep cliffs (Fig. 13-4).

In contrast to these puny streams, valley glaciers in the Himalayas and Alaska are more than 100 km long and more than 1 km thick, with many tributaries merging into an integrated system draining thousands of km² (Fig. 13-3). At the foot of the St. Elias Mountains in Alaska, several of these valley glaciers emerge, spread over the plain, and coalesce to form Malaspina Glacier (Fig. 13-5), a lobate mass covering nearly 2000 km²—a **piedmont glacier**. Behind the stagnant frontal area, ice emerging from the valley glaciers above is forced into horizontal fold patterns of fantastic scale and complexity (Fig. 13-6).

The great ice barrier of the Ross Sea, the size of Texas, is also formed partly by coalescence of valley glaciers at the mountain front, but here most of the ice is floating rather than lying on a coastal plain (Fig. 13-7).

Small masses of radially spreading ice blanket much of Iceland, Spitzbergen, parts of Scandinavia, and the Canadian Arctic islands (Fig. 13-8). These are **ice caps**.

The largest glaciers of all, the **continental glaciers**, are now found only at high latitudes—Greenland and Antarctica—but they were formerly much more widespread. All of Greenland except a narrow coastal strip is covered by ice flowing from two high centers. The ice mass covers 1.73×10^6 km² and averages 1.6 km thick, with a maximum thickness of 3.4 km and a volume of 2.8×10^6 km³. Much of Greenland's rock surface is below sea level, though the ice surface

Figure 13-5
Folds in the Malaspina glacier, Alaska. Seward glacier in the right rear. The scale of folding is in kilometers, in places as much as 16 km. [Photo by Austin Post, U. S. Geological Survey.]

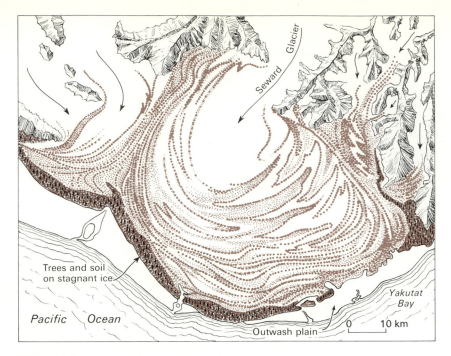

Figure 13-6
Map of the Malaspina glacier, Alaska. The arrows indicate the flow of the valley
glaciers that feed the Malaspina. [After R. S. Tarr and Laurence Martin, 1914.]

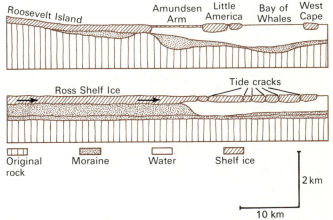

Figure 13-7
(Left) Ice in the Bay of Whales, Antarctica. The Bay
of Whales is a re-entrant in the Ross Shelf Ice, protected in part by islands. The bay ice, 10 to 15 m
thick, is folded by the pressure of the advance of the
much thicker (120–150 m) shelf ice around the
protecting islands. The individual folds are several
meters high. Similar folds are formed by the drag of
the shelf ice over its morainal deposits. [Air photo
by T. S. Poulter.] (Above) Sections through the Ross
Shelf Ice and the Bay of Whales. The ice thickness
has been determined by seismic methods. [After T. C.
Poulter, Stanford Research Institute.]

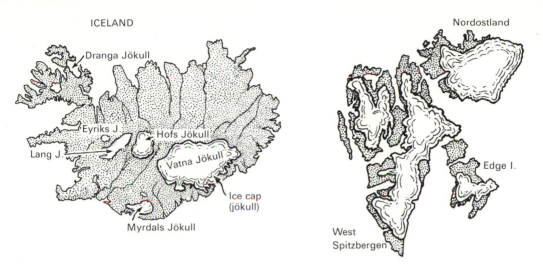

Figure 13-8
Ice caps of Spitzbergen and Iceland. [After Stieler's *Hand Atlas*.]

is far above. Parts of the coast are mountainous, and inland ice flows through valleys in tongues to spill icebergs into fiords.

Antarctica supports a much larger continental glacier—in fact, one that contains 90 percent of the earth's ice. It covers an area of 12.6 million km² — larger than the United States and Mexico — with an average thickness of more than 2 km; its volume, estimated at 25 million km³, if melted, would suffice to raise sea level over the earth by 40 m. At the South Pole the ice surface is nearly 3 km above the sea, but most of the rock surface is close to sea level. The glacier overrides the coast in many places, and huge bergs deform the shelf ice. In places the ice escapes between mountains in great glaciers; the famous Beardmore Glacier, ascended by several early explorers in their quest of the Pole, is 500 km long and 20 km wide, and flows from the interior plateau far out into the Ross Shelf Ice.

GLACIER FLOW

Ice appears rigid and brittle; under a blow it shatters like glass. Under short-term stress, it appears very strong, but if it were, even the highest mountains above snowline would eventually be buried in snow, firn, and ice. Under long-term stress, ice is a weak rock; when only a few meters of ice have formed beneath firn and snow, it begins to flow downslope. Laboratory experiments show that pure ice at the freezing point begins to flow under a shear stress of about 1 kg/cm², but many glaciologists think that under natural conditions it may flow at stresses only a tenth as great. The rate of flow increases with higher stresses at a rate proportional to the third or fourth power of the stress. Glaciers are perfect examples of the influence of size and time on strength (Chapter 2).

Load adequate for notable creep is reached at a depth of about 20 m in temperate glaciers, and crevasses in them are seldom deeper; in polar glaciers, where temperatures are far below the pressure-melting point, ice is stronger, and crevasses as deep as 60 or 70 meters are found in stagnant ice. Deeper crevasses are kept open only by continual movement or by meltwater pouring into them.

Glacier motion is imperceptible to the eye, but is readily demonstrated by driving rows of stakes across the surface (Fig. 13-9). Velocity is greatest over the thickest part of the glacier—generally near the middle, where frictional resistance is least. The speed varies seasonally. In winter the higher part of the glacier is more heavily loaded with snow and hence moves faster; in summer the more rapid melting at the glacier snout lowers friction and the lower part speeds up.

The speed may suddenly increase through the whole glacier in a **surge** (Fig. 13-3). Surges are explained by large-scale melting at the base because of frictional heat when the velocity exceeds

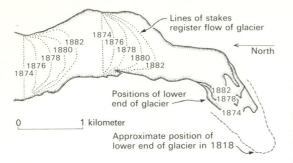

Figure 13-9
Records of flow and frontal shrinkage of the Rhone glacier, Switzerland. [After Albert Heim.]

some critical value. The Kutiah glacier in the western Karakorum advanced as much as 113 m/day in 1953; during three months it advanced fully 10 km.

Speeds of 50 m per day have been measured on a few Alaskan and Greenland glaciers, but such velocities are exceptional (Fig. 13-3). The Beardmore glacier moves at variable rates between 300 and 1400 m/year. The Athabaska glacier in the Canadian Rockies, where it is 200 meters thick and on a slope of 6.3°, moves at the rate of 65 m/yr; where it is 320 m thick but the slope is only 3.5°, it travels at only 30 m/yr. The Argentiére glacier in the French Alps sometimes moves

as much as 41 mm/hr, but its annual average is only 200 m.

At the other extreme, the Malaspina glacier is virtually stagnant near its snout.

The mechanism of flow is complex. Microscopic studies show that some crystals bend, some glide along lattice planes, others are broken and granulated, others melt and refreeze. Thus, though the mass as a whole flows, some crystals break, as in the metamorphic flowage of many other rocks. Even at the front of a moving glacier, crystals rarely exceed 2 or 3 cm in diameter, in contrast with the huge crystals in stagnant fields.

Many crystals are strung out in parallel sheets, as in other foliated metamorphic rocks.

Ice in deep crevasses generally shows layering that superficially resembles the true stratification found in snowfields. The layers roughly parallel the floor and curve upward along the walls. Measurements show that adjacent layers move at slightly different speeds; close to the wall each successive layer inward flows slightly faster than its neighbor. The layers are thus not bedding, but shear surfaces. Measurements in tunnels cut into moving glaciers show that the shearing is by no means uniform. Much of the motion, ranging from as little as 12 percent in a thick temperate glacier to as much as 90 percent in a thin polar one, is concentrated into a narrow zone just above the glacier bed, where a thin layer of water is present.

Figure 13-10
Concentrated zone of shearing over a roche moutonée at the base of the Grindelwald glacier, Switzerland. [After Hans Carol, *Journal of Glaciology*, 1941; by permission of the Glaciological Society.]

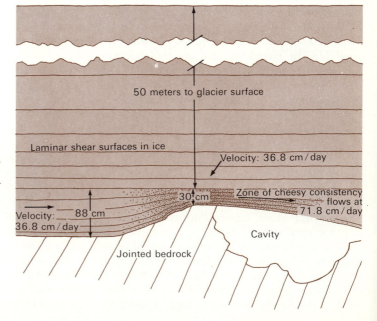

Figure 13-11
Overfolded silt band at snout of Taylor glacier, South Victoria Land, Antarctica, showing how ice overriding the basal slip surface has dragged the silt band into a flat fold. [Photo by Warren Hamilton, U. S. Geological Survey.]

In the polar glaciers of Antarctica this meltwater is attributed to heat supplied from below, for it is present even in areas of no perceptible motion. In the temperate Blue glacier of the Olympic Mountains, Washington, the bottom 2 m of a pipe inserted vertically in the glacier was bent through an angle of 55° in a single year; bending at higher levels was slight.

A crevasse gave access to the floor of the Grindelwald glacier of Switzerland at a point where the ice was about 50 m thick. Here an upward projection of bedrock obstructed the flow. A shear zone separated brittle ice above, moving at 36.8 cm/day, from a mass below, about 30 cm thick, that had the consistency of cheese and a velocity of 71.8 cm/day, nearly twice that above.

The whole basal meter of the ice upstream from the obstacle was squeezed into this thin cheesy layer. As it passed over the obstruction it immediately recrystallized into firm ice sufficiently strong to roof a cavity several meters wide. (Fig. 13-10). Had the glacier been much thicker, such an open cavity could not have been maintained.

Carbon dioxide is as much as twenty times more soluble in snow and ice than in water. As a result glaciers in the Canadian Rockies dissolve limestone from the upstream side of bed projections and redeposit some of it in the lee side.

The shear banding of glaciers is commonly emphasized by streaks of rock debris dragged into the glacial mass (Fig. 13-11). The foliation, very like that of many gneisses and schists, proves that the ice is no longer brittle but reacts to stresses by flowage.

The upper parts of most glaciers are covered with snow, even through the summer. Farther down, snow accumulated in winter but is melted and evaporated by the end of the summer, exposing the underlying firn. Still farther down-glacier the firn ends: evaporation and melting have exposed the underlying ice. This lower limit of firn in summer is the **firn line**; above it is the zone of net accumulation, below it the zone of net loss. It is the line at which the average velocity of glacial flow is generally greatest.

Above the firn line the transverse profile of the surface of a valley glacier is always concave, sloping toward the glacier axis; below the firn line it is convex. (Fig. 13-12). Above the firn line a particle of ice tends to sink into the body of the glacier as it becomes buried by increasing thicknesses of snow, firn, and ice; below the firn line it rises toward the surface as the overlying ice melts. The shear zones of the glacier are similarly curved. Toward the snout, the shear surfaces tend to steepen and become thrust surfaces over the more stagnant lower frontal ice.

Figure 13-12
Portage glacier, Alaska, showing concave profile above and convex profile below firn line, which is just below the steep glacial slope in middle distance. [Photo by U. S. Air Force.]

GLACIER LOADS

Frost Weathering and Avalanching

Glaciers acquire rock debris in several ways. Valley glaciers generally cover only small parts of the mountains from which they flow. Gentle slopes hold extensive snowfields, but steep slopes are swept bare by wind and avalanche, exposing great expanses of crags and cliffs above the glaciers. Frost action strongly shatters these rocks (Fig. 13-4). During the day meltwater from snowbanks seeps into the crevices; that same night it may freeze, breaking off fragments. Small grains, loose chips, and even great blocks of rock thus freed roll down the slopes and accumulate in talus piles at the glacier edge. Landslides and rockfalls from cliffs undermined by the glacier crash down along with snowslides. The power of these slides is truly stupendous: snow fences at Fionnay, Switzerland, designed to withstand pressures of 12 tons per square meter, were swept away by a slide on a slope only a little steeper than 15°.

In this way a glacier becomes charged with rock debris, especially along its edges, making **lateral moraines**. Where two valley glaciers join, their inner moraines unite to form a **medial moraine**; if tributaries are many, several moraines may streak the surface of the trunk glacier (Figs. 13-6, 13-13). The moraines are not merely surficial; they extend into the body of the glacier and eventually to the very bottom, for rocks are denser than ice. Glaciers thus drag rocks along their floors, pick up fractured material from them, and rasp their beds. Crevasses that open as the glacier moves over changes in slope of the floor also permit surface material to reach the bottom.

Meltwater Shattering

At the head of most valley glaciers is a deep arcuate crevasse or a series of them, the **bergschrund** (Fig. 13-14), formed as the plastic downhill motion of deeper layers of the glacier dragged the brittle surface ice away from the headwall. The bergschrund yawns open in summer but is generally filled or bridged with snow in winter.

Figure 13-13
Valley glacier in Alaska Coast Range near Skagway, Alaska. Note medial moraines springing from spurs between tributaries. [Air photo by Tad Nichols, Tucson, Arizona.]

Figure 13-14
Gravity movements at the head of a glacier, Sunset Amphitheatre, Mount Rainier, Washington. Length of view one and a half kilometers. Note (1) the sinuous bergschrund formed where the glacier moved away from the headwall, (2) the dark streaks and fans of broken ice and rock left by avalanches from the higher chutes, and (3) the precariously balanced blocks at the edge of the 60-m high cliff in shadow. [Photo by Miller Cowling, 116th Photo Section, Washington National Guard.]

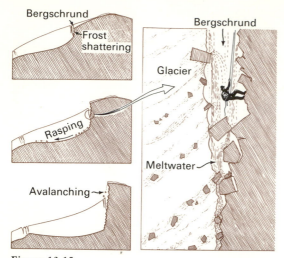

Figure 13-15
Progressive stages of erosion at the head of a valley glacier (left), with a detail of the bergschrund (right). [In part after W. V. Lewis, *Journal of Glaciology,* 1938.]

Adventurous observers have descended bergschrunds on ropes to find that at depth the crevasse has a rock wall on the upstream side and an ice wall on the other. The rock wall is cracked and riven with joints, some gaping as the blocks they bound are loosened (Fig. 13-15). Many blocks are broken free but are still nearly in place; others lean out against the ice in precarious imbalance; still others are entirely separated from their parent ledges and are embedded in the ice. During the summer day, meltwater pours over the bergschrund cliff, filling the crevices; the almost nightly freezing loosens new blocks that tumble into the chasm; thus the glacier grows headward.

Crevasses deeper than 20 or 30 m are normally closed by plastic flow but the continual drag of the glacier away from the headwall permits sporadic opening of the bergschrund to greater depths. On hot summer days great quantities of meltwater cascade into the bergschrund but never fill it; the meltwater must open its own channel at the bottom of the glacier, where, of course, the melting point is lowest. The meltwater carries heat down, keeping the drainage open. Daily temperature changes do not reach deep into the glacier though; frost shattering must be very much less than at the surface.

Cirques

The moving glacier carries away the rock shattered in the bergschrund, exposing new surfaces to attack. Many valley glaciers have melted completely away, exposing their headwalls. Most such glaciated valleys head in semicircles of steep cliffs that bound a rock basin holding a pond or lake. Such cliffed valley heads are **cirques** (Fig. 13-26) (Mt. Assiniboine). The cliffs at cirque heads are shattered for their full height, and generally meet the grooved and polished valley floor at a high angle. Such cliffs, jagged and shattered products of frost plucking in the bergschrund, contrast strongly with the smoothed valley floor, polished by the glacial debris.

The Glacial Rasp

On the floor of a glacier, the ice, heavily charged with rock debris, moves over the bed like a gigantic rasp (Fig. 13-16), gouging into the rock beneath, scraping off irregularities and polishing and grooving it. In places, entrained debris breaks

Figure 13-16
Rock surface showing glacial grooving, polishing, and, at the left, plucking, near Mount Baker, Washington. The ice flowed from the upper left to the lower right. [Photo by H. A. Coombs, University of Washington.]

Figure 13-17
Lunate chatter marks on the back of a roche moutonée near the lower end of Florence Canyon, Sierra Nevada, California. [Photo by F. Matthes, U. S. Geological Survey.]

off lunate wedges of rock from the floor—**chattermarks**—resembling the tracks of a horse going downglacier (Fig. 13-17).

Debris in Continental Glaciers

A continental glacier covers so much of the ground that frost shattering and avalanche accumulations are limited to the rare pinnacles that project above the ice. Yet the Greenland ice is as heavily charged with debris as any valley glacier. It must have been entrained at the bottom of the glacier, but the mechanism by which this was done is still poorly understood.

Erosion by Continental Glaciers

Rock floors exposed by the melting of continental glaciers are scratched, deeply grooved, and locally chattermarked (Fig. 13-17). Continental ice sheets are generally thicker than most valley glaciers, and the added weight makes them very effective rasps. The preglacial valleys of the Great Lakes and the Finger Lakes of New York were scoured to great depths by the Pleistocene glaciers that once occupied the country.

A glacier's efficiency in erosion depends on four factors: (1) the resistance to abrasion of the floor, (2) the abundance and hardness of the rock debris entrained in its basal layers, (3) the thickness of the ice, and (4) the speed and duration of the flow. Thick continental glaciers flowing over weakly coherent rocks cut down rapidly; valley glaciers are most effective on cirque floors, where the ice is thickest and most heavily armed.

Where great thicknesses of ice have been funneled into mountain valleys at the seashore, as in the fiords of Norway, Alaska, Greenland, and New Zealand, deep troughs are carved several hundreds of meters below sea level. Bedrock sills at shallower depth generally mark the approximate maximum advance of the ice, where it broke up into icebergs.

Till

The effectiveness of glacial erosion is attested by the great amount of debris released at the snout on melting. Hummocky ridges of unsorted mixtures of boulders, sand, and silt are piled along the ice front (Fig. 13-18) as **till**. The proportions

Figure 13-18
Till deposited by a valley glacier, West Walker River, Nevada. The largest boulders are about half a meter in diameter. [Photo by Eliot Blackwelder, Stanford University.]

of boulders to fine material in till varies widely: some till is mainly boulders, but that from thin ice caps resting on shale may be chiefly clay and silt.

Rock fragments in till differ from those in stream and beach deposits in being largely sub-rounded or even sharply angular rather than rounded. Some have been crushed, and many are little-modified joint blocks. Some are faceted, with nearly flat, grooved, and polished surfaces formed as they were dragged along the floor.

Rock Flour

The glacial rasp produces much **rock flour**—extremely finely ground silt—so abundant as to make the meltwater milky as it gushes from the glacial snout. Such streams are invariably over-loaded, and their braided channels quickly deposit rubble and coarse sand on a floodplain at the glacier's end. The finer debris goes on to form bottom deposits in lakes and seas. Rock flour from granodiorite builds great terraces of white silt along many rivers of Washington, British Columbia, and Alaska.

Rock flour is also widely blown about in late summer when the shrinking glacial streams expose their braided channels to the wind. Unstrati-fied deposits of **loess** (a loam consisting of silt particles settled from the wind) are widespread

near many glaciers and formerly glaciated areas (Fig. 13-19). Most loess contains innumerable roughly vertical tubules left by the rotting of grass stems and roots. Although only feebly con-solidated, loess stands well in nearly vertical banks.

GLACIAL DEPOSITS

The debris dumped by glaciers or glacial streams is **glacial drift**; the unstratified material freed directly by melting is **till**; that reworked by streams is **stratified drift**.

Moraines

Most till forms moraines, a term applied both to the topographic forms of the deposits and to the till composing them, as well as to the debris on or in an active glacier. Most large moraines form at the glacier front; if the rates of flow and melt-ing are roughly constant, so that the snout re-mains nearly stationary for a long time, the moraine may become large and hummocky. As the ice front often varies with minor climatic changes, more than one morainal ridge commonly forms. The outermost is the **end moraine**; those formed during halts in the glacial retreat are **recessional moraines**. Small recessional moraines

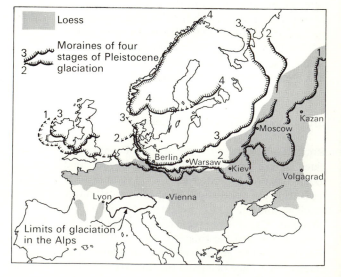

Figure 13-19
Relations of loess-covered areas to the end moraines of four glacial advances in Europe. [Adapted from R. F. Flint, *Glacial Geology and the Pleistocene Epoch*, John Wiley & Sons, 1947, and from R. A. Daly, *The Changing World of the Ice Age*, Yale Univ. Press, 1934.]

Figure 13-20
Recessional moraines in lower Victoria Dry Valley, South Victoria Land, Antarctica. The surficial deposits are patterned by front polygons. A frozen lake lies against the snout of the glacier. [Photo by U. S. Navy for U. S. Geological Survey.]

may be almost buried by later outwash (Fig. 13-20).

The terminal and recessional moraines of most continental glaciers are broadly lobate and can be followed for many kilometers except where breached by outwash from later stages (Fig. 13-19). Some debris left by the melting ice is strewn as patches of till over the deglaciated area (Fig. 13-21). Most such **ground moraine** is not aligned in ridges but is scattered at random as the most widespread residue of a continental ice sheet.

A later glacial surge may mold the ground moraine into systematically clustered hills, each shaped like half an egg cut lengthwise (Fig. 13-21). These **drumlins** range widely in size; many are as much as a kilometer long, 150 m wide, and fifty m high. Excavations show bedrock cores in some drumlins, but most are wholly till.

294

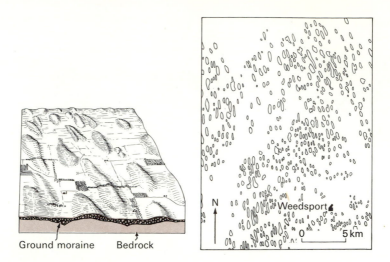

Figure 13-21
Sketch of ground moraine and map of drumlins near Weedsport, New York. [From Weedsport quadrangle, New York, U. S. Geological Survey.]

Stratified Drift

The coalescing alluvial fans of many braided streams spread from the margins of continental glaciers to build **outwash plains**. Many are pitted by countless undrained holes, most only a few meters across but others a kilometer or more long and 30 or 40 m deep. These **kettles** mark the places where blocks of ice, stranded during the glacial retreat, were surrounded or buried by outwash gravels and ultimately melted to leave the depressions (Fig. 13-22).

Many outwash plains are dissected by streams

Figure 13-22
Small kettle lake in outwash gravels of the Baird glacier, Alaska. [Photo by A. F. Buddington, U. S. Geological Survey.]

(no longer overloaded after glacial retreat), and remnants stand as terraces. Long ridges of stratified sand and gravel wind across some formerly glaciated areas; most are only a few tens of meters high and 100 to 200 m wide, but may be several kilometers long. These **eskers** merge downstream into outwash plains or abandoned deltas. They must have been built by subglacial or crevasse streams while the ice was stagnant; the slightest bodily movement by the glacier would have strewn them into ground moraine.

Many lakes are dammed on one side by ice, and water also fills holes in the wasting glacier. Such lakes are of course unstable and fluctuate in level as the glacier moves. Some in Alaska are completely drained nearly every year. Their sediments are thus commonly interlayered with stream deposits and till. Patches of such deposits are **kames**, or if standing above the modern stream, **kame terraces**.

GLACIAL LAND FORMS

Areas recently uncovered by glacial recession have landforms that are very different from those of unglaciated areas. The polished and grooved bedrocks of glacial floors differ greatly from those of stream channels. Glaciated areas contain many lakes in rock basins scooped out by the ice or dammed by moraines. The basins of the Great Lakes, the Finger Lakes of New York, Lakes Maggiore, Leman, and Luzern, as well as thousands of others were excavated by glacial scour. Streams draining such areas are generally ungraded, with many waterfalls, lakes and rapids, such as Niagara and Sault Sainte Marie.

The upglacier sides of most hills and rock knobs overridden by ice are rounded, polished, and grooved; the downglacier sides are jagged from the dragging out of joint blocks. Rock knobs so sculptured are **roches moutonnées** (sheep rocks).

Cross profiles of mountain stream canyons are V-shaped, but those of formerly glaciated valleys are characteristically U-shaped because of grinding by the glacial ice. Glaciated valleys are also straighter and smoother than most stream valleys; spurs between tributaries have been worn away by the ice (Figs. 13-23, 13-24, and 13-25). A viscous glacier does not turn sharp curves as

readily as water, so that the glaciated valley tends to be straighter.

The long profile of a glaciated valley is commonly interrupted by abrupt steps; both above and below such steps, the smooth U-shaped

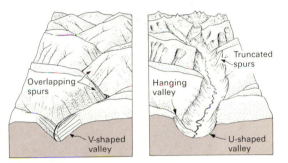
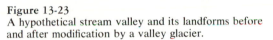

Figure 13-23
A hypothetical stream valley and its landforms before and after modification by a valley glacier.

Figure 13-24
A glaciated valley in the Sierra Nevada, showing typical U-shape, faceted spurs, and hanging valleys. [Photo by Warren Hamilton, U. S. Geological Survey.]

Figure 13-25
The valley of the Roaring Fork, Colorado, showing hanging valleys, faceted spurs, and a tiny underfit stream, wholly out of scale with the width of the valley. [Photo by C. B. Hunt, U. S. Geological Survey.]

valley may continue with flatter gradients. Several such "cyclopean steps" may interrupt a single valley, alternating with polished rock floors and lake-filled basins.

Most glaciated valleys head in cirques; in some places, these are so closely spaced that the divides have been reduced to knife-edged "combs" and triangular "horns" (Fig. 13-26). The Matterhorn is an excellent example.

Stream valleys and glaciated valleys also differ in the junctions of their tributaries. All tributaries in stream valleys join at grade. As Playfair pointed out in 1802, this is a compelling argument that streams form their own valleys. During glaciation, tributary glaciers also generally join at grade, though not all do; thin tributary glaciers cannot lower their channels as rapidly as a thicker more powerful main glacier. When the glaciers melt away the tributary valleys stand high above the main valley; they are "hanging valleys."

FORMER EPISODES OF GLACIATION

In most of Scandinavia, southern Canada, Labrador, and parts of the northern United States, soil profiles are either poorly developed or missing. Instead, rounded hilltops expose smoothly polished and striated surfaces as fresh as those beneath a modern ice sheet. Boulders, many huge, are strewn over the surface (Fig. 13-27). Most of these "erratic" boulders differ from the local bedrock; many in Iowa are of gneiss or granite resting on limestone or shale. Neither gneiss or granite is to be found within many kilometers, so that the boulders are too large to have been brought in by a flood. Sporadic chunks of copper like that mined on the Keeweenaw Peninsula, Michigan, are occasionally found in eastern Iowa. Boulders of an unusual kind of granite called "rapikivi," found in place only in Finland,

Figure 13-26
Mount Assiniboine, a glacial horn near Banff, Alberta. Note the many cirques, some with hanging glaciers. [Photo by Alberta Department of Mines and Resources.]

Figure 13-27
Glacial boulders resting on a surface polished by a Pleistocene valley glacier, Sierra Nevada, California. [Photo by Eliot Blackwelder, Stanford University.]

are scattered widely over Esthonia and Poland. A large nickel deposit in what was formerly northern Finland (now part of the USSR) was found by tracing scattered ore fragments northward to their bedrock source near Petsamo.

The Pleistocene Glaciations

In both Europe (Fig. 13-19) and North America (Fig. 13-28), many abandoned moraines have been mapped for many kilometers. North of them lie scattered exposures of striated rock floors strewn with till and erratic boulders. Innumerable lakes occupy gouges in the glacial floor. Debris-dammed older drainage and partly filled kettles in outwash aprons record the old glaciers. Outwash fans spread from gaps in the moraines; beyond them lie terraces of silt and sheets of loess many meters thick. All these features—in central Europe, the British Isles, and central and eastern North America—record formerly widespread glaciers where none now exist.

In mountainous areas not covered by the continental ice sheets, such as the Alps, the American Cordillera, and the circum-Pacific mountains, innumerable cirques, U-shaped valleys, superb waterfalls tumbling from hanging valleys, clear mountain lakes nestled in rock-scoured basins—all tell of formerly more extensive valley glaciers. The mountain glaciers of the Southern Hemisphere were more extensive than now at times estimated from radiocarbon dating to be contemporaneous with the expanded glaciers of the Northern Hemisphere. The earth now has about 26 million km^3 of ice; during the Pleistocene maximum it had about 70 million.

ADVANCE AND RECESSION OF
PLEISTOCENE ICE SHEETS

At many places in North America and Europe, roadcuts expose two or more different layers of till, one superposed on the other (Fig. 13-29). The upper till contains boulders of almost unweathered granite and gneiss, some with polished facets and striae; beneath this is a mature soil profile developed on deeply weathered till that grades downward into less-altered material. In the upper part of the lower till, the outlines of boulders can still be recognized, but clay is all that remains of them, and they can be cut with a knife. These relations show that the lower till had been weathered for a long time before burial by the younger. In places the B-horizon of the weathered till is a tough, sticky clay commonly called "gumbo," and since palimpsests of boulders show it is derived from till, it is called *gumbotil*.

Moraines old enough to have weathered to gumbotil generally have integrated drainage, except where overridden by younger moraines. Careful study of the superposition and degree of

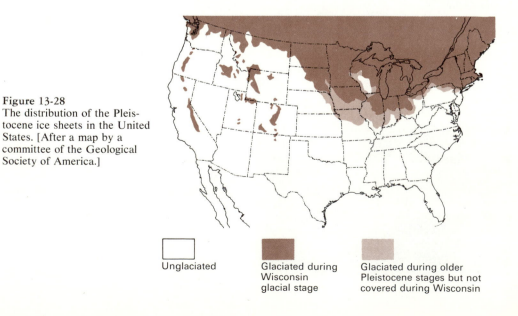

Figure 13-28
The distribution of the Pleistocene ice sheets in the United States. [After a map by a committee of the Geological Society of America.]

Unglaciated

Glaciated during Wisconsin glacial stage

Glaciated during older Pleistocene stages but not covered during Wisconsin

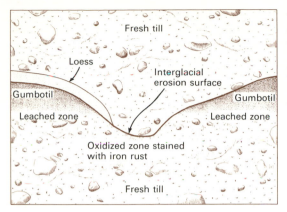

Figure 13-29
Superposed tills, southwestern Minnesota. In the upper part of the lower till, all boulders of granite and gneiss are thoroughly rotted, and only resistant quartzite is still undecomposed. The upper till contains abundant fresh boulders of granite and gneiss.

weathering of different glacial deposits, of the fossil content of associated stream-laid deposits, and of radiocarbon ages of enclosed tree trunks have led to the recognition of several episodes of Pleistocene glaciation: four major ones in North America, six in Europe, and ten or twelve in Iceland. Radiometric and fossil evidence make probable the time correlation of the last two glaciations in Europe and North America, but correlation of the earlier ones is quite uncertain. Sandwiched between deposits left behind by several of the ice advances are others in which preserved plant remains tell of past climates warmer than today's.

DATING PLEISTOCENE DEPOSITS
BY VARVED LAKE SEDIMENTS

As we have seen, lake deposits abound in formerly glaciated regions. Torrents of turbid meltwater pouring into these lakes dropped their coarser sand and silt in deltas; the finest clay and silt settled slowly and spread throughout the lakes. Present glacial lakes are deep green, owing to the dispersion of light by the abundant particles of fine suspended matter.

In winter the glacial lakes freeze over, and most small tributary streams freeze solid. During this quiet period, the suspended clay particles beneath the ice, along with the fine algal matter that accumulated during the summer, settle slowly to the bottom to form a dark fine-grained layer. By spring most glacial lakes are nearly clear. Thus the summer layer of deposits is coarser and consists almost entirely of rock waste; the winter layer is finer and much richer in organic matter. Such cycles can be seen in many existing lakes; cores drilled into the bottom sediment show this two-fold layering.

These thin laminae of alternating dark and light material, each pair the deposit of a single year, are called **varves**, from a Swedish word meaning "seasonal deposit." The typical pair is less than a centimeter thick (Fig. 6-5), but some are much thicker.

By counting varves, we can determine the number of years represented by a glacial lake deposit. The varves also record climatic variations; for example, an exceptionally warm year yields an exceptionally thick and coarse summer layer. By matching sequences of comparable variations in thickness, it is often possible to correlate the upper layers of a southerly lake with the varves near the bottom of a more northerly lake and thus time the interval of recession.

Such studies show that the last ice sheet retreated from the site of Stockholm, Sweden, about 9000 years ago, that southern Ontario was ice-covered 13,500 years ago, and that about 4300 years elapsed while the glacier front retreated from a site near Hartford, Connecticut, to St. Johnsbury, Vermont, a distance of about 320 km.

Radiocarbon gives another method of measuring ages during the past 35,000 years. Ages so determined for trees overrun by glaciers are generally, but not invariably, consistent with varve counts.

Drainage Changes Caused by Glaciation

The drastic climate that buried so much of Europe and America under glacial ice also had striking effects in latitudes not reached by the continental glaciers.

LAKES OF THE GREAT BASIN

In Pleistocene time, Nevada and western Utah were not barren semideserts as they are today. Lake sediments containing fossil leaves and fresh-water animals show that many intermont valleys held fresh-water lakes rimmed by trees and luxuriant grass. The highest wave-cut shore-

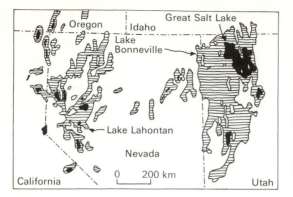

Figure 13-30
Map showing the extent of the great Pleistocene lakes in the western United States. Present lakes, some ephemeral, are shown in black. [After G. K. Gilbert and O. E. Meinzer.]

line of the largest of these vanished lakes, Lake Bonneville, makes a conspicuous horizontal terrace along the Wasatch Mountain front more than 300 m above Great Salt Lake. When filled to this level Lake Bonneville overflowed to the Snake River and thence by way of the Columbia to the sea. The outlet was over unconsolidated alluvium into which the torrent quickly carved a channel to bedrock, about 100 m below the original divide. This resistant barrier held the lake at a nearly constant level for a long time, and great deltas and terraces much wider than those at the highest level were formed along the shores. Moraines from valley glaciers in the Wasatch extend to the old shorelines. Some rest on beach deposits of the lake and in turn are cut by beaches; the glaciers were thus contemporaneous with expansion of the lake.

As aridity increased, the glaciers and streams dwindled, and evaporation began to exceed inflow to the lake. The lake fell to lower levels, some recorded in fainter shore features; Great Salt Lake and the Bonneville salt flats remain as the last pools of this once vast inland sea (Fig. 13-30).

GREAT LAKES AND
MISSOURI VALLEY AREA

In the north-central United States, the continental glaciers overrode a stream-carved landscape, damming many north-flowing streams to form

ephemeral lakes, some of which were destroyed by further advance. The present channels of the Missouri below Great Falls, and of the Ohio below Pittsburgh, follow closely the approximate edge of the glaciers for many miles in new channels determined by the blockage of former north-flowing streams. The Milk and Yellowstone rivers formerly flowed to Hudson Bay, as shown by till-filled channels.

As the ice front retreated, new lakes were impounded between the glacier and higher ground to the south. The whole valley of the Red River

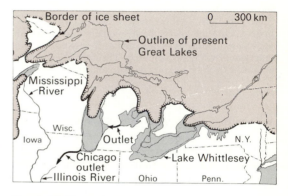

Figure 13-31
Meltwater lakes formed during the recession of the Pleistocene ice sheet from the Great Lakes. The outlines of the states and present lakes are shown for reference. [After Frank Leverett and F. B. Taylor, 1915; Frank Leverett and F. W. Sardeson, 1922; and W. S. Cooper, 1935.]

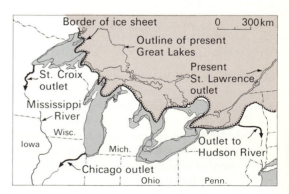

Figure 13-32
Same as Figure 13-31 at a somewhat later stage of ice retreat. [After Frank Leverett and F. B. Taylor, 1915.]

of the North was once a lake bed extending far to the north over the sites of Lakes Winnipeg, Manitoba, and Winnipegosis for a length of more than 1100 km. This former lake has been named Lake Agassiz, in honor of a distinguished early worker in glacial geology.

The present Great Lakes are the last in a long series, the main elements in whose history were worked out more then 60 years ago, though further details are of course still being uncovered. The former lake shores, recognized by beach ridges, deltas, spits, and bars, stand high above existing lakes. When followed northward the shorelines generally slope upward and end abruptly against a moraine or outwash apron of the former glacial front that dammed the lake.

As shown in Figures 13-31 and 13-32, a series of lakes occupied the southern end of the Lake Michigan basin, draining to the Mississippi through the Illinois River. Lake Whittlesey, a large lake that once occupied the basin of Lake Erie and part of Huron, drained westward across what is now Michigan to join the Mississippi by the same route. As the ice retreated northward it cleared a lower valley, and Lake Whittlesey drained down the Mohawk Valley to join the Hudson. The lake then shrank; the outlet across Michigan dried up. Still later, when the ice exposed the St. Lawrence outlet, both the Chicago and Mohawk channels were abandoned, and the present Great Lakes were established.

GRAND COULEE

Glacial ice diverted the great Columbia River in eastern Washington to create a landscape unique on earth (Figs. 13-33, 13-34). A huge glacial lobe advanced from the north to dam the westerly course of the river, forcing it to spread in a myriad of channels across the loess-covered basaltic plateau that here slopes southward at more than a meter per kilometer. Armed with huge quantities of gravel and basalt blocks torn from its channel, the river swept new channels through the loess and cut deep into the basalt below. As the ice advanced and retreated, the diversions shifted from one place to another, finally leaving a network of braided channels across the plateau. Though the Columbia alone must have been a mighty stream, it was augmented by enormous floods from Lake Missoula, a glacially dammed mountain lake that debouched northeast of

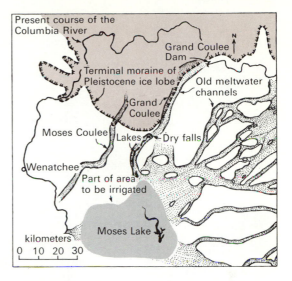

Figure 13-33
Map of central Washington, showing Grand Coulee and other features formed by the glacial diversion of the Columbia River. [In part after J. Harlan Bretz.]

Spokane and which was alternately blocked and opened by glacial fluctuations. These floods were so great that they ignored channels scores of meters deep and swept over hills that stand a hundred meters above the wide lowlands.

Between these great floods the river ran mainly in a single channel, now called the Grand Coulee. It is a huge canyon, 150 to 300 m deep and 1500 m to 25 km wide, cut in the flood basalt of the plateau. Midway was a gigantic waterfall nearly 120 m high and 5 km wide (Fig. 13-34). Today the falls are dry; retreat of the ice allowed the Columbia to reoccupy its preglacial channel. Grand Coulee is now the route of water pumped from the Columbia and used to irrigate thousands of hectares of rich but arid land.

Pre-Pleistocene Glaciations

Many ancient sedimentary formations show all the characteristics of till except that they are tightly cemented. These are **tillites**. They are unsorted, contain striated and faceted stones, and are associated with varved shales and slates or with sandstones and conglomerates having characteristics of outwash deposits. Some rest on polished and grooved rock floors.

Figure 13-34
Dry Falls, a former huge waterfall in the glacially diverted Columbia River. [Courtesy of the State of Washington Department of Conservation and Development.]

Tillites occur in the Ordovician and Silurian rocks of the Hoggar Mountains of the Sahara and in the Silurian or early Devonian of South Africa and Argentina, but the most widely distributed pre-Pleistocene tillites are those of the Carboniferous and Permian of all the Gondwana continents. Glaciers were also widespread in the late Precambrian, and evidence has been reported from every continent except South America. Incidentally, on the southwestern outskirts of Adelaide, South Australia, Precambrian and Carboniferous tillites crop out within a few kilometers of each other!

CAUSES OF GLACIAL CLIMATES

For much more than a century, geologists and climatologists have puzzled over the causes of recurrent continental glaciations. Hypothesis has followed hypothesis, but all seem to explain either too much or too little, and none has won general acceptance; yet the suggestions have a certain interest.

Facts to be Explained

1. Continental glaciers in Greenland and Antarctica occupy about 10 percent of the lands today. At several times in the Pleistocene they covered an area three times as great.

2. During the Pleistocene, climatic zones of the northern hemisphere generally paralleled their present positions, fluctuating southward during glacial stages and northward during interglacial. In low latitudes—at least for the last two glacial stages—heavy rainfall was contemporaneous with glaciation at higher latitudes, and the snowlines were lower than now (Fig. 13-35).

3. Estimates of the duration of the several glacial and interglacial stages of Pleistocene time do not suggest periodicity.

4. Evidence of continental glaciation comparable to that of the Pleistocene is found in late Paleozoic and in late Precambrian rocks but only in far smaller amounts in other strata.

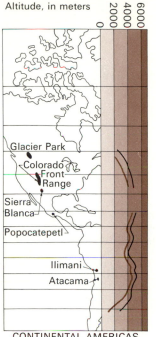

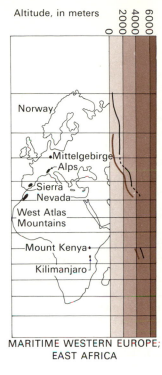

CONTINENTAL AMERICAS

MARITIME WESTERN EUROPE; EAST AFRICA

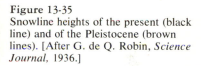

Figure 13-35
Snowline heights of the present (black line) and of the Pleistocene (brown lines). [After G. de Q. Robin, *Science Journal*, 1936.]

Geologic Hypotheses

Some theorists maintain that the ice accumulations of the Pleistocene are due to the unusually high stand of the continents. This can hardly be an important factor, however, for a major center of accumulation during the last glaciation was on the lowlands of the Canadian Shield near Hudson Bay. Furthermore, at several of the interglacial stages the temperature was warmer than at present, and evidence of vertical movements of the continent is insufficient to account for these changes. Moreover, the mountains of today, a relatively nonglacial time, are as high as any that existed during either glacial or interglacial times.

Changes in carbon dioxide content of the air (greenhouse effect) and in amounts of volcanic dust have been suggested as controlling earth temperatures. Certainly the average temperature over the whole earth has been lessened by several degrees after such explosions as those of Katmai and Krakatau, and remained so for several years. But volcanoes were not demonstrably more active in Pleistocene than in earlier times. The greenhouse effect of carbon dioxide is certainly a factor in earth's climate, but, again, no independent evidence of its unusual deficiency during the Pleistocene has been brought forward. And it would be partly counteracted by the opposite tendencies of water vapor.

The American geophysicists Ewing and Donn have suggested that the drifting of crustal plates has placed the nearly isolated Arctic Ocean over one pole at the same time that the continent of Antarctica covered the other. In this position oceanic circulation hardly affected the Arctic; its ice-free surface allowed ready evaporation and a source of snow for continents in the northern hemisphere. Widespread snow reflects much of the solar energy; the snowfields expanded, the temperature dropped, and glaciers grew. Even though most of the snow for the northern hemisphere glaciers came from the Atlantic and Pacific oceans, the open polar sea had triggered the beginning of glaciation. After the temperature dropped and the Arctic Ocean froze over, the trigger for glaciation was not needed. Glacial retreat was possible but not necessarily immediate, because cooling the ocean temperature lessened evaporation and the feeding of the glaciers. The theory gives no cause for the warmer interglacials, for the map pattern must have been about the same then as during the glaciations. It seems not to account for retreat of the glaciers at all. In support of the idea, though, is paleomagnetic evidence that the Gondwana glaciations took place in succession as the Gondwana continents drifted over the South Pole, nicely accounting for their sequence.

It has been pointed out that strengthening of the earth's magnetic field should cool the earth considerably, as it would increase the shield against the solar corpuscular radiation—"solar wind." Although the field strength is well known to have varied widely, no one seems to have developed a specific hypothesis involving this mechanism.

Astronomic Hypotheses

Three astronomic mechanisms have been suggested to account for glaciation: (1) that the solar system from time to time encounters clouds of cosmic dust; (2) that the earth varies periodically in its distance from the sun, and hence in the heat it receives; and (3) that the sun's heat radiation itself varies.

Dark nebulae are known to be partly cosmic dust. Were the earth to enter such a nebula, the dust might either screen out the sun's radiation or screen in the earth's radiation to outer space, depending on the size of the particles. There is now no way of testing this suggestion, even though the cloud could not have passed out of range of telescopes since the last glaciation, only 10,000 years ago. No such cloud has been identified.

The earth's distance from the sun is influenced by three concurrent cycles: (a) changes in eccentricity of the earth's orbit, with a period of 92,000 years; (b) changes in the angle of tilt of the earth's axis to the ecliptic, in a period of about 40,000 years, and (c) changes in the positions of the equinoxes, with a period of about 22,000 years. The collective effect of these variables is to produce periodic changes in the distance between each point on the earth's surface and the sun, and thus in the heat it receives. This hypothesis, first suggested more than a century ago, was elaborately worked out by the Yugoslav astronomer Milankovitch between 1920 and 1938. If such changes had brought about glaciation, the effects on northern and southern hemispheres, though not exactly opposite, would

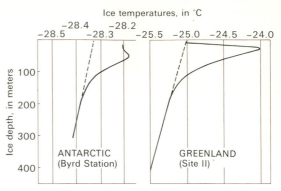

Figure 13-36
The temperature measured at depth in drill holes in Antarctic and Greenland ice sheets. Both curves are nearly straight below about 150 m. The curves between 150 and 60 m show warming in both hemispheres during the past 40 to 80 years. Above 60 m the curves show slight cooling in both hemispheres during the past 40 years, approximately. [After G. de Q. Robin, *Science Journal,* 1966.]

certainly not have been parallel, as they obviously have been from temperature measurements in the Antarctic and Greenland ice (Fig. 13-36). Furthermore, the effect at the equator should have been negligible, but the snowline fell there as notably as in higher latitudes (Fig. 13-35). Age determinations of oceanic sediments in both hemispheres show contemporaneity of Pleistocene glaciation, though the Antarctic glaciers were active much earlier, probably in the Miocene and perhaps even earlier. The temperature effects on the isotopic composition of ice (less ^{18}O at lower temperatures of freezing) indicate virtually perfect parallelism of temperature back 73,000 years in the deep drill holes in Greenland and Byrd Station in the Antarctic. This time covers the entire span of the latest glacial episode.

A still further argument against the significance of the Milankovitch cycles is that, since they are continuous, we should have regularly spaced glaciations through time, perhaps as many as twenty episodes per million years. Instead we must go back to the Permian to find glaciation comparable to that of the Pleistocene.

Astronomers have shown that solar radiation fluctuates, and so too the heat received by the earth. Short term variations of as much as 3 percent seem well established. Meteorologists estimate that Pleistocene temperatures were 7° or 8°

cooler than the present average. Whether from solar change or dust in the air, a drop nearly this great took place between May 1958 and April 1963, showing that the balance is delicate. The American astronomer Charles Abbot, an authority on solar radiation, favored this explanation for the cooling of the Pleistocene.

But more than a temperature decrease is needed for continental glaciation. The mean annual temperature of freezing lies much nearer the equator than the existing ice sheets. Obviously an increase in precipitation is also needed. The present configuration of land and sea, essentially that of the Pleistocene, is such that increase in precipitation would require increased evaporation. The British meteorologist Sir George Simpson built an ingenious hypothesis on this idea.

Simpson reasoned that if solar radiation increased, the air would warm, and cloudiness and precipitation would everywhere increase. More snow would fall, the ice caps would expand, and the cloud cover would lessen summer melting. But if the temperature continued to rise, the melting would eventually exceed snowfall and the glaciers would disappear (Fig. 13-37). At the high point on the radiation curve, the world climate would be milder and wetter than now. When radiation began to decrease, the sequence would reverse; ice would advance at first, then retreat when heat became insufficient to supply precipi-

tation. Thus, paradoxically, one rise and fall of temperature from present conditions would supply two glaciations, separated by a warm wet episode; the four American glacial eposides would thus require two radiation cycles; the six of Europe, three.

Unfortunately for this theory, however, temperatures at the equator should be warmer throughout, whereas both the ratios of oxygen isotopes in tropical shells and the equatorward shift in ecologic niches of marine invertebrates indicate quite the opposite—a worldwide depression of snowline and temperature.

No hypothesis seems wholly satisfactory to account for the expanded continental glaciations, nor can we say whether we are now in an interglacial interval. We noted that for several decades most glaciers have been receding in both hemispheres, and sea level has risen a few centimeters. Stillstands during the sea-level rise that accompanied glacial melting are recorded by niches formed by beach erosion at depths of −60 m and −160 m in the Gulf of Mexico. Should the retreat continue until all land ice is melted, the sea would rise about 60 m; at the peak of the latest Pleistocene glaciation, it must have stood about 130 m lower than now—a figure supported by the thick alluvium in the valleys of most rivers near the coast. During an earlier episode the sea level may have been depressed as much as 210 m. We may expect that, some centuries hence, either our coastal cities and low-lying coastal plains will be drowned beneath a shallow sea or the sites of Chicago, Copenhagen, and Warsaw may again be overrun by glaciers. We have no way of finding which way the trend will be.

EFFECTS OF GLACIAL LOADS ON THE CRUST

We noted in Chapter 2 that the earth's crust is in near isostatic condition and that it is unable to sustain large variations in load over wide areas. It would thus be expected to respond to variation of the Pleistocene ice sheets, and the record is clear that it did.

Wherever glaciers spilled over or around mountains, some measure of their thickness is possible. In New England, for example, the ice must have been more than 1200 m thick, for the highest peaks were overridden. In southern Canada,

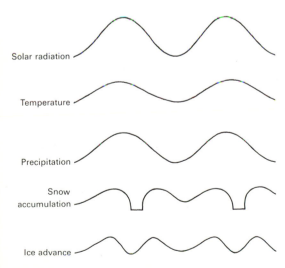

Figure 13-37
Simpson's theory of glaciation based on variations in the sun's radiation.

nearer the source, the ice must have been much thicker, as the glacial surface slopes in the direction of ice movement. Precambrian boulders from the Hudson Bay lowland were carried to elevations of 1300 m in the Alberta Rockies. Boulders that could only have come from Sweden were carried over 1750-meter mountains in Norway.

Seismic soundings of ice thickness in Greenland and Antarctica, when plotted against surface slopes, show that ice thickness is systematically related to distance from the front (Fig. 13-38). Applying such profiles to the Fennoscandian ice cap, which spread over much of northern Europe, it seems probable that at its maximum it was more than 3 km thick (Fig. 13-39).

The density of ice is only about a third that of ordinary rocks; a load of ice 3 km thick is thus equivalent to a load of rock 1 km thick, and when spread over a wide area should have depressed the crust isostatically. That it did so is evident from the ancient shorelines of glacial lakes and of the Baltic. These were, of course, level when formed; after the ice melted, the crust responded isostatically to the unloading, so that the shorelines now rise toward the center of the former ice cap (Fig. 13-40).

The rebounds from the glacial loads are not yet over, either in Europe or North America. The north shore of the Gulf of Bothnia is rising at the

Figure 13-39
Reconstruction of the European ice sheets at their maximum extent. The contours are based on analogy with the slopes of the present Greenland and Antarctic glaciers. [After G. de Q. Robin, *Science Journal,* 1966.]

rate of 89 cm per century; Stockholm at 40 cm per century; Churchill, on Hudson Bay, at 91 cm per century, and Manitoulin Island in Lake Huron at over 200 cm per century.

Similar uplift after release of glacial load took place in the Great Lakes region, in Labrador, and even in the basin of Lake Bonneville, where the load was not ice but merely 300 m of water. The rebound of the lake center was as much as 60 m. The continental shelves of Antarctica stand several hundred meters lower than those of other continents; it has been suggested that this is because of the ice load weighing down formerly normal shelves.

THE DURATION OF THE PLEISTOCENE EPOCH

Despite the abundant exposures of Pleistocene deposits and the wealth of stratigraphic data available, one of the most controversial items in geochronology is that of the duration of the Pleistocene epoch. For those who accept the Milankovich curve as definitive, the oldest of the four generally recognized glaciations of North America took place less than 600,000 years ago. The variations in $^{18}O/^{16}O$ ratios in fossil shells from

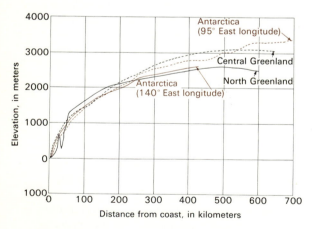

Figure 13-38
Similar surface profiles of Greenland and Antarctic ice sheets, showing that despite differences in rate of snow accumulation, in temperature, and in rock floors, the surface slopes are systematically related to distance from the glacier front. [After G. de Q. Robin, *Science Journal,* 1966.]

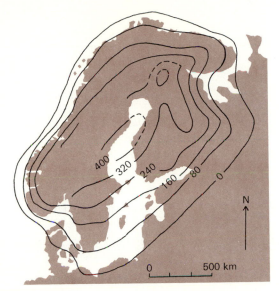

Figure 13-40
Postglacial uplift in Fennoscandia. The heavy lines connect points of equal uplift (in meters) of the highest strand line of the sea that flooded the area just after the glacier melted. [After R. A. Daly, *The Changing World of the Ice Age,* Yale Univ. Press, 1934.]

the sea floor indeed indicate that there were four cold intervals during this time, but no one knows whether these were indeed the main glacial epochs or merely minor fluctuations within one or two of them. The Bishop Tuff, a rhyolitic ash flow along the east foot of the Sierra Nevada, rests on glacial till thought to represent the second of the major glaciations of the midcontinent. If this correlation is correct—a doubtful matter— an age of at least a million years would not be unreasonable for the earliest North American glaciation.

But the Pleistocene was not defined by its relation to glaciations but from a sequence of marine strata in southern Italy, far from any glacial deposits. And it is generally conceded that the base of this type section long antedates the first of the Alpine glaciations. European vertebrate faunas somewhat younger than the marine fossils of the type section are closely similar to others in eastern Africa, with which human remains are associated. Potassium-argon dates from associated ash beds indicate an age of 1.85 m.y. On this basis, the Pleistocene began still earlier, perhaps about 2 m.y. ago.

Questions

1. What is the evidence that recrystallization takes place in the transformation of snow to firn and ice?

2. Why are most glacial crevasses less than 60 m deep?

3. Explain the process by which the head of a valley glacier acquires (a) new snow and ice, (b) rock debris.

4. Explain how a continental glacier acquires its rock load.

5. Draw a longitudinal profile through a valley glacier and label the following features: cirque, terminal moraine, meltwater tubes, bergschrund, shear banding in ice, snowfield, rasped bedrock, plucked and shattered bedrock.

6. How does rock flour released from a glacier differ from the fine-grained material formed during weathering?

7. Draw a hypothetical sketch map showing the location of all the following: (a) a lobate end moraine, (b) a recessional moraine, (c) pitted outwash, (d) ground moraine, (e) a drumlin cluster, (f) a plain underlain by varved clays, (g) an esker, (h) kame terraces, (i) an abandoned stream course.

8. How can a glacier in contact with sea water lower its bed below sea level?

Suggested Readings

Ahlman, H. W., *Glaciological Research on the North Atlantic Coasts* (Research Series No. 1). London: Royal Geographical Society, 1948.

Charlesworth, J. K., *The Quaternary Era, with Special Reference to Its Glaciation* (2 vols.). New York: St. Martin's Press, 1957.

Coleman, A. P., *Ice Ages, Recent and Ancient*. New York: Macmillan, 1926.

Daly, R. A., *The Changing World of the Ice Age*. New Haven: Yale University Press, 1934.

Flint, R. F., *Glacial and Quaternary Geology*. New York: John Wiley and Sons, 1971.

Gilbert, G. K., *Lake Bonneville* (U.S. Geological Survey, Monograph 1). Washington, D.C., G.P.O., 1890.

Matthes, F. E., *The Geological History of the Yosemite Valley* (U.S. Geological Survey, Professional Paper 160). Washington, D.C.: G.P.O., 1930.

Zeuner, F. E., *Dating the Past—An Introduction to Geochronology*. London: Methuen, 1958.

Scientific American Offprints

809. William O. Fields, "Glaciers" (September 1955).

823. Gilbert N. Plass, "Carbon Dioxide and Climate" (July 1959).

834. Edward S. Deevey, Jr., "Living Records of the Ice Age" (May 1949).

835. Ernst J. Öpik, "Climate and the Changing Sun" (June 1958).

843. Harry Wexler, "Volcanoes and World Climate" (April 1952).

849. Walter Orr Roberts, "Sun Clouds and Rain Clouds" (April 1957).

861. Gordon de Q. Robin, "The Ice of the Antarctic" (September 1962).

CHAPTER 14

Ground Water

Where does water in wells come from? Why, in some dry areas, do shallow wells deliver good water while at others none is found at depths of thousands of meters? The great limestone caverns of Kentucky and Virginia extend for distances so great that many have not yet been explored. How were they formed? All these questions relate to the part of the precipitation that infiltrates the soil—the **ground water**. It fills pores and crevices in soil and rock, surfaces in springs, swells or shrinks the volume of streams by seeping into or out of them. It also supplies the water to wells.

SOURCE OF GROUND WATER

Most soils and rocks contain voids and openings into which rain or meltwater can seep: tiny pores between the mineral grains, small tubules left by decay of grass roots, larger openings made by burrowing animals, and shrinkage cracks in drying clays. Even well-consolidated rocks are riven by faults, joints, and intergranular openings. Some infiltrating water remains near the surface, adsorbed by the soil colloids or held in the smaller voids by capillarity—the force that pulls water up a slender tube and holds it there against the pull of gravity. Some water, however, percolates deeper, ultimately reaching a zone where all rock pores are completely filled with water. Above this

surface, most openings in soil and rock are open to the air; below it they are filled with water. At still greater depths, the pores and cracks are so closed by compaction or so cemented by minerals that they are virtually watertight; the rocks beneath are dry, as has been found in many deep mines.

THE WATER TABLE

If we sink wells to the zone of saturation, measure the altitude of its upper surface, and then contour the surface defined by these measurements, we have defined a smooth, even surface (if the rock is homogeneous) generally with a gentle slope. This is the **water table**. Because rocks are variably porous, the actual interface between air and water is not smooth but minutely irregular, for capillarity can pull water higher in small tubes than in larger. (The capillary rise in a well only a decimeter across is negligible.) Although impossible to measure, the pressure on the water in the capillary zone above the water table must be less than atmospheric, for the capillary force opposes the pull of gravity. The pressure relations are as shown in Table 14-1.

Lohman has defined the water table as the surface at which the pressure on the water is precisely atmospheric; below it the pressure is

Table 14-1
Water table and pressure relations above and below it.

Zone	Pressure relations	Contents
Zone of aeration	Pressure in air is atmospheric	Air and discontinuous water in capillary spaces
	Pressure in water less than atmospheric	
Zone of continuous capillary saturation	Pressure less than atmospheric	Water
The water table; pressure atmospheric		
Zone of unconfined ground water	Pressure greater than atmospheric	Water

SOURCE: Slightly modified from S. W. Lohman, U.S. Geological Survey, 1972.
NOTE: Compare with Figures 14-1 and 14-2.

higher, above it lower. This definition, however, is rarely usable; we usually talk about the water surface penetrated by wells. The volume of rock whose pore spaces are completely filled with water also includes a zone of variable thickness above the water table. In fine-grained rocks the thickness may be considerable, measured in meters; in coarse rocks it may be virtually missing. This is the **zone of capillary saturation**. Between it and the surface is the **zone of aeration**, wherein weathering is active (Fig. 14-1). The water table is generally at a depth of a few meters, but in marshes, lakes, seas, and rivers it is at the surface. In some deserts it may lie at depths of more than a kilometer. The relief of the water table is, in general, a greatly subdued replica of the surface topography, rising under the hills and sinking under the valleys, though less regularly than the land surface.

POROSITY AND PERMEABILITY

Porosity is the ratio of pore volume to total volume, expressed as a percentage. Porosities of clastic sediments range as high as 80 or even 90 percent, but porosities of most rocks range between 12 and 45 percent, depending on the **shape** of the grains, their **sorting** and **packing**, and the degree of cementation (Fig. 14-2).

Mineral and rock grains vary in shape from thin plates and irregular chips to nearly perfect spheres. Their packing, whether tight or loose, greatly affects the porosity; fine material may largely fill spaces between larger grains. Most sediments are loosely packed when deposited but gradually become more compacted because of loading by younger beds and cementation. Uniform spheres, whether 1 mm or 2 m in diameter, when most tightly packed, have 26 percent pore space. Greater porosities indicate either that packing is loose or that the grains are themselves porous. Shape obviously affects porosity, but the presence of nonspherical grains may either raise or lower it, depending on the packing.

The capacity of a rock or soil to *hold* water is determined by its porosity; the capacity to *yield* it to the pump depends on pore size and intercommunication rather than on porosity alone.

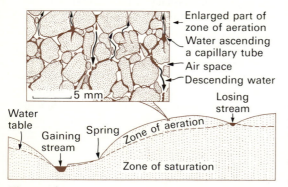

Enlarged part of zone of aeration
Water ascending a capillary tube
Air space
Descending water
Losing stream
Water table
Gaining stream
Spring
Zone of aeration
Zone of saturation
5 mm

Figure 14-1
Cross section showing the water table and its relations to a spring and to the streams. The greatly enlarged inset shows the movement of the water (black) in the zone of aeration.

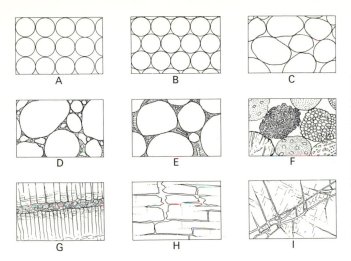

Figure 14-2
Porosity in rocks. (A and B) Variations in porosity due to differences in packing arrangement of spheres. (C) Sand with high porosity due to good sorting. (D) A sandstone with low porosity due to poor sorting. (E) Low porosity due to cementation. (F) Very high porosity due to well-sorted grains that are themselves porous. (G) Porous zones between lava flows. (H) Limestone rendered porous by solution along joints. (I) Massive rock rendered porous by fracturing.

Much water is held in capillary films; rocks such as shales and clays may yield little water to a well even though highly porous. Laboratory tests indicate that a fairly definite limit to free flow is at a pore diameter of about 0.05 mm. **Permeability, hydraulic conductivity**, the capacity of a porous medium to transmit a liquid, is thus the property determining the yield of a water-bearing material. A gravel with 20 percent pore space is much more permeable than a clay with 35 percent.

AQUIFERS AND AQUITARDS

No rock, except perhaps asphalt, is completely impermeable to water, given time enough. But some are so slightly permeable that little water moves through them even under high pressures. Such rocks have been loosely called impermeable; more properly they are **aquitards** (retarders of water), in contrast to materials both permeable and porous that yield water readily to wells— **aquifers** (bearers of water).

Most aquifers are beds of sand, gravel, sandstone, conglomerate, or other permeable rocks through which the water moves in laminar flow because the openings are small. Some highly prolific aquifers, however, are leached and cavernous limestones or beds of lava with tubes and other openings between them, some as much as a meter high; in these, water may flow in turbulent underground streams. A few aquifers are narrow, sinuous bodies of gravel that fill former stream courses, but these are not as common as the popular expression "underground stream" implies; the only true underground streams are in cavernous limestones and lava tubes.

PERCHED WATER

An aquifer may rest on an aquitard that, in turn, overlies porous, unsaturated rock above the normal water table. Such water is **perched**, retarded from downward movement by the aquitard beneath (Fig. 14-3).

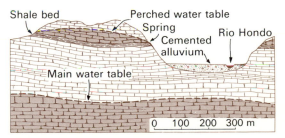

Figure 14-3
Cross section of aquifers in porous limestone containing an interbedded shale aquitard, southeastern New Mexico. Note that the Rio Hondo is also perched above the water table on its own caliche-cemented channel sands. [After A. G. Fiedler and S. S. Nye, U. S. Geological Survey.]

312

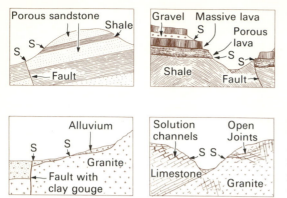

Figure 14-4
Cross sections showing likely sites for springs (S).

GROUND-WATER MOVEMENT

The water below the water table is not stagnant; like all water, it flows downhill if there is a gradient. The high areas of the water table tend to flatten, and the low ones to fill up or discharge water. If not replenished by rain, the water table would ultimately flatten completely at the level of discharge into a stream or the sea; indeed, in many areas of high permeability—in limestone and volcanic terranes—the water table slopes almost imperceptibly.

Where the watertable intersects the ground surface, springs form (Fig. 14-4). Most perennial streams are **gaining streams**, situated along troughs in the water table toward which the ground water flows (Fig. 14-1). Streams that flow from well-watered areas into deserts (for example, the Nile, Euphrates, Tigris, and Colorado) lose water by infiltration. These are **losing streams**; they lie on ridges in the water table and lose water to it (stream on the right in Fig. 14-1). Irrigation also feeds water to the water table. The water table has risen a hundred meters or more in fifty years beneath large areas of irrigated land in southern Idaho.

The rate and direction of water movement are determined by the permeability and the **hydraulic gradient**, the ratio between the *difference of elevation*, or **head**, H, and the horizontal distance, L, between the two points. Most ground-water gradients are low: 0.001 (1 m/km) or 0.01 (1 m/100 m).

Variations of Intake and Water Tables

For any particular hydraulic conductivity, the hydraulic gradient adjusts to the water supply. In dry spells the ground-water divides are lowered, the hydraulic head and discharge lessened. In arid western Texas, eastern New Mexico, and northern Mexico, the water table is nearly flat and commonly as much as 300 m below the surface. Large areas lack permanent streams, though intermittent ones may flow on perched ground-water bodies or in natural flumes cemented by caliche deposited as the seasonal streams evaporate under the desert sun (Fig. 14-3).

DARCY'S LAW

The modern concepts of ground-water movement were discovered in the mid-nineteenth century by the French hydrologist Henry Darcy, while studying the water supply of the city of Dijon. His work showed that nearly all ground-water movement is by laminar flow, with both velocity and discharge varying directly as the hydraulic gradient.

Darcy's Law states that $Q = KIA$, where Q = quantity of water moving in a unit time through a unit cross section A, and K = the hydraulic conductivity (a measure of the permeability). It is measured in volume transmitted through a unit cross-sectional area per unit of time under a hydraulic gradient, I (the unit change in head through a unit length of flow path).

If Q is measured in cubic meters per day, A in square meters, and I in meters (loss of head per meter of flow distance), K is measured in meters per day. In natural aquifers I is always a very small fraction; few water tables slope more than a few meters per kilometer.

Darcy's law is used to determine the hydraulic conductivity. If we know the discharge, Q, of water from a well; the area, A, of the openings in the well pipe; and the difference in elevation between the water in the well and the water table, I, the hydraulic conductivity is the only unknown in the equation. Measurements show that some aquifers have hydraulic conductivities several thousand times as great as others. Well yields differ correspondingly.

Rates of Ground–water Movement

The movement of ground water through uniformly permeable material is shown diagrammatically in Figure 14-5. Some flow lines go much deeper than others, but all ultimately reach the streams.

Mean rates of movements can be calculated from Darcy's Law after the hydraulic conductivity and gradient have been measured. Rates can also be measured directly by introducing dye or salt in one well and timing its arrival at others. American ground-water authority O. E. Meinzer thought the rate of 15 m/yr in the Carrizo Sandstone of Texas to be typical. Rates of 3 to 6 m/day are common, but rates as high as 118 m/day have been measured in highly permeable materials.

Darcy's Law implies that in material of low permeability, the gradient of the water table increases steeply with recharge; with high permeability, gradients are always low.

Drawdown by Pumping

A pumping well is a point of artificial discharge that disturbs the water table. We have already seen how the water table adjusts to loci of natural discharge, such as springs and gaining streams; it reacts to well discharge similarly by forming a **cone of depression**, thus greatly increasing the hydraulic gradient close to the well (Fig. 14-6).

In the example of Figure 14-6, the well drew water from the moderately permeable alluvium of the Platte River, Nebraska, where the undisturbed water table slopes eastward at about 1 m/km. The water table was monitored in more than 80 wells, in lines radiating from the well. The lowering, or **drawdown**, at the well was 7 m after 48 hours of pumping; 80 m from the well it was 30 cm, and at 400 m was barely measurable. Flow lines toward the well must have extended at least this far horizontally and also for some distance below the level of drawdown (Fig. 14-5).

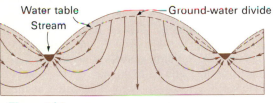

Figure 14-5
Approximate flow pattern of ground water in uniformly permeable material. [After M. K. Hubbert, *Journal of Geology*, 1940.]

CONTAINED WATER: ARTESIAN WELLS

A highly permeable aquifer may be overlain by an aquitard, such as a shale. If the rocks have been tilted and eroded, the aquifer may crop out in the nearby hills or mountains and be deeply buried beneath the aquitard in the lower country. There the water is *confined*, but it is readily recharged in the high country (Fig. 14-7). Water enters the aquifer at A, so that the water table

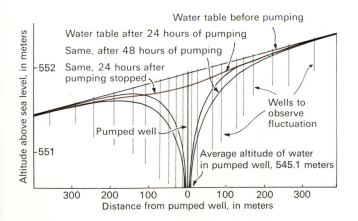

Figure 14-6
Cross section, with vertical scale greatly exaggerated, showing the water table before, during, and after pumping from a well. Observation wells are indicated by vertical lines. [After L. K. Wenzel, in *Hydrology*. Courtesy of Dover Publications.]

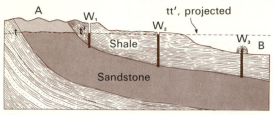

Figure 14-7
Cross section showing a series of wells penetrating a confined aquifer. The water table in the recharge area is tt′.

(t–t′) stands higher than the ground surface at B, where the aquifer is confined. A well sunk near B will allow the water to rise under the head of water from the intake area and flow freely without pumping—an **artesian well**.

The name "artesian" was originally restricted to flowing wells; it is now applied, however, to any well in which the water rises higher than the top of the aquifer penetrated (W_1 and W_2 in Fig. 14-7). (The term "artesian" is derived from the Roman name for a French province along the English Channel, where artesian conditions were common.)

Many artesian wells are of great value, furnishing copious supplies of water even in very arid country. Huge artesian supplies are present in the northern Sahara and in the desert of the Artesian Basin of Queensland, Australia. Semiarid southwestern South Dakota offers a dramatic example. The early railroads crossing this region needed much water for their steam engines. In 1905 N. H. Darton of the U.S. Geological Survey, having geologically mapped the pertinent region, recommended drilling at Edgemont, just south of the Black Hills, where a presumably productive aquifer should be met at a depth of about a kilometer. The aquifer was profitably intersected at a depth within 10 meters of that projected!

Many Arctic areas furnish impressive evidence of the power of artesian pressures. In the delta of the McKenzie River, the permafrost is commonly 30 m or more thick. As all pores in frozen ground are filled with ice, it is an effective aquitard. Artesian pressures derived from highland recharge areas are sufficient to punch up the permafrost and overlying tundra into large blisters, called **pingos** (Fig. 14-8), some nearly a kilometer across and more than 100 m high. The injected water may then freeze to form a huge biscuit of ice beneath the hill. Some similar topographic forms are sporadic beneath the Beaufort Sea; they have not been drilled, but it seems likely that these are pingos also.

Figure 14-9 illustrates the head conditions in a confined aquifer. Before discharge from an artesian well, the water would have stood in open casings in the well and in the observation wells B and C to the height marked "artesian head before discharge." This is equivalent to the line t–t′ of Figure 14-7. Well A, which does not penetrate the aquitard, draws its water from an isolated aquifer, and its hydraulic regime is independent of the artesian system. The load of all the rock between aquifer and surface rests on the aquifer, partly supported by the rock framework of the aquifer and partly by the hydrostatic pressure (artesian head) acting on the base of the aquitard. When the well is allowed to flow, the

Figure 14-8
A pingo on the McKenzie delta, Northwest Territories, Canada. [Photo courtesy of Robert F. Leggett, Canadian Research Council.]

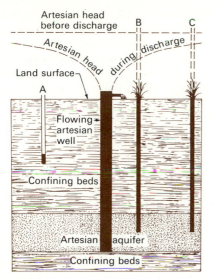

Figure 14-9
Diagrammatic section through an artesian well and two observation wells, B and C, illustrating head conditions in and near an artesian well before and during discharge. [After S. W. Lohman, U. S. Geological Survey.]

pressure in the nearby aquifer falls, just as in the cone of depression of an unconfined aquifer, but the two are not identical either in speed of response or area affected. Observation wells show a declining head toward the discharge well. The load of rock is there less supported by the artesian head; it bears more heavily on the aquifer, compressing it and squeezing water from its pores. In an unconfined aquifer the drawdown is caused by gravity acting on the water alone; in a confined aquifer, however, the rock pressure is so high that the drop is felt over a wide area almost instantaneously. Under otherwise similar conditions, like pressure drops in confined aquifers have been recorded over areas several thousand times the area of measurable water-table depression in unconfined aquifers.

Effects of Water Extraction

The reduction of water pressure in confined aquifers has often permitted significant compaction of the rocks and dramatic subsidence of the surface. In Mexico City, for example, the cathedral and surrounding area has subsided almost 6 m following pressure reduction in a confined aquifer. Similarly, an area of more than 3000 km² near Los Banos, California, is subsiding for a like reason. The pressure reduction increases grain-to-grain support within the aquifers, packing the grains tighter together and compressing them. In 1964 the subsidence was as much as 7 m and was proceeding at the rate of 40 cm/yr. This subsidence must be taken into account in the planning of the large irrigation works to be built in the area.

Dewatering of aquifers beneath Houston, Texas, San Jose, California, and dozens of other places has caused extensive subsidence. The collapses at Carletonville, South Africa, mentioned in Chapter 2, are also caused by removal of fluid support of cavern roofs.

Though subsidence is measured in centimeters rather than meters, the entire economy of Venice is threatened by a combination of withdrawal of artesian pressure and the slow but significant rise in sea level due to glacial retreat since the city was laid out in medieval times. The high winter tides—again trivial in measure—and the subsidence combine to flood St. Mark's Square frequently and threaten the future of the city.

Aside from damage due to subsidence, excessive withdrawal of water, either by artesian flow or by pumping, may destroy the value of an artesian system. If a confined aquifer is overpumped, the water may become unconfined, even though it is separated from the surface by an efficient aquitard. This has been the fate of many artesian systems that have been overexploited.

Subsidence of as much as 9 m has gone on at the Wilmington oil field, in Long Beach, California, because of reduced pressure in the oil sands due to pumping of oil. This is a highly industrialized area, practically at sea level, and great expense has been entailed in building dikes, replacing water and sewer lines, and pumping. Further subsidence is being combatted by pumping sea water into the oil sands along the periphery of the field, thereby restoring the artesian pressure. This is a delicate operation, for similar recharge of oil-sand pressure at the Baldwin Hills oil field in Los Angeles caused renewed movement on old faults and bursting of a dam with great resulting damage.

The most striking oil-field subsidence due to withdrawal of oil is in the Lake Maracaibo, Venezuela, fields, where the volume of subsidence has been nearly equal to the volume of oil removed. Subsidence of the Po delta, 50 km west of Venice, owing to extraction of natural gas, became so great that the field is being closed down because of the excessive costs of repairs to levees, roads, and other structures.

Along many coasts, fresh ground water discharges directly to the sea, in places for long distances offshore. In the Black, Baltic, Caspian, and Adriatic seas, the discharge of fresh water is locally so large that fishermen obtain their supplies simply by dipping buckets hundreds of meters offshore. Discharge to the Gulf of Carpenteria off Australia is also very great. Fresh water may extend well below sea level at the shoreline (Fig. 14-10). The lighter, higher column of fresh water seems to be in static balance with the denser sea water, as though floating within it. A column of sea water 300 m high weighs the same as a column of fresh water 308 m high. If, therefore, the water table near shore stands 8 m above sea level, fresh water might theoretically be recovered at a depth of 300 m below. This is true only if flow is laminar, so that mixing is minimal; in the porous and broken volcanic lava edifices of the Hawaiian Islands, however, lava passages are large enough to permit considerable mixing. A mixed brackish zone thus intervenes between fresh water and sea water of normal salinity. The friction of flow through the rock retards spreading out of the fresh-water lens, though of course if it were not constantly replenished by rainfall, the steep interface would soon flatten.

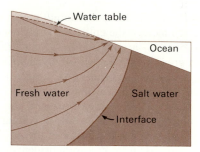

Figure 14-10
Cross section showing fresh-water flow lines in relation to the contact with underground salt water. [After M. K. Hubbert, *Journal of Geology*, 1940.]

GROUND WATER IN CARBONATE ROCKS

Rainwater, especially the slightly acid ground water of humid regions and the more acid meltwaters from snow, effectively dissolves limestone, enlarges cracks and pores, and forms tunnels, irregular passages, and large caverns (Fig. 14-11).

Figure 14-11
Solution-etched limestone surface giving direct access to ground water. Near head of Valentine Creek, Glacier National Park, Montana. [Photo by M. R. Campbell, U. S. Geological Survey.]

Figure 14-12
A large sink, with alluvial floor, in limestone, Kars region of Yugoslavia, the type area of karst topography. [Photo by Th. Benzinger, Stuttgart.]

In many limestone areas such solution openings are so extensive that the surface drainage quickly goes underground through **sinks** (Fig. 14-12) and discharges through caves (Fig. 14-13). An artesian system in a Paleozoic limestone that crops out in the Black Hills discharges to wells nearly five times the volume of water it receives from surface streams. The difference is due to direct rainfall on the limestone and its rapid infiltration into the ground before forming streams. Sinks and caves of course develop only slowly. Water seeping through limestone enlarges the passages by dissolving the rock; gravitational collapse ensues when the hole is large enough, allowing it to trap more water and further dissolve limestone. The sink may eventually discharge into solution caverns bottomed by a less soluble aquitard. In time, the entire limestone terrane becomes honeycombed with interconnected sinks and caverns. Where CO_2-saturated ground water seeps into the roof of a cave open to the air, it loses part of its carbon dioxide, partly evaporates, and thus deposits some of its dissolved calcium carbonate

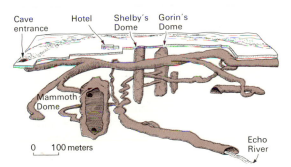

Figure 14-13
Diagram showing part of the Mammoth Cave system, Kentucky. [After A. K. Lobeck, *Geomorphology*, McGraw-Hill Book Company, 1939.]

as **dripstone** hanging from the roof or rising from the floor in bizarre forms (Fig. 14-14).

The ease with which water moves through cavernous limestone and the flatness of the water table have often been demonstrated by pumping. At the Los Lamentos mine, about 150 km south-

Figure 14-14
Dripstone in an Indiana cave. [Photo by Arch Addington.]

east of El Paso, Texas, two years of constant pumping failed to lower the water table appreciably, preventing the rich mineral deposit below from being mined. Attempts to mine deposits at Tombstone, Arizona, Eureka, Nevada, and Orange Free State, South Africa, have been similarly thwarted.

Karst Topography

Some limestone or dolomite regions have few or no surface streams: the runoff passes quickly underground through sinks, flows through caverns, cascades at intervals to lower levels, and finally reaches the water table (Fig. 14-15). Where valleys cut down to the water table, giant springs may gush forth. Such areas of underground drainage are **karst** regions (from the Kars region of Yugoslavia).

The rivers are fed mainly by large springs and may disappear into another chasm within a few kilometers, only to reappear in a neighboring one (Fig. 14-15). Along the Dalmatian coast of Yugoslavia, some karst drainage is discharged offshore in such volume as to form visible bulges at the surface of the Adriatic in quiet weather.

Other karst regions include the Causses Plateau

of southern France, parts of the Cumberland Plateau of Kentucky and Tennessee, and the Shenandoah Valley of Virginia. Probably the most spectacular karst region on earth is in the Vogelkop peninsula of New Guinea, where sinks hundreds of meters deep are so close together that, like dough cut by a biscuit cutter, only knife-edge ridges stand between them.

Karst drainage has influenced human activity for ages. Crops are poor on the dry plateaus, lush in the well-watered valleys. In southern France limestone caverns sheltered the ancient creators of the Lascaux paintings. Great springs emerging from limestone caverns have determined the sites of the major towns at least since the time of the Romans.

Piping: Pseudo-karst Formation

Subsurface drainage and underground caverns are not confined to carbonate terranes. In many arid regions where the water table lies deep, ground water filtering downward may wash out finer grains from a weakly consolidated sedimentary rock, in a process resembling that in which a clay-rich B-horizon soil is formed. The small tubes or "pipes" extend farther and farther headward from the gullies into which they empty. This is especially common in "badlands" (Fig. 14-16), where the pipes may extend for several hundred meters from the deeper gullies. Collapse of the pipes forms karst-like topography and sinkholes. These pipes, however, do not extend below the water table, as do many in true karst regions.

SOLUTION AND CEMENTATION BY GROUND WATER

In general, solution predominates above the water table; deposition and cementation below, though in carbonate rocks solution may go on far below the water table. But carbonates are not the only minerals that dissolve; the solution load of all streams include many ions besides those in calcite and dolomite. Even quartz, a relatively insoluble mineral, slowly dissolves. Garnet may be pitted and etched, and pyroxene and amphibole completely dissolved from the more permeable parts of sandstone beds, though they remain in tightly cemented parts of the same bed. Fossil

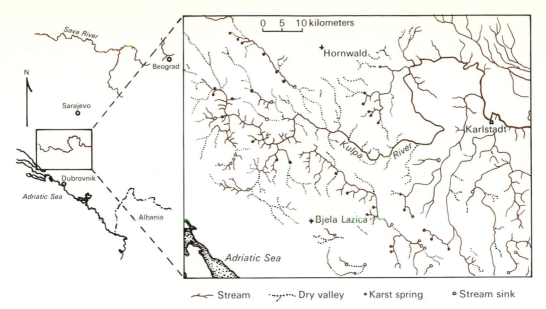

Figure 14-15
Map of part of the Kars region, Yugoslavia. The drainage is largely underground, as shown by sinks in small, partly dry valleys, and great springs in larger and deeper valleys. [After N. Krebs, 1928.]

① Shale and sandstone of Cretaceous Mancos Shale
② Tan silt and clay, sandy in places, of Quaternary age
③ Pipe system
④ Block left as natural bridge
⑤ Debris blocks undermined and sapped by pipes
⑥ Flow of ephemeral drainage

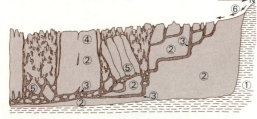

Figure 14-16
(Top) Pseudo-karst topography in shale of the Chinle Formation (Triassic), near Round Rock, Arizona. (Bottom) Diagram showing inferred subterranean conditions near the Book Cliffs, Utah. [After G. G. Parker, U. S. Geological Survey.]

shells are commonly leached out, leaving open cavities that may later be filled to form "casts" that faithfully preserve the fossil form. Casts of such soluble minerals as halite and pyrite are also common.

In Chapter 4, we described the cementation of sand to form sandstone. Calcite cements many sandstones, but the reason is obscure; surely water far below the water table cannot evaporate and deposit calcite as it does in forming caliche and dripstone. Possibly the decreased pressure as it approaches the surface allows carbon dioxide to escape and calcite to crystallize. Some sandstones are only locally tightly cemented around shell fragments or other nuclei to form ball-like masses—**concretions** (Fig. 15-15). Experiments show that slow precipitation of any material from solution generally results in enlargement of existing crystals rather than development of new crystallization nuclei. Ground water also deposits many other substances: silica as opal, chalcedony, or quartz, and iron oxide as limonite or hematite.

Ground water that circulates to depths of several kilometers may be heated close to the boiling point, and many other minerals such as feldspar, mica, clays and chlorite may be dissolved at depth and deposited at cooler areas. The water may return to the surface without losing much of its heat, forming hot springs (Fig. 14-17). Hot

Figure 14-17
Steamboat Springs, Nevada, where heated ground water returns to the surface from depth. [Photo by W. D. Johnston, Jr., U. S. Geological Survey.]

springs abound in volcanic areas, but are not confined to them. Some hot springs are valuable sources of power: those of Lardarello, near Pisa, supply most of the electricity that runs the Italian railroad system. Hot springs and steam jets (fumaroles) are also exploited for power in New Zealand, Iceland, and California.

Heated ground water is a powerful solvent of rocks, and when it returns to the surface most of its dissolved load is deposited. Thus the geysers of Iceland, Yellowstone Park, and New Zealand build mineral terraces precipitated from their hot pools (Fig. 14-18). Some terraces are of opaline silica, others of calcite. Algae of various colors participate in the process of precipitation and give brilliant colors to some pools.

Figure 14-18
Hot spring terraces of calcium carbonate, Mammoth Hot Springs, Yellowstone National Park. [Photo by Tad Nichols, Tucson, Arizona.]

Figure 14-19
Water witcher or dowser of the sixteenth century.
[Styled after old wood cuts.]

WATER WITCHING

Before the nineteenth century, men thought that ground water flowed in definite rivers like surface water; a lucky well would reach one of these streams and produce copiously; an unlucky one missed. Since no one can see beneath the surface, the digging of wells was always uncertain. Many farmers seeking help with this vexing problem consulted "water witches" or "dowsers." Dowsers supposedly possessed supernatural powers that enabled them to discover the "underground streams"; the belief still persists locally. A dowser walks about with a tightly held forked stick (Fig. 14-19), which is supposed to dip sharply when he is over the underground stream. His success, if any, has no known scientific basis, but his probability of success is high, for aquifers are widespread.

COMPOSITION OF GROUND WATER

Ground water beneath swamps, peat bogs, and rain forests is always slightly acid because of organic acids released from decaying vegetation. When it contacts limestone, or even moderately calcium-rich rocks such as granite, it may dissolve enough calcium to become **hard water**. The same thing happens even with rainwater. The amount of calcium ion in hard waters in humid regions is only a small fraction of one percent, but is enough to inhibit lathering of soap and to encrust boilers. Such water may also contain fully as much sodium, but this is not readily precipitated and ordinarily goes unnoticed.

In arid regions, water in the capillary fringe near the surface may evaporate, precipitating calcium carbonate as crusts of caliche at the very surface of the ground. Complete evaporation precipitates sodium carbonates and sulfates that are toxic to plants and nearly ruin agricultural land. If such "alkali soils" are drained and the solvents washed away by heavy irrigation, they may be reclaimed for agriculture.

Some ground waters, even in humid regions, contain enough salt to make them undrinkable and injurious to plants because they have come in contact with beds of salt or with sea water entrapped between the grains of marine sediments — **connate water**. Few connate waters retain the precise composition of sea water because of dilution by ground water, solution of salt from sedimentary beds, and reaction with the wall rocks. Most of them are more concentrated than sea water. Salinity generally increases with depth; one reason is that brines are denser and displace the less saline waters.

GROUND–WATER SUPPLIES IN THE UNITED STATES

Water is a vital resource. Even in humid western Europe and eastern North America, municipal and industrial demands tax all available supplies. The quantity and quality of recoverable ground water are therefore of prime social importance. Ground water within the United States to a depth of 0.8 km is estimated at fully 700 times the amount in all fresh water lakes and streams. In almost any area, wells will yield some water, but rocks of low hydraulic conductivity yield negligible amounts.

Much of our ground water comes from unconsolidated Pleistocene and Holocene surface formations and from lava flows and cavernous limestones of various ages. Some comes from consolidated sedimentary rocks. The principal unconsolidated aquifers are: (1) river gravels, sands, and glacial outwash, (2) the coarser beds of deltaic and other coastal plain deposits, and 3) sands and silts of floodplains.

About half the ground water used in the United

States is derived from unconsolidated alluvium in the intermontane valleys of the west, in the valleys of the Mississippi and its tributaries, and on the Great Plains and the Atlantic coastal plain. Sandstone aquifers supply much ground water in the Mississippi Valley and in Texas. Basalt flows are important aquifers in the Pacific northwest and in Hawaii. In New England glaciofluvial gravels yield much water, but even more is derived from jointed gneiss and schist.

The abundant joints, lava tubes, and brecciated tops and bottoms of basalt flows make them almost or quite as permeable as cavernous limestones. The most productive springs in the United States are in flood basalts of the Snake River Canyon. These springs discharge 140 m³/sec, much of it derived from the drainage of the Lost River and other streams in the mountains to the north. These disappear as they reach the basalt plain. Cavernous limestones are important aquifers in Florida, the Cumberland plateau, and the Shenandoah Valley. Some in Turkey have discharges even greater than the Snake River springs of Idaho.

Several productive aquifers are described in the following pages.

Ground Water of Long Island, New York

Because it is small and infiltration is easy, Long Island has no large streams. Of the 100 to 125 cm of annual rainfall, about 20 percent goes to surface runoff, 40 percent to ground water, and the rest to evapotranspiration. Lacking adequate surface supplies, the nearly 5 million inhabitants depend on ground water and water piped from the mainland. Three fourths of the ground water comes from glacial outwash sands and gravels, the rest from weakly consolidated Cretaceous sands below.

Of the Cretaceous aquifers, the most productive is the basal clean quartz sand, 30 to 80 m thick, overlain by shale. Pleistocene sediments as much as 120 m thick unconformably overlie the Cretaceous. Moraines form two ridges, one near the north shore, the other near the middle of the island. The chief aquifers are the outwash sands between and south of the moraines.

In 1903 the water table underlying Brooklyn sloped from about +4.5 m to sea level, but by 1943 excessive pumping had reduced it to below sea level, so that salt water invaded the aquifer. The State of New York now requires that water pumped for cooling and air conditioning be returned to the aquifer from which it was withdrawn. By 1946 more than 200 recharge wells were operating in the urban area of the island. In the rural area several very large recharge basins have been built so that storm runoff and industrial waste water can seep into the ground. The warm recharged water has raised the ground-water temperature a few degrees, lessening its value for cooling, but the recharge wells and basins have served their purpose: the salt water has been displaced by fresh.

The intensive study of the water budget here has shown that urbanization greatly affects ground-water recharge. Of two adjacent streams, one was urbanized during the study while the other remained rural; the new roofs, pavements, and storm sewers of the urbanized area cut off at least 2 percent of the recharge. The urbanization of the Atlantic region from Portsmouth, New Hampshire, to Richmond, Virginia, has had a measurable effect on ground-water recharge throughout.

Basins in Southern California

Several basins in semiarid southwestern California contain alluvial deposits across which intermittent streams from the more humid mountains flow to the sea, charging the alluvium with ground water on the way (Fig. 14-20). The Santa Ana basin is filled to a depth of 150 to 350 m by compound alluvial fans from the San Gabriel Mountains that extend almost entirely across the basin. At their heads the fan gravels slope as steeply as 9°; basinward their slopes lessen, and the gravels interfinger with sands and silts. Over parts of the fans, relatively impermeable soils have formed and been buried as the fans grew, forming aquitards in an otherwise highly permeable mass. Waters low on the fan were once artesian, but a century of overexploitation has rendered them no longer artesian.

Many of these basins are broken by faults so recent that they cut Holocene alluvium. Impermeable clay gouge along some of these forms effective barriers to ground-water movement. At the fault shown in Figure 14-20, the water table

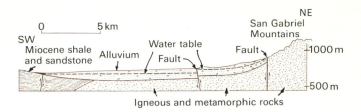

Figure 14-20
Section across the Santa Ana ground-water basin, showing the water-bearing Pleistocene and Holocene alluvium and the effect of a fault on the position of the water table. [After California Division of Water Resources Bull. 45, 1934.]

stands 120 m higher on one side than on the other, making a considerable difference in pumping costs on the two sides.

In this populous but arid area, the winter flood waters are diverted to spreading grounds of coarse gravel, thus helping to recharge the aquifers. Here, as in Brooklyn, excessive pumping near the sea reversed the water table gradient so that sea water invaded the aquifer; the problem was met by a line of recharge wells along the coast, the effect being to build a ridge in the water table, thus reversing its coastal slope and halting the salt-water incursion—an effective if costly solution.

It is noteworthy that lowering of the water table by pumping may have beneficial as well as harmful effects. In an area where a fully charged aquifer would result in bypassing of surface or subsurface flow to the sea, pumping makes pore volumes available for recharge, and steepening of the hydraulic gradient allows the aquifer to accept water during the rainy season.

Dakota Sandstone Artesian Aquifer

The great Dakota Sandstone basin forms the largest and most important source of artesian water in the United States, extending over much of the Dakotas, Nebraska, and parts of adjacent States. At least 15,000 wells have been drilled into this Cretaceous sandstone, which is generally about 30 m thick and is overlain by scores or even thousands of meters of other beds, largely shale. As shown in Figure 14-21, much of the recharge takes place in the upturned zones along the edges of the Black Hills and Rocky Mountains. But the sandstone is not a simple aquifer, and recent studies by Frank Swenson of the U.S. Geological Survey have shown that, except near the mountains, much of the recharge is by leakage from the Pahasapa Limestone, a cavernous limestone of Carboniferous age that unconformably underlies

the Dakota. The Dakota Sandstone includes a widespread shale bed near the eastern margin of the basin, dividing it into two members, in which the compositions and pressures of the artesian waters vary independently. Pressure measurements in wells prove that the flow from both the surface intake and from the Pahasapa below is locally hindered by less permeable parts of the formation. Because of heavy use, pressures in the aquifer have decreased progressively since the first well was drilled in 1882.

The Dakota Sandstone is not the only valuable aquifer in this region; we have already mentioned the Paleozoic sandstone whose depth and productivity were so accurately foretold by Darton.

Water Mining on the Llano Estacado

The Llano Estacado of west Texas and eastern New Mexico is formed of Tertiary gravel and sand spread in a great alluvial apron from the southern Rocky Mountains. The canyon of the Canadian River on the north has separated the

Figure 14-21
Section through the Dakota Sandstone artesian aquifer, from an intake area in the Black Hills of South Dakota to northern Iowa. Vertical scale greatly exaggerated. [After F. A. Swenson and N. H. Darton.]

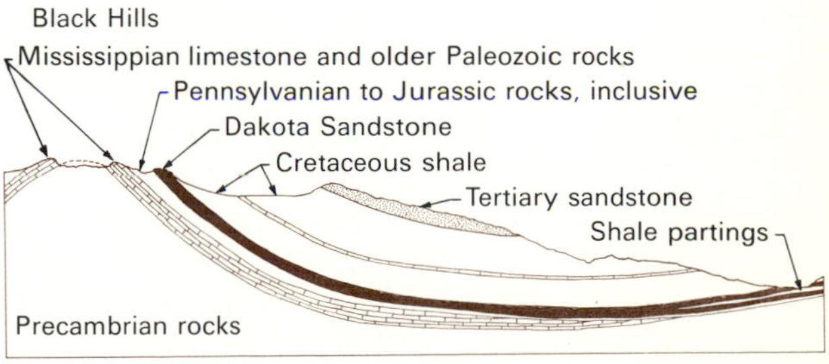

Llano Estacado from the source area of the gravels, and they now form an isolated mass, resting on an aquitard of Cretaceous shale. The water in the gravels is perched. The gravels vary in thickness between 120 and 180 m. The rainfall is about 50 cm annually, but very little reaches the water table because of high summer temperatures and heavy cultivation; recharge is nearly negligible.

The Llano Estacado was virtually unpopulated in 1880, and had fewer than 20,000 people in 1900. Soon afterward it was found that ground water could be used to grow remarkable crops of cotton; agriculture increased phenomenally. Pumpage has increased to the point at which more than 56×10^{10} m^3 of water are being withdrawn from storage annually, and the prosperous agricultural area now supports a population of about 400,000 people. But the recharge is so little that if pumping were to stop today, it would require 4000 years for the water table to return to its original level. No one can say how long the pumping can continue at present rates, but the water table had fallen more than 30 m in some areas by the early 1960's and as much as 13 m in a decade. It is obviously more costly to raise the water from greater and greater depths. As much as 44 percent of the water originally beneath Lubbock County, Texas, had been withdrawn by 1962. Clearly, it will eventually become uneconomic to deepen the wells, and some farms will be abandoned, thus prolonging the life of the reservoir, so that one need not expect a sudden collapse of the area's economy, even though a steady decline over the long run is certain. The water is being mined just as literally as is the coal of Pennsylvania or the copper of Arizona, and just as inevitable is the ultimate exhaustion of the water resource.

ECONOMIC AND LEGAL ASPECTS OF GROUND-WATER USE

Where neither surface nor ground water suffice for all, disputes have arisen between individuals, communities, states, and even nations. Many such disputes have been carried to the courts. Applying, but narrowing, the English Common Law rule that the owner of the land surface also owns everything beneath it (a rule not recognized in Roman Law or in the codes deriving from it), United States courts have held that all the owners of land overlying a ground-water basin jointly own the water. Water may not be exported without compensation, and water rights are prorated according to acreage. Furthermore, the principle of "best use" has been established for settling disputes, as between cattlemen wishing to preserve feeble springs and truck gardeners wishing to pump ground water from the same area for more productive use.

In some states permission of water-control authorities is required for the drilling of large wells, and return of used water to the aquifer may be required. Ground water is an important public commodity, and its use increasingly demands regulation by well-informed officials.

Questions

1. Draw sketches showing several geologic conditions that could result in the formation of a spring.
2. Draw one well-labeled cross section showing all of the features listed below:
   (a) An area of rounded hills with two through-flowing streams: one with a floodplain, the other downcutting far above grade.
   (b) Two deep but dry ravines.
   (c) The position of the normal water table.
   (d) A perched water table.

(e) A swamp.

(f) Two wells of equal depth: one with water, the other dry.

3. A group of large fresh-water springs emerges on the sea floor about half a mile off the mountainous coast of Ecuador. Show by a well-labeled diagram how this is possible.

4. Amphibole and garnet grains are abundant in well-cemented concretions from a sandstone, but the remaining, poorly cemented sandstone contains only a few etched grains of these minerals. How do you account for this?

5. Compare the drawdown at the test well near the Platte River (Fig. 14-6) with that of the Los Lamentos Mine (p. 317), and account for the difference.

6. How can you tell an area of karst topography from the hummocky surface of a large landslide or debris flow?

7. Large springs are common in areas underlain by basalt, but almost nonexistent in areas of granite. Why?

8. How can ground-water basins be artificially recharged from waste water at the surface?

9. If water is neither moving through an aquifer nor being discharged from it, can there be a hydraulic gradient? Explain, using a diagram.

Suggested Readings

Hubbert, M. King, The Theory of Ground Water Motion. *Journal of Geology,* v. 48, 1940, pp. 785–944.

Lohman, S. W., *Geology and Artesian Water Supply of the Grand Junction Area, Colorado* (U.S. Geological Survey, Professional Paper 708). Washington, D.C.: G.P.O., 1965.

Meinzer, O. E., *Ground Water in the United States: A Summary* (U.S. Geological Survey, Water Supply Paper 836-D). Washington, D.C.: G.P.O., 1939. [Pages 157–229.]

Meinzer, O. E. (ed.), *Hydrology; Physics of the Earth. No. 9* (National Research Council). New York: Dover Publications, 1942.

Scientific American Offprint

818. A. N. Sayre, "Ground Water" (November 1950).

CHAPTER 15

Deserts and Wind Erosion

Deserts are barren because not enough water is available to support a continuous vegetational cover. No true desert receives more than 25 cm of precipitation per year. But many areas receiving less, such as the tundra of northern Alaska, with only 15 cm, are lush with plant growth because evaporation is low and permafrost prevents water from draining downward. To a visitor from a well-watered region, the desert at first seems to have been molded by different forces from those of his homeland. The contrasts, however, do not reflect different agencies, but merely result from differences in their effectiveness caused by different climatic conditions (Fig. 15-1).

About a sixth of the lands of the earth are desert. The most arid lie in the Horse Latitudes — the subtropical belts of high atmospheric pressure, where descending air masses are dry, clouds few, precipitation low, and evaporation high. Annual evaporation from the Persian Gulf is as much as thirty times the rainfall (Fig. 10-4). Other deserts lie in the Trade Wind belts, where the winds have swept long distances over land, with little opportunity to pick up moisture. Other great deserts in higher latitudes, such as those of Central Asia and parts of western North America, lie in "rain shadows" behind high mountains that force the prevailing winds to rise, cool, and give up most of their moisture in crossing. Greenland and Antarctica are barren because nearly all their water is ice, not available for plant growth.

CLIMATIC CONTROLS

In most great deserts, such as the Sahara, the average annual rainfall is less than 18 cm. A year or even several years may pass with no rainfall, though dew is common. Some students divide arid regions into **steppes**, where scattered bushes and short-lived grasses furnish a scant pasturage, and **true deserts**, where vegetation is sparse or absent. So defined, most deserts of North America are steppes. Of course, all gradations exist.

INTERIOR DRAINAGE

Only the greatest rivers, such as the Nile, Indus, Tigris, Niger, and Colorado, can persist through wide deserts to the sea. These and lesser desert streams are called "losing streams" because they lack adequate tributaries and evaporation rates are high. Some merely sink into the ground; others flow into a closed basin, like the Dead Sea, or into an alkali mud flat that may seasonally hold a shallow lake but be dry for the rest of the year, a **playa** (Fig. 15-2). Such **unintegrated drainage** is characteristic of deserts.

The water table generally lies far deeper in deserts than in humid regions. Rainfall there, as in humid areas, is greater at high altitudes than at low. After a rain, rills or even rushing torrents rise in the mountains but quickly diminish on

Figure 15-1
Broad alluvial plains between desert mountains, Salton Desert, California. Belt of small sand dunes in foreground. The straight black line is a railroad. [Air photo by R. C. Frampton and J. S. Shelton.]

Figure 15-2
North Alvord Playa, southeastern Oregon. The irregular dark patches on the white playa floor are wet ground. The straight mountain fronts to the upper left are fault scarps. [Air photo by Richard E. Fuller, Seattle, Washington.]

Figure 15-3
Wadi Araba, Eastern Desert, Egypt. The dry watercourse of a braided stream.
[Photo by Tad Nichols, Tucson, Arizona.]

the plains below. Though most stream courses are dry for most of the year, effects of stream erosion nearly everywhere dominate the landscape (Figs. 15-1, 15-3). Barren mountains scarred by gullies and sun–baked plains underlain by stream deposits are characteristic.

Most desert storms are local, and the streams they feed flow for only a few hours. Most sediment is thus not transported far, but in a short distance is dumped on alluvial cones at the mouths of mountain canyons. Cones of adjacent canyons grow and merge to form great alluvial aprons along the mountain fronts (Figs. 12-37, 15-4, Top). These compound alluvial aprons, or **bahadas**, flatten gradually toward the valley and merge imperceptibly with it, as the streams, weakened by evaporation and infiltration into the permeable ground, carry only fine material. A

basin of interior drainage surrounded by bahada slopes is a **bolson**.

Water infiltrating warm desert ground quickly dissolves calcite and, later deposits it just below the surface upon evaporating from the capillary fringe. The resulting lime-cemented rock is **caliche**. In the central Australian Desert and in the Kalahari Desert of South Africa, the silica-cemented **desert armor** at or near the surface has a like origin. Thus many bahada slopes and valley floors become plated with well-cemented rocks within a few years.

In wet seasons shallow bolsons may be filled to overflowing, thus leaving a string of playas when the dry season ensues. Although Great Salt Lake is perennial and thus not strictly a playa, it fluctuates widely with wet and dry cycles, so that the flat western part of its bed — the Bonne-

Figure 15-4
Three stages in the erosion of desert mountains. (Top) Panamint Range, California, showing bahada slopes at the foot of a moderately eroded fault block. (Center) Ibex Mountains, California, showing broad pediment embaying deeply eroded range. (Bottom) Cima Dome, California, showing a broad graded surface with a few residuals of former large mountains. [Photos by Eliot Blackwelder, Stanford University.]

ville Salt Flats—has all the character of a true playa. It is surfaced with halite, various carbonates, and sulfates of sodium—salts typical of "alkali flats" in many arid regions.

EROSIONAL PROCESSES

Weathering

Rocks disintegrate and decompose in deserts, though more slowly than in humid regions, because of the paucity of moisture and especially of organic acids in the soil. We noted in Chapter 10 that an Egyptian obelisk moved to New York weathered more in 90 years than it had in 3500 years in Egypt.

Because weathering is slow and the barren ground unprotected from rainwash and wind, fine-grained residual soils are rare. The fertile soils of Egypt and Mesopotamia, where alluvium has supported civilization for centuries, were not weathered in place but in the humid headwater regions of the rivers. Limestone, which dissolves readily and generally underlies lowlands in humid regions, stands in bold ridges in the deserts, partly because joint blocks broken off dissolve slowly and so protect the slopes, and partly because water in the ground evaporates near the surface, redeposits its dissolved calcite, and thus seals openings in the rock. One important factor in desert weathering is the shattering of porous rocks by the crystallization of salts from evaporating water. The growing crystals tend to split the rock into splinters.

Rainsplash and Rillwash

Desert plants are so scattered that they protect little surface from the direct impact of the pelting rain. Even though the mass of an individual drop is slight, on unprotected ground it splashes fine rock fragments into the air, to fall or roll downhill. Anyone who has seen the mud and sand splashed onto a garden path after even a light rain can readily imagine the effects of splash during a heavy hailstorm or rainstorm when perhaps 5 cm of rain falls in a single hour. This is nearly 50 kg/m^2. During a heavy rain small rills quickly form and sweep mud, sand, and, as the rills join and enlarge, even gravel and boulders downhill. Desert streams carry vastly more material than corresponding rills in humid regions, where vegetation binds the soil. Much of this load is left stranded after only a short journey, helping to fill older channels. Because of the low water table, most desert streams are losing streams and run dry in a few hours. Between storms, newly exposed rock is accessible to weathering—moisture being supplied by dew; loosened chips and grains fall into the rills, so that if the dry spell continues, the smaller channels may lose their identity.

Mudflows and Sheetfloods

From time to time, perhaps only once in a decade, or even a century, intense rains pour down on deserts. In September 1970, one dropped 29 cm of water on southern Arizona in 24 hours. Similar downpours are more frequent in humid than in arid regions, but their effects are far less because of vegetation cover. These so-called "cloudbursts" may drop as much as 10 cm of rain in an hour. Most are very local; a few kilometers away the sky may be blue. Cloudbursts quickly cut gullies and strip off loose debris; rapidly gaining in both volume and load, they race downcanyon as a wall of debris-laden water, so charged with mud and sand as to form a turbid sludge far denser than water alone. Such mudflows are capable of buoying up huge boulders, some as much as 10 m through. One poured down Cajon Pass in California as a **flash flood**, overwhelmed a freight train, carried the engine more than a mile downstream, and buried it so deeply that it could only be found by a magnetometer. Mudflows may travel completely out of their source area before reaching an alluvial fan on the mountain front, there to grind to a halt as the flood spreads and its waters sink into the alluvium. One that originated in Titus Canyon, California, left a steep frontal wall 2 m high where it stopped in Death Valley. Excavations show mudflows to be virtually unsorted. They greatly resemble unstratified glacial drift, and indeed some ancient mudflows have been misidentified as tillite.

Loose silt, sand, and rock fragments so abound on desert fans that water flowing over them is soon loaded to capacity; it is unable to scour deeply into the surface. Diverted by cobbles and clumps of plants, the water spreads widely in a plexus of small braided channels, or it may cover the whole surface as a **sheetflood**. When it sinks in or evaporates, a coating of mud and silt is left to dry in the sun.

Downslope Movements

Downslope movements produce somewhat different results in deserts than in humid areas. Although weathering is relatively slow, many joint blocks are so weakened that they fall apart when they tumble from the cliffs. Talus therefore forms only beneath rocks resistant to weathering, such as quartzite, chert, or limestone. Both steep and gentle slopes may be mantled with fallen boulders and chips, but most of these are only "one boulder thick," with bedrock visible beneath. The boulders are ultimately swept away by cloudbursts or reduced by slow weathering to grains small enough to be carried away by rills.

Although during cloudbursts runoff is rapid and great, thick masses of water-soaked soil and rock like those responsible for the Gros Ventre slide (Chapter 11) rarely develop in deserts because the storms are brief. In fact, the large ancient landslides and debris flows in Arizona are considered strong evidence of a formerly more humid climate.

Relation of Slopes to Structure

The lack of soil and vegetation affects desert erosion in still another way. As the loose surface material is not root-bound, it does not creep as a mass. Thus changes in steepness and roughness of the surface are not softened and blurred as they are in moist climates. Slope steepness is apparently determined by the size of the bedrock joint blocks; steep where blocks are large, gentle where small. Even minor differences in particle size are accurately reflected by changes in slope developed on different rocks (see Figs. 5-2 and 12-32). Abrupt changes in slope at the contacts of different rock masses are the rule in the desert, in marked contrast with the blurred and transitional slope changes characteristic of regions of active soil creep. In the deserts abrupt changes in gradient rather than transitions link hillslopes and valley floor; most steep slopes remain steep, perhaps "case-hardened" by quick evaporation of capillary water, with consequent deposition of dissolved matter. Even on wide desert plains, gradients depend on the size of the weathering particles (Fig. 15-5).

EVOLUTION OF DESERT LANDFORMS

Crustal deformation may disrupt pre-existing drainage in deserts as well as elsewhere; so also may growing alluvial fans, mudflows, volcanic eruptions, or even roof-collapse of limestone caverns. Basins range from enormous areas such as the Caspian and Aral depressions, the Dead Sea trough, or the basin of Great Salt Lake, down to wind-carved hollows a few meters across. In a region of interior drainage, the base level of the streams constantly rises as the alluvium collects. We now outline the evolution of a landscape of interior drainage.

As the base level rises, the stream gradients flatten and the streams aggrade their alluvial fans. As the stream drops its load it becomes unstable, overflowing now in one direction, now in another. In this way, the terminal lake in the Lop Nor depression of central Asia has moved more than 100 km since the thirteenth century. Rainfall is always higher in the mountains than over lowlands, so the only gaining streams are there; when they reach the fans they quickly lose capacity, deposit their loads, and build up the fans, so that they eventually grow headward into the mountain valleys.

But the upbuilding of the fans is opposed by the slow weathering of their material and its removal by intermittent streams that head, not in the mountains, but on the fans themselves. Feeble as such streams are, they strongly influence the landforms. A fan can continue to grow only as long as the mountain-fed streams continue to bring more material to it than is removed; obviously, the greater the fan area grows, with respect to the source of fan material, the greater the area of rain attack on the fan. A dynamic equilibrium is thus approached, at which material

Figure 15-5 (*facing page*)
Erosion in friable sandstone, Coal Canyon, Arizona. Note the abrupt change in slope at the foot of each hill and the uniform gradients of the hill slopes. [Photo by Tad Nichols, Tucson, Arizona.]

deposited on the fan is just equalled by that removed to the basin floor.

The topography of desert areas where crustal deformation has been fairly recent (as shown by folding or faulting of late Tertiary or Pleistocene deposits) comprises three principal land forms: (1) relatively steep mountain slopes of bare rock and loose fragments, (2) bahada slopes of coalescing fans, and (3) playa floors covered with fine silts, clays and various salts residual from evaporation (Fig. 15-4, Top).

As the mountain front retreats, the bedrock is slowly regraded to slopes virtually identical with those of the alluvial fans, forming a broad, gently sloping surface, a **pediment**, strewn with a thin, discontinuous veneer of gravel in slow transit downslope. At this state the desert landscape comprises four main elements: (1) the mountains, with almost unchanged slopes, (2) the pediment, or planed-off bedrock surface, abruptly flatter than the mountain slope, (3) the bahada, now composed of old fan deposits regraded to a lower surface that blends imperceptibly upslope into the pediment and downslope into (4) the playa (Fig. 15-4, Middle). This stage of landscape development is widespread in southern Arizona and New Mexico, where deformation has not been quite as recent as in the Great Basin to the north.

It is not only time since deformation, however, that determines the relative extent of bahada and pediment. If the lowlands are small relative to the uplands, fans will continue to grow faster than they are worn down, and the dynamic equilibrium necessary for pediment formation will be long delayed; conversely, broad lowlands and small highlands favor early formation of extensive pediments.

The ephemeral streams may eventually fill the basin with their deposits and overflow the rim into an adjoining basin, thereby integrating the drainage. As erosion goes on, streams draining to lower basins grow headward just as they do in humid areas, capturing successively higher basins and regrading them to lower levels. The main streams thus grow longer and longer, and

in time their long profiles become graded throughout, even though the waters of no single storm may ever flow their entire lengths.

This process of drainage integration and regrading of higher basins results in the scouring out of older deposits of the higher basins, cutting slopes appropriate to the size of their component particles. Because such deposits are generally poorly consolidated, pediments develop quickly. At this stage playas are absent from most of the region, bahadas much less extensive, mountains have shrunk, and pediments cover most of the area. This is the stage represented widely over the deserts of South Africa and, in the United States, south of the Gila River in southwestern Arizona.

In a still later stage, the mountains, though retaining their steep slopes, have shrunk to small isolated hills rising abruptly from a pediment, like islands in a sea (Fig. 15-4, Bottom; Fig. 10-12). These are **inselbergs** (German, "island mountains"). Presumably, if no structural or climatic change intervened, continued erosion would ultimately produce a wide rock plain whose flat surface would be subject only to wind erosion. Parts of the Kalahari Desert of southwest Africa approach this condition, but no large area has been recognized as representing such a hypothetical end stage.

WORK OF THE WIND

The landforms just described indicate that water-molded surfaces dominate the desert landscape. Many people have the impression, fostered by melodramas, that deserts are chiefly wastes of sand dunes. This is far from true; sand covers less than one-fifth of the desert areas of the earth; yet because vegetation is sparse or absent, wind erosion is much more effective in deserts than elsewhere, and locally is the major agent.

Every gust of dry winds wafts dust into the air. In the open, on hot summer days, dust devils swirl fine debris high above plowed fields. Sporadic tornadoes uproot trees, lift soil, and destroy houses. Dust is in the air everywhere; even in humid regions closed rooms require dusting every few days.

Sorting by the Wind

Anyone who allows a handful of dry soil to dribble slowly from his hand during a wind notices that some falls almost vertically but that much is strung out downwind, and the finest dust is carried away completely. By repeated trials, the coarse grains can be rather cleanly winnowed from the fine, even in a gentle breeze, just as wheat is winnowed from chaff on primitive threshing floors.

This illustrates a general condition: any object falling through a fluid, such as air or water, at first accelerates but eventually reaches a speed that remains constant—the **terminal velocity** of fall. Two forces are acting: (1) the pull of gravity on the body and (2) the resistance of the fluid to its passage. The gravitational pull depends directly on the difference between the density of the object and that of the fluid it displaces. The resistance of the fluid depends on its viscosity, the diameter and shape of the body, and its velocity. Resistance increases with increasing speed. Thus gravity, which in a vacuum would produce constant acceleration, is ultimately balanced by the resistance of the viscous fluid; acceleration ceases and speed is constant at the terminal velocity.

Experiments show that terminal velocities of different-sized spheres falling through a fluid vary tremendously. Those of particles smaller than about 0.01 mm vary almost exactly with the square of their diameters, a relationship deduced by Sir George Stokes in 1851. Such particles fall slowly enough that the air accommodates to their passage by laminar flow; larger spheres displace such a volume of air that their inertia and the turbulence they cause both become important. Figure 15-6 shows the general relations between size and terminal velocity of particles in air, though flakes and other irregular grains fall more slowly than spheres of the same mean diameter and density. The figure shows that the fine particles in our handful of soil were blown farther than the coarse because they fell more slowly.

This helps us to understand the transporting power of the wind. All winds are turbulent; gusts and eddies swirl in every conceivable direction, superposed in their general motion. Close to the ground the ratio of the speed of upward gusts to average forward speed is extremely variable, but averages about 1 to 5. Thus some particles whose terminal velocities are lower than $\frac{1}{5}$ of the wind speed will be carried upward by gusts and remain suspended until they are caught in a downdraft or sink under their own weight to the ground. While in suspension they will of course travel along with the wind. Particles whose terminal velocity exceeds $\frac{1}{5}$ of the wind speed are not wafted aloft. Measurements in dune areas indicate that sand begins to move when the average wind velocity reaches about 5 m/sec (16 km/hr). Figure 15-6 shows that if the maximum updrafts have $\frac{1}{5}$ of this speed, grains less than about 0.2 mm in diameter should be winnowed out of surface sands.

This deduction is amply confirmed by observation of actual dunes. Sieving dune sands through graded screens shows that grains in the range 0.3

Figure 15-6
Graph showing the variation of terminal velocity with grain size of falling particles in air. [After R. A. Bagnold, *The Physics of Blown Sand and Desert Dunes*, William Morrow and Co., 1942.]

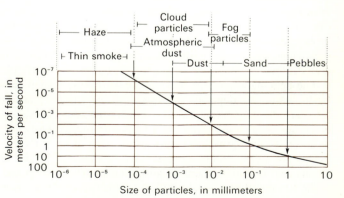

Figure 15-7
Fluted and polished Eocene limestone, sandblasted by the prevailing winds, Western Desert, Egypt, near Kharga Oasis. [Photo by Tad Nichols, Tucson, Arizona.]

Motion of Particles with the Wind

Sand grains, momentarily carried forward by gusts, strike the ground at a low angle. If the surface is rocky, they bounce into the air and move on in a series of hops. If the surface is loose sand, a grain may "splash" to a stop on the surface and eject several others into the wind stream. Even grains too large to be caromed into the air will be pushed along slowly by successive impacts of many grains. The thickness of the layer of sand entrained depends on wind speed and grain size. If the sand is pebbly, the wind may winnow the sand away, leaving the pebbles to accumulate over the desert surface as a residual layer, one pebble thick, making a **desert pavement** (Figs. 15-8, 15-9). The more exposed pebbles in the pavement generally have smooth, sandblasted facets. Pebbles undermined as sand is blown from beneath them may roll over, exposing more sand and a different pebble surface to the sandblasting. In this way sandblasted facets may develop on several sides of the pebble, as seen in Figure 15-8. Eventually, wide areas are covered by such deflation pavements, one pebble thick (Fig. 15-9).

Figure 15-8
Desert pavement in Death Valley, California. Note the facets on several boulders and pebbles. [Photo by Eliot Blackwelder.]

to 0.15 mm greatly predominate. Even the finest dune sands contain almost no grains smaller than 0.08 mm.

Wind-blown sand rarely rises higher than about 2 m, even in a severe storm, and most moves within a few dm of the ground, as shown not only by measurements in storms but also by abrasion of telegraph poles and rock outcrops (Fig. 15-7). The great clouds that darken the sun in areas like the floodplain of the Nile and the Dust Bowl of the High Plains, are clouds of dust, not sand. In sandy deserts away from the floodplains, the air is generally clear, even in high winds, just above a carpet of sand-laden air only a meter or so thick.

Figure 15-9
Deflation armor, or desert pavement, in the valley of the Little Colorado River, Arizona. [Photo by Tad Nichols, Tucson, Arizona.]

Similar processes go on in the snowy deserts of Antarctica and Greenland. The hardness of snow crystals increases greatly with decreasing temperature; at −78°C they are as hard as orthoclase. At some stations in Adelie Land, the wind velocity averaged over an entire month is as much as 110 km/hr, gusting to 160. Drifting snow is thus a powerful erosional agent, and boulders facetted by snow blast are as common there as sandblasted boulders in a sandy desert (Fig. 15-10).

The surface of a sand dune is so rough that the turbulent wind whips the most exposed grains aloft in momentary whirls. But where the average diameter of the grains is less than about 0.03 mm (much below sand size), even the most exposed grain projects so slightly from a smooth surface that the wind cannot pick it up except at very high speeds. Dunes never form on a surface composed exclusively of such fine grains, and only a very high wind can set them in motion.

In wind-tunnel experiments, a British military engineer, Brigadier R. A. Bagnold, found that a smooth surface of loose, dry Portland cement was stable and the air above dustless even though the wind was strong enough to move pebbles 4 mm in diameter. This stability of even-surfaced fine material accounts for the general lack of dust storms on the large playas, whose surface material is both fine grained and well sorted. Such material is stabilized by the small size of the grains and their strong cohesion if capillary water is present. Only when the playa surface has dried up and is covered with curled flakes of dried mud does it yield much dust, even to a strong wind. After the mudcurls have blown away (to accumulate as clay dunes to leeward), the playa is nearly dust-free unless disturbed by animals or wheels.

Measurements made while sand is drifting show that the wind near the ground moves much less swiftly over loose sand than over bare rock, even though the velocity at a height of 2 m is identical. Studies of grain movements suggest why this is so. Momentarily suspended grains that bounce along the rocky floor are so highly elastic that little energy is needed to keep them rebounding. Much more energy is needed to keep similar grains moving over loose sand, because the grains lose momentum on splashing into it and disturbing other grains. Thus grains that bounce over a rocky floor slow down or stop

Figure 15-10
Snow-blasted boulder, Taylor Dry Valley, Antarctica. [Photo by Warren Hamilton, U. S. Geological Survey.]

when they strike loose sand. Sand dunes grow because of their peculiar ability to collect sand grains from intervening barren areas instead of permitting them to spread evenly over a broad surface.

SURFACE FORMS OF MOVING SANDS

Small-scale Features

When wind becomes strong enough to lift sand grains, the surface of a moving mass of sand is bombarded by grains that splash into it and eject some of the grains they hit; it is estimated that between 20 and 25 percent of sand transport takes place by surface creep. Though the leaping grains differ widely in range and trajectory, since

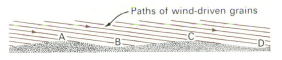

Figure 15-11
The beginning of rippling on a sand surface. [After R. A. Bagnold, *The Physics of Blown Sand and Desert Dunes*, William Morrow and Co., 1942.]

most are of about the same size, they will strike a flat surface at about the same angle and with like momentum. Saltating grains can move grains as much as six times their own diameter.

A not-quite-flat surface of sand is shown in Figure 15-11. A small hollow has formed at B. The paths of the leaping grains are represented by the equally spaced parallel lines. The forward drift of the sand, the aggregate movement of both jumping grains and those hit by them should be roughly proportional to the number of grains striking a unit area. Fewer grains per unit area will strike the upwind slope of the hollow (AB) than strike an equal area of the downwind slope (BC). More grains will be driven up the slope BC than are driven down the slope AB; the hollow will deepen. The slope BC also receives more impacts per unit area than a horizontal area of equal size. Grains carried up the slope will therefore accumulate at C, on the lip of the hollow. A second slope (CD) is thereby formed. On it, as on AB, grain motion is at a minimum; a second hollow must form downwind. In this way the originally flat surface of the sand becomes rippled (Fig. 15-12).

Once rippling begins, more grains are ejected from slopes facing the wind than from the shel-

Figure 15-12
Wind ripples on the surface of a dune near Newport, Oregon. [Photo by Parke D. Snavely, Jr., U. S. Geological Survey.]

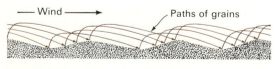

Figure 15-13
Uniform transfer of grains and wind-formed ripples. [After R. A. Bagnold, *The Physics of Blown Sand and Desert Dunes*, William Morrow and Co., 1942.]

tered slopes (Fig. 15-13). The ripples thus tend to become similar in size and spacing. Furthermore, as their crests rise, they enter streams of stronger wind; the larger and heavier grains thus tend to linger on the crest as the lighter are more readily moved. This concentration of coarser grains on ripple crests is the exact opposite of grain-sorting in water-formed ripples; when preserved in consolidated rocks, it helps distinguish the two depositional environments.

Rippling takes place during gentle wind; when wind speed rises (in wind tunnel experiments to about three times that needed to start motion), the ripples are destroyed, apparently because the difference in speed of wind over crest and hollow becomes negligible. If the wind dies away gradually, the hollows tend to fill, as the wind in these protected places becomes too feeble to maintain them. Hence winds that slacken slowly may leave a nearly flat, though mildly rippled surface.

WIND EROSION

Unlike streams and glaciers, winds are not confined between banks but blow freely uphill and down. Dust clouds may travel far before the dust settles out. The only base level for wind erosion is the local water table, and even this may be slowly lowered by evaporation as wider areas are eroded down to the capillary fringe.

Most large undrained depressions in the deserts of North America and Asia—for example, Death Valley and the Caspian depression—have been formed by crustal deformation, not by wind erosion. But in Wyoming, Colorado, New Mexico, Texas and Egypt, the wind has carved hollows—some more than 100 m deep and covering several tens of square kilometers. In the Kalahari Desert of South Africa, many shallow "pans" have been cut below the general rock surface. These are undrained, so they could not have been formed by stream erosion.

The most striking wind-carved depressions make up a chain of oases extending about 650 km westward from the Nile Delta into the Lybian Desert (Fig. 15-14). Although perhaps started by some other process (for example, ground-water solution), strong evidence shows that the depressions have been enlarged and deepened by the wind. Their northern margins are steep escarp-

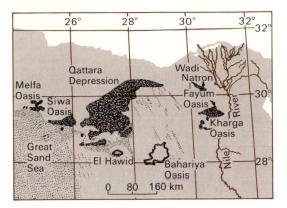

Figure 15-14
The large depressions and sand dune areas of Egypt and Libya. The depressions (dark pattern) are below sea level. [After U. S. Air Force Aeronautical Charts.]

Figure 15-15
Concretions left on the rim of the Kharga Oasis, Western Desert, Egypt, by deflation of finer sand grains not so firmly cemented. [Photo by Tad Nichols, Tucson, Arizona.]

ments, greatly dissected by stream-carved ravines. Some depressions bottom well below sea level. The floors of successive depressions rise gradually southeastward and merge with the general level of the desert plain. Long chains of sand dunes built in part of material blown out of the oases, are strung across their slopes and beyond the oases for hundreds of kilometers to the southeast.

Flat-lying concretionary sandstone underlies this part of the desert; no evidence suggests the depressions are fault troughs. The dunes to leeward attest to the transporting ability of the winds, but the water-carved slopes draining into the depressions show that the wind need not have abraded the rock very much; the rocks were already broken down by weathering and streams. The winds left behind concretions too coarse to blow away (Fig. 15-15).

When such basins are lowered to the water table, ground moisture and vegetation greatly slow further downcutting. Many depressions have springs of fresh water around a central salt marsh or playa sealed off by clay.

Though impressive, such wind-blown depressions are rare; most deserts show little sign of deep wind erosion, although in narrow passes grooves a meter or two deep and a few tens of meters long have been carved in sedimentary and even more resistant rocks. The main role of the wind, however, is to remove loose material from the surfaces of the alluvial fans whose very presence testifies to the dominance of running water in the making of the desert landscape.

Large Accumulations of Sand

Sand that blows into an area where either the nature or configuration of the ground or the vegetation interferes with the wind, it accumulates. Two kinds of accumulations are obviously related to topography: **climbing dunes** form where the wind rises over a sharp topographic break (an example is the sand sea banked against the northeast wall of Panamint Valley, California; another is the piling of sand to heights of more than 900 m against the west face of the Andes in northern Chile); **falling dunes** form where sand sweeps over a cliff and tumbles into a sheltered hollow. Almost as clearly conditioned by topography are the dunes formed where the wind, after sweeping sand through a topographic gap, diverges across a wide plain with consequent lesser speed.

Sand also accumulates on flat plains, forming persistent dunes that migrate slowly across country for scores or even hundreds of kilometers. Though the mechanism is complex, we can outline a few of the factors involved. Among these, vegetation is paramount; other important factors are the effect of sand accumulation itself on the pattern of wind currents and the relation between sand supply and prevailing winds.

DUNES IN BARREN DESERTS

No desert is entirely without vegetation, but in some the plants are so small and widely scattered as to exert only trivial influence on wind speed. Where enough sand accumulates to create a wind shadow, the wind speed is obviously greater on the windward than on the lee side (Fig. 15-16). Sand is then selectively eroded from the windward side and deposited to leeward. When the speed to leeward slackens so much that the average grain in suspension is not carried to the foot of the slope, it is deposited in the wind

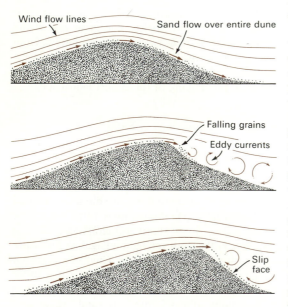

Wind flow lines
Sand flow over entire dune

Falling grains
Eddy currents

Slip face

Figure 15-16
Evolution of a sand dune with a slip face. [After R. A. Bagnold, *The Physics of Blown Sand and Desert Dunes*, William Morrow and Co., 1942.]

shadow high on the leeward slope. Eventually the accumulation becomes unstable and the pile slides down to form a **slip face**, thus building an even more efficient wind shadow. As Figure 15-17 shows, a slip face is composed of many small "landslips" extending from top to bottom of the dune. They generally slope at an angle of repose of 34°.

Dune accumulations a few meters high rise into speedier air streams and therefore become increasingly unstable. The wind funnels through any gaps along their crests, counteracting the tendency of grains to roll sidewise into the gaps, and thus a large dune of irregular height transverse to the wind tends to split in two.

In deserts with very sparse vegetation and constant wind direction, many dunes are of the crescentic variety called **barchans** (Fig. 15-18). The points of the crescent point downwind; the curving bow faces the wind. About two thirds of the lee face is a slip face. Even during strong winds the immediate lee of the dune is quiet, but sand streams away at the ends in great quantities. Since no sand escapes from the slip face, sand arriving from upwind must escape from the wing tips or the dune will grow higher. Sand streaming from the wings commonly starts new dunes farther to leeward. Barchans are confined to areas of nearly constant wind direction.

In general, the larger a barchan, the slower its travel. A small dune will thus overtake a large one and enclose a hollow between its tips. Barchan fields can thus become very complex, especially where vegetation or topography modifies their advance.

Barchans are unstable if wind direction varies greatly. If sand is supplied chiefly from one direction but winds from other quarters are stronger, the sand may move very irregularly and the dunes are strung out in long chains at an angle to both winds (Fig. 15-19). For these dune chains the Arabic word seif (sword) has been applied. Seifs can grow to great size; some in Iran rise more than 200 m above their bases and are a kilometer wide; in Libya and on the Rub al Khali, Arabia, they are nearly as large. Some individual seif ridges are as much as 100 km long, and groups of seifs extend for more than 300 km downwind from the oases of the Western Desert, Egypt. The sand between the seifs is better compacted than that of the dunes and forms the **serir**, or hard sand desert of the Arabs (Fig. 15-20).

Figure 15-17
A transverse dune in the White Sands National Monument, New Mexico, showing irregular slip face.
[Photo by Tad Nichols, Tucson, Arizona.]

Figure 15-18
Barchan dunes near Laguna, New Mexico. The dunes are several hundred meters long. [Photo by Robert C. Frampton and John S. Shelton.]

Figure 15-19
Seif ridges of the Sahara, Africa. [Photo by U. S. Air Force.]

DUNES IN CONFLICT WITH VEGETATION

Even an open vegetative cover greatly influences dune forms. Plants establish themselves more readily in sags in the dunes than on their more active crests, and are there closer to ground water. Hence **transverse dunes**, at right angles to the wind, do not split into barchans if plants in the low spots stabilize the sand that would otherwise form the barchan's wings. Many transverse dunes are a kilometer or more long and reach heights of 4 to 5 m before breaking up (Fig. 15-21).

Figure 15-20
The serir, or hard sand desert, Kharga Oasis, Western Desert, Egypt.
[Photo by Tad Nichols, Tucson, Arizona.]

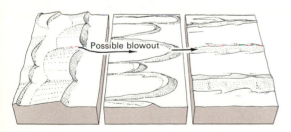

Possible blowout

Figure 15-21
From left to right: transverse, parabolic, and
longitudinal dunes; the arrows indicate possible
transitions between the three forms. [After John T.
Hack, 1941.]

Where vegetation is able to establish itself
widely over the sand, however, long transverse
dunes do not form; instead, two other varieties
predominate: "blowouts," or **parabolic dunes**,
and **longitudinal dunes** (Fig. 15-21).

Parabolic dunes, some of elongate "hairpin"
shape, have their points facing upwind, the reverse of the barchans. Some form by blowouts
of formerly stabilized sand, by accumulation
downwind from patch sources, or where excessive cultivation or trampling by animals destroys
the plant cover and exposes patches of sand.
Some form where sand from a dry stream bed is

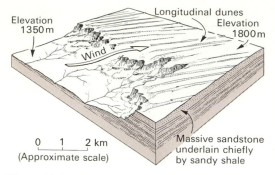

344

Figure 15-22
Longitudinal dunes on the Moenkopi Plateau,
Arizona, formed by climbing dunes reaching the cliff
crest and supplying sand through the notches in
discrete trains across the plateau. [After photo by
the U. S. Soil Conservation Service.]

Figure 15-23
Cross-bedding in the Navajo Sandstone, Zion
National Park, Utah. [Photo by Tad Nichols,
Tucson, Arizona.]

swept up a ravine onto higher, vegetated ground,
where it overwhelms the plants in the line of
maximum supply. The ravine funnels both wind
and sand so that the center of the dune advances
faster than the wings. In extreme cases the dune
may be shaped like a hairpin, or even break up
into longitudinal sand ridges.

Where the sand supply is spottily distributed
or comparatively scanty and the wind direction
constant, dunes may become downwind ridges.
Such dunes also form where climbing dunes
reach the top of the slope and the sand is chan-
neled through notches in the crest (Fig. 15-22).
These are the dominant forms in the Navajo
country of northern Arizona, where many are
several kilometers long and as much as 10 m high.

SUMMARY OF FACTORS
INFLUENCING DUNE FORMS

We have suggested some of the many factors influencing the dune forms. Others are doubtless
involved, and much remains to be learned, but
four factors seem clearly important: wind speed,
constancy of direction, sand supply, and vegetation. Abundant sand and strong winds produce
transverse dunes in both barren and brushy
deserts; less sand and weaker vegetation favor
barchans and longitudinal dunes. Moderate
winds may produce parabolic dunes where vegetation grows rapidly enough partly to anchor the
slowest moving parts.

ANCIENT DUNE SANDS

Highway and railroad cuts through dunes show
that they differ from other sediments in several
ways, among which the most conspicuous is their
cross-bedding. The internal stratification of dunes
is extremely complex, as is suggested by the slip
faces and their complicated progress. A well-
cemented sandstone that preserves similarly complex patterns is illustrated in Figure 15-23. This
sandstone, the Navajo Sandstone of Jurassic age,
contains other features that confirm its dune
origin, such as wind-faceted pebbles and frosted
or sandblasted grain surfaces.

Frosted surfaces on sand grains are common in
windblown but rare in waterlaid sandstone, both
because small grains move faster in winds than

in streams and because their effective mass is far higher because the density of the fluid is much lower. Accordingly, the impact of sand grains on each other, though not enough to shatter them completely, does tend to pit or frost their surfaces. Such textures have enabled recognition of ancient dunes in geologic formations of many ages. But this criterion is not infallible, as some waterlaid sand grains are slightly corroded by calcite cement and so appear frosted.

LOESS

Great areas of southern Germany, Russia, Turkestan, and China in the Old World (see Fig. 13-23), and of the Mississippi Valley and Columbia Plateau in the New, are blanketed with fine-grained, loosely coherent material called loess. Despite its weak cohesion, loess stands in nearly vertical walls because it contains vertical tubules (left by rotting out of grass roots) and shrinkage joints. These also make it highly permeable to ground water. Its porosity may reach 60 percent. Many loess-covered areas are very fertile farmlands—for example, the Palouse country of eastern Washington. Some loess deposits in western China are many meters thick, but elsewhere most are only 2 or 3 m thick.

Microscopic studies show that loess is composed of angular particles—mostly less than 0.05 mm in diameter—of quartz, feldspar, hornblende, and mica, pieces of fine-grained rocks, and a little clay. Most of the grains are fresh or only slightly weathered.

These features suggest that loess is a deposit of dust and silt that settled from the air in grassy country. Most loess lies downwind from areas that were glaciated during the Pleistocene epoch (see Chapter 13), but some deposits are in the lee of deserts. The great loess deposits of China, for example, lie downwind from the Gobi and other deserts, and are probably being added to at the present time, just as dust from the Dust Bowl must, in the middle 1930's, have added to the soil of the more humid lands of the eastern United States. Nigeria has vast loess deposits derived from the Sahara, though much Saharan dust goes on to settle in the Atlantic Ocean (Fig. 17-19). Sahara dust identified in deep-sea sediments, 400 km southwest of the Canary Islands, amounts to about 0.6 mm per 100 years.

Questions

1. Why are the southwestern slopes of the Hawaiian Islands arid, whereas the northeastern sides receive heavy rainfall?

2. Why do dune sands vary so little in grain size? Why are they generally free of clay?

3. Why does sparse sand on a rock floor accumulate into dunes instead of spreading out uniformly over the whole surface?

4. Draw a cross section through the area shown in Figure 15-1, and label the following features on the section: (a) area that is being reduced in height by rainwash and gullying, (b) pediment, (c) area of stream deposition and braided streams, (d) area of active sand dunes, (e) area where wind-carved pebbles and desert pavement might be found, (f) area in which water might be obtained from wells.

5. Why are pediments not formed in humid regions?

6. Why are windborne sands more likely to be frosted than stream sands?

7. What differences can you find between pediments and stream-cut terraces?

8. The so-called cloudbursts rarely exceed 10 cm in total rainfall over a period of an hour or two. Heavy rains that last for much longer periods of time are common in humid regions. Explain the cause of the generally more drastic results associated with cloudbursts in deserts.

9. Rainfall is higher along the recently uplifted shores of the Baltic Sea than it is along the Italian coast. Why, then, are sand dunes so much more abundant on the North German and Polish coasts than near Naples?

10. The walls of many canyons in the steppes of southeastern Oregon are draped with remnants of debris flows and landslides that are being actively gullied by rillwash from cloudbursts. What does this imply about climatic changes in the area? What additional features would you look for to prove the point?

Suggested Readings

Bagnold, R. A., *The Physics of Blown Sand and Desert Dunes*. New York: William Morrow, 1942.

Bryan, Kirk, *Erosion and Sedimentation in the Papago Country, Arizona* (U. S. Geological Survey, Bulletin 730). Washington, D. C.: G.P.O., 1932.

Denny, Charles S., "Fans and Pediments." *American Journal of Science* v. 265, 1967, pp. 81–105.

Gautier, E. F., *Sahara, the Great Desert* (translated by D. F. Mayjew). New York: Columbia University Press, 1935.

Hack, J. T., "Dunes of the Western Navajo Country." *Geographical Review* v. 31, 1941, pp. 240–263.

Scientific American Offprints

841. Victor P. Starr, "The General Circulation of the Atmosphere" (December 1956).

847. Joanne Starr Malkus, "The Origin of Hurricanes" (August 1957).

CHAPTER 16

The Oceans–Producers and Collectors of Sediments

The seas now cover 70.8 percent of the earth, and rocks containing marine fossils or having other characteristics indicating a marine origin crop out over nearly three-fourths of the lands. As most of these outcrops are bounded by erosion surfaces, Steno's reasoning shows that even more of the continental surface has, at one time or another, lain beneath the sea. It therefore seems possible, even probable, that no spot on earth has been land throughout geologic time.

To interpret the records of the marine rocks, we must know something of the sea, its processes and organisms, and its widely varying topographic features, and ecosystems.

CIRCULATION OF THE SEA

Ocean water is about 800 times as dense and several thousand times as viscous as air of the same temperature; its movements, therefore, though complex and unpredictable in detail, are sluggish compared to those of the atmosphere. Ocean waves may move at speeds of scores of kilometers per hour, but the water composing them rarely, if ever, travels more than 8 km/hr. Until recently it was thought that no ocean current exceeds the 28 km/hr tidal current in the Moluccas. The equatorial countercurrents, which run at depths of a few score meters beneath the equator in all three oceans, are now known to have speeds up to 45 km/hr in the Pacific, 36 km/hr in the Atlantic, and 18 km/hr in the Indian Ocean. Most ocean water, though, moves only a kilometer or so a day, and much appears to be nearly stagnant.

Surface Currents

The surface currents of the ocean closely accord with wind patterns. Indeed, wind drag is their chief driving force. Because of the Coriolis Force (Chapter 10), the water is driven somewhat to the right of the wind currents in the northern hemisphere and to the left in the southern. The Trade Winds thus drive the surface water westward as well as equatorward in both hemispheres. Thus is produced a strong **equatorial drift**, a drift ridden by the famous raft *Kon Tiki* in its journey from Peru to the mid-Pacific islands. The water of the equatorial drift tends to diverge both northward and southward, leaving an eastward-moving countercurrent between, traveling at the high speeds just cited. In this way a continuous upwelling of deeper water is brought about, both along the equator and along the western shores of all the continents (Fig. 16-1).

In the middle latitudes of both hemispheres, westerly winds set up a similar eastward drift. No land masses interrupt it in the southern hemisphere; a continuous West-Wind Drift

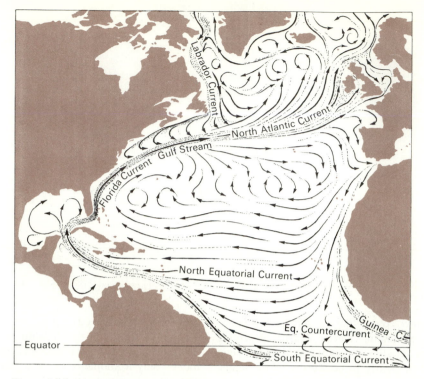

Figure 16-1
The currents of the north and equatorial Atlantic Ocean in February–March.
[After H. U. Sverdrup, M. W. Johnson, and R. H. Fleming, *The Oceans.*
Copyright 1942 by Prentice-Hall, Inc.]

rings the Antarctic (Fig. 16-2). In the northern
hemisphere the drift is obstructed by the continents, which deflect it both north and south.
The continents similarly deflect the equatorial
currents. In the landlocked North Atlantic Basin,
a vast clockwise eddy, of which the Gulf Stream
is a part, is thus established. The South Atlantic
and other oceans have comparable systems.

THE GULF STREAM

"There is a river in the Sea—the Gulf Stream,"
said Maury, the great American oceanographer
of the nineteenth century. Modern studies confirm
his picture of a well-defined stream of warm
water that courses with the speed of a river across
thousands of kilometers of ocean (Fig. 16-1).

The Gulf Stream system consists of three
segments: the Florida Current, the Gulf Stream
proper, and the North Atlantic Current. The
Florida Current pours through the Straits of

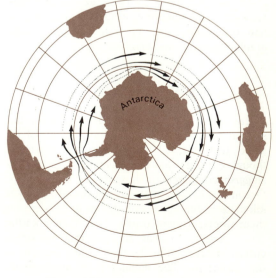

Figure 16-2
The West-Wind Drift in the Antarctic.

Florida at an average speed of nearly 5 km/hr, carrying between 20 and 40 million m³ of water per second—more than a thousand times the normal flow of the Mississippi. The North Equatorial Current, driven by the Trade Winds and deflected northward by South America, piles water up in the Gulf of Mexico, so that it is as much as 17 cm higher than in the Atlantic, as shown by precise leveling across Florida; this slope powers the Florida Current.

The Florida Current, emerging from the Straits, is joined by a northward drift along the east coast of Cuba; the combined stream sweeps northeast along the continental shelf, augmented by water from great eddies in the western Atlantic. It leaves the continental shelf off Cape Hatteras and reaches its maximum volume off Chesapeake Bay, where it has grown to 70 million m³/sec—more than 4000 times the average flow of the Mississippi. It varies in width between 50 and 150 km at this latitide, and different threads of current travel at speeds between 5 and 12 km/hr. Sometimes the stream meanders over a belt 500 km wide with a wavelength of 300 km. Some meanders spin off completely to form large independent eddies that rapidly mix with the cold surrounding water. The volume varies with the season, but the current averages about 80 km in width and 500 m in depth, although it reaches bottom at much greater depths locally. Inshore from the Gulf Stream, numerous counterclockwise eddies set up the southward drift along the Atlantic beaches.

As it sweeps past the Grand Banks, near Newfoundland, the Stream spreads out, subdivides, slows, and becomes less definite; this is the North Atlantic Current. Its northern branches spread far into the Norwegian Sea and even into the Arctic Ocean; their warm water notably ameliorates the climate of northern Europe. Southern branches carry water back to the trade wind belt, whence it is driven back across the ocean in the Equatorial Drift, completing the circuit.

The Deep Water Circulation

Measurements of temperature and salinity (the concentration of dissolved salts) at various depths show that the deep sea water is stratified according to density. Warm water is of course less

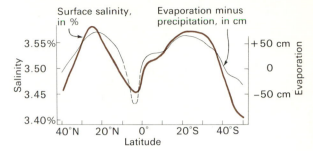

Figure 16-3
Graph showing the close correlation between changes of surface salinity and evaporation-minus-precipitation at different latitudes. Note that near the equator, though evaporation is very high, the salinity is low because of the very heavy precipitation. [After George Wüst.]

dense than cold water of the same salinity; in tropical seas, the water is less dense than elsewhere. But density also increases with salinity. Salinity is high in the Horse Latitudes, where evaporation is high (Fig. 16-3), and low in areas of heavy rainfall and where great rivers discharge. When sea water freezes, ice excludes nearly all the salt; both Arctic and Antarctic waters are highly saline as well as cold. They are also rich in oxygen dissolved by storm-tossed waves; they sink to the bottom and spread widely across the ocean floor, so that the bottom water is nearly everywhere relatively rich in oxygen. As it slowly mixes with adjacent water masses, the oxygen becomes depleted by the metabolic activity of organisms, and the water is enriched in carbon dioxide, a change of considerable importance in sedimentation (Chapter 17).

Thus measurements of temperature, salinity, and oxygen content at various depths and localities enable oceanographers to trace the movements of water masses of various sources with remarkable accuracy; we describe two examples.

THE MEDITERRANEAN DENSITY CURRENT

The Mediterranean Sea, because of the hot, dry climate, is a huge evaporating pan whose surface water acquires a density 10 percent higher than normal sea water. Thus even though the surface water is relatively warm (about 13°C in winter), it is dense enough to sink to the bottom. At the Strait of Gibraltar it is out of equilibrium with the less saline but cooler Atlantic water and pours

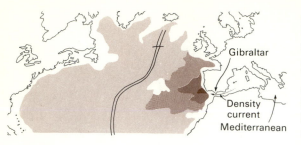

Figure 16-4
Salinity of the North Atlantic at a depth of 1750 m.
The map is drawn as if all the water above that depth
were stripped away. Extremely salty water from the
Mediterranean pours through the Straits of Gibraltar
and flows down the sloping sea bottom until it meets
water of higher density, upon which it spreads as a
flat sheet. The successively lighter shades indicate
the gradual dilution by mixing with other water
bodies. [After "The Anatomy of the Atlantic" by
Henry Stommel. Copyright © 1955 by Scientific
American, Inc. All rights reserved.]

across the sill of the Strait in a huge density
current that delivers about two million m³/sec
and attains speeds of 9 km/hr locally. A surface
current from Atlantic to Mediterranean balances
the outflow. During World War II, German submarines, with engines turned off to avoid detec-
tion, rode into the Mediterranean in the upper
current and out in the lower.

The Mediterranean current scours the Strait
clean, and at depths of 600 m forms sand waves
5 m high on the floor of the Atlantic. It flows
down the continental slope until, at a depth of
about 1700 m, it reaches water of slightly higher
density but of lower temperature and salinity,
over which it spreads in a great sheet covering
much of the North Atlantic (Fig. 16-4).

At the other end of the Mediterranean, similar
exchanges take place with the Black Sea, which
pours less saline water through the Bosporus
while an undercurrent of Mediterranean saline
water enters the Black Sea. Here it forms a
stagnant water mass virtually depleted in oxygen
below depths of about 150 m in the center and
250 m near the shore.

THE ANTARCTIC BOTTOM WATER

As noted earlier the water around the Antarctic
continent is extremely cold and saline; its density
is 1.0274. This water sinks to the ocean floor,
forming a wedge whose upper surface is generally

about 3500 m below the surface (Fig. 16-5). Note
how the Mid-Atlantic Ridge channels the Antarctic Bottom Water into the western half of the
Atlantic Basin until it reaches the low Romanche
Trench, through which some can escape to the
eastern Atlantic. Note, too, how this water extends far into the northern hemisphere. In the
South Pacific the Antarctic water moves fast
enough to scour the bottom, exposing sediment
as old as early Pliocene. The water also seems to
be eroding the Falkland Platform in the South
Atlantic, exposing Tertiary sediment. A comparable but weaker Arctic bottom water flows
from the Arctic Ocean.

TIDES, WAVES, AND CURRENTS

Water is a powerful erosive agent both on land
and in the sea, but even though the ocean currents move immensely greater volumes of water
than all the rivers of the continents, they erode
but little, though they seem to prevent deposition
over wide areas. Their energy is largely dissipated by friction with other water masses instead
of with the sea floor. Currents erode locally, but
the main geologic effect of the ocean currents is
climatic. It is the smaller but swifter currents
caused by winds, tides, or earthquakes and the
turbid flows of sediment-laden water that are the
main agents of erosion and transportation in
the sea.

Tides

The ancients knew that the ebb and flow of the
tides varies with the phases of the moon. So complex is the real earth, as compared with the idealized earth of the astronomers, that no general
theory permits tidal forecasts for any point on
earth. Tides are, of course, predicted with great
accuracy for all the principal ports, but these
predictions are not derived from theory but
from the tidal records of previous years.

Since Newton's time, the forces that produce
the tides have been understood much as shown
in Figure 16-6. If D is the distance from the center of the earth to that of the moon, M the mass
of the moon, and r the radius of the earth, the
moon's attraction for a mass, m, at the earth's
surface on the side toward the moon is greater
than its attraction for a like mass at the earth's

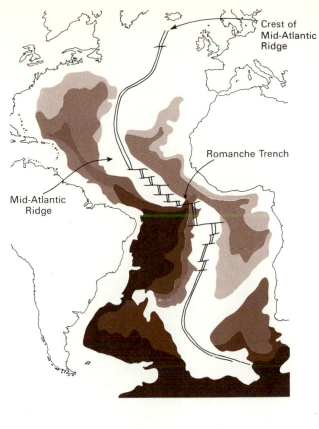

Crest of
Mid-Atlantic
Ridge

Romanche Trench

Mid-Atlantic
Ridge

Figure 16-5
The Antarctic Bottom Water, as it
would appear if all the water above a
depth of 3500 m were stripped away.
The darkest shade indicates undiluted
Bottom Water; the lighter shades,
successive stages in its dilution. The
Mid-Atlantic Ridge is shown only
schematically. [After "The Anatomy
of the Atlantic" by Henry Stommel.
Copyright © 1955 by Scientific
American, Inc. All rights reserved.]

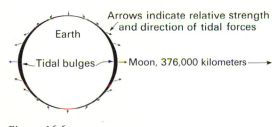

Earth

Arrows indicate relative strength
and direction of tidal forces

Tidal bulges

Moon, 376,000 kilometers

Figure 16-6
The lunar tides. The tidal bulges are greatly
exaggerated. [In part after H. U. Sverdrup, M. W.
Johnson, and R. H. Fleming, *The Oceans*, Copyright
1942 by Prentice-Hall, Inc.]

center; $Mm/(D - r)^2$ is greater than Mm/D^2.
Theoretically, if this surficial mass is a fluid, it
will bulge upward. By like reasoning, mass on the
opposite side of the earth will be less attracted;
it is "left behind" in a corresponding bulge. At
all other points there are differences of lesser
degree, with the net orientation of gravitational
pull as shown by the arrows in Figure 16-6. Because the earth rotates with respect to the moon
once in 24.84 hours (not 24, for the moon advances eastward in its orbit), two tidal bulges
should pass over any given point during this time.

The sun has similar attraction but because it is
so far away its tide-producing force is only 0.46
that of the moon. At new and full moon, sun and
moon lie in a virtually straight line, and their
influences are additive, producing the highest
tides, the **spring tides**. At the moon's first and
third quarter, the tidal influences counteract each
other, and the tides are smaller—**neap tides**. The
tides are widely variable, but the world average
spring tide, about 6.5 m, is nearly twice the world
average neap tide. The tides affect all the earth,
not merely the water bodies, but the deformation
of the rocky crust is so small that it can be detected only by a sensitive gravimeter.

Although the simple diagram of Figure 16-6
illustrates the tide-producing forces, it fails completely to explain the tidal vagaries: many ports
have but one tide a day; in others the tide lags
many hours behind the moon's overhead passage;
in still others the two daily tides have greatly
different heights throughout the lunar cycle.
These observations and many others make clear
that the tides are not direct responses to the
vertical pull of the moon's gravity (which is far
too small to exert a lifting effect on so much
water) but to its horizontal component, causing

water to flow laterally. It is thus the configuration of the ocean floor that modifies the local tidal action and produces the differences cited.

The principal geologic interest of the tides is in their erosional power. Some tidal currents are indeed prodigious, particularly those in estuaries with shores converging inland. In the Bay of Fundy, the vertical tidal range is as much as 20 m; the possibility of harnessing the power of the huge water masses that surge in and out twice daily has been seriously studied. So far, the costs, and the irregularity in power from hour to hour, have made the project seem uneconomic. Yet the currents attain speeds of 15 km/hr during both rise and fall, and have scoured basins more than 50 m deep.

In St. Malo Bay, Brittany, the tidal range is 12 m, and currents reach speeds of 13 km/hr. Between the Channel Islands and the French coast, currents sometimes reach 20 km/hr, as they also do between the Shetland and Orkney islands. The speediest tidal currents known are in the Moluccas, about 28 km/hr. Although the Mediterranean tidal range is only about 3.5 cm, the timing of high and low tides at the ends of the Strait of Messina—between the Scylla and Charybdis of Homer—is such that currents as fast as 10 km/hr are common. During storms in the South Pacific, the lagoon of Tuamoto atoll at times becomes so overfilled with water washing over the reef that currents of more than 20 km/hr develop in the narrow outlet at low tide.

In some rivers, high tides may reverse the flow as they rush upstream in continually breaking waves called **bores**. At spring tides the bore of the Hangchow River is as much as 5 m high, traveling at 26 km/hr; the bore in the Amazon is comparable, as is that of the Yellow River. The estuary of the Scheldt, although it has a seaward sill less than 10 m deep, is scoured by the tide to depths of nearly 60 m. In the muddy Gironde estuary, high tides cause sedimentation upstream, and low tides somewhat compensate this by scour. On balance, though, deposition exceeds scouring, so that Nantes, a considerable seaport two centuries ago, is no longer important commercially.

Experiments show that water moving a kilometer per hour will carry medium sand and, at 5 km/hr, gravel more than 2 cm in diameter. At such speeds, tidal currents are important agents of transport and erosion; dredging off the Mull of Galloway, Scotland, proves that coarse gravel moves there at depths of more than 23 m.

Current meters placed on the sea floor indicate that some tidal motion extends to great depths. The presence of coarse gravel in the Romanche Trench and the abrasion of submarine cables where they cross the Mid-Atlantic Ridge attest to the effects of tidal currents. Over the continental shelves, tidal currents should be strongest near the edges, as the volume of water moving in and out is there greatest; this may account for the sediments there being commonly coarser than those nearer shore. The stronger currents should winnow out the finer sediment and drop it off onto the continental slope. Photographic measurements of light transmission (by cameras lowered on cables) show that water even at great depth is commonly cloudy from fine sediment. Tidal currents seem active on all the submarine ridges, for most yield bare rock to the dredge, and nearly all show coarser sediments than neighboring basins. Tidal currents rushing in and out through offshore bars commonly build submarine deltas at both ends of the channels.

Tidal currents thus significantly affect the size, sorting, and distribution of sediments over most of the sea floor. But only locally are they important in shaping the shore itself; there waves and wave currents are far more effective.

Two wholly different kinds of waves that have no relation to tidal forces have unfortunately become popularly known as "tidal waves." One, the seismic sea wave, or tsunami, has been described in Chapter 7; the other is the "sea surge"—a high water wave caused by prolonged and unusually violent onshore winds. Such a hurricane-driven water mass overwhelmed Galveston, Texas, in 1900, with great loss of life. A great storm of January 31 and February 1, 1953—probably the greatest in the North Sea since 1571—raised a huge sea surge far above normal sea level, in places as much as 3 m, and crashed against the shores (Fig. 16-7). More than 2100 people were drowned in England and Holland, and shore cliffs were battered back in places for many meters. One of the most tragic episodes in history resulted from a surge on November 12 and 13, 1970, that struck the Ganges delta and drowned more than 300,000 Bangladesh peasants. The amplitude of the surge was only 1.5 m, but it was added to a tide of nearly 6 m.

A similar storm in October 1972 struck the island of Funafuti in the Ellice group and; covering the airfield with 2 m of water, it built a ram-

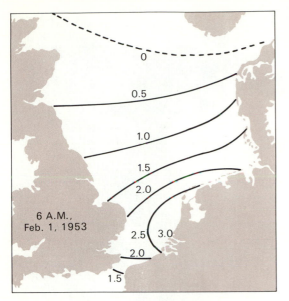

Figure 16-7
A synoptic chart of sea surface disturbance during
the surge of January 31–February 1, 1953, in the
North Sea. Disturbance heights are in meters above
mean sea level. [After P. Groen and G. W. Groves,
in M. N. Hill (ed.), *The Sea* (vol. 1), Interscience
Publishers, New York, 1962.]

part of broken coral reef 18 km long, 35 m high
and about 40 m wide, a considerable contribution
to the size of the island. A similar wall was built
by a surge at Jaluit atoll in January 1958.

Waves

GEOLOGIC EFFECTIVENESS

Waves and the currents they generate wash
every shore. Anyone watching the perpetual play
of waves on the shore—whether the ripples that
waft sand grains up and down the beach or the
great breakers that hurl tons of water and
boulders against the cliffs (Fig. 16-8)—cannot
but be impressed with the ability of waves to
modify the shore. Armed with cobbles, pebbles,
and sand, the waves saw away, undercutting
cliffs and notching every rock exposed at their
level. In weakly consolidated materials erosion
is rapid; the low cliffs of glacial gravels on the
Holderness coast of Yorkshire have been eaten
back at a measured rate of 2 to 4 m a year for
more than a century. Repeated soundings and

studies of bottom sediments in many places also
show that great quantities are moved at shallow
depths.

WAVE MECHANISM

The mechanism by which winds produce water
waves is complex. Theoretically it should take
a wind of 18 to 25 km/hr to make water waves,
but actually, winds of only 4 km/hr do make them,
probably because of turbulence. The motion of
the water particles depends on wave length, wave
height (Fig. 16-9), and water depth. Waves in
water deeper than one-fourth the wavelength—
deep-water waves—are not affected by water
depth. Particles in them move in roughly circular
orbits but retain nearly the same position; a
floating object rises and falls but does not move
forward at anything like the wave speed. Indeed,
if the water masses moved at wave speed, no
ocean would be navigable. Surface particles move
in orbits whose diameters about equal wave
height, but the orbits diminish quickly with depth
(Fig. 16-10). In waves 100 m long and 5 m high—
about as large as North Atlantic waves ever
get—traveling about 47 km/hr, the surface particles move at only about 7.3 km/hr; at 20 m
depth, the particle velocity is only 2 km/hr, and
at 100 m it is negligible (Fig. 16-10). Because of
this quick falloff of orbital velocity Vening-
Meinesz's submarine (Chapter 8) was stable
enough at shallow depths for gravity observations with his relatively crude instruments. But
in great swells 360 m long, 10 m high, and traveling at 95 km/hr, even though the surface particle
velocity is still only about 7 m/hr, at 100 m it is
about 1½ km/hr. Hurricane waves sometimes
reach heights as great as 35 m; they can thus disturb the waters to still greater depths.

Under steady winds, the waves grow in size
and speed up to a limit imposed by friction. The
maximum wave height known was estimated at
35 m, but few attain a height of 15 m. The wind
energizes waves in two ways: by its push against
the wave form and by the frictional drag of the
air on the water surface. The first depends on the
difference in speed between wind and wave;
normally, wind speed is greater and the wave is
pushed. Frictional drag depends on the difference between the speed of the wind and that of
the far slower-moving particles of surface water.
Thus the speed and size of waves depend on wind
speed, wind duration, the **fetch** (the distance

Figure 16-8
Breakers crashing against a headland, Boiler Bay, Oregon. [Photo by Oregon State Highway Department.]

Figure 16-9
Schematic diagram of a progressive wave.

across which the wind acts with the same velocity and direction), and, finally, the state of the sea when the wind began.

WAVE REFRACTION

When deep-water waves enter shallow water, normally at depths between $\frac{1}{4}$ and $\frac{1}{2}$ wavelength, the bottom interferes with the orbital motion of the particles, the orbits become first elliptical and eventually linear—the particles move back and forth in the line of wave travel. The wave speed in shallow water no longer depends on wavelength but becomes proportional to the square root of the depth. Most waves begin to "feel bottom" at depths of less than 150 m, but during great storms they extend to three times this depth.

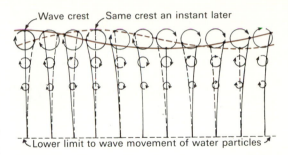

Figure 16-10
Circular movement of water particles in a deep-water wave of small height. Solid lines show the position of the water particles at one instant; broken lines the same particles $\frac{1}{4}$ period later. [After U. S. Hydrographic Office, Publ. No. 11.]

The inshore end of a wave approaching the shore obliquely over a uniformly sloping bottom feels bottom soonest and thus slows down sooner than the rest of the wave. The wave bends, tending to approach the shore more nearly head on. It is exceptional, along straight coasts, for the obliquity of approach to exceed 10° (Fig. 16-11).

Because wave energy is carried in paths at right angles to the wave crest, refraction modifies its distribution. Uniformly distributed in deep

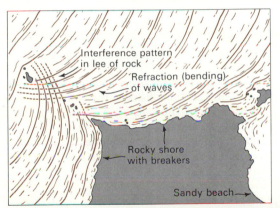

Figure 16-11
Vertical air photo and diagram showing bending of the waves (wave refraction) around a point of land. [After U. S. Hydrographic Office, Publ. No. 234.]

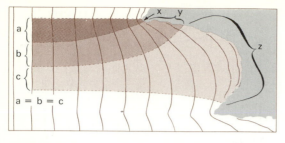

Figure 16-12
Distribution of wave energy as conditioned by coastal form. Each of the equal segments, a, b, and c, of the offshore wave has the same amount of energy. The refraction leads to concentration of energy on the headlands and to its dispersion in the re-entrants. Energy is everywhere carried in directions normal to wave crests; that in section a is concentrated at x, but that in c is spread over the much greater span z.

water, the energy tends to be concentrated against sections where shallow water extends farthest from shore. Wave energy is thus concentrated against the headlands and dissipated more widely in bays (Fig. 16-12). This differential attack tends to straighten the shoreline by selectively eroding the headlands and filling the coves with transported debris.

BREAKERS

When waves reach water so shallow that water is no longer available to fill out their wave forms, they steepen and begin to break. They may continue, though, for several hundred meters without curling over at the crest—the ideal wave for the surf-rider. The wave becomes a **wave of translation**, in which wave speed equals water speed; the wave curls over and breaks with a crash, tumbling forward onto the beach as a **breaker**. Waves break where the stillwater depth is between one and two times the wave height. Exceptionally, where the shore shape favors convergence of the translation wave, two or even three breaker zones may coexist, as on the Landes of Gascony, but this is extremely rare.

The energy of the breaker is dissipated in throwing the bottom material violently into suspension, thereby scouring a trough along the breaker line. Some material goes to build bars both landward and seaward of the breaker line; some is carried landward and swept up the beach, where, in less turbulent water, it is sorted according to size, shape, and density.

WAVE EROSION

Waves erode the shores in three ways: (1) by impact and hydraulic pressure; (2) by **corrasion**, the sawing and grinding action of the sand, gravel, and cobbles hurled against the cliffs or rolled and dragged across the foreshore; and (3) by solution, a minor process, even on limestone shores.

Hydraulic Pressure

If we calculate the pressure of even a moderate wave 3 m high and 100 m long, we find it capable of exerting a push of 118 kg/m². Great storms in 1963 exerted measured pressures of 72 tons per square meter against harbor works at Dieppe and Le Havre, France. At Wick, in northern Scotland, waves have broken away masses of concrete weighing as much as 2600 tons from the harbor breakwater. The wave pressure acts not only directly but also by driving air into crevices with such force as to push a breakwater block weighing 7 tons toward the sea, at Umuiden, Holland.

Storm waves wear away weak strata or joint cracks in a cliff and dislodge the blocks between them, tunneling out sea caves, scores or even hundreds of meters long; some may tunnel entirely through small promontories and produce spectacular arches. Some cavern roofs collapse, leaving isolated **stacks** in front of the cliffs (Fig. 16-13).

Corrasion

We have seen that great storm waves can tear huge blocks from the cliffs and hurl them against the shore. Even moderate waves can move boulders and cobbles, and feeble ones sand. In a Cornish mine that extends beneath the sea, the menacing grinding of boulders overhead is distinctly heard through a rock roof 3 m thick. The breaker zone is a veritable grinding mill. At Cape Ann, Massachusetts, angular fragments from granite quarries have become rounded on the beach within a single year. Shakespeare's Cliff, part of the Chalk Cliffs of Dover, more than 100 m high, is so rapidly undercut as to produce frequent large landslides. One in 1810 was such as to cause a strong earthquake at Dover, several kilometers away. The cliffs of the volcanic island

Figure 16-13
Sea caves and a sea stack at low tide near Santa Cruz, California. [Photo by Eliot Blackwelder, Stanford University.]

Figure 16-14
Cliff in rhyolite undercut by waves, Kendall Head, Moore Island, Maine. [Photo by E. S. Bastin, U. S. Geological Survey.]

Krakatau in the Sunda Straits were cut back over 1500 m between 1883 and 1928, an average of more than 30 m/yr. These are spectacular examples of wave erosion in weak rocks, but even the most resistant rocks are impressively notched (Fig. 16-14).

Despite these examples of significant marine erosion, cliff retreat at the shore is vastly inferior to fluvial erosion. Careful estimates by C. K. Wentworth indicate that even in the Trade Wind belt, fluvial erosion in Hawaii is at least seven times as effective as marine. Studies of shore abrasion and sediment volume in the Black Sea indicate that only 5 percent of the sediment was derived from the waves, 95 percent from the streams. If this were not generally true, we should find waterfalls along most of the coasts of the world.

Although most wave energy is spent between the breaker line and the shore, corrasion is not confined to that zone. Where wavelengths are great, sediment particles move vigorously at greater depths. Great waves 150 m long and 7 m high should theoretically produce speeds of 25 cm/sec in water 100 m deep, enough to move fine sand. When account is taken of tidal and other currents, the bottom water should be agitated enough to keep clay particles in motion to depths of 200 m—on some coasts to more than 300 m. These theoretical conclusions are generally, but not invariably, confirmed by the character of sediments dredged and also by the fact that sand is frequently found on the decks of North Sea fishing craft even when they are on water more than 100 m deep. Stones weighing five hundred-grams have been found in lobster pots at depths of 50 m off Land's End, England, during storms.

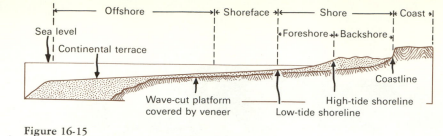

Figure 16-15
Elements of the shore zone and profile of equilibrium. [After D. W. Johnson, *Shoreline Processes*, John Wiley & Sons, 1919.]

The depth to which sediment is moved significantly by waves is **wave base**.

Outside the breaker zone sediment particles moved by the waves reverse their motion with each wave; as most of the shallow seafloor slopes seaward, the seaward motion is favored. As wave motion diminishes with depth, coarser particles tend to come to rest before finer ones, which continue their drift to deeper water. The moving particles wear away each other and the bottom so that not only in the breaker zone, where the wear is greatest, but throughout, the shore zone is corraded down to wave base.

Inshore from the breaker zone the shoreward-moving water masses may dash coarse fragments to heights from which the feebler backwash cannot carry them, even though favored by a steep slope, because some of the water filters into the porous beach and returns seaward far more slowly than it moved landward. A **storm beach** of coarse gravel may thus be built above the reach of normal waves. For example, the crest of Chesil Beach, Dorset, England, is as high as 13 m above calm water level. Such beaches are steep; coarse material can only be moved seaward on steep slopes: shingle beaches in Morocco slope as steeply as 31°; sand beaches of the eastern Pacific have slopes ranging from 24° to as little as 1°, depending on the average grain size of the shore sediment:

Mean grain size, in millimeters	Mean beach slope, in degrees
$\frac{1}{16}$–$\frac{1}{8}$	1
$\frac{1}{4}$–$\frac{1}{2}$	5
1–2	9
4–64	17
64–256	24

The heights and slopes of beach ridges are not constant, but change with wave changes. For

example, prolonged onshore winds during World War II allowed the shingle ridge at Fécamp, France, to be regraded to a level 2 m lower.

Solution

Although solution effects are trivial compared to those of corrasion, they are not negligible. The Irish geologist Joly has shown by experiment that minerals such as feldspar and hornblende are from three to fourteen times as soluble in sea water as in fresh. In landlocked bays of Indonesia, a corrosion groove as much as several meters deep has been dissolved at the shoreline, not just in limestone but in volcanic rocks. Similar solution is doubtless going on on exposed shores elsewhere, but is masked by corrasion.

THE PROFILE OF EQUILIBRIUM

Partly from wave theory, partly from observations of wave action, partly from dredging and sounding, and partly from study of dried up lakes, such as Bonneville (Chapter 13), and uplifted sea floors (Fig. 16-16), geologists have derived the concept of the **marine profile of equilbrium**. This profile is a smooth, sweeping curve, concave upward (Fig. 16-15). It is steep in the breaker zone and flattens quickly seaward; its slope is adjusted to the average particle size and wave characteristics. The slope just suffices to keep the material in slow transit seaward. As the waves fluctuate in size, the seaward slope flattens with weak waves and steepens with big ones, on the whole varying around the most common conditions.

That such profiles are not merely hypothetical is proved by soundings that show the sea floor near the coast to be generally concave upward

Figure 16-16
Uplifted marine terraces, Palos Verdes Hills, California. [Photo by John S. Shelton and R. C. Frampton.]

despite the wide range in sea level during and since the Pleistocene. Several such profiles are seen in Figure 16-16, where they have been uplifted successively to form **marine terraces**.

WAVE CURRENTS

Even though refracted, most waves break against the shore at a slight angle, so that the swash of the breaking wave has a component of motion parallel to the shore. This produces a current along the beach—the **longshore drift**—which may consistently flow in one direction or reverse with changes in the winds. Such a drift may be swift, especially if a longshore wind or tidal currents reinforce it. The breaking wave carries sand and gravel obliquely up the beach; the swash of the retreating wave also has a smaller longshore component (Fig. 16-17). Marked pebbles made from bricks have been traced along a beach for as much as a kilometer a day; similar movements doubtless go on in the whole agitated zone. Measurements off Jutland, Denmark, have shown that the longshore drift, to depths of 20 m and for as much as 2 km offshore, is virtually equal across the whole zone. At Westport, England, measurements show that the sea is eating into the shore

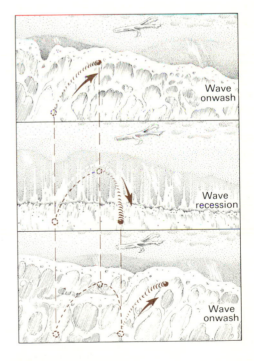

Figure 16-17
Longshore drift (to the right) resulting from oblique wash of the waves. Each arcuate segment of the pebble's path represents the movement due to one wave.

along a 5-km stretch at rates between 2 and 8 m/yr; the longshore drift is carrying more than 3.8 million m³/yr. The Columbia River discharges silt particles identifiable by their radioactivities (derived from the atomic reactors at Richland, Washington); these allow the sediment to be readily traced offshore. The sediment is being moved northward by longshore currents at 12 to 30 km/yr, and seaward at rates between 2½ and 10 km/yr.

ARTIFICIAL INTERFERENCE WITH SHORE PROCESSES

The nice adjustment of beach profiles and shore outlines to the average power of waves and currents acting on shore detritus is emphatically shown wherever the regimen is disturbed. Even slight changes may bring drastic disturbance. Dredging of sand and gravel offshore is commonly followed by erosion of the shore. More striking effects follow the building of groins on many bathing beaches. Groins, low walls extending seaward from the high tide line, are built to prevent sand removal by longshore currents. Where they have been successful in holding the sand, beaches farther downdrift, being deprived of their normal supply of traveling sand, have been severely scoured and changed from sandy beaches to gravel or cobble ones. Many parts of the famous Waikiki Beach in Honolulu have been scoured of their sand and destroyed as bathing sites by updrift groins.

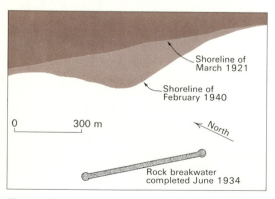

Figure 16-18
The Santa Monica beach before and after building of the breakwater. The beach has continued to advance since 1940, but later surveys are not available.

Breakwaters of rock or concrete, built to protect anchorages from storms, commonly cause beach modification. At Santa Monica, California (Fig. 16-18), a 600-m breakwater was built about 600 m offshore to make a small-boat anchorage. The longshore drift here is mainly to the southeast. The breakwater reduced the waves striking the shore behind it, thus reducing the longshore drift; the beach in the lee of the breakwater began to build forward, advancing into the sea about 160 m in 6 years. Although the beaches to the southeast have not been greatly eroded, it is noteworthy that great storms cause more damage there than elsewhere nearby.

EROSIONAL AND DEPOSITIONAL FEATURES OF SHORES

The Shores of Hilly Coasts

Waves attack promontories selectively because of wave refraction and cut cliffs; the debris is distributed by long-shore drift, producing characteristic shore features: bayhead and baymouth bars, land-tied islands, spits, and bars.

Wave refraction on bay shores produces currents that generally tend to sweep detritus toward the bayhead, where it accumulates as a bayhead bar (Fig. 16-19). Sediment carried alongshore from eroding headlands tends to continue straight across coastal indentations; the detritus from a cliffed point may form a smoothly curved spit. Toronto harbor is thus protected by a long spit growing westward. As spits grow into deeper water, wave refraction causes their ends to curve landward (Fig. 16-19). Where the outflow from a bay is small, the spit may grow entirely across the opening, to form a baymouth bar (Fig. 16-20). The lagoon behind may be brackish or fresh and may slowly become a marsh filled by silt from the land and sand blown inland from the beach.

Islands near shore, like breakwaters, weaken longshore currents in their lee, so that a mainland beach grows out toward the island and may eventually so divert the longshore currents as to extend to the island and tie it to the shore (Fig. 16-21). Bottom irregularities may deflect the drift, even on straight coasts, so that the beach builds seaward, leaving behind a series of beach ridges, some half a dozen meters high, separated by swales. Perhaps the best example is at

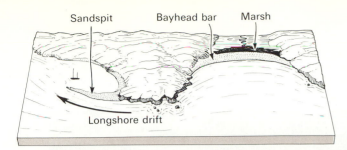

Sandspit   Bayhead bar   Marsh

Longshore drift

Figure 16-19
An embayed coast, showing a sandspit springing from a headland and a bayhead bar that encloses a marsh.

Figure 16-20
Baymouth bar and lagoon, St. Mary's Lake, Glacier National Park. [Photo by Eugene Stebinger, U. S. Geological Survey.]

Dungeness, in southeastern England, which has advanced into the sea at an average rate of 5 m/yr since the time of Elizabeth I.

Shores of Plains

Along low-lying shores like those of the Gulf of Mexico and the southeastern Atlantic States, the wave attack is spread rather evenly instead of being concentrated on headlands. The sea deep-ens so gradually that the waves feel bottom far offshore and there stir up sediment to form an off-shore bar just landward of the breaker zone (Fig. 16-22). The waves continually attack the bar, which, in order to survive, must move slowly inland as long as sea level is stable. In many places they have been driven landward until they touch the mainland, advance inland, and expose low cliffs, as in northern Florida and southern North Carolina.

Figure 16-21
Land-tied island, Hancock County, Maine. [Photo by E. S. Bastin, U. S. Geological Survey.]

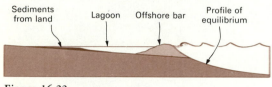

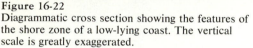

Figure 16-22
Diagrammatic cross section showing the features of the shore zone of a low-lying coast. The vertical scale is greatly exaggerated.

The lagoon behind an offshore bar is salty, but large rivers may freshen it and perhaps raise the lagoon level to such a height that the bar is broken at low tide, forming a tidal inlet. Great hurricanes also break the bars from time to time. Because of longshore drift, the tidal inlets shift, some fill in, and new ones appear. Salt grass grows in the protected lagoon and, as plant material and silt are added by inflowing streams, and by winds and tidal currents, the lagoon gradually becomes a swamp.

The famous beaches of Florida and New Jersey are offshore bars. Even more striking examples are the long sand bars—the Frisches Nehrung and the Kurisches Nehrung—that fringe the southern Baltic from Danzig to Memel, enclosing an almost continuous brackish lagoon. Magnificent examples of offshore bars fringe the Carolina coast, enclosing Albemarle and Pamlico sounds and meeting at stormy Cape Hatteras. Their cuspate shape has been attributed to junction of Gulf Stream eddies, but it may partly result from bottom irregularities. Plains coasts have no coarse gravel, and the beaches are almost wholly of sand. Between tides the beaches dry out, in almost any climate, and sand may blow inland to form great dunes, as it has in many lands.

LIFE ZONES OF THE SEA

Life is ubiquitous in the sea. Photographs from a bathyscaphe at the bottom of the Mariana Deep—far below levels to which light penetrates and where the water temperature is near freezing—show living organisms. In fact, fully 300 species of animals, chiefly mollusks, worms, and holothurians (sea cucumbers), have been dredged from the abyssal ocean floor. Marine environments differ greatly from place to place, and so do their organisms. Recognition of these faunal facies greatly improves interpretation of marine formations.

Marine biologists classify the different parts of the sea as shown in Figure 16-23. The two main divisions are the **benthic zone**, or sea-bottom environment, and the **pelagic zone**, or open-water environment. Both are divided, at a depth of

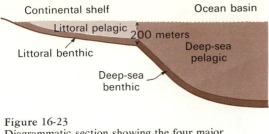

Figure 16-23
Diagrammatic section showing the four major
divisions of the life zones of the sea. [After H. U.
Sverdrup, M. W. Johnson, and R. H. Fleming, *The
Oceans*, Copyright 1942 by Prentice-Hall, Inc.]

about 200 m—that of the edge of most continental shelves—into a littoral system landward and a
deep-sea system seaward. The 200-m depth is
critical, because it is near the limit to which light
can penetrate. On the deep-sea floor are neither
light nor seasons and no photosynthesis; life is
confined to scavengers living on sinking organisms and to a few species of bacteria that depend
on organic compounds brought in by currents.

Marine life, both plant and animal, is classed
in three large groups: **benthon** (Greek, "depth of
the sea"), or bottom-dwellers; **nekton** (Greek,
"swimming"), or swimming forms; and **plankton**
(Greek, "wandering"), or floating organisms.

The benthon includes all the attached, creeping, or burrowing organisms of the bottom: sea
weeds, grasses, sponges, barnacles, mollusks,
corals, bryozoa, worms, lobsters, crabs, and the
minute unicellular foraminifers, most of which
secrete shells of calcium carbonate.

The plankton includes all the surprisingly
varied organisms, chiefly microscopic, that float
with the ocean currents: diatoms of several
species with opaline shells; the Coccolithophores,
a group of uncertain classification, but usually
considered algae; many foraminifers (of only a
few species), and the silica-secreting Radiolaria.

The nekton includes the squids, fishes, seals,
whales, and many other animals that are economically important but of little geologic importance.

All living things, except for a few bacteria,
ultimately depend on **photosynthesis**—the process
by which plants utilize radiant energy from the
sun to form organic compounds directly from
carbon dioxide and water. Plants also require
minute amounts of many substances, among them
phosphates, nitrogen, iron, and manganese. Vari-

ations in the abundance of these in different parts
of the sea are reflected in the varying densities
of plant populations and in the abundance and
kinds of animals that feed on the plants.

On death, a planktonic or nektonic organism
sinks, perhaps to feed some scavenger in the
depths. During and after sinking, much of it is
decomposed by bacteria, using up oxygen and
discharging carbon dioxide to the water, thus
favoring solution of calcium carbonate shells.
Aragonite is missing below about 3500 m depth,
and calcite below about 4500 m. Such shells as
those of foraminifers and coccolithophores are
therefore missing from deep-sea sediments, even
though they are abundant at the surface.

Decomposition returns phosphorus, nitrogen,
and other nutrients to solution, making them
once again available for plant growth when currents bring them into the photic zone by upwelling. Thus areas of upwelling of deep water are
areas of colder than average water, of rich plant
life, and prolific fisheries. Among these the
Newfoundland Banks, and the eastern Atlantic
and Pacific oceans are most productive.

CORAL REEFS

Although some corals can live in cool water, the
reef-building species require water at least as
warm as 20°C, of normal salinity, and nearly free
from mud. The coral animal, locked in its limy
case, depends on the waves and currents to bring
its food; it therefore thrives best on the windward
and offshore sides of islands. Many species of
coral live on reefs, but the reefbuilders do not
grow below a depth of about 50 m, nor much
above low tide. Because of their narrow depth
range, reefbuilding corals are sensitive indicators
of crustal movement.

Although both living coral and broken fragments form the most conspicuous framework of
the reefs, algal growths bind them together, and
many other calcareous organisms—mollusks,
worms, foraminifers and a host of others—make
up far more of the reef than does coral.

Coral reefs have three forms: fringing reefs,
barrier reefs, and atolls. **Fringing reefs** are confined to the very border of the land. A few are
more than a kilometer wide but most are less than
40 m.

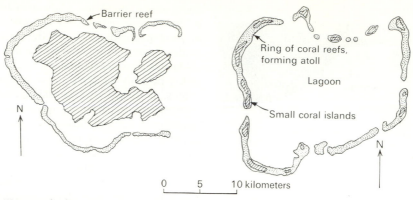

Figure 16-24
(Left) The barrier reef of Vanikoro Island, Caroline Archipelago. (Right) Peros Banhos Atoll, Chagos Archipelago. [After Charles Darwin, 1842.]

Barrier reefs are separated from shore by lagoons, some shallow and some scores of meters deep. Many volcanic islands of the Pacific and Indian oceans are ringed by white barrier reefs a short distance offshore (Fig. 16-24). The Great Barrier Reef of Australia roughly follows the Queensland coast for 2000 km south from Torres Strait at distances ranging from 40 to nearly 300 km offshore.

An **atoll** is a ring-like reef enclosing a lagoon, commonly dotted with isolated coral heads and floored with algal mud and coral sand.

The Origin of Atolls

Darwin considered fringing reefs, barrier reefs, and atolls to form in sequence (Fig. 16-25): if an island with a fringing reef subsided, the reef would grow upward (if growth could keep pace with sinking), first forming a barrier reef and, when the central peak drowned, an atoll. In support of this, it has been pointed out that the shores of many islands with fringing or barrier reefs are embayed as they might be if a stream-dissected island had sunk slightly. Moreover, the outer slopes of many atolls descend steeply to great depths; on Bikini and Eniwetok atolls, borings and seismic soundings show that the reef structure extends to great depths and rests on a volcanic basement. Fossils from the basal part of the Eniwetok reef are of Eocene age; the atoll has sunk 1200 m in 50 million years.

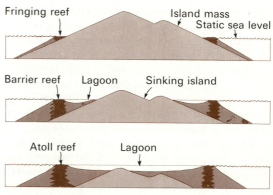

Figure 16-25
Cross section showing three stages in the formation of an atoll, according to the subsidence theory of Darwin.

Drowned reefs with perfectly preserved atoll shapes have been sounded near the Philippines; the sinking was too fast for coral growth to compete. Greatly elevated reefs in Indonesia, Fiji, and Samoa show that the ocean floor can rise as well as sink.

Other features testifying to subsidence are the **guyots**, flat-topped submerged mountains shaped like the frustum of a cone. Many have yielded basaltic pebbles to the dredge; clearly they are extinct volcanoes that once stood above the sea and were truncated by the waves before being submerged to various depths, some as great as 3 km.

EROSION BELOW WAVE BASE

It was long thought that the only erosion below wave base is that caused by locally constricted currents. With the discoveries since World War II, it has become apparent that considerable erosion is going on in many submarine environments. Among the various agents, the most widely active are **turbidity currents**.

Turbidity Currents

Aviators in the Arctic report that some silt-laden streams of meltwater discharged from the Greenland ice cap do not halt or spread out on entering the sea but dive and flow on beneath the surface, because their density is greater than that of the enclosing water. This phenomenon has also been recognized in many reservoirs and lakes. Such turbidity currents are very effective at all depths in the sea; a spectacular example is that of the Grand Banks flow.

The Grand Banks Flow

The American oceanographers Heezen and Ewing studied the turbidity current of 1929 that overwhelmed 280,000 km² of the North Atlantic floor. An earthquake with an epicenter at the edge of the Grand Banks off Nova Scotia snapped eight transatlantic cables immediately and five others in succession downslope, the last one 13 hours and 17 minutes later (Fig. 16-26). All the broken cables lay on the continental slope south of the Grand Banks or on the gently sloping ocean floor below and south of the slope.

Each break was timed precisely by the interruptions of the teletype machines, and accurately located by measurements of the electrical resistance of the cables—a method long used in cable repair. Heezen and Ewing concluded that the first eight cables broke as the earthquake triggered a landslide in the weakly consolidated sediments of the continental slope; the slide continued downslope, destroying one cable after another over a distance of nearly 500 km. The current was traveling about 100 km/hr near the base of the slope but had slowed to a little less than 25 km/hr when it cut the last cable. The higher speed is much greater than that of most mountain torrents and certainly was adequate to do much erosion. The slide material broke up into individual grains and settled out over a wide area of the ocean floor in accordance with the grain size of the particles, the coarse material settling our first and the finer later, to form **graded bedding**. This inference has been verified by the presence of excellent graded bedding in drill cores from the ocean floor below the slide. The volume of this slide was at least 1 million m³ and may have been ten times that.

Other Evidence of Submarine Erosion

Since the invention of the underwater camera and oriented recording current meters, current ripples have been identified on the Blake Plateau, where the Gulf Stream sweeps across it; on the Scotia Ridge in the South Atlantic, where the cold Antarctic water pours into the South Atlantic at depths of 3000 m; and on the very bottom of the Pacific off Eniwetok Atoll (Fig. 16-27). A generation ago, features such as these seen in marine

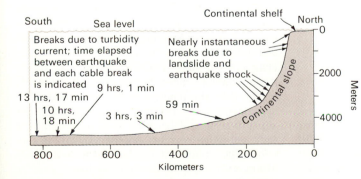

Figure 16-26
Profile of the sea floor off the Grand Banks, showing the positions of the transatlantic cables broken by the landslide and turbidity flow started by the earthquake of November 18, 1929. [Modified from B. C. Heezen and M. Ewing, 1952.]

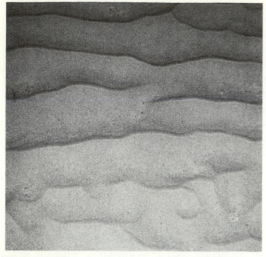

Figure 16-27
Current ripples in calcareous sand at a depth of 2000 m on the southwest slope of Eniwetok Atoll, Marshall Islands. [Underwater photo by C. J. Shipek, U. S. Navy Electronics Laboratory.]

sandstone would have been cited as proof of shallow-water deposition; now we know they can form at any depth.

Submarine cables, heavily armored by steel wires and thick insulation are so abraded by sediment-laden currents where they cross the Mid-Atlantic Ridge at depths of 2000 m that they must be replaced in 25 years; surely the rocks are being equally eroded. Erosion of submarine mountains may not be as rapid as that of land mountains, for there is no frost rifting, but it cannot be trivial.

The drilling of the ocean floor by the *Glomar Challenger* has revealed many disconformities—some spanning whole periods—in completely undisturbed strata at great depths; the bottom currents must be strong enough to prevent deposition, even if not to erode, over very wide areas of the sea floor.

SUBMARINE CANYONS

It was long thought that, except in some small areas of concentrated current action, significant erosion does not go on below wave base. With the advent of sonic sounding, however, it has become evident that most of the continental shelves are dissected by steep-sided canyons and extensive submarine valleys. Figure 16-28 shows many canyons indenting the shelf margin. Note that the only river clearly linked with a submarine canyon is the Hudson. This exceptional trench extends from the estuary almost across the shelf before fading out; the canyon is a deep chasm extending down the slope virtually to the continental rise, where a flat fan contains some of the material eroded.

Other continental shelves, notably those bordering California, Alaska, and southwestern Europe, are similarly trenched. Most submarine canyons incise the slope, begin far from shore, extend downslope for 100 to 200 km, and cut 1 to 2 km deeper than the slope alongside. Some canyons are comparable to the Grand Canyon in size.

The Congo Canyon extends virtually to the base of the continental slope from the mouth of the river. The Nazare Canyon, off Portugal, heads far from any river mouth. The Monterey Canyon, off California, heads near but not at the mouth of the Salinas River; it is joined by the Carmel Canyon about 40 km from its head, and extends to a depth of more than 3500 m, where it ends in a topography like that of an alluvial fan, with gentle slopes. Its long profile is seen in Figure 16-29. Some irregularities of the profile may be due to sounding errors, but the general form is trustworthy. The average grade is almost 4° for the first 65 km from shore; that of the lower Salinas River is less than 0.1°. The submarine canyon is cut chiefly in Miocene and younger strata, but has granite on one wall near the middle; the Carmel Canyon is largely cut in sheared granite. The canyon head is just outside the beach zone and is cut in Holocene and Pleistocene sands and muds.

Most submarine canyons head on the continental shelf, have steep gradients, concave long profiles, and few tributaries. The few canyons heading near river mouths include most of the largest. Like the Monterey Canyon, many others end in fan-like cones on the ocean floor, in which channels are sunk a few score meters. Where these channels curve, the outer bank is notably higher than the inner; it seems clear that these are constructional features like the natural levees along river floodplains, built up by the overflow of sediment-laden currents sweeping down the canyons. The volume of the fan at the foot of the Monterey Canyon is much more than a hundred times the volume of the canyon itself.

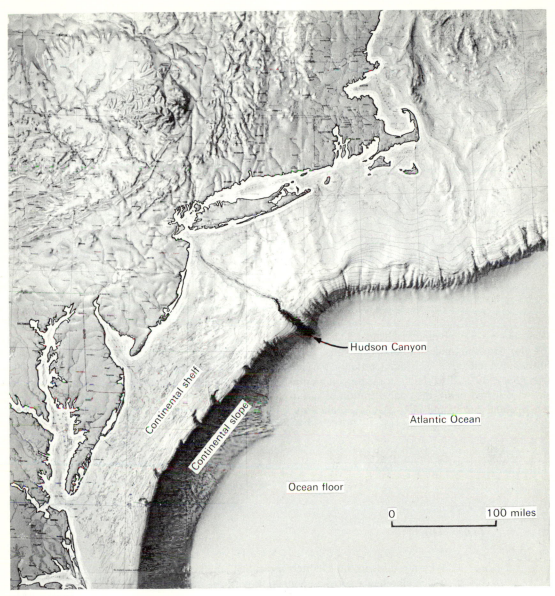

Figure 16-28
Relief model showing the land and submarine topography of the northeastern United States and adjacent continental shelf. Submarine canyons abound on the steep continental slope (partly in shadow). The irregularly ridged topography in the upper right corner is submerged morainal topography. [Courtesy of the Aero Service Corporation, Philadelphia, Pennsylvania.]

Hypotheses of Origin

The many hypotheses advanced to account for submarine canyons belong to two groups, subaerial and submarine. The main subaerial hypothesis is based on the lowering of sea level during the Pleistocene glacial episodes. The amount of lowering is somewhat uncertain. Fossils from borings in the Mississippi delta suggest a lowering of about 120 m; this is slightly greater than most estimates. Thus the upper canyons could have been cut during the lowering of sea

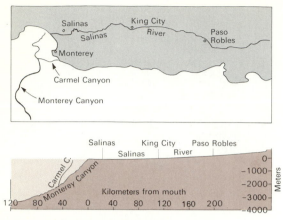

Figure 16-29
Map and profile of the Salinas River, the Monterey submarine Canyon, and its tributary, Carmel Canyon. The vertical scale is exaggerated 20 times.

level, but no conceivable lowering could account for the Hudson and Monterey canyons being formed by subaerial erosion. Certainly the sea was never lowered to the base of the continental slope; if it had been, the concentration of salt in the residual sea would have been so great as to extinguish whole families of organisms. Furthermore, many canyons head far from rivers, but at points where longshore currents converge, as in Santa Monica Bay, California, into which no significant streams enter.

The alternative hypothesis is that the canyons were formed under conditions such as exist today; in short, they are being cut down now, by turbidity currents and more commonly by currents carrying fine sand. Consonant with this is the fact that the Scripps Canyon, constantly monitored by the nearby Scripps Institution of Oceanography of the University of California, episodically clears itself of sediment by slides. It is also clear that the longshore drift both from the north and south toward the head of the Santa Monica Canyon, California, does not accumulate to prograde the shore, but disappears down the canyon; obviously it must be eroding the channel. Scuba divers have observed conspicuous sand streams in the sea at Scripps Canyon and at Cape San Lucas, Baja California. On the Atlantic shelf off Cape Hatteras, the depth of the canyons is due partly to upbuilding of their walls by sediment, as shown by the drooping of bedding toward the canyons. Of course, some drooping may be due to creep, but the actual structure seems more likely to be sedimentary.

Questions

1. Why are tides in the Mediterranean Sea smaller than those in San Francisco Bay, which also has a narrow connection with the ocean?

2. If the effect of wave refraction is to concentrate attack on headlands, why is a cuspate form of sandspit such as those off Cape Hatteras not destroyed?

3. If a wave 120 m long has a period of 8 sec, what is its velocity in meters per second?

4. Roughly estimate the ratio between the velocity of a storm wave far from land and that of the swiftest ocean current.

5. What would be a reasonable difference in elevation between seaward and landward edges of the rock floor of an erosional marine terrace 800 m wide? What is the basis of your estimate?

6. The Congo submarine canyon heads in the Congo estuary, which is studded with alluvial islands. The depth of sedimentary fill in the estuary is not known. Draw

two long profiles of the base of this fill: (a) assuming that the estuary is the slightly emergent head of a canyon formed beneath the sea, and (b) assuming that the whole submarine canyon was formed subaerially, but has since been warped below sea level. Explain.

7. At the equator, water on the bottom of both the Pacific and Atlantic is only a few degrees above freezing. What is the source of this cold water, and why does it not mix quickly with the warm water at the surface?

8. What probable effects did the waxing and waning of the Pleistocene ice sheets have upon: (a) the salinity of the ocean? (b) the number of turbidity currents on the continental slopes? (c) the circulation through the Straits of Gibraltar?

Suggested Readings

W. Bascom, *Waves and Beaches*. New York: Anchor Books, 1964.

Darwin, C. R., *in* Mather, K. F., and S. L. Mason (eds.), *Source Book in Geology*. New York: McGraw-Hill, 1939. [Pages 354–357.]

Davis, W. M., *The Coral Reef Problem* (Special Publication 9). New York: American Geographical Society, 1928.

Steers, J. A. (ed.), *Applied Coastal Geomorphology*, Macmillan, 1971.

Steers, J. A., *Introduction to Coastline Development*, Macmillan, 1971.

R. A. R. Tricker, *Bores, Breakers, Waves and Wakes*, New York: Elsevier, 1964.

K. K. Turekian, *Oceans*, Englewood Cliffs, N.J.: Prentice-Hall, 1968.

Scientific American Offprints

807. Bruce C. Heezen, "The Origin of Submarine Canyons" (August 1956).

810. Henry Stommel, "The Anatomy of the Atlantic" (January 1955).

813. Walter Munk, "The Circulation of the Oceans" (September 1955).

814. Robert L. Fisher and Roger Revelle, "The Trenches of the Pacific" (November 1955).

828. Willard Bascom, "Ocean Waves" (August 1959).

830. Herbert S. Bailey, Jr., "The Voyage of the 'Challenger'" (May 1953).

839. James E. McDonald, "The Coriolis Effect" (May 1952).

845. Willard Bascom, "Beaches" (August 1960).

860. V. G. Kort, "The Antarctic Ocean" (September 1962).

CHAPTER 17

Sedimentary Rocks as Historical Documents

Because they cover two-thirds of the lands, and because fossils are restricted to them, sedimentary rocks are the chief records of earth history and furnish the most complete chronology. This chapter reviews a few examples of the sedimentary record; a full review would require scores of books the size of this.

THE VOLUME OF SEDIMENTARY ROCKS

For many years, geochemists have been estimating the volume of sedimentary rocks formed throughout geologic time on the assumption that the salt in the oceans is a measure of the amount of average igneous rock that has been weathered to produce the much less sodic sedimentary rocks (Appendix IV). Such an assumption leads to estimated volumes of somewhere between 3×10^8 and 7×10^8 km³. We now know from the discovery of vast masses of salt in the sedimentary column and from the similarity of sediments of widely differing ages, that the salt dissolved in the sea is as cyclic as the sedimentary rocks. In fact, it seems likely that the seas of the earth have always —in "legible" geologic history—been of comparable saltiness, and sedimentary rocks roughly as abundant as those of today. If so, the amount of sediment produced throughout geologic time is nearly an order of magnitude larger than that

estimated from the comparison of salt and average igneous rock: the total is nearer 40×10^8 km³, equivalent to a layer of crust 27 km thick beneath the continents. This is nearly equivalent to all the continental crust above the M Discontinuity. In fact, if, as seems likely from sonic soundings of the sea floor, oceanic sediment averages 1 km thick, the present sea floor carries as much sediment as the salt-ratio method yields, and this all seems to be of Jurassic age and younger (Chapter 8).

Abundance of Different Rock Varieties

Scores of varieties of sedimentary rocks have been described, but more than 99 percent of the total volume is made up of only three: shale (including siltstone), sandstone (including graywacke and conglomerate), and limestone (including dolomite). Different geologists have computed the relative proportions of these, as reported in measured stratigraphic sections. The data are poor, for several reasons; many silty sandstones and all graywackes, for instance, contain much silt and clay, as do many limestones. The result is that shale is greatly underrepresented in measured sections (Table 17-1) compared to the quantity expected on the assumption that the composition of all sediments should be that of the "average" igneous rock.

Table 17-1
Proportions of sedimentary rocks computed
by different geologists.

Rock	Percentages measured in outcrops	Percentages expected under differing assumptions of authors
Shale	42–58	70–83
Sandstone	14–40	8–16
Limestone	18–29	5–14

Perhaps a disproportionate amount of the clay
has been carried to the deep sea, thus accounting
for the discrepancy.

Factors Influencing Diversity of Sediments

Sediments are affected by conditions in the source
area, during transportation and in the place of
deposition.

IN THE SOURCE AREA

Most rocks, on weathering, produce some clay,
but no amount of weathering can release clay
from a pure quartz sandstone or a pure limestone.
Climate is a major factor in weathering; contrast,
for example, the frost-riven fragments of fresh
granite on an Arctic mountain, the clay-rich soil
derived from granite in Virginia, and the high-
alumina laterite, also derived from granite, in a
Surinam lowland. Climate-controlled vegetation
strongly influences the breakup of rocks and the
rate and mechanism of erosion. Relief is also a
major factor, as seen by the contrast between
simple downslope movements on the Matterhorn
and solution, acting almost alone, on a coral
atoll. Relief also influences the maturity of the
weathering profile, and thus the completeness
of chemical decomposition. Many sedimentary
rocks are stamped with hallmarks of their source,
giving strong clues to local history.

   The fragments in a talus pile—angular and unsorted by shape and only slightly coarser toward
the base than higher up—exemplify material
moved by gravity alone. Talus is rarely preserved as a consolidated rock, but where it has
been, as along the buried mountain slope underlying the Titus Canyon Formation (Oligocene)
in the Death Valley region, California, its angular
fragments and obvious derivation from the bed-

rock cliff against which it rests, leave no doubt
as to its origin. Contrast this with the Navajo
Sandstone (Fig. 15-23), with its sweeping cross-
bedding, well-rounded grains of uniformly sized
quartz free from clay and other silicates, sparse
faceted pebbles, and rare lenticles of limestone—
features entirely duplicating those of modern
deserts. Or contrast the mixture of unsorted
rock flour, sand and boulders at a glacier's snout
with the organic mud of a tide flat, or the wave-
rounded sand and gravel of a coastal spit. These
examples, of which scores more could be given,
show how the transporting agency places its
stamp on many rock characters—the Present is
the key to the Past.

STRATIFICATION

The most striking feature of most sedimentary
rocks is their stratification (Figs. 11-5, 15-24,
17-1, 17-2, 17-3). Stratification gives some of the
clearest evidence of conditions of deposition.
Layers may be microscopically thin or many
meters thick. In coarse clastic rocks, grains of
varying size lie in layers, parallel or nearly so to
the base of the stratum. Slight differences in size
and color of the grains from one layer to another
emphasize the stratification. In fine clastic rocks,
grain size may seem uniform from one layer to
another, but a color-banding, due to slight differences in content of organic matter or cement may
bring out the bedding. Many carbonate rocks
form very thick, massive strata; others include
many thin layers of clay or shale. Four varieties
of stratification in clastic rocks merit discussion,
as they record significantly different depositional
processes. These are: parallel bedding, current
bedding, graded bedding, and massive bedding
(Fig. 17-4).

Origin of Stratification

The word sediment implies settling from a fluid,
but in geology it generally implies current transport as well. Even though a crystal may form on
the sea floor, its component ions have been
brought together by currents. Most water currents are turbulent, with different threads ranging
widely in velocity and direction and subjecting
suspended particles to wide fluctuations, even
though these average out in time.

Figure 17-1
Scour-and-fill bedding in the Cretaceous strata of the Wasatch Plateau, Utah. A filled stream channel in the alluvial deposits of an ancient coastal plain is seen at the top. [Photo by Warren Hamilton, U. S. Geological Survey.]

Figure 17-2
Bedding in the volcanic ash of the John Day Formation (Oligocene), near Mitchell, Oregon. Continuous thin beds are ash falls (Chapter 9). The weak and discontinuous layering records intermittent weathering and stream flow between the pyroclastic eruptions. [Photo by Oregon State Highway Department.]

Figure 17-3
Massive limestone of the Horquilla Limestone, (Pennsylvanian), Big Hatchet Mountains, New Mexico. Note thin beds at the upper right, passing leftward into much thicker masses of white limestone. This limestone is a former reef, composed of algal-rich fossil fragments. [Photo by Robert A. Zeller, Hachita, New Mexico.]

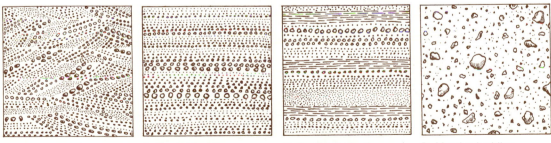

A: Current bedding   B: Graded bedding   C: Parallel lamination   D: Massive bedding

Figure 17-4
Varieties of stratification of clastic rocks. (A) Current bedding, in which the minor laminae (cross-bedding) lie at notable angles to the major bedding surfaces. (B) Graded bedding, in which the laminae are nearly parallel, each grading from coarser at the base to finer at the top. (C) Parallel lamination, in which thin layers of variable grain size lie virtually parallel. (D) Massive bedding, in which no systematic arrangement by grain size is recognizable within individual strata.

CURRENT BEDDING

Where a sand-laden current slows, as on the inside of a river bend or on passing from a shallower to a deeper reach of a stream or at the face of a delta, the sand grains of the traction load are deposited to form point bars, sand bars, and foreset bedding, respectively. The traction load is carried forward and dumped over the edge of the material deposited earlier, building the deposit forward parallel to the foreset slope but at a considerable angle with the water surface. Deposition is also favored by an upstream bottom eddy generated in the lee of the embankment as the

Figure 17-5
Diagram showing eddy formed in the lee of a prograding embankment, with upstream current counter to the traction load that is sliding down the front. [Modified from A. V. Jopling, *Journal of Sedimentary Petrology* v. 35, 1965.]

main stream loses contact with the bottom (Fig. 17-5). Each increment of the advancing foreset slopes at the angle of sliding friction as modified by these eddies. This may be a few degrees or as steep as 40°. This process is not limited to streams; many photographs of ocean bottoms show dunes formed by strong current action, and many marine sedimentary rocks show dune bedding (Fig. 17-6). Current bedding thus may form in any fluid carrying a traction load: sand dunes form in air (Fig. 15-23), in streams (Figs. 12-28, 17-7), or in the sea (Fig. 17-6). They can readily be distinguished by their scales and the sorting of their component grains.

Cross-bedding is thus often a clue to the environment of deposition. Fluctuations in current speed and volume may scour out depressions in the bottom, to be filled by later deposits. Thus scour-and-fill structures are commonly associated with cross-bedding (Figs. 17-1, 17-7).

GRADED BEDDING

Graded bedding consists of layers, each with sharply marked base, on which lie the coarsest grains of the bed (Figs. 17-4B, 17-8). In each layer the coarse material grades gradually upward to finer and finer material, which is succeeded

Figure 17-6
Current bedding in dolomitic marine sandstone of the Vinini Formation (Ordovician), Cortez Mountains, Nevada. The dark layers at top and bottom are shale.

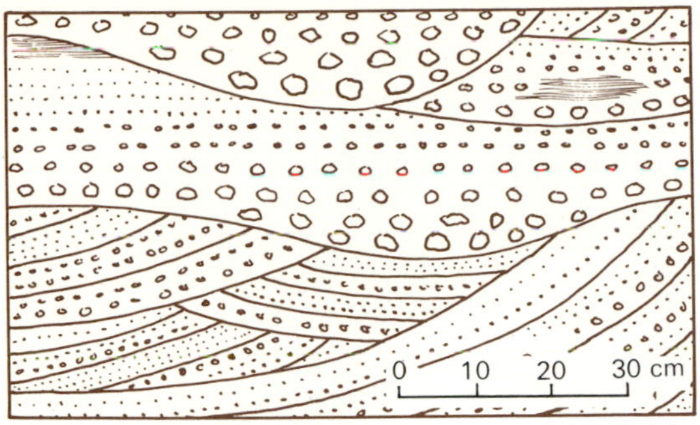

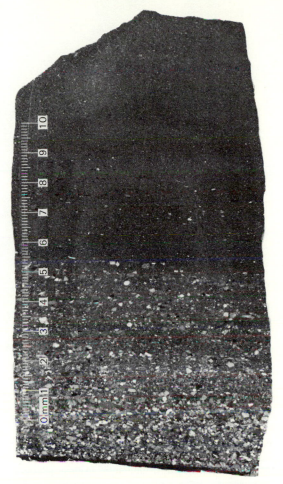

Figure 17-7
Scour-and-fill structures and cross-bedding characteristic of fluvial deposition.

sharply by the base of the next layer (Fig. 17-8). Such grading can be readily produced by stirring up sediments of a wide size range in a beaker full of water and allowing it to stand. The coarse fragments are first to settle out, but the finest may remain in suspension for hours. Much finely laminated material is laid down by pulses of such sediment-laden water, and each lamina may be graded.

Many, if not most, turbidites are composed of sandstone containing considerable feldspar, rock fragments, and a clay matrix—a "dirty" sandstone, or **graywacke**. The poor sorting, even with excellent grading, is doubtless due to the incorporation of all sizes of material in the slide parental to the turbidite flow. In most graywacke the fine grain of the matrix is original, but in many it is due to submarine weathering of rock fragments in place, for the bottom waters are generally well oxygenated.

The varves of glacial lakes (Chapter 13) are excellent examples of graded bedding; it is also found in stream backwaters into which freshets overflow; in sheltered bays, where storm waves that had entrained much sediment are driven; and at great depths in the ocean, where turbidite deposits grade upward from sand to silt and clay. Graded beds are commonly interbedded with others having parallel lamination or current bedding, laid down in the intervals between turbid flows. This is to be expected, and is seen in Figure 11-5.

Graded bedding is commonly associated with submarine landslides, as these are likely to generate turbidity currents. Sliding of weakly consolidated sediment may gouge and plow the bottom (Figs. 17-9, 17-10, 17-11).

Figure 17-8
Graded bedding in a limestone turbidite from the Dimple Formation (Pennsylvanian) near Marathon, Texas. Note coarser fragments at the base, grading upward to the extremely fine-grained material at the top. [Photo by Alan Thomsen, Shell Oil Company, Houston, Texas.]

The association of graded bedding with evidence of submarine landslides has persuaded many geologists to consider it a criterion of deep-water deposition. But varved lake sediments and overbank floodplain deposits are also truly graded, and subaqueous slides can form at shallow and intermediate depths as well as in very deep water, as shown by the slide in the Lake of Zug (Chapter 11) and the sediments illustrated in Figure 17-10. Graded bedding of itself indicates only that it formed where bottom currents were not active enough to sort the material. Such currents may have been active at the site before

Figure 17-9
Bottom of a turbidite stratum showing bottom marks indicating flowage into the underlying layer. The sediment was moving toward the bottom of the picture. Scale 15 cm long. Tyee Formation (Eocene), Oregon Coast Range. [Photo by Parke D. Snaveley, Jr., U. S. Geological Survey.]

Figure 17-10
Slumping of lake sediments of the Lisan Formation (Pleistocene) in the Dead Sea graben, Israel. Presumably the sliding was caused by differential loading. [Photo by Tad Nichols, Tucson, Arizona.]

A

B

C

D

Figure 17-11
Features of sedimentary rocks. (A) Thinly laminated sandstone interbedded with a few thicker beds in Coaledo Formation, near Chehalis, Washington. [Photo by Parke D. Snavely, Jr., U. S. Geological Survey.] (B) Slump bedding and interbed disturbance due to sliding of unconsolidated sediments. Miocene strata near Dos Pueblos Creek, Santa Barbara County, California. [Photo by M. N. Bramlette, U. S. Geological Survey.] (C) Massive bedding in two-foot mudflow on beach at Cape Thompson, Alaska. [Photo by Reuben Kachadoorian, U. S. Geological Survey.] (D) Ripple marks on Silurian siltstone, west shore of Moose Island, Maine. [Photo by E. S. Bastin, U. S. Geological Survey.]

the graded bed was deposited and may recur later, as shown by the common interbedding of graded and current-bedded sandstones. The only unambiguous criterion of very deep origin for graded bedding is the finding of deep-water benthic fossils.

PARALLEL LAMINATION

The size of grain that can be kept in suspension or traction depends on current speed. A mountain torrent can roll great boulders, but as its velocity decreases, so does its competence. Part of the traction load comes to rest, and smaller grains drop from suspension to become part of the traction load; a layer of sediment is thus laid down. If the current continues to slow, finer and finer material is deposited. As we saw in Figure 12-12, finer grains are more difficult to entrain than coarser, when the size is smaller than about 0.02 mm; a new pulse of sediment-laden water may carry in coarser grains without eroding the finer. The resistance of clay-sized particles to re-entrainment is particularly marked, as has been proved in many irrigation canals. This may in part explain the parallel lamination in many shales and siltstones, though the lamination in shales is also in part due to compaction of randomly oriented flakes of clay by load of younger sediment, tending to flatten the originally unoriented particles and align them to form bedding planes.

MASSIVE BEDDING

Strata that show no internal lamination have **massive bedding**, without sorting by either size or density. Mudflows, whether in the desert or on the slopes of a volcano, may be wet enough to

flow but too viscous to permit sorting (Figs. 17-4,C, 17-11,C, 17-32). Some deposits of volcanic lapilli are also without bedding, doubtless because of turbulence in the eruptive cloud (Fig. 17-12). Some well-sorted sediments are also massive; the coarser because of reworking of previously well-sorted sediments and the finer because organisms have so reworked a mud as to destroy all trace of bedding.

TEXTURES

The texture (Appendix III) of a rock is commonly a clue to its depositional environment. Boulders, even of resistant rock, commonly become rounded in a mountain stream within a few hundred meters of their source, but sand grains, with relatively much greater surface per unit mass, are rounded much less quickly. In a poorly sorted sandstone it is common to find the coarser grains well-rounded even though the finer ones remain angular (Fig. 17-13). Even long wear on a beach may fail to round sand-sized quartz grains very well—in fact, many geologists think that highly rounded quartz grains are never formed in a

Figure 17-12
Massive-bedded volcanic ash flow, Crater Lake National Park, Oregon. Except near the top of the exposure, sorting by size is lacking. The pinnacles—some more than 60 m high—were left standing after rain erosion, where scattered larger lava fragments served as protecting caps. [Photo by Oregon State Highway Commission.]

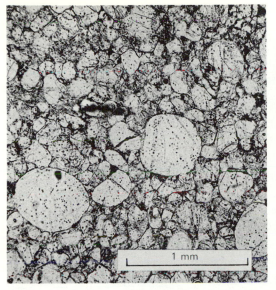

Figure 17-13
Photomicrograph of a sandstone from the Valmy
Formation (Ordovician), Shoshone Range, Nevada.
Although this sandstone is more than 99 percent
quartz, the sorting is very poor; the coarser grains are
well rounded, the finer sharply angular.

single cycle of erosion and deposition, but require
reworking from one sandstone to another, perhaps several times. As a sand grain weighs more
in air than in water, wind transportation is much
more abrasive than water transport; dune sand
grains are thus generally better rounded than
those of streams and beaches.

LAND-LAID SEDIMENTS

Distinguishing features of many land-laid sediments have already been described. Here we
examine some modern deposits with a view to
recognizing any analogs in ancient rocks. Coarse
stream deposits characteristically include scour-
and-fill structures, current bedding, and local
graded bedding; overbank deposits are finer.
When the muds dry on retreat of a flood, they
generally crack into polygonal patterns (Fig.
17-14) characteristic of clay deposits that are
alternately wet and dry. Though such cracks form
on tidal flats as well, the intertidal interval is much
too short to make them as deep as in the Nile
floodplain of Figure 17-14. If, as is common, the
floodplain mud is graded, the polygons curl up-

Figure 17-14
Mudcracks on the floodplain of the Nile at Khartoum,
Sudan. [Photo by Tad Nichols, Tucson, Arizona.]

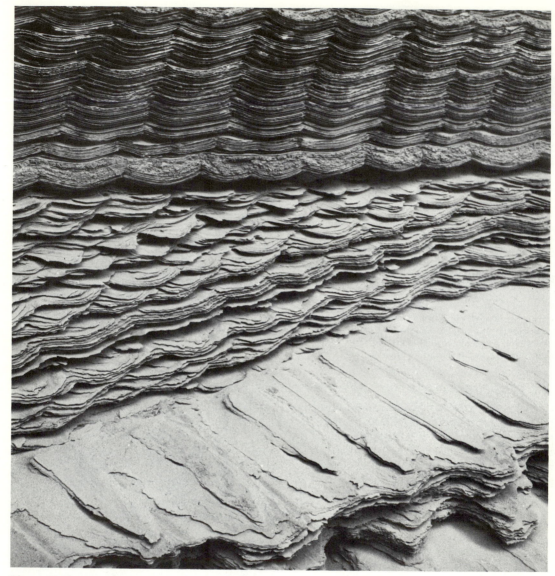

Figure 17-15
"Climbing ripples" in the alluvium of the Colorado River delta, California. Flume experiments show that climbing ripples form only when the stream is choked with rather well-sorted sand and is rapidly aggrading. [Photo by Tad Nichols, Tucson, Arizona.]

ward, for the more silty and sandy lower part of the graded bed contracts less on drying than the more clayey surface layers.

Streams so choked with uniform sand as to be rapidly aggrading form a distinct kind of progressive ripples, each higher layer being slightly advanced downstream with respect to the lower ones (Fig. 17-15)—**climbing ripples.**

Streams such as the lower Mississippi, the Amazon, and the Nile carry little coarse sediment; their floodplains are chiefly made of silt and fine sand—the sand along the deeper channels and silt and clay in overbank deposits.

Sporadic lenses of fine gravel occupy the deeper channels of the lower Mississippi, and similar lenses of coarser silt, sand, and fine gravel have been outlined in well cores from depth in the floodplain and delta. At depth Pleistocene fossils are found: fresh-water mollusks and remains of plants like those along the river bank. Clearly these are fluvial deposits of the ancient Mississippi. From studies of many cores, the positions of a long succession of delta distributaries were determined (Fig. 17-16). Well data from several oil fields in Oklahoma have been used to develop a precisely comparable pattern of sandstone distributaries in Pennsylvanian rocks of an ancient delta.

Long chains of gravel-capped mesas made up of the Arikaree Formation (Miocene) form a river-like pattern in parts of the High Plains of Colorado and adjacent states. The finer sediments have been more readily eroded away; there can be no question that the Arikaree was deposited by aggrading streams rising in the Rocky Mountains to the west. Fossils found in it include bones of antelope, camels, and horses, shells of land snails, and pollen of grasses—all of which might be expected on an open floodplain.

Few lakes are large enough to allow enough fetch for waves to become large or persistent

currents to be established; coarse sediment is generally confined to the shore zone and bottom sediments are fine grained and evenly laminated, not unlike the glacial lake varves of Figure 6-5.

MARINE SEDIMENTS

Marine sediments are now and were at all times in the geologic past far more voluminous than land-laid beds. Except for a trivial amount of meteoritic dust, the much greater but unmeasurable increment from submarine volcanism, and an uncertain, but probably minor addition from tidal scour and slides from submarine ridges, all marine sediment comes from the lands. Even the carbonate of marine shells, which bulk large in marine sediments, was originally leached from the land.

Every year the rivers sweep about 20 billion tons of rock waste to the sea, there to join the debris won by the ocean's own attack upon the shores. More than four-fifths is clastic material in suspension or traction; the rest is dissolved. The fate of the clastic material is clear:

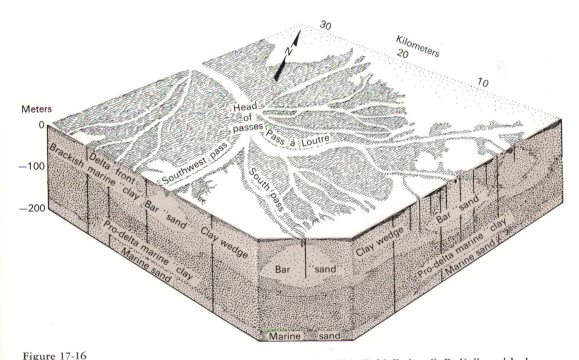

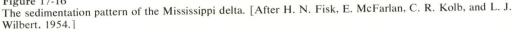

Figure 17-16
The sedimentation pattern of the Mississippi delta. [After H. N. Fisk, E. McFarlan, C. R. Kolb, and L. J. Wilbert, 1954.]

382

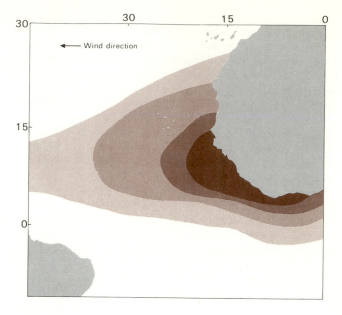

30   30   15   0

← Wind direction

Figure 17-17
Phytoliths (silica derived from grass fires) and fresh-water diatoms in a dust storm of January 17 to 19, 1965, off the west coast of Africa. These land-derived materials abound, not only in the dust but also in the deep-sea sediments for many hundred kilometers from the shore. Pattern indicates abundance of diatoms and opaline silica particles recovered during the storm by H. M. S. *Vidal.* [After D. W. Folger, W. H. Burckle, and B. C. Heezen, *Science* v. 155, pp. 1243–1244, March 10, 1967. Copyright 1967 by the American Association for the Advancement of Science.]

sorting, already begun in the stream regimens, continues in the undersea regimen. Samples dredged from the sea floor show partly consolidated sediments identical in grain size and mineral content with the river detritus. Because of continual wave agitation, stratification and sorting of the coarser clastic sediments are commonly more pronounced in marine than in fluvial sediments. As suggested both by the deficiency of shale among the stratified rocks and by the composition of the sediments of the deep oceans, this agitation is generally enough to permit a large share of the finer clay to move off the continental shelf and to the deep sea. This inference has recently been fortified by light-meter measurements of transmitted light at depth in the sea, which have shown that water masses are cloudy with suspended sediment down the continental rise and far out in the ocean basin. Even a slight load of finely suspended matter, such as those samples summarized in Table 17-2, constitutes a prodigious aggregate in process of settling to the sea floor. Windblown dust also is spread far from shore, especially off the desert belts (Fig. 17-17), but also even off the humid coasts of the northeast Pacific.

The fate of the dissolved material is not so obvious. Though many huge masses of salt and anhydrite occur in older strata, they seem now to

Table 17-2
Suspended sedimentary particles in bottom waters.

Water body	Suspended load, in milligrams per liter
Bering Sea	2–4
North Atlantic	0.05–1
East Pacific	up to 5
Baltic	7–10
Indian Ocean	0.39–2.2

be deposited only in a few small lagoons along some desert coasts or on tidal flats, such as the Rann of Kutch in westernmost India, which the monsoons annually flood to shallow depths. Evaporites are being deposited today in vastly smaller amounts than in much of geologic time. Present river water holds much higher proportions of silica and calcium than does sea water (Appendix IV)—so much higher that the composition of the sea would change greatly in a geologically short time were these elements not somehow removed. Its removal is largely organic; much calcium and silica are incorporated into the shells of animals and into plant and animal skeletons. Some geologists think that the lesser proportion of limestone being formed in modern seas is due to the relatively recent evolution of the planktonic unicellular Foraminifera. These drifting animals now carry much calcium carbonate to the deep sea; before they evolved, during the Cretaceous, this lime remained on the continental shelves, fixed by sessile rather than drifting organisms.

Marine Sedimentary Realms

A core problem in stratigraphy—and through stratigraphy to tectonic geology—is the recognition of the marine realm in which individual strata were deposited. The great surge of oceanographic research since World War II has yielded much information, but we are still far from a thorough understanding of sedimentational regimes of the sea.

The Strand Zone

The **strand zone** lies along the shore; its widely varied sediments include those of beaches, deltas, bars, and tideflats.

BEACHES

Beaches are wave-deposited cobbles, gravels, and sands on and near the shore. During storms, much or all of a beach deposit may be swept away or, depending on the set of the waves, piled to heights several meters above its normal level; between storms only a thin surface layer is agitated. On mountainous coasts, beaches are short and discontinuous and may be composed of poorly sorted debris largely supplied by downslope movement; but along plains coasts, beaches may extend unbroken for hundreds of kilometers and are generally composed of well-sorted sand. On coasts of high relief, the beach sands may contain readily weatherable minerals, such as feldspar, supplied by rock fragmentation without much weathering; in Hawaii whole beaches are composed of sand-sized grains of olivine, a mineral that normally weathers readily but is available there from cinder cones formed where basaltic lava rich in olivine phenocrysts entered the sea. Such minerals are missing from plains coasts, where beach sands consist wholly of quartz and other resistant minerals, such as rutile, magnetite, and garnet.

Some beaches, particularly in the tropics, are composed largely of shell fragments. Along the Brazilian coast these are being cemented into **beach rock** almost as quickly as they form; sea spray dissolves a little calcite from the shells and it reprecipitates below as a cement. Many sand beaches, like stream deposits, contain concentrations of dense and resistant minerals such as magnetite, cassiterite, zircon, gold, and diamond. Many have been mined for rare metals in India, Ceylon, Australia, and elsewhere. The beaches of Southwest Africa have been great producers of diamonds; those of Nome, Alaska, have furnished much gold, both from the modern beach and from ancient raised beaches a few kilometers inland.

Although raised beaches are common, few beaches have been recognized among the consolidated rocks. They are obviously vulnerable to erosion if uplifted and to reworking if submerged only modestly. Nevertheless, Cambrian beaches with associated sea cliffs and tumbled sea stacks have been recognized in Montana, and Carboniferous and Jurassic beaches in Great Britain.

DELTAS

Where a stream enters the sea, the surface slope that provides its current vanishes. If wave action is strong, river and sea water quickly mix, but along quiet coasts the fresh water may float for a while above the salt. The powerful flood of the Amazon spreads widely over the Atlantic, and fresh water may be dipped from the surface far at sea. If heavily silt-laden, though, a stream may sink and continue to flow as a turbidity current.

Whether the river water sinks or floats, its sediment is widely distributed over the sea floor. Different rivers bring vastly different sedimentary loads, and all vary seasonally. The Colorado, for example, carries much coarse sand and rapidly aggrades its delta plain (Fig. 17-15); the Mississippi and Rhine carry little except clay, silt, and very fine sand, aggrading their delta plains only in floods.

Most streams that enter the sea build deltas, but if the load is small, the coast exposed, or the tidal range high, the detritus may be quickly dispersed, and no delta is formed. Even the great

Columbia has none, though the Fraser, 300 km to the north, empties into a protected area and has a vigorously growing delta. Deltas furnish most of the sediment to longshore currents—far more than is mobilized by wave attack.

Deltas in shallow water, especially those supplied with sand, commonly form foreset, bottomset, and topset bedding, but the Mississippi, building its muddy delta into the deep water of the Gulf, shows nearly level bedding; foreset bedding is there a meaningless term. On either side of the distributaries, coarse silt and fine sand extend far out into the Gulf as underwater natural levees, continuous with those farther upstream.

Prograding of a delta varies with river discharge, with shifts of distributaries, and with storms. Some segments advance while others retreat under wave attack. Deltas furnish a good example of the difficulty of geologic classifications, for all marine deltas contain some marine beds, some fluvial and some brackish lagoonal deposits—all complexly interfingering.

Delta subsidence Surveys and topographic changes show that many deltas are sinking, that of the Mississippi at the rate of $2\frac{1}{2}$ m per century in places. Some consider this an isostatic response to the load of the delta, but this is obviously not the whole story, for the delta sediment is far less dense than any mantle material and so the sinking responsible for the "birdfoot" pattern may be in part due to compaction of the shales and perhaps to tectonic forces rather more than to isostatic response.

Intimate associations of such variable sediments record many ancient deltas: in New York and Pennsylvania, a Devonian delta more than $2\frac{1}{2}$ km thick has been recognized, bordering the Old Red continent (p. 397), and in Louisiana, borings for petroleum have outlined an ancient Mississippi delta of comparable size.

TIDAL FLAT DEPOSITS

Tidal flats in estuaries of sluggish streams or in coastal lagoons often build up rapidly. Some near Wilhelmshaven and Cuxhaven on the North Sea coast of Germany gather 2 or 3 m of mud a year. Some of this mud is brought in by rivers, but much is swept in by the tides from the Rhine delta. The mud is thickly populated with worms and other mud-ingesting organisms, so that no stratification is preserved. Some of these mud deposits contain so much organic matter that they are used as fertilizer.

Like beaches, lagoonal muds are rarely preserved in the geologic column but some Devonian shales in Germany have many features reminiscent of modern lagoonal sediments, and they may have had a like origin.

SEDIMENTS OF THE
CONTINENTAL SHELVES

The marine profile of equilibrium (Chapter 16) is merely an idealized concept, but it applies as a sound generalization for such areas as the northern Gulf of Mexico, the Gulf of Paria, the shelf off Somaliland, and in many other places where accurate sounding and bottom sampling have been done. In all these areas, nearshore sand deposits grade outward to a mud bottom in deep water. It may be significant that these places were not occupied by the Pleistocene glaciers and are abundantly supplied with riverborne sediments. Whatever the reason, this systematic arrangement of grain sizes is not ubiquitous, though it does appear that the nearshore sea bottom is generally concave upward and flattens seaward.

Farther offshore, though, the bottom in many places deviates widely from the ideal smooth profile; sediment in transit does not everywhere become finer seaward. Part of the deviation may be explained by the rise of sea level, perhaps by 150 m since the disappearance of the Pleistocene glaciers—a rise so recent that the sea may not have had time to adjust its profile to the new level except in areas of heavy sediment contribution or where no morainal sediments were flooded. A well-studied example of a complex pattern of bottom topography and sedimentation patterns is that of the Atlantic coastal plain off the United States.

Atlantic continental shelf sedimentation For nearly 40 years it has been known that the bottom sediments on the Atlantic continental shelf are not systematically sorted. Figure 17-18 shows the grain-size distribution of material that adhered to the tallow on a sounding lead, as compiled by the U. S. Coast and Geodetic Survey. More sand than gravel is near shore, and sediments of the outer shelf are in many places coarser than those inshore. Silt and clay are rare on the shelf at

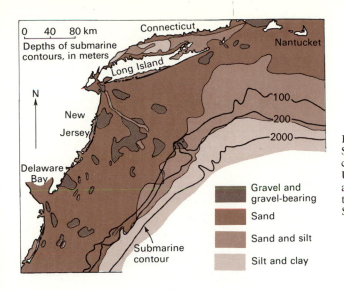

Figure 17-18
Size distribution of sediments on the continental shelf off the northeastern United States. [After F. P. Shepard and G. V. Cohee, 1936, from charts of the U. S. Coast and Geodetic Survey.]

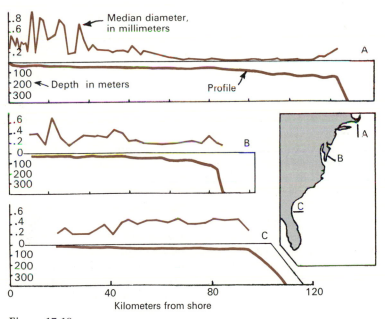

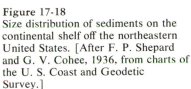

Figure 17-19
The distribution of sediment sizes along three profiles of the Atlantic shelf of the United States. [After H. C. Stetson, 1938.]

depths of less than 200 m and abound on the continental slope below.

Three profiles surveyed and sampled across the shelf show notable regional differences (Fig. 17-19). Above each profile is plotted the median diameter of particles recovered from the cores.

Profile A, off Cape Cod, shows the sea floor deepening rather systematically except near the shelf edge. The coarse sediment nearshore is irregularly distributed; it is reworked morainal material. Profile B, off New Jersey, an unglaciated area, shows finer material near shore, but

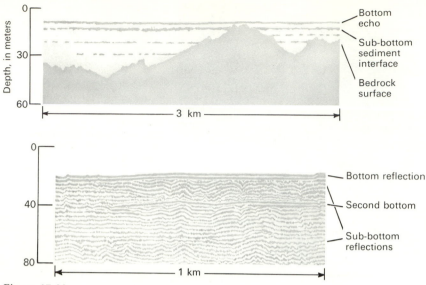

Figure 17-20
(Top) A representative continuous reflection profile over an uneven bedrock surface
buried in sediment in Long Island Sound, New York. [After W. C. Beckmann, A. C.
Roberts, and B. Luskin, *Geophysics* v. 24, 1959.] (Bottom) Reflection profile in
Vineyard Sound, Massachusetts, showing younger sediments smoothing the surface over
an irregular underlying topography. [After J. B. Hersey, in *The Sea*, Interscience
Publishers, New York, 1963.]

coarser at depths where A shows silt. Here, too,
grain sizes are irregularly distributed, but more
than 40 km from the coast they tend to diminish
seaward except at the shelf edge. Silt is lacking,
perhaps because the sea is reworking coastal
plain rocks composed chiefly of sand. Profile C,
off Florida, slopes regularly seaward, but here
again sediment more than 40 km offshore is
coarser than that nearshore. The Gulf Stream
sweeps finer sediment into oceanic depths. Photographs of the bottom show dunes, and in many
places the dredge brings up rock as old as Cretaceous—the currents are actively eroding.

Despite the poor sorting of the sediments, the
smoothness of the profiles indicates considerable
reworking of them. Sonic profiler records show
this, even for sheltered areas (Fig. 17-20).

Sediments of the Continental Slopes

Off Spain, Brittany, and Ireland, the continental
slope is too steep to retain sediment, and the sea
floor is barren rock. Generally, though, the slopes
are mantled with fine silt and mud lying in strata
parallel with the slope, but locally broken by
faults or by slumping.

Sediments of the Continental Rise

Coring and sonic sounding reveal the profound
influence of deep ocean currents over the continental rise. A southward-trending deep current
sweeps along the rise off the eastern coast of the
United States and Canada. In many places it
lies directly beneath the Gulf Stream, flowing in
the opposite direction. Photometric measurements
show that it is carrying much fine sediment. The
swiftest part of the current follows the base of
the continental slope, and presumably because of
the larger supply, deposits more sediment here
than elsewhere (Fig. 17-21), thus accounting for
the gentle seaward slope of the rise.

Even the relatively feeble movement of the
Antarctic Bottom Water, moving northward
around the Bermuda Rise, seems to govern deposition of the fine sediments there.

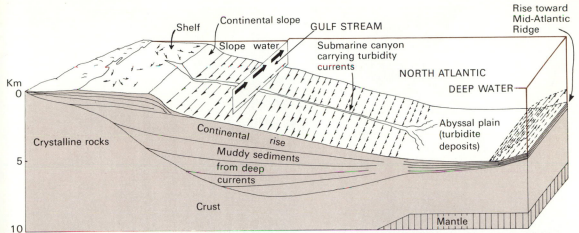

Figure 17-21
Diagrammatic section of the sedimentation pattern off the east coast of the United States. [Greatly modified from B. C. Heezen, C. D. Hollister, and W. F. Ruddiman, *Science* v. 152, pp. 502–508. April 22, 1966. Copyright 1966 by the American Association for the Advancement of Science.]

Table 17-3
Theoretical dispersal of sedimentary particles by currents.

Diameters of particles, in millimeters	Rate of settling in water at 27°C, in cm/hour	Distance a particle would be transported by a 1 cm/sec current while sinking 1 km
0.10 (fine sand)	2000	1.8 km
0.030 (coarse silt)	180	20 km
0.005 (fine silt)	5	720 km
0.001 (clay)	0.2	18,000 km

SOURCE: Data from Neeb, *Snellius* Report.

Sediments of the Atlantic Abyssal Plains

Cores drawn from the Atlantic abyssal plain west of the Mid-Atlantic Ridge show thin graded beds, averaging considerably coarser than the sediments of the continental rise. Similar beds were found at the surface soon after the great turbidity current from the Grand Banks (Chapter 16); it seems likely that the graded beds are also turbidity current deposits. Though the Grand Banks current was not confined to submarine canyons, the observed motion of coarse sediment down them, the cones at their lower ends, and the pattern of levees across the cones all suggest that many turbidites reach the ocean floor through submarine canyons (Fig. 17-21).

Many turbidite sequences contain submarine slides, involving great boulders, some measured in scores or even hundreds of meters—masses too large to break up into turbidity currents. These are very abundant in the Appenines but are also found in many other ranges. A relatively minor example is seen in Figure 17-22, another in Figure 17-33.

Distribution of Clays

Experiments by Neeb show that quartz grains of various sizes settle in sea water at the speeds indicated in Table 17-3.

Clay particles have very large surface areas relative to their weight. The surface ions of the

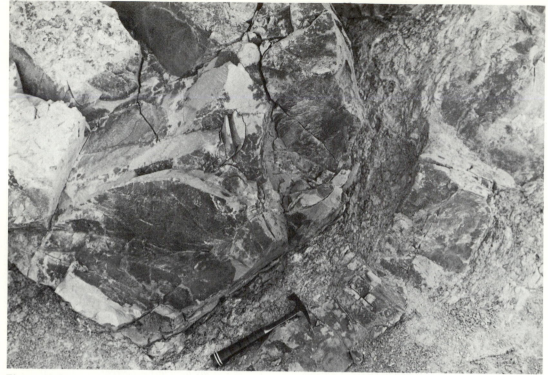

Figure 17-22
Large boulder at the base of a submarine slide, showing how the soft sediment beneath was squeezed and deformed while still unconsolidated. Tesnus Formation (Mississippian and Pennsylvanian), Marathon region, Texas. [Photo by Alan Thomsen, Shell Oil Company, Houston, Texas.]

clay crystals have unsatisfied positive charges that attract negatively charged ions such as OH^- from the river water. Each particle thus acquires a shell of negatively charged ions. As the negative charges repel each other, the clay particles do not stick together in fresh water but remain dispersed. On entering the sea, the negatively charged ions unite with the abundant positive ions of Na^+ and Mg^{++} in sea water; thus neutralized, the clays flocculate to form clots or aggregates, which sink faster than individual clay flakes.

Clay thus settles somewhat more quickly in the sea than in fresh water, but the flocculated clots must generally not be large and the settling velocity not great, for the sea-floor sediments record wide dispersal. The high density of the very cold water of the deep ocean would also delay sinking.

Though Neeb's computations neglect floc-culation of the clay, we see that fine silts and clay can be carried almost any distance by very modest currents. In fact some studies indicate that the average clay particle may take several centuries to sink to the abyssal floor. Deep-sea clay may thus have a much more distant source than the coarser particles accompanying it. The Swedish Deep Sea expedition of 1947–1948 dredged much clay from the deep eastern Pacific, more than 3000 km from the nearest land.

At first glance it would seem that with such slow settling velocities, clays should be almost uniformly distributed over the sea floor in areas of equal depth. Both silica and calcium carbonate are organic sediments that vary with latitude, but the amount of clay in a deep-sea core should, it might appear, be a measure of its age. Radiometric dating of cores, however, shows that this assumption is not true, doubtless because fast

Table 17-4
Radiometrically determined rates of pelagic clay accumulation.

Ocean area	Rate, in millimeters per thousand years
North Pacific	4–7 (Holocene)
Pacific, off Mexico	11 (Holocene)
Atlantic continental rise, off Florida	50–500 (Holocene)
Caribbean southwest of Hispaniola	28 (Pleistocene and Holocene)
Atlantic, Argentine Basin	17–34 (Holocene)
Atlantic, Argentine Basin	60–110 (Late Pleistocene)
Southeastern Pacific	80 (Holocene)

From various sources.

currents bring more sediment over a given area than slow. Some radiometrically determined sedimentation rates are given in Table 17-4.

It should be noted that several students have suggested that not all deep sea clay is detrital; despite its low solubility, they think that some of the aluminum in the clay was in ionic solution in the ocean water and was precipitated in the presence of silica brought in organically.

COMPARISON OF ANCIENT AND MODERN OFFSHORE SEDIMENTS

Little silt and clay is now coming to rest on the continental shelves except near such muddy streams as the Hoang Ho, Mississippi, Orinoco, and Nile. Elsewhere clay seems to remain suspended until it drops over the shelf edge into a quiet basin or onto the ocean floor. In many places—not merely down submarine canyons—even sand is swept by currents onto the ocean floor. Beyond the immediate beach zone, at least in areas of slow deposition, the grain size of sediment depends more on local currents than on distance from the shore.

The most widespread sediments of Phanerozoic age are marine shales, clays, and mudstones containing fossils of shallow-water organisms. Many can be traced laterally into sandstones in one direction and into limestones in the other. Similar associations are found today off the coast of

northern Australia, in the Gulf of Mexico, in the southern Caribbean, the Yellow Sea, the South China Sea, and the Persian Gulf. But we know of no such extensive marine environments of mud deposition in shallow water as that of the Cretaceous geosyncline that covered the interior of North America from the Gulf of Mexico to the Arctic and from the present Mississippi-Hudson Bay embayment to the west side of the Cordillera. We must infer the conditions of deposition of the Cretaceous shales from their own character and fossil content (all from the photic zone) rather than from comparison with present-day depositional environments.

It is easy to match the turbidites of the Dimple and Haymond Formations of Texas (Figs. 11-5, 17-8) with the turbidite cores from the Atlantic floor. The weakness of this analogy is that although turbidites are common in mountains of all ages, no mountain seems to include large areas of ocean floor; more likely flysch is derived from basins like those of Indonesia rather than like those of the Atlantic.

A few dominantly shale formations, notably the Repetto Formation (Pliocene) of California, and some others in Timor, Cuba, and elsewhere, contain fossils that indicate water depths of several thousand feet, like those in basins off the coast of southern California. The Repetto basin ultimately became filled with sediment dumped in from adjacent land, and its upper beds contain only shallow-water fossils.

Deep-sea deposits, such as organic oozes and red clay, described in the next sections, are exceedingly rare on the continents, although some cherts may represent siliceous ooze and some red shales may possibly represent deep-sea clay of narrow basins.

Pelagic Sediments

Pelagic (deep-sea) sediments cover nearly three-quarters of the ocean bottom, as has been known since the cruise of the first oceanographic ship, the *Challenger*, nearly a century ago. They are of four main varieties, distributed in the pattern generalized in Figure 17-23. In addition, considerable areas of pyroclastic deposits are known in broad outline. Calcareous ooze is most widespread; land-derived sediments are next most

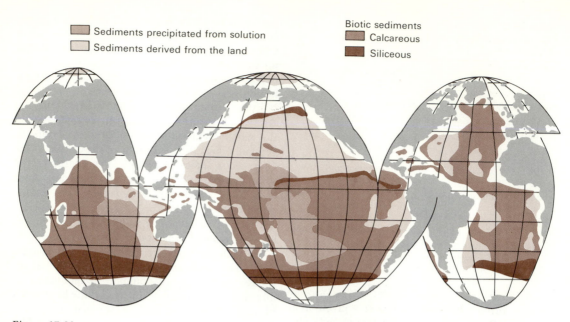

Figure 17-23
Areal distribution of the pelagic sediments. Sediments whose major components have been formed from ions in solution in sea water are less abundant than the sediments derived from the continents. Each variety contains some components characteristic of some of the others. The scale prohibits showing many small areas of distinctive character within broader regions of another variety. [After Gustav Arrhenius, in *The Sea*, Interscience Publishers, New York, 1963. Use of Goode's interrupted homalosine equal area projection by permission of the University of Chicago Press, copyright by the University of Chicago.]

abundant, strewn widely by turbidity and other currents; siliceous oozes are next lower in total amount, but locally predominate; and the minor sediments formed by precipitation from solution dominate only where sedimentation of other types is extremely slow—generally on current-swept hills on the ocean floor.

Calcareous Ooze

Calcareous ooze (Fig. 17-24) covers most of the floor of the western Indian Ocean, the mid-Atlantic Ocean, and much of the equatorial and southern Pacific Ocean. It consists chiefly of calcium carbonate shells of planktonic organisms, chiefly foraminifers, and coccoliths (calcareous algae?), but with some clay, siliceous shells, minute fragments from the disintegration of meteorites in their travel through the air, and minerals crystallized from the ocean waters. Calcareous ooze, the shallowest of the pelagic

sediments, seems to be forming where skeletons of calcareous planktonic organisms rain down on the floor so abundantly as to mask the terrigenous sediments and oceanic precipitates; also faster than the cold deep water can dissolve the minute limy shells.

Red Clay

Red clay covers most of the very deep ocean floor. Much is too fine grained to be identified microscopically, but X-ray study has shown the chief minerals to be quartz, mica, several kinds of clay minerals, chlorite, and several of the hydrous aluminum silicate minerals of the zeolite group, all generally stained red with iron rust. Despite the name, not all the "red clay" is actually red; some is pinkish, and some is gray. The iron is oxidized because the bottom water is generally rich in oxygen dissolved in the stormy waters of high latitudes in both hemispheres, whence the

Figure 17-24
Foraminiferal ooze dredged from 450 m off Central
America; enlarged about 15 times. [Courtesy of Patsy
J. Smith, U. S. Geological Survey.]

bottom water sank. Although some of the clay
and zeolite minerals are thought to form in place
from ions in solution, most of the pelagic mud is
clastic material from the continents, a source
consistent with the settling velocities in Table
17-2. Though there are no quartz-bearing rocks
on the oceanic islands, more than a fourth of the
mud of the North Pacific is quartz, presumably
dropped from the upper atmosphere, for its relation to the continents is inconsistent with the
pattern of either surface winds or oceanic currents.

Volcanic ash was once considered the chief
source of red clay. It contributes, of course, but
does not seem to be an adequate source for the
great bulk. Kaolinite seems never to form in the
sea, whereas chlorite forms both on land and in
the sea. The distribution of the two minerals
strongly suggests that the pelagic muds are
chiefly of continental origin, for kaolinite is far
more abundant in the equatorial Atlantic (into
which many of the greatest rivers empty) than it
is either to the north or south (Fig. 17-25). Chlor-

ite, abundant in high-latitude soils, also abounds
in high-latitude pelagic sediments.

The red clays contain some organisms like
those of both the calcareous and siliceous oozes
and intergrade with them. Red clay dominates
either where limy plankton is sparse (in high
latitudes) or where the shells of such organisms
are dissolved in water made rich in bicarbonate
ions by decaying organisms; at great depths only
the thickest shells, or none at all, remain.

Siliceous Oozes

Siliceous oozes are restricted to relatively narrow
strips in the equatorial and North Pacific, and to
a somewhat broader band in all the southern
oceans. Those in the equatorial belt are composed
chiefly of remains of *Radiolaria*, minute single-
celled animals; those in high latitudes are chiefly
of shells of single-celled plants, the *Diatomacea*.
Radiolaria flourish in the nutritious, phosphate-
rich water of the equatorial upwelling. The rising
water is also cold and rich in bicarbonate, so that
most of the abundant limy organisms (that also
flourish in the upwelling) dissolve before sinking
to the bottom. In high latitudes the mixing of
deep water with cold surface water (and the
abundant oxygen dissolved in the excessive turbulence due to the prevailing westerlies) furnish
abundant food for diatoms, which there dominate
the plankton.

Manganese Nodules

Scattered very unevenly over parts of the sea
floor where other sedimentation is slow are
biscuit-shaped nodules consisting of oxides and
hydrated oxides of manganese, iron and nickel,
with variable amounts of silica. The manganese
content may be as high as 55 percent, but the
average is far lower, perhaps 14 percent. En-
thusiastic oceanographers sometimes contend
that these nodules are potential ores of manganese,
but it is hard to see them soon competing with
the rich ores of Brazil, South Africa, India, and
the USSR. The manganese is generally attributed
to submarine springs associated with volcanoes,
but underwater leaching of submarine volcanic
rocks may be more important.

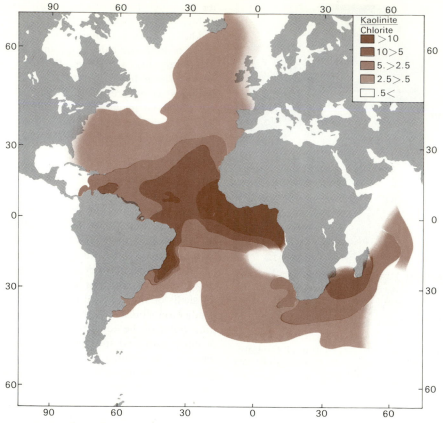

Figure 17-25
Ratio of kaolinite to chlorite in the pelagic sediment of the Atlantic Ocean, as inferred from X-ray studies of the grains less than 0.002 mm in diameter. [Modified from P. E. Biscaye, *Geological Society of America Bulletin*, 1965.]

Pelagic Sediments in the Geologic Record

Claystones so similar to pelagic red clays (fine-grained, iron-rich, containing manganese nodules and fragmental fossil foraminifers and Radiolaria) that some geologists regard them as true pelagic deposits, have been found in the Alps, in Barbados (West Indies), and Timor (Indonesia). The rocks are closely associated with coarse sediments generally considered shallow-water deposits, so that a truly pelagic setting has been often questioned. Should the interbedded coarser deposits turn out to be turbidites, however, the matter would be in a new light. The development of the electron microscope has permitted the recognition of coccolith ooze—definitely an abys-

sal sediment—as parental to some thick Jurassic limestones in the Austrian Alps (Fig 17-26). Doubtless other examples will be recognized elsewhere.

The Cretaceous chalk of England and France was once thought to be a consolidated deep-sea ooze, but its fossils are such as lived in the photic zone, not more than 200 m deep. The Cretaceous rocks of the Caspian region include thick chalks containing coccoliths, like some of the deep-sea oozes, associated with some blue-mud shales, not unlike some deep-sea muds. Shallow-water fossils have not been seen. The strong geophysical arguments against a continental plate being submerged to depth sufficient for true pelagic deposition make the interpretation doubtful but certainly do not rule it out.

Figure 17-26
Transmission electron micrograph of Coccolith limestone, partly recrystallized. Oberalm Beds, Upper Jurassic, Austrian Alps. [Courtesy of Robert Garrison, University of California, Santa Cruz, California.]

DOLOMITE AND LIMESTONE

The carbonates make up about 20 percent of the sedimentary rocks. As we have just noted, calcareous ooze covers nearly half the ocean bottom, but it is apparently rare on the continents. Most limestones are formed in shallow water, as is shown by the coarse clastic texture of some, by the cross-bedding and other indicators of strong currents in others, and by the shallow-water fossils in life positions of nearly all. Among modern shallow-water sediments are many analogs of these limestones, but modern dolomites are few, and those that are forming do not resemble the most abundant of the ancient ones.

Chemical Conditions of Carbonate Deposition

Calcite is among the most soluble of minerals. Though only slowly soluble in pure water it dissolves readily in water containing bicarbonate ions. Though long disputed, it is now generally agreed that normal sea water is about saturated with calcium carbonate at 20°C. Warming the water decreases the carbon dioxide content and thus favors precipitation of calcium carbonate, whereas decomposition of organic matter, which produces carbon dioxide, increases the solubility. We have just seen that cold deep-sea water is undersaturated with calcium carbonate.

Life processes modify the carbon dioxide content of the sea and thus the deposition and solution of carbonates. Many organisms secrete shells of calcite and aragonite (a mineral of the same composition but of different crystal structure). Photosynthesis by marine plants extracts carbon dioxide from the water and favors precipitation of the carbonates. Over the shallow Bahama Banks several processes contribute to lowering its solubility: (1) as water flows onto the Banks to compensate for the high evaporation, it is warmed, lowering its carbon dioxide content;

(2) photosynthesis further lowers it; (3) the abundant bacteria in the shallow carbonate muds include species that combine nitrogen with hydrogen to produce ammonia, a weak alkali that neutralizes some of the free hydrogen ions and also favors precipitation of calcium carbonate. These processes thus cooperate to bring the water to saturation; fine needles of calcium carbonate crystallize and settle to the bottom, there to accumulate as limy mud, crowded with remains of the microorganisms of the Banks (Figs. 17-3, 17-27).

Many lime-secreting plants and animals flourish in the littoral zone of tropical seas. Shells, whole or fragmented, make up many beaches and accumulate in shallow water to be cemented into limestone. Many ancient limestones are also made of fossil fragments and presumably were formed in like manner. Putrifying dead organisms form amino acids that tend to dissolve calcite.

Figure 17-27
Fossil reef in the Ellenburger Limestone (Late Cambrian–Early Ordovician), Central Mineral Region, Texas. [Photo by Preston E. Cloud, University of California, Santa Barbara, California.]

Calcite dissolves and reprecipitates so readily in water that many microscopic textural features are corroded; doubtless many limestones showing no obvious organic fragments were nevertheless originally composed of shell fragments.

Dolomites

Dolomite, the double salt of calcium and magnesium carbonate, presents one of the great geologic enigmas. Few organisms secrete shells containing appreciable magnesium carbonate, and in the few that do the ratio of magnesium to calcium is far short of that in dolomite, where it is nearly 1:1. Nor is inorganic dolomite being deposited now in the sea. Yet dolomite, though extremely rare among the Tertiary rocks, is both widespread and abundant in rocks of the long interval from middle Precambrian through the Mesozoic. Does the Principle of Uniformitarianism fail us here?

Around some igneous intrusions, large volumes of limestone have been altered to dolomite, as shown by their lateral passage, either gradual or abrupt, into unaltered limestone away from the intrusive. This dolomite shows definite association with faults and fissures; it is clearly a product of reaction of limestone with ascending hot solutions. No such origin is possible for the great sheets of early Paleozoic dolomite that can be traced for scores or even hundreds of kilometers — some of them interbedded for the whole distance with unaltered calcite limestone and shale.

On wide mud flats in Australia, the Netherlands, West Indies, Andros Island (Bahamas), and along the Persian Gulf, dolomite is forming as a replacement of slightly older aragonite and calcite muds. In each place the topographic setting is similar; the mud flats, just above normal high tides, are flooded only when spring tides or onshore winds are favorable. These localities are all in the tropics, where evaporation is high. Chemical experiments and field observation agree in showing that when sea water is evaporated, the first mineral to crystallize from the bittern is calcite. This is followed by gypsum ($CaSO_4 \cdot H_2O$). Evaporation on the mud flats eventually produces a bittern in which the ratio of Mg to Ca may be 30:1 or 40:1 — a highly concentrated, magnesium-rich solution of high density that sinks into the mud and there replaces about half the calcium ions with magnesium, producing a dolomite with some calcite in solid solution with it. It seems that we have here modern examples that go far to explain the frequent association of dolomite with beds of evaporites, such as salt, gypsum, or anhydrite.

Such examples do not, however, explain the origin of the huge dolomite sequences that are not associated with evaporites and for many kilometers are interbedded with quite unaltered limestone. This problem has puzzled geologists for more than a century. In 1904 a Royal Society expedition investigating the atoll problem drilled a core well on the atoll of Funafuti, in the Ellice Islands. All material cut by the drill (about 360 m deep) was coral reef rock like that at the surface; it contained coral heads, algal deposits, and shells of many reef-dwelling animals. Although the organic structures were well preserved, the magnesium content increased with depth, and below about 190 m the rock was nearly all dolomite. It was suggested that the replacement was by magnesium-enriched water from the shallow evaporating lagoon, but whatever the cause, the replacement was highly irregular and undoubtedly controlled by local conditions. This was demonstrated by the fact that a hole drilled to 640 m on Bikini atoll in the late 1940's found no dolomite whatever. The lowest beds cut at Bikini are of Miocene age and must have been in contact with sea water for 15 m.y. without magnesian enrichment.

We are thus left with no satisfactory theory for the origin of extensive dolomites not associated with evaporites. It is true, however, that most of these strata show signs of shallow-water origin, making plausible some former surficial concentration of magnesium-rich bitterns, even though the gypsum that may originally have been associated with them has since been dissolved. Striking it is, though, that in Nevada, thick Silurian and Devonian dolomite strata deposited in a basin a thousand kilometers across gradually change and become limestone within 50 km of the old shore line — precisely opposite to what might be expected. The dolomite problem is indeed an enigma yet unsolved.

STRATIGRAPHIC SYNTHESES

The Principle of Uniformitarianism is the key to an understanding of the past; were we able to determine the environmental conditions of each

396

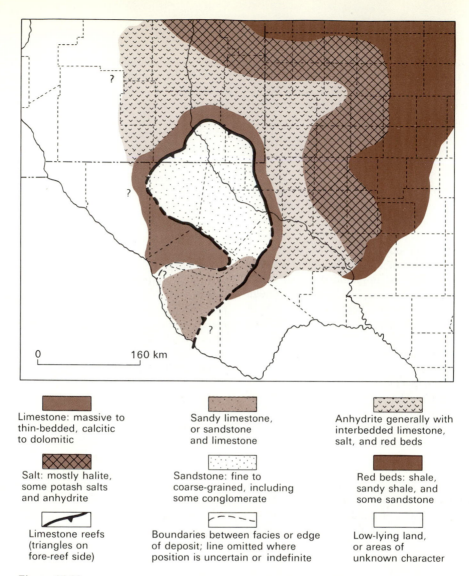

Limestone: massive to
thin-bedded, calcitic
to dolomitic

Sandy limestone,
or sandstone
and limestone

Anhydrite generally with
interbedded limestone,
salt, and red beds

Salt: mostly halite,
some potash salts
and anhydrite

Sandstone: fine to
coarse-grained, including
some conglomerate

Red beds: shale,
sandy shale, and
some sandstone

Limestone reefs
(triangles on
fore-reef side)

Boundaries between facies or edge
of deposit; line omitted where
position is uncertain or indefinite

Low-lying land,
or areas of
unknown character

Figure 17-28
The Paleogeography of part of Middle Permian time in western Texas and southeastern
New Mexico. [After P. B. King, 1942, with permission of the American Association of
Petroleum Geologists.]

formation and its relation to others, we could construct a segment of ancient geography. Tracing a succession of ancient geographies is the task of **historical geology,** most of which is beyond the scope of this book, but we illustrate the reasoning behind it by three examples, applying some of the data reviewed here and in previous chapters about sedimentational environments. Many economic mineral deposits are confined to specific geologic settings; an understanding of their origin is commonly necessary for their intelligent ex-

ploitation; the key to the past may unlock the door, not only to scientific understanding, but to economic success as well.

The Permian Basin of the American Southwest

Permian rocks are widely exposed in the "Permian Basin" of western Texas and southeastern New Mexico—a huge area, rich in petroleum, natural gas, and potash salts. Hundreds of geologists over the past 60 years have studied Permian geography, aided by many thousand wells and excellent outcrops in this arid region. The results are summarized in Figure 17-28. The weakly stippled area is the site of the Delaware Mountains, with the Delaware Basin to the east. The mid-Permian rocks are well exposed in the mountains, but are buried by younger rocks in the Basin, where, however, they are well known from well borings as fine-grained sandstone interbedded with very dense limestone. The even lamination of both rocks suggests deposition in quiet water.

On all sides except the south, the sandstone area is bordered by thick massive fossiliferous beds of limestone with which the thin limestones of the basin interfinger. Near the zone of interfingering are huge beds of breccia containing large blocks of the same kind of limestone as the thick beds. The breccias lie at the foot of long layers that slope upward from the basin at angles of 10° to 30° for several hundred meters before flattening out and merging into a sequence of thin-bedded, well-laminated limestones forming a flat bench nearly surrounding the basin. These limestones are highly fossiliferous and total more than 800 m in thickness.

These relations are interpreted as follows: the thick massive limestone beds are a fossil reef that grew on a gently sloping bottom. The breccias are relics of storm-broken reef that rolled down the reef front into the basin, whose thin sandstones and limestones are deep-water strata. The upper set of flat-lying limestones was formed in the quiet water of back-reef lagoons (Figure 17-29). Thus the sloping curve of the massive reef limestone is considered an original feature of the growing reef, and the amount of its rise

is a rough measure of the difference in depth of basin and back-reef lagoons. This implies that the basin deposits were formed at a depth of about 400 m.

Many corollaries of this interpretation support it. If the reef were a fringing reef (whose most conspicuous outcrop is Guadalupe Peak; Fig. 17-30) bordering the ancient land nearly surrounding the basin, we can understand why the back-reef limestones become poorer and poorer in fossils inland from the reef, and finally interfinger with beds of anhydrite, halite, and red shale, as might be expected on a low-lying coastal plain in an arid or semiarid climate. Still farther north, in the Texas and Oklahoma panhandles and eastern New Mexico, the limestones give way to red sandstones and shales containing reptilian fossils—clearly stream deposits of a semi-arid climate. Southward, relations are obscured by overlying Cretaceous rocks, but well records suggest that the sand of the basin deposits was derived from what is now the Big Bend country of Trans-Pecos Texas.

Interpretation of this ancient geography has proved to be of great economic significance. The highly porous reef limestones are, in some areas, prolific reservoirs of petroleum, and the salt deposits are locally rich in potash salts, valuable for fertilizers and other chemicals. Determining the relations of the several formations has been a fascinating detective game, by no means over, and the prizes have been vast mineral wealth as well as knowledge.

The Old Red Continent in Europe and America

Prominent in the history of geology is the Old Red Sandstone of England and Scotland. More than a century ago, the British geologist Hugh Miller attracted thousands of amateurs to geology by his reconstructions of the conditions of origin of this formation. As the name implies, the Old Red is made up chiefly of sandstone stained red by iron rust. The sandstone consists of quartz and much relatively unaltered, poorly sorted feldspar, and with lenses of coarse boulder conglomerate and breccia alongside wedges of fine silt and clay partings. Some clay beds contain cubical casts resembling salt crystals in form. In these and associated beds many fish skeletons are preserved. These fish had lung-like breathing organs and are

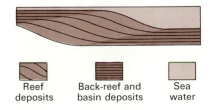

Reef deposits | Back-reef and basin deposits | Sea water

Figure 17-29
The relation between basin, reef, and back-reef deposits as interpreted from the Middle Permian rocks of the Permian Basin. [After P. B. King, 1942, with permission of the American Association of Petroleum Geologists.]

Figure 17-30
The great reef deposit of Guadalupe Peak. Oblique aerial view of El Capitan and Guadalupe Peak from the south. Southern Guadalupe Mountains, Culbertson County, Texas. [Photo by U. S. Army Air Force.]

inferred to have resembled the so-called lungfish of modern Australia that can survive long periods of drought by burying themselves in moist mud. Moreover, some of the complexly cross-bedded sandstones with sparse faceted pebbles closely resemble modern dune sands with desert pavements. The current bedding in the coarser layers, the dominantly mechanical rather than chemical weathering in the source area, the evidence of wind at work and of playas in which salt sometimes crystallized—all these together seem to imply that the Old Red is a desert deposit, recording conditions like those of today in the Mohave Desert or in western China.

To the south, in northern Devonshire, some of these red sandstones are interbedded with thin limestones that contain marine fossils, and still farther south the sandstone grades into shale that also contains marine fossils. This is the type area of the Devonian System, and one of the early triumphs of systematic stratigraphy was the recognition that these marine beds correlate with the continental Old Red farther north. Between them the shoreline must have fluctuated in position over a belt some scores of kilometers wide.

The Old Red Sandstone crops out in the Hebrides, and similar rocks, containing the same kind of fish and plant fossils are found in southern Norway. Interfingering marine and continental rocks like those of northern Devon are found in the Ardennes, in Poland, and in Czechoslovakia. In the Alps to the south, the Devonian strata are all marine (Fig. 17-31). Across the Atlantic the

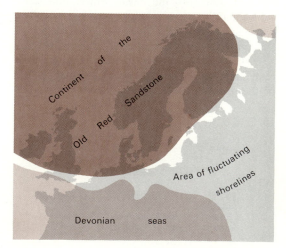

Figure 17-31
The Old Red Sandstone continent and its bordering seas. [After *Stratigraphic Geology* by Maurice Gignoux. W. H. Freeman and Company. Copyright © 1955.]

Devonian rocks of New Brunswick strongly resemble the Old Red Sandstone; to the northwest they interlens with marine beds just south of the St. Lawrence. We see here strong support for the thesis that the two sides of the North Atlantic were once joined, before that ocean formed.

A Miocene Geography in California

In the San Joaquin Hills, southeast of Los Angeles, California, a large area is underlain by a thick diatomaceous shale, the Monterey Shale, which contains, besides myriads of diatoms, many foraminifers and, in some sandy beds, marine mollusks. The very thin and evenly laminated strata testify either to deep water or a sheltered position, for they show little or no sign of bottom disturbance, either by burrowing animals or waves. Northeastward the shale is interbedded with thick, poorly sorted sandstones whose composition suggests derivation from the granitic rocks nearby to the north and east. A few of these sandstones extend well out into the shale body and suggest, by their molluskan fossils and bedding that they are nearshore deposits. Some interfinger, near the center of the hills, with a very unusual rock, the San Onofre Breccia, made up almost wholly of the debris of glaucophane schist and related metamorphic rocks. (Glaucophane is a blue amphibole, readily recognized with a hand lens or under the microscope.) In the breccia are great boulders and angular blocks a meter or more across (Fig.

Figure 17-32
An old mudflow, part of the San Onofre Breccia. [Photo by Wright M. Pierce.]

0   10 meters

400

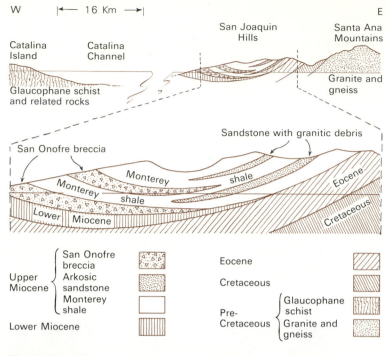

Figure 17-33
Map and sections showing the relations of the San Onofre Breccia to the Monterey Shale. [After A. O. Woodford.]

17-32). Nearly all the recognizable fragments are of glaucophane schist, and the matrix is glaucophane-rich silt. Although it interfingers with glaucophane-bearing marine rocks to the northeast, it does not contain marine fossils. It is poorly or not at all sorted; most of its 750-m thickness is massive. It seems to represent a series of mudflows that slid from a land of high relief and flowed out into a protected seaway.

A most significant feature of the breccia is its composition. To the east no glaucophane schists, readily recognizable as they are, are known nearer than Wales, Corsica, and the Alps. But such rocks are well exposed to the west on the Palos Verdes peninsula and on Catalina Island, not far offshore. The San Onofre Breccia unquestionably came from a landmass to the southwest, now almost buried beneath the Pacific (Fig. 17-33). The San Joaquin Hills thus expose

deposits of a northwesterly trending arm of the sea, sheltered by a highland to the southwest, that received sediment from both northeast and southwest. Perhaps the climate was semiarid; not only do the mudflows suggest this, but there is almost no clay in the Monterey Shale. Did the paucity of streams on a semiarid landscape hamper sorting on both sides of the channel? Critical exposures are lacking to evaluate this suggestion.

The former highlands to the southwest are all but sunken, with only local remnants still visible on the islands and submarine ridges. We are not simply dealing here with a Miocene flooding of the continental margin, but with a series of much more complex earth movements; some areas once high above the sea have sunk far below; other areas have gone through completely opposite movements.

Questions

1. Why is the systematic distribution of sediments in lakes not paralleled by that in the ocean?

2. Assuming that turbidity currents build fan-like deposits on the ocean floor at the mouths of submarine canyons, how would you expect the form of such a deposit to differ from that of a normal fan? How would its composition and internal structure differ?

3. Why are dune sands generally better sorted than water-laid sediments of the same average grain size?

4. If relief in the source area is an important factor in sedimentation, how would you expect the marine sediments of California to compare with those off Texas?

5. How would you expect first-cycle sandstones to differ from reworked sandstones?

6. The St. Lawrence carries much more water to the sea than the Colorado. Why does it end in an almost unsedimented estuary whereas the Colorado has been building a huge delta since Pliocene time?

7. Aside from the temperature control of the life cycle of reef corals, would you expect greater, lesser, or similar proportions of calcium-carbonate deposits and noncarbonate deposits in arctic or tropical waters? Why?

8. How would you expect graded bedding formed by a steadily weakening current to differ from that formed by settling of a turbid flow?

9. Why does the paucity of clay in the San Onofre Breccia suggest an arid landscape?

Suggested Readings

Gignoux, Maurice, *Stratigraphic Geology*. San Francisco: W. H. Freeman and Company, 1955. [Chapter 1.]

Kuenen, P. H., *Marine Geology*. New York: John Wiley & Sons, 1950.

Marr, J. E., *Deposition of the Sedimentary Rocks*. Cambridge: Cambridge University Press, 1929.

Shepard, F. P., *Submarine Geology*. New York: Harper and Bros., 1948.

Society of Economic Paleontologists and Mineralogists, *The Filling of the Marathon Geosyncline*. Publication 64-9, 1964.

CHAPTER **18**

Mineral Resources

THE INDUSTRIAL REVOLUTION

Our world differs more from the world of 1800 than that did from the world of Alexander the Great. In 1800 nearly four-fifths of the population of Great Britain and Italy lived on farms, as did more than nine-tenths of those in the rest of Europe. Today a Zulu miner travels third class to his labor compound in the Rand goldfield in greater comfort than Louis XIV did in his state coach between Versailles and Paris. By our standards, the Zulu's lot is hard and his pay pitiably small, yet he is better clothed and fed than most of the people of George III's England. He is fortunate, indeed, compared to a slave in the mines of Laurium, whose life expectancy beneath the Athenian lash was only four years, but whose labors produced the silver that sustained the Golden Age of Greece.

Most of us now reject the slave-holding philosophy of Plato and Pericles, but it is not primarily ethical principles that account for the differences between the ancient and the modern worlds. Material goods are perhaps as unevenly distributed today as under most of the cultures of the past, but the standard of living—of the Western World, at least—is almost incredibly higher.

The change began with two events of the eighteenth century. Neither attracted as much notice at the time as the intrigues of Bonnie Prince Charlie or the campaigns of Frederick the Great. But about 1730, a Shropshire Quaker, Abraham Darby, discovered how to use coke in iron smelting, and in 1768 James Watt invented the steam engine. Over the next few decades cheap iron and mechanized power created the industrial age. Without machinery, population would long since have outstripped food supply the world over, as Malthus predicted in 1798, and indeed, as it has in many parts of the less-developed world.

Now, as always, agriculture is the basic industry. But a wholly agricultural economy imposes sharp limits on division of labor and the increased productivity this allows. As transport improved, first with iron rails, then with locomotives and steam-driven ships, a specialization formerly unknown made possible tremendous savings in labor. By 1830 a twelve-year-old girl, operating a machine loom in a Lancashire mill, could turn out 35 yards of cloth daily—in a year, enough to clothe about 1200 persons.

The little girl's existence was doubtless as dismal as any Norman serf's, and even now the "better life" is beyond reach for many. But the ideal of a "better life for all" would be pathetically ludicrous had there been no industrial revolution. Even now the food supply of much of the world is below subsistence levels; it would be appallingly worse if we were to revert to the economy of 1800.

These facts are commonplace and generally accepted. What is not so widely understood is that all these changes in living standards ultimately depend on the world's diminishing and nonreplenishable assets—its mineral resources. And the continuity of our standard of living is far from assured; our energy supplies are already being sorely strained; without ample energy the capacity of our economy is bound to fall in view of the population explosion.

THE MINERAL BASIS
OF CIVILIZATION

Throughout history mineral resources have played a greater role than is usually recognized. Today this role is second only to agriculture's. The relationship between mineral wealth and national power, even in ancient times, can be clearly traced, though most historians ignore it.

The Greeks who turned back the Persian hosts at Marathon carried metal swords and shields; many of the enemy had only leathern shields and stone weapons; the Greek fleet at Salamis was built by the Athenian profits from the silver and lead mines of Laurium, discovered only a few years earlier. These profits also paid the mercenaries who fought Athen's battles in the Peloponnesian Wars; with the exhaustion of the mines came Athen's downfall. Philip burst from the wild Macedonian mountains, and his son Alexander the Great swept over the world, financed by the flush production of gold—roughly a billion and a half dollars in modern equivalent—from the new-found mines on Mount Pangaeus. When Scipio drove the Carthaginians from Spain and won for Rome the gold, iron, copper, silver, and mercury of the peninsula, he sealed the fate of Carthage.

These are but a few examples from preindustrial days. Today, mineral resources and national power are even more closely linked. Gold and silver could hire mercenaries and support campaigns, but useful goods could not be created from them; they merely gave control of the few goods then available to one group rather than another. They still possess this conventional value, but living standards and national power depend on them only incidentally. Most useful goods depend on the mineral fuels and the industrial metals—iron, copper, aluminum, lead, and others. It was no accident that Britain was able to maintain the *Pax Britannica* throughout the nineteenth century; her industrial and military supremacy came from the happy fortune that her "tight little island" held a greater known mineral wealth per hectare than any similar area on earth, together with a population intelligent and aggressive enough to exploit it.

At one time or another in the nineteenth century, Great Britain was the world's largest producer of iron, coal, lead, copper, and tin. From these came her machines, her mills, and the great cities founded on them. Before 1875 she had built more miles of railroad than any of the much larger continental countries. Her flourishing internal markets and her manufactured products, carried to all the world in the British merchant marine, brought her the greatest wealth any country in history had ever enjoyed. True, the cheap foodstuffs she received in return eventually ruined the island's agricultural economy, but her favorable trade balance enabled British capital to control Malayan tin, Spanish iron, and many of the best mines and oilfields of Mexico, Chile, Iran, Australia, Burma, and the United States. These holdings saw her through one world war and maintained her credit through a second. When her flag followed her mining investments into South Africa and the Boers were defeated, she gained control of more than half the world's production of new gold, and ultimately of great deposits of copper, chromite, diamonds, asbestos, and manganese. The independence of South Africa has only slightly diluted control; the African mineral holdings are among the most valuable sources of British capital to this day.

Nowhere better than in the United States can be seen the cardinal significance of minerals to living standards and national power. Before 1840 manufacturing was inconsequential, and only high tariffs made it possible to compete with the advanced British industries. The small, scattered iron deposits along the eastern seaboard did, it is true, supply enough of the local demand to cause the British parliament in 1750 to forbid their further exploitation; after independence, growth was slow: as late as 1850 iron production was only about half a million tons annually. In 1855 the "Soo Canal" brought the rich Lake Superior iron deposits within economic reach of the Pittsburgh coal; by 1860 iron production had trebled, and by 1880 it had passed that of Great Britain.

The Civil War was won by the greater productivity of Northern industry, the greater weight of armament and supplies flowing over a far superior railway net. The Tredegar iron works at Richmond was the only one worth mentioning in the Confederacy, and it could not compete with the overwhelming output of the Pennsylvania furnaces.

Its huge internal market, prodigious endowment of many of the minerals basic to manufacturing, and favorable agricultural heritage have all contributed to making the United States one of the two most powerful countries in the world at present. During the Battle of the Bulge in World War II, the Allies hurled more metal at the Germans than was available in all the world in Napoleon's time. Cannon and tanks, as well as plowshares and tractors, are made of metals, and all are transported by mineral fuels.

ENVIRONMENTAL IMPACT OF A MINERAL ECONOMY

A most important element in man's many impacts on the environment is his exploration and exploitation of its nonrenewable resources, its mineral deposits. Each year man digs from the crust rocks equivalent to 8 tons per person on the globe; of this, 2.5 tons is waste, 3 tons is construction materials, 1.7 tons is mineral fuels, and the remainder is mineral concentrates from which 140 kg of metals and 160 kg of nonmetallic minerals are obtained. One can imagine the size of the excavation if this were all obtained from a single mine: it would be at least 1 km wide and deep and 32 km long—a considerably greater excavation than that for the Panama Canal.

The most striking thing about this exploitation of the earth's mineral resources is the rate of its growth. Since 1900 mineral consumption has multiplied 40 times, a per-capita growth of 10 times. The amount is 30 times all the mineral production from the beginning of the Industrial Revolution to 1900. Since 1950, a mere quarter century, the production has been more than half that of all pre-1900 history.

But this expansion cannot possibly continue indefinitely. Huge as it is, the earth is still a finite body. We rediscover this fact with every major war, when supplies of "strategic minerals" —a different list for each belligerent—are diminished or cut off completely. We found it true in 1974 when energy supplies in the form of fossil fuels were inadequate and restraints were put on highway speeds just as the super-highway net of the United States was approaching completion!

SALIENT FEATURES OF MINERAL RESOURCES

Economic mineral resources are concentrated in relatively small areas; they are exhaustible and irreplaceable. These facts—of the greatest social and political implications—are often, to the peril of the public welfare, overlooked by those unfamiliar with the mineral industry. Their social effects are so profound that no citizen should fail to be aware of them and of the geologic factors determining them.

Sporadic Distribution

As emphasized later in the chapter, mineral deposits of economic value are all essentially "freaks of nature." An abnormal pituitary may make a man a giant, although he is otherwise normal. Similarly, mineral deposits result from normal geologic processes under various unusual conditions. Only a few geologic environments favor the formation of mineral deposits.

The favored spots are by no means evenly distributed over the earth. Nearly a third of the world's nickel produced in 1971 came from less than a score of mines in the Sudbury district of Ontario. This district at one time furnished nine-tenths of the world supply; it is still highly productive, but new mines have had to be opened in western Australia, New Caledonia, and a few other spots. The only production from the United States is from southwestern Oregon, which supplied only a little more than 10 percent of the national consumption. A single mine at Climax, Colorado, furnished more than 60 percent of the world production of molybdenum. Nearly a quarter of the copper produced in the United States since 1880 has come from an area of less than 10 km^2 at Butte, Montana, an area that has supplied about 8 billion dollars worth of metals to the American economy. The Rand goldfield of South Africa produces half the new gold of world from an area about 80 km by 30 km.

The mineral fuels are less localized, but they underlie only trivial parts of the continents. Less than 25 percent of Pennsylvania, a leading coal state, is underlain by coal. The East Texas oil field, the greatest thus far found in the United States, with the possible exception of the still-unexploited Prudhoe Bay field, Alaska, covers an area about 15 by 65 kilometers, but for several years it yielded about a quarter of the nation's oil. The most extensive oil field in the world, the El Nala anticline in Saudi Arabia, is almost continuously productive for 175 km but it is less than 10 km wide.

The economic implications of this unequal distribution of the mineral fuels are great. Modern chemistry may be able to make a nylon purse out of a sow's ear (perhaps better than a silk one), but only with the expenditure of energy. Today, this means mineral fuels, with minor water and atomic power. The dreams many countries have of emulating the industrial development of the United States are foredoomed to failure because they lack adequate sources of cheap fuel. It is this fact that explains the tremendous interest of all nations in the prospect of developing cheap atomic power. If such power could be obtained safely, the economics of many nations might come abreast that of the United States. Some countries, such as the Scandinavian, have higher educational and health standards or, like Argentina, a higher agricultural output per capita, but none is now so fortunate as the United States in its combination of high average education (hence a skilled work force), great agricultural productivity, and a nearly balanced supply of mineral resources. But the supply of minerals—thus far so nearly balanced—is finite, and many are already in short supply. No nation, with the probable exception of the USSR, is completely self-sufficient in all industrial minerals.

Exhaustibility

All mineral deposits are limited in size; they represent unusual concentrations. Once the ores are mined or the oil withdrawn, all that is left are holes in the ground. This is the fate of all mines, even of the greatest.

True, the mines of Almaden, Spain, have yielded mercury since the days of the Carthaginians and still hold the richest known reserves of this metal, but these deposits are nearly unique. The mines of Cornwall, the "Cassiterides" of the ancients, supplied tin to the Phoenicians and provided varying amounts of it throughout the years until about fifty years ago. The recent high price of tin has, it is true, made economical the reopening of some mines, but it is obvious that the venture is marginal unless the price of tin continues to rise, and in any event merely postpones the inevitable depletion of the deposits. The world's greatest single oil well, the Cerro Azul No. 4, in the Tampico field, Mexico, yielded nearly 60 million barrels of petroleum in a few years, then suddenly gave forth only salt water.

The still–fertile Valley of the Nile has been the granary of the Mediterranean throughout most of recorded history, and huge areas in China and India have been farmed nearly or quite as long. The forests of Norway that built the Viking ships still produce lumber. But the mines of Freiberg, where the world's leading mining academy of medieval times was founded, have long been abandoned. Belgium and Wales, with their cheap coal and high metallurgical skills, are still centers of smelting, but nearly all the mines on which the industry was founded have been exhausted for generations. Potosi, which supplied tons of silver to the Viceroyalty of Peru, the fabulous Comstock Lode of Nevada, and the copper mines of the Upper Peninsula of Michigan are not quite dead, but they are pale shadows of their former greatness. *There is no second crop of minerals!* And lest the economic, political, and social meaning of this statement be underevaluated, more metal has been mined in the past 30 years than in all of preceding history.

This expansion cannot go on forever. The earth is a finite planet, and growth of the mineral industry must eventually stop, and decline set in. Those economists who claim that with increasing prices it will be possible to work ores of lower and lower grade are, of course, right for the very short haul, but the working of lower-grade ores always demands more energy. And energy resources, like all things on the planet, are also finite; growth in their use cannot continue indefinitely.

What the exhaustibility of mineral resources means socially can be seen in the long roll of western "ghost towns," in which a few families or none now live where thousands lived before. More pathetically, it is seen in the "distressed" coal mining towns of England, Wales, and Ap-

Table 18-1
Value (1972 dollars) of world mineral production in the past, with an estimate for the year 2000.

Year	Population, in millions	Mineral production, in billions of 1972 dollars	Per capita value, in 1972 dollars
1770	700	1.3	1.80
1900	1550	13.3	8.50
1970	3680	166	45
2000	6500	520	80

SOURCE: Data from Sutulov, 1972.

Table 18-2
World production of minerals and mineral fuels.

Year	Coal, tons × 10^6	Oil, barrels × 10^6	Gas, cubic meters × 10^9	Pig iron, tons × 10^6	Steel, tons × 10^6	Aluminum, tons × 10^3	Copper, tons × 10^3	Nickel, tons × 10^3
1900	763	166	small	40.8	27.8	7.3	470	6.26
1970	2983	16,690	900	434.5	5904	10,320	6688	666.6

palachia, where, although the mines are not, strictly speaking, exhausted, the higher costs of deeper mining, of pumping ground water from greater depths, and of longer hauls from coal face to portal have weakened their competitive position in world trade. Unemployment, lower wages, and lower living standards have followed.

The fantastic growth of the mineral industries since the start of the industrial revolution is shown, along with an estimate of growth to the end of the century, in Table 18-1. The estimate for the year 2000 is neither extremely low nor high. It is obvious that such growth cannot continue at the rates implied for any extended time.

Optimists point to the development of nuclear fuel as a future energy source, and this may indeed considerably relieve the energy shortage, but not until a "breeder reactor" is developed, for the supply of uranium is not great, and considerable problems of thermal contamination are involved. The growth in production of some of the most critical mineral fuels since 1900 is shown in Table 18-2.

ECONOMIC IMPORTANCE OF MINERAL RESOURCES

Although mineral resources account directly for less than 4 percent of the Gross National Product, they play a vital role in industry, for all heavy manufacturing depends on them. Nearly two-thirds — in some years nearly three-fourths — of the value of mineral production in the United States is supplied by the energy sources — the **mineral fuels**, petroleum, gas, and coal. Mineral resources also include the **metalliferous deposits**, the sources of our metals, and the **nonmetallic deposits**, such as building stone, cement rock, clay, sand and gravel. Of these the nonmetallic deposits normally are somewhat more remunerative than the metals. From 1940 to 1971 the total annual value of mineral products in the United States ranged between 6 and 30 billion dollars, though part of the increase is due to inflation, not corrected for.

The Mineral Fuels

Fuel is the most important mineral resource, essential for heat, power, metal refining, and as sources of many useful chemicals and of nitrogen fertilizers. The mineral fuels are especially important to the geological profession, for the search for oil is the principal work of about half of all geologists.

The Industrial Revolution was based on coal. It is still a basic fuel, though petroleum has largely replaced it in the field of transport and natural gas has made great inroads in power plants and metallurgy — accounting for only

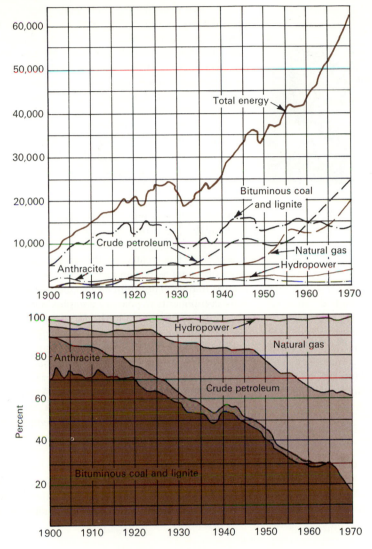

Figure 18-1
Percentage of total production, in British Thermal Units equivalent, of mineral and hydropower energy sources in the continental United States, 1900 to 1970. [From U. S. Bureau of Mines Mineral Yearbooks.]

about 12 percent of the energy consumption in 1940 and increasing to more than 33 percent in 1971. Since the total energy consumption has more than doubled during the same time, the natural gas industry has grown by a factor of about 8 (Fig. 18-1). As a result, coal, which supplied about half our energy requirements in 1940, supplied less than 18 percent in 1971. In absolute amount, however, the output was nearly the same because of the great growth in energy supply.

In 1970 world production of coal, exclusive of the Communist states, was 2983 million tons, of which the United States produced 555 million tons. World production of petroleum that year was 16,690 million barrels; the United States' share was 3517 million barrels. The great producers were the USSR, Venezuela, and the Arabian and Iranian principalities around the Persian Gulf.

Coal

Coal is a brownish-black to black combustible rock that forms beds ranging from a few centimeters to as much as 30 m or more thick, interbedded with other sedimentary rocks (Fig. 18-2). In the western Powder River Basin of Montana,

Figure 18-2
Coal beds on Lignite Creek, Yukon region, Alaska.
[Photo by C. A. Hickcox, U. S. Geological Survey.]

one coal bed is fully 60 m thick. Many beds may be found in the same stratigraphic sequence; in West Virginia, 117 different beds are of enough economic interest to have been named. Such sequences of **coal measures** commonly include many alternations of marine and nonmarine beds; the coal beds are almost all nonmarine. They are composed chiefly of flattened, compressed, and somewhat altered remains of plants: wood, bark, roots (some in position of growth), leaves, spores, and seeds.

COAL RANK

Most coals have been formed of plant remains that accumulated in swamps where downed trees and fallen leaves were protected from an oxidizing environment. A continuous series exists,

from brown peat, in which the plant residues are only slightly modified, to a hard black, glistening type of coal that contains no recognizable plant remains. The principal members of this series are **peat**, **lignite**, **subbituminous coal**, **bituminous coal**, and **anthracite**. The more obvious features of all except peat, which is not considered a coal, are shown in Table 18-3.

In most coal fields the beds low in the stratigraphic section are of higher rank than those higher, but this is not invariable. Most students attribute rank to the weight of overburden or to tectonic pressure, though the exceptions indicate that the original composition of the beds has some effect on their later metamorphism. Carboniferous coals of the Moscow Basin are still lignite, never having been deeply buried.

When coal is heated in the absence of air, it gives off water vapor and hydrocarbon gases, the **volatile matter**. The ultimate product is **coke**. The plant components in peat are complex compounds of carbon, oxygen, and hydrogen. In the air they oxidize and rot away, yielding chiefly carbon dioxide and water, but if air is excluded in the stagnant water of a swamp or by burial beneath younger strata, they slowly alter into many solid products, as well as some gases. Among the solid products is finely divided black elemental carbon, a substance whose presence distinguishes coal from peat. The higher the proportion of elemental ("fixed") carbon and the lower the volatiles, the higher the rank of the coal. Clay and sand washed into the swamp while the coal accumulated remain as ash when the coal is burned, lowering the heating value and increasing waste.

Table 18-3
Distinctive features of coal of various ranks.

Kind of coal	Physical appearance	Characteristics
Lignite	Brown to brownish black	Poorly to moderately consolidated; weathers rapidly; plant residues apparent
Subbituminous coal	Black, dull or waxy luster	Weathers easily; plant residues faintly shown
Bituminous coal	Black, dense, brittle	Does not weather easily; plant structures visible microscopically; burns with short blue flame
Anthracite coal	Black, hard, usually with glassy luster	Very hard and brittle; burns with almost no smoke

Table 18-4
Estimated total potential coal resources of the world*, in billions (10⁹) of metric tons.

Continent	Known on basis of existing mapping and exploration	Probable additional in less–explored areas	Total potential
Asia and European USSR	6360	3670	10,000
North America	1560	2620	4,180
Europe, excluding USSR	560	270	830
Africa	70	140	210
Oceania	55	65	120
South and Central America	18	9	27
	8623	6774	15,367

SOURCE: Data from Paul Averitt, U. S. Geological Survey.
*Mostly less than 1200 m deep but includes small amounts between that depth and 1750 m in beds 30 cm or more thick.

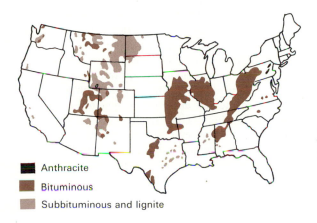

Anthracite

Bituminous

Subbituminous and lignite

Figure 18-3
The coal fields of the United States. [After Bituminous Coal Institute, 1948.]

Most bituminous and anthracite coals are of Carboniferous age. Such coals abound in Europe and eastern North America. In Europe most lie in a belt extending from Wales across Belgium, Luxembourg, and Germany into the Ukraine. This is the industrial heart of Europe.

COAL RESERVES

Coal is so abundant (Table 18-4) that generally only the thicker, more accessible, and higher-rank deposits are now being mined. An estimate of coal reserves means nothing unless the limits of thickness, depth (for it is expensive to mine deep underground), and quality of coal are stated. We avoid details but present Table 18-4, a careful estimate prepared by Paul Averitt of the U. S. Geological Survey. This estimate includes all anthracite and bituminous beds 36 cm or more

thick and all subbituminous and lignite beds 75 cm thick to a depth of 1750 m. Much of the coal of Montana, North Dakota, and Alaska is lignite; most of that in Wyoming, Washington, and Colorado is subbituminous; of the rest, nearly all is bituminous except for small tonnages of anthracite, in Pennsylvania, Colorado, Alaska, and West Virginia (Fig. 18-3). As seen in Figure 18-3, coal-bearing rocks underlie 14 percent of the contiguous United States; the total tonnage is between a fifth and a sixth of the world total.

OIL AND GAS

Earth oil (petroleum) and natural gas are found in similar environments, commonly together. Commercial accumulations exist only in special geologic conditions and are almost, but not quite, limited to sedimentary rocks. Producing areas are known as "pools," though oil and gas, like ground water, generally fill pore spaces in rocks rather than open caverns.

Formation of a commercial oil pool requires (1) a source rock, (2) a permeable reservoir rock which will yield the oil rapidly enough to make drilling worthwhile, (3) an effectively impermeable cap rock, and (4) a structure permitting the cap rock to retain the oil. We defer discussion of the source rock until we have briefly dealt with the other requirements.

Essential to a reservoir rock is the presence of connected pores through which a liquid can move. **Permeability** (Chapter 14) is thus the prime requirement, though **porosity** — storage capacity — is obviously important. Most reservoir rocks are sandstones, though some are carbonate rocks, either granular, with interstitial pores, or jointed and cavernous. Shattered brittle rocks, such as chert and basalt, and very rarely even granite and schist, are productive reservoirs in a few fields. Extremely fine-grained sandstones, many formerly nonproductive because of low permeability, have recently been made producers by "hydra-fracturing" — a process of injecting sand-charged liquids under such high pressure as to fracture the rock and inject the newly formed crevices with coarse sand to keep them open.

A **cap rock** must be virtually impermeable to oil and gas. Most are shale, but nonporous sandstone seals some reservoirs, and others are sealed by asphalt residues left by oil that escaped to the surface where its more volatile fraction has evaporated. Such residues are **oil seeps**; in Spanish countries, **breas**. Many commercial pools have been found by drilling through them.

Oil and gas are of course far less abundant than ground water, and being less dense, they float on it; their accumulation requires a **cap rock** over a **favorable structure** wherein the floating oil and gas are confined, generally in the highest part of the reservoir. Since gas is less dense than oil, it ordinarily rises to the top, although, under high pressures at depth much dissolves in the oil and is released only when the pressure is released. Some favorable structures are diagrammed in Figure 18-4.

Most oil fields occupy the crests of elongated **anticlinal folds**, often, but incorrectly, called domes (Figs. 18-4,A; 18-5). Examples are the Salt Creek Dome, Wyoming, the Kettlman Hills North Dome, California, the tremendous El Naha anticline in Saudi Arabia, and the prolific Bahrein Island anticline in the Persian Gulf. Less common are the **salt dome** fields, which lie above or beside plugs of massive salt that have plastically risen from depth and bowed up the surrounding strata, because of the weight of sediment pressing down on a deeply buried salt deposit. These abound along the Texas and Louisiana coast of the Gulf of Mexico, both on and offshore; the famous Spindletop field, one of the first found, is representative. Similar fields have been found

Figure 18-4
Structures favorable to the accumulation of oil and gas.
(A) Anticlinal fold with reservoir sand underlain by shale (possible source rock) and overlain by shale (cap rock). Note that the reservoir sand to the right is capped by asphalt, making a second trap. (B) Salt dome with oil at crest and along flanks. (C) Trap in sandstone, formed where an unconformity is overlain by shale. (D) Porous limestone reef reservoir in impermeable limestone and shale. (E) Fractured schist reservoir beneath domed shale.

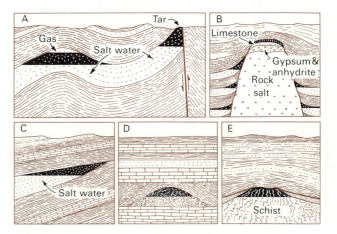

Figure 18-5
Anticline with oil well sited on its crest. East Los Angeles, California, 1931. [Photo by M. N. Bramlette, U. S. Geological Survey.]

along the Caucasus, in the Carpathian foothills of Rumania and offshore Nigeria. Finally, **stratigraphic traps** (Fig. 18-4,C) are reservoirs that are capped, generally unconformably, by blankets of relatively impermeable sediment. The greatest of all American oil fields, East Texas, is a stratigraphic trap formed by shale unconformably overlying and covering the outcrop of a gently dipping sandstone bed. The great Leduc oil field of Alberta lies in the porous interior of a fossil coral reef, covered by impermeable sedimentary rocks (Fig. 18-4,D). The small but interesting Edison Field, California, has fractured schist as the reservoir, into which oil must have seeped from adjacent sedimentary rocks (Fig. 18-4,E). Some traps, like the immense Hugoton gas field of Kansas and Oklahoma, contain gas but no oil.

Source Rock

The conclusions about reservoirs, cap rocks, and favorable structure are well-established generalizations based on numerous examples. Oil is found in permeable rocks confined by virtually nonpermeable ones and in structural positions determined by hydraulic laws. But the sources of earth oil are less surely inferred. Most (but not quite all) geologists think that *petroleum*

originates exclusively in sediments. Nearly all oil pools are in sedimentary rocks. The rare exceptions, such as the Edison Field just mentioned, could have received oil by migration from adjacent sedimentary rocks. Most pools are separated from the nearest igneous or metamorphic rocks by hundreds or thousands of meters of barren sedimentary rocks containing not a trace of oil.

Many oil fields, too, are in or near thick accumulations of marine or deltaic sedimentary rocks that include cubic kilometers of shales that generally contain several percent—exceptionally as much as 80 percent—of organic matter, chiefly compounds of carbon originating from the bodies of former plants or animals. These organic shales were probably deposited as marine ooze, for modern oozes contain similar compounds.

The carbon dioxide and water that went to form the organic compounds must have been synthesized by plants, perhaps largely the very abundant diatoms. The organic compounds thus produced may then have been repeatedly worked over in the digestive tracts of many kinds of animals on the sea floor before final burial. The bodies of dead animals also contribute.

Even after burial, the organic residues in the ooze are further altered by bacteria, with partial elimination of oxygen. Some oozes probably accumulated in stagnant basins too poor in oxygen to permit bacterial transformation of the hydrogen-carbon residues into water and carbon dioxide again. Many geologists think that most oil is formed in this way, for many oilfields are associated with strata suggesting such environments. Yet the transformation of organic matter into liquid oil is still poorly understood. It must have taken place under a load of strata, and the pressure and rise of temperature resulting from burial may have been important factors.

The tentative prehistory of petroleum just outlined is based on much geological and chemical research. Three additional generalizations seem justified, though not all are universally accepted. First, little free oil has been found in sedimentary rocks of fresh-water origin, despite the vast quantities of "oil shale" (which contains no free oil but from which liquid oil can be distilled) in the Eocene lake beds of Utah, Wyoming, and Colorado. Second, oil and coal are not the liquid and solid products, respectively, of the alteration of

412

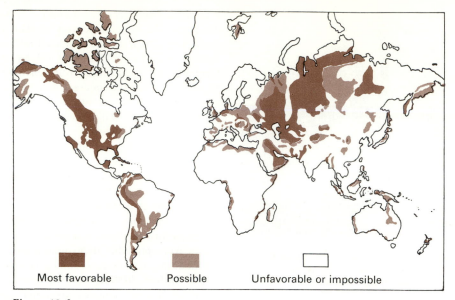

Figure 18-6
Oil map of the world, based on the estimated likelihood of finding oil in each region.
[Modified from Arabian American Oil Company, Middle East Oil Developments, 1948.]

peat, even though both have been found in the same sedimentary sequence (but at different levels) in the mid-continent area. Third, the oil-forming process seems to be very slow, as no oil has been found in Holocene sediments, though similar compounds have been extracted from them. The hope expressed by some chemists that earth oil may now be forming at about the same rate as it is being consumed seems quite unjustified geologically.

Oil Map of the World

Because of the economic and strategic importance of oil, a knowledge of oil geology is essential for the formulation of intelligent policies, both foreign and domestic. Where is oil now being produced? What reserves are still in the ground? Where may additional supplies be found?

First, we mark off on a map the areas where only igneous and metamorphic rocks crop out, calling these impossible or wholly unimportant because of the absence of marine organic sedimentary rocks. We can add to them, as unfavorable, areas where only thin or nonmarine sedimentary rocks overlie crystalline rocks (Fig. 18-6). Then we out-

line as favorable the places where oil is being produced, and add those where thick marine rocks have yielded evidence of oil, such as seeps. Finally, we may distinguish an intermediate or possible group of areas containing thick marine strata but no positive indications of petroleum.

Although such a map is valuable, it leaves many questions unanswered. Which areas have produced the most oil? Which have the most left? Up to 1948 the United States had produced more than half the world's oil, but by 1970 it was producing only 21 percent. And what about proved reserves? The United States has much oil. Careful estimates indicate that of an original endowment of between 165 and 200 billion barrels, 99.9 billion barrels have been produced, leaving probable reserves of between 65 and 100 billion barrels. At present rates of production this would last for between 19 and 28 years.

No one expects complete exhaustion of domestic petroleum in such short times, because not all the potential oil will have been discovered, and increased imports will also prolong the productive cycle. Nevertheless, it is clear that in 50 years we will see a striking reduction of internal combustion engines unless ways are found to obtain liquid hydrocarbons from coal. That oil is becoming harder to find in the United States is dramatically shown by the declining numbers of successful wells—from 15,329 in 1967 to 11,858 in 1971. The proved reserves of the Middle East are far the greatest in the world, more than 200 billion barrels. Those of Venezuela are also large, and perhaps equal to those of North America, as the oil industry has not yet explored all of the country fully. The fields of Iran, Kuwait, Bahrein, and Saudi Arabia are intrinsically so rich and are being managed so much more efficiently than were those of the United States in the early days of wasteful competition—town-lot well spacings—that their value is almost unimaginably great. Never in history have there been such prizes of concentrated wealth. Even in 1957, early in their development, more than $3\frac{1}{2}$ million barrels flowed daily from a few hundred wells. Compare this with America's hundreds of thousands of wells with an average production of about 10 barrels a day, and consider the vast wealth that has come from them. Clearly, developments with regard to the Middle East's oil will be crucial in the next generation.

No inventory of potential oil provinces would be complete without mention of the continental shelves. These submerged extensions of the continents are repositories of huge volumes of sedimentary rocks, most of them undoubtedly marine. Though the technological difficulties of both exploration and exploitation of these underwater reserves are tremendous, production in 1964 amounted to nearly 50 million barrels of oil a year off the coasts of California, Texas, and Louisiana. Though initially less productive, reservoirs developed later in Cook's Inlet, Alaska, and in the North Sea will surely add greatly to the world's supplies. With improved technology we may look forward to considerable additions to reserves from subsea sources.

Oil Shale

Vast quantities of oil are locked up in the so-called oil shales of Wyoming, Colorado, and Utah. But the oil can be otained only by distilling the shale, leaving a spongy residue of virtually the same volume as was distilled. Disposal of this waste is one of the many problems that thus far have prevented commercial operations for the extraction of the oil, though distillation in place underground was successfully tested in 1973. Presumably when the price of oil rises high enough, production will begin, though it is very doubtful that the expected yield will be attained. If it were, it would postpone the energy crisis for several decades.

Oil-Finding

The first earth oil put to human use oozed from oil seeps. Noah's Ark, like the present native boats of the Near East, was probably calked with asphalt from a Mesopotamian seep. The first well successfully drilled for oil outside of China (in Pennsylvania, 1859), was sunk beside a seep, as were many successful wells later. Most such wells have been small producers, but Cerro Azul No. 4, perhaps the greatest in history, was an exception. This well, near Tampico, Mexico, "blew in" on February 10, 1916, for 200,000 barrels a day, the column of oil rising nearly 200 meters into the air. It produced almost 60 million barrels before suddenly yielding only salt water. The reservoir was in limestone and apparently contained a real pool of oil and gas floating on salt water in interconnected caverns.

Seepages are obvious clues to oil, but their absence does not deter exploration. By 1883, random drilling in Pennsylvania and West Virginia had shown that many of the producing wells were grouped along anticlinal crests. Noting this, I. C. White advocated the anticlinal theory of accumulation and pointed out the physical reasons for it; it was generally accepted as a guide to exploration. Between 1900 and 1910 most oil companies built up geological staffs and gradually came to rely, in exploration, not on random drilling, but on systematic geologic search for anticlines and similar favorable structures. By 1928 most anticlines observable from surface geology in the United States had been drilled and hundreds of oil fields discovered.

But the surface beds in many areas conceal the structures of the rocks beneath if an unconformity intervenes. Nor can they give clues to stratigraphic traps—a reservoir type generally overlooked until the early 1930's. Systematic exploration demanded some means of determining geologic structure at depth.

GEOPHYSICAL METHODS

The first geophysical method applied was the **gravity survey**. A few huge salt domes (Figure 18-4, B) along the Texas coast had yielded much oil from their flanks or crests. As salt is less dense than most other rocks, measurements of gravity over the flat coastal country should reveal similar buried domes by their smaller gravitational attraction. Systematic surveys during the 1920's did, in fact, reveal many anomalously low values of gravity, and by 1930 the coasts of Louisiana and Texas were dotted with salt dome oilfields. Gravity methods are less useful in rough terrain, however, and the method is little used today.

Explosion seismology, introduced about 1924, has proved more widely applicable. Of the several techniques, **reflection shooting** has been most successful; the travel times of seismic waves from small explosions in shallow holes are recorded on portable seismographs, allowing identification of various strata by their differing elastic properties. By comparing travel times to different points, the structure of buried strata can be determined, even through an unconformity.

This method has been highly successful. For example, the Lowden field in Illinois was so accurately located by seismic work that the oil company was able to lease in advance of drilling all of the 8000 hectares that have since proved productive. This is a major field, with ultimate recovery estimated at 200 million barrels. The seismograph is the most used instrument in present-day prospecting.

SUBSURFACE METHODS

By the late 1920's so many wells had been drilled in and near oilfields that the subsurface geology of the country surrounding them could often be determined almost independently of the surface rocks. This is obviously extremely important where unconformities, lensing of beds, or facies changes are encountered. The major problem is that of correlating beds between wells. If this can be done, favorable structures may be further explored and unfavorable ones avoided.

Of the many kinds of **well logs**—records of the characteristics of strata cut by a well—four are particularly widely used: lithologic, paleontolog-

ical, electrical, and gamma-ray. **Lithologic correlations** are based on microscopic study of well cuttings of cores from the drill holes. An illustration showing such correlations in the Yates field, Texas, is given in Figure 18-7.

Where the rocks are not so readily distinguished as the red shale and anhydrite of the Yates Field, paleontologic methods are at times applicable. As any larger fossils are generally ground to bits in drilling, the principal paleontologic materials are microfossils, chiefly foraminifers (Fig. 17-25), diatoms, and spores. These tiny fossils are extremely abundant in many strata and are small enough to escape being ground up. Some fossil zones only 3 to 6 m thick can be distinguished by their microfossils and readily traced from one field to another nearby, even though the facies may change from sand to shale. Hundreds of micropaleontologists are now working in the oil industry to make correlations by means of fossils.

Electric logs are made by lowering electrodes into the uncased well to measure the electrical properties of the beds penetrated. These characteristics vary markedly because of differences in the composition of the rocks and in the kind of fluid, oil, salt water, or fresh water contained in the pores. Figure 18-8 illustrates the structure of a wide, flat anticline as developed from electric logs.

GAMMA-RAY LOGS

Most rocks carry trace amounts of radioactive materials—uranium or thorium—that emit gamma rays in the course of their decomposition. By lowering a detector into the wells, variations in the intensity of the radiation can be recorded. The radiation is characteristic for each sedimentary unit, and this method of correlation has become routine.

SUMMARY OF EXPLORATION METHODS

This outline of some of the many techniques employed in exploration for oil shows that, although the presence of oil in a particular place at a particular depth cannot be foretold by either geological or geophysical techniques in advance of drilling, the search is not merely a matter of luck. Oil is localized in response to physical laws, in structures and stratigraphic traps that are geo-

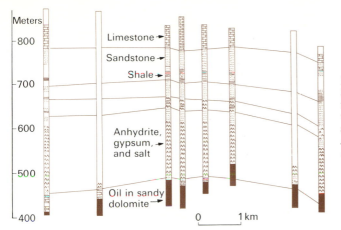

Limestone →
Sandstone →
Shale →

Anhydrite,
gypsum,
and salt

Oil in sandy
dolomite →

0   1 km

Figure 18-7
SW–NE section across the Yates oil
field, western Texas, showing well-to-
well correlations. [After G. C. Gester
and H. J. Hawley, *Structure of Typical
American Oil Fields*, American
Association of Petroleum Geologists,
1929.]

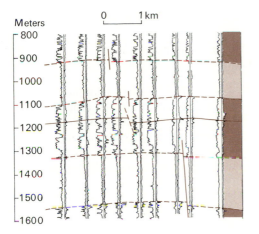

0   1 km

Figure 18-8
SW–NE section across Odom dome, southern Texas,
showing correlations (broken lines) from electric
logs. The irregular lines to the right of well-location
lines record the resistance of the rocks to an electric
current; the lines to the left measure the amount of
natural current produced by the rocks themselves.
They effectively mark the positions of the three
sandstone strata shown by the dark brown pattern at
the right. [After Society of Exploration Geophysicists,
Geophysical Case Histories, 1949.]

logically determinable. The problem of oil-finding
is geological, and its solution depends on application of many principles of geophysics, stratigraphy, and paleontology. All large oil companies
therefore maintain elaborate geological depart-
ments, and many maintain research laboratories
as well.

ORE DEPOSITS

Ore deposits are rock masses from which metals
are obtained commercially. Every ore body has
been formed by the selective concentration of one
or more elements in which the rock is greatly enriched as compared with the average of the earth's
crust. How great this enrichment must be can be
realized from the fact that more than 99 percent
of the earth's crust is made up of only 10 elements (Appendix IV, Table IV-3). Of these, aluminum, iron, and magnesium are industrial metals,
but only very small proportions of these are sufficiently concentrated to constitute ore. Some industrial metals—copper, zinc, lead, tin, and
others—are found in only minute amounts in the
earth's crust (Table 18-5).

The abundance of an element in the "average
igneous rock" is called the **clarke** of the element
(in honor of the American geochemist F. W.
Clarke). In any ore body the concentration of the
ore element must obviously be higher than the
clarke of the element; this concentration is the
clarke of concentration of the ore body. Thus an
aluminum ore of low grade, say 25 percent Al,
has a clarke of concentration of 3, for aluminum
is three times as abundant as in the average
igneous rock. Similarly, an iron ore with 50
percent Fe has a clarke of concentration of 10;
a copper ore of 1 percent Cu, a clarke of concentration of more than 100. Most gold mines operate
on ore that assays 10 parts per million, a clarke
of concentration of 20,000. The clarke of concentration of a metal in an ore body is thus a rough

Table 18-5
Abundance (clarkes) of some useful metals in the average igneous rock.

Element	Weight percentage	Element	Weight percentage
Aluminum	8.3	Cobalt	0.002,3
Iron	5.0	Lead	0.001,6
Magnesium	2.0	Tungsten	0.001,5
Titanium	0.44	Uranium	0.000,4
Manganese	0.1	Antimony	0.000,1
Chromium	0.02	Mercury	0.000,01
Vanadium	0.015	Silver	0.000,01
Nickel	0.008	Gold	0.000,000,5
Copper	0.007	Platinum	0.000,000,5
Zinc	0.005	Radium	0.000,000,001,3
Tin	0.004		

measure of the value of the metal. High clarkes of concentration must reflect either very unusual geologic conditions or the carrying of ordinary geologic processes to unusual perfection. We shall see examples of both aberrations.

Gold is among the few native elements found in ore bodies; most ore minerals are compounds of the valuable element with others (Table 18-6). If the combination is such that the useful metal may be economically separated, the deposit is **ore**. Separation of the ore minerals from their worthless associates—the **gangue minerals**—and the extraction of the valuable metal from the ore minerals is **metallurgy**.

The definition of ore is purely economic: it is rock that can be worked commercially for extraction of a useful metal. The exact percentage of the metal in the ore is irrelevant; it must be enough for a profit. A particular low-grade, uneconomic ore may become economic when metal prices increase (as happened to many mercury deposits during both world wars and to many gold mines in 1971); with improved metallurgical techniques (as with many deposits of copper, lead, and zinc with the invention of the flotation process for separating ore minerals from gangue and from each other), or with subsidies (as did the very low-grade Rhineland iron deposits under the Nazi self-sufficiency program). Conversely, a price decline or higher mining and metallurgical costs have often rendered once-valuable ore worthless.

Among the many factors that determine whether a particular mineral deposit is ore are:

1. The size, shape, and depth of the deposit. (All these greatly affect the cost of mining.)
2. The amenability of the rock to metallurgical treatment. (Fine-grained mineral aggregates may need to be ground very fine for clean separation from the gangue and from each other—an expensive operation.)
3. The distance to metallurgical centers or to market. (The exploitation of the rich Brazilian iron ores was long delayed because of distance from coking coal.)
4. The cost of labor. (Highly mineralized rock in the Broken Hill mining district, New South Wales, changed from ore to non-ore in 1972, when miners demanded large increases in pay and several mines were shut down as they became unprofitable.)

Dramatic examples of the influence of these factors on mining are many. The great copper mine at Bingham Canyon, Utah, today mines ore containing as little as 0.50 percent copper (5 kilograms of copper to the ton), yet fifty years ago masses of pure copper weighing several tons were occasionally found in the Michigan mines but were not ore because they could not be effectively blasted and it cost too much to chisel them out. A generation ago rock containing 50 percent iron could not be mined on the Minnesota "iron

Table 18-6
The common ore minerals.

Metal	Mineral	Elements contained	Percentage of metal in ore mineral	Chemical composition
Gold	Native gold	Gold	50 to 100 (alloyed with silver)	Au and Ag
Silver	Native silver	Silver	100	Ag
	Argentite	Silver, sulfur	87.1	Ag_2S
Copper	Native copper	Copper	100	Cu
	Chalcopyrite	Copper, iron, sulfur	34.6	$CuFeS_2$
	Chalcocite	Copper, sulfur	79.8	Cu_2S
	Enargite	Copper, arsenic, sulfur	48.3	$3Cu_2S \cdot As_2S_5$
Lead	Galena	Lead, sulfur	86.6	PbS
Zinc	Sphalerite	Zinc, sulfur	67	ZnS
	Franklinite	Zinc, iron, oxygen, manganese	12±	$(Fe,Mn,Zn)O \cdot Fe,Mn_2O_3$
Iron	Hematite	Iron, oxygen	70	Fe_2O_3
	Magnetite	Iron, oxygen	72.4	Fe_3O_4
Aluminum	Bauxite (actually a mixture of several minerals)	Aluminum, oxygen, hydrogen	35–40	$Al_2O_3 \cdot 2H_2O$ (varies)

ranges" because blast furnaces were designed to use only higher grade ores. With the exhaustion of the high-grade ores, furnace practices were changed so that the lower-grade materials can be used. At Birmingham, Alabama, the happy combination of abundant local coal and limestone flux for use in metallurgy permits mining of ores with as little as 25 percent iron. Such material would be useless rock, not ore, in Montana or California.

Ore Formation

The processes that concentrate minerals into economic deposits are both mechanical and chemical, acting alone or in combination. The processes are those we have already noted: weathering, solution, transportation and sedimentation at the surface, volcanism, metamorphism and flow of solutions below the surface. Ore bodies are rocks, and at one place or another

nearly every rock-forming process has produced a valuable mineral deposit. This is indicated in Table 18-7, a brief tabulation of a few geologic types of mineral deposits, to which many others could be added. The few selected are economically important and illustrate different ways in which various elements have been selectively concentrated.

MAGMATIC SEGREGATIONS

As noted in Chapter 9, many large floored intrusions are stratified, and more or less regularly banded, with the mafic minerals near the base. In the Palisades sill the mineral so concentrated by early sinking is olivine, of no economic value, but in some places economic minerals are similarly concentrated. Among these are the chromite and platinum ores of the Bushveldt, South Africa, and the chromite of the Stillwater region, Montana. An outcrop of segregated chromite is illustrated in Figure 18-9.

Table 18-7
Types of mineral deposits.

Type	Manner of formation	Representative deposits
Magmatic segregation	By settling of early formed minerals to the floor of a magma chamber during consolidation.	Layers of magnetite, chromite, and platinum-rich pyroxenite in the Bushveldt lopolith, South Africa.
	By settling of late-crystallizing but dense metalliferous parts of the magma, which either crystallize in the interstices of older silicate minerals, or are injected along faults and fissures of the wall rocks.	Copper-nickel deposits of Norway and most of those of the Sudbury district, Canada. Injected bodies of magnetite in Sweden (greatest in Europe) and in New York.
	By direct magmatic crystallization.	Diamond deposits of South Africa (nonplacer).
Contact-metamorphic	By replacement of the wall rocks of an intrusive by minerals whose components were derived from the magma.	Magnetite deposits of Iron Springs, Utah; some copper deposits of Morenci, Arizona.
Hydrothermal (deposits from hot, watery solutions)	By filling fissures in and replacing both wall rocks and the consolidated outer part of a pluton by minerals whose components were derived from a cooling magma. These differ from contact-metamorphic deposits in having fewer silicate minerals and more obvious fissure control.	Copper deposits of Butte, Montana, and Bingham, Utah; lead deposits of Idaho and Missouri; zinc deposits of Missouri, Oklahoma, and Kansas; silver deposits of the Comstock Lode, Nevada; gold of the Mother Lode, California, Cripple Creek, Colorado, and Lead, South Dakota.
Sedimentary	By evaporation of saline waters, leading to successive precipitation of valuable salts.	Salt and potash deposits of Stassfurt, Germany, and of Saskatchewan, Canada, New Mexico, Utah, Michigan, Ohio, and New York.
	By deposition of rocks unusually rich in particular elements.	Iron deposits of Lorraine, France, Birmingham, Alabama, and some in Minnesota and Labrador.
	By deposition of rocks in which the detrital grains of valuable minerals are concentrated because of superior hardness or density.	Placer gold deposits of Australia, California, Siberia, Nome, Alaska, probably of the Rand, South Africa; titanium of Travancore, India, and Australia; diamond placers of Southwest Africa.
Residual	By weathering, which causes leaching out of valueless minerals, thereby concentrating valuable materials (originally of too low a grade) into workable deposits.	Iron ores of Cuba, Bilbao, Spain, some of those of Michigan and Minnesota; barite deposits of Missouri.
	By concentration due to weathering, together with further leaching and enrichment of the valuable mineral itself.	Bauxite (aluminum) ores of Arkansas, France, Hungary, Jamaica, and British Guiana.

Floored intrusions of a variety of gabbro called norite, with concentrations of nickel sulfides and copper sulfides near their bases, are found in localities as widely scattered as Norway, Canada, and South Africa. The great Norwegian geologist J. H. L. Vogt pointed out in 1893 that the bulk compositions of these intrusives resemble those of smelter charges of sulfide ore—that is, a small percentage of sulfide and a great preponderance of silicates of iron, aluminum, magnesium and calcium. When such a sulfide ore is smelted, the molten sulfides (the matte) sink to the bottom of

Figure 18-9
Banded chromite, Seldovia district, Alaska. [Photo by
P. W. Guild, U. S. Geological Survey.]

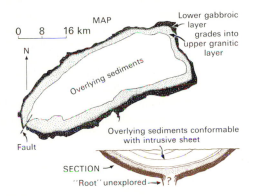

Figure 18-10
The Sudbury lopolith, Ontario, Canada. [After A. P.
Coleman and E. S. Moore, 1929.]

the crucible, while the slag of silicates rises to the
top. Vogt suggested that the same mechanism
might operate in nature to produce the observed
relations.

Among such deposits are the greatest nickel
deposits of the world, those of Sudbury, Ontario,
which largely occur along the base of the lopolith
shown in Figure 18-10. Here nickel sulfides are
commonly molded against the silicate minerals,
which possess their own crystal forms. This suggests strongly that the sulfides remained molten
after the silicates crystallized and accommodated
themselves to the intergrain spaces between the
silicates.

CONTACT METAMORPHIC DEPOSITS

In Chapters 4 and 9 we noted that the wall rocks
of some intrusives contain quite different minerals
from those of the same beds at a distance. In
many places features such as bedding and even
fossils are still identifiable in beds of limestone
right up to an intrusive contact, although the limestone near the contact may be almost completely
converted to garnet, pyroxene, amphibole, and
epidote. Such mineral changes in an already existing rock at plutonic contacts prove that material
has been transferred from the magma to wall rock
during the cooling of the intrusive. If the new
minerals were merely concentrated from the limestone itself by solution and removal of other constituents, the volume must have shrunk. Locally
such shrinkage has been found, but it is exceptional. Most contact zones show no volume
changes; minor details of bedding and delicate
fossil structures are faithfully preserved in the
new mineral phases.

Clearly, new material must have been added
to the wall rock, and carbon dioxide and other
substances removed, in order to change the
carbonates to silicates without volume change.
Supporting this interpretation is the observation
that gases escaping through fissures at many
volcanoes have deposited iron-rich minerals,
such as hematite and magnetite, on the fissure
walls.

Only ions or gases and relatively dilute solutions could so intimately permeate the wall
rocks as to bring about their transformation
without disturbing delicate textural features; it
is equally evident that part of the former material
must have been removed by the same solutions.
Volume-for-volume replacement is widespread
in geology, not just at intrusive contacts, though
these are distinctive because of the high-temperature mineral phases formed.

Contact-metamorphic deposits yield many
useful products: garnet for sandpaper, corundum
(Al_2O_3) for abrasives, rubies and sapphires for
gem use, copper, iron, zinc, and lead. The iron
deposits of Iron Springs, Utah, are representative
(Fig. 18-11). These deposits are pod-shaped,
having replaced a limestone, in places to its full
thickness; elsewhere only in part. The ore is a
mixture of magnetite and hematite, with a little
apatite and smaller amounts of garnet, pyroxene,
and quartz. Of these minerals only quartz is

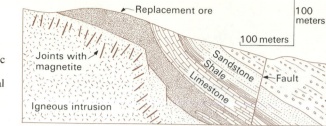

Figure 18-11
Cross section of contact metamorphic iron deposit at Iron Springs, Utah. [After J. H. Mackin, Utah Geological Survey, 1947.]

found in the unaltered limestone. The rest have been introduced from the magma concurrently with removal of calcium and carbon dioxide from the limestone.

HYDROTHERMAL DEPOSITS

Most ores of copper, lead, zinc, mercury, silver, and many of gold and tungsten are classed as hydrothermal deposits; that is, they were formed by deposition from hot water solution, as proved by the following facts:

1. Closely similar deposits have been observed to form in hot springs and fumaroles.
2. Many deposits are localized along faults and fissures that cut older rocks, and therefore must have crystallized from fluids that penetrated cracks in the host rock.
3. The minerals of many of these ores are identical, though the deposit can be followed from one kind of wall rock to a very different one. The ores must have formed in an environment nearly independent of, and therefore later than, the diverse environments that prevailed during formation of the differing wall rocks.
4. Although the wall rocks commonly show drastic mineral changes, delicate pristine structures are preserved, showing that the alterations were brought about by solutions so fluid that their passage did not mechanically disturb the rocks.
5. The well-developed crystal faces found on many of the minerals imply growth from solution, for such faces can readily be produced in the laboratory and can be seen to form from solutions in nature.

That the solutions inferred from these features were of magmatic origin is less certain, though there surely is plenty of evidence that plutonic magmas contained water in solutions in excess of that present in the crystallized rocks. Isotopic studies suggest that much of the solutions were probably meteoric waters—ground water put into circulation by the heat of the pluton—for the proportions of ^{18}O and D (deuterium) to ^{16}O and H, respectively, are closer to those in ground water than to those waters contained in vesicles in igneous rocks. One would, of course, expect this; ground water continues active circulation during the long time of cooling of a deep-seated pluton, so that the altered minerals would finally equilibrate with the ground water rather than the magmatic solutions, even if originally wholly formed from magmatic solutions.

Among the minerals that have been seen to form within fumaroles, geysers, and hot springs are magnetite, galena, sphalerite, and cinnabar (HgS, the principal source of mercury). In a few places, such as the sulfur mines of Sicily and the mercury deposit at Sulphur Bank, California, economic mineral deposits are directly associated with hot springs, and their valuable minerals were obviously deposited by ascending hot waters. The similarities in both ore minerals and wall rock alteration of these springs with those of other localities where hot springs are not now active are so close that most geologists agree that the latter—the so-called hydrothermal deposits—have largely been formed from hot solutions. Along the walls of these metallic deposits, just as along those of modern hot springs, the rocks have been altered to the clay mineral chlorite and other hydrous minerals, no doubt by the action of the long-vanished hot solutions.

Paradoxically, one of the strongest evidences that hydrothermal deposits are of magmatic origin and are not formed merely by circulating ground water set in motion by magmatic heat, lies in the fact that many plutons are *not* accompanied by such deposits, though other igneous masses that cut the same strata and similar structures have great suites of them. Both plutons should have developed similar patterns of groundwater circulation, but only one has ore deposits associated. This seems to leave little doubt that the metals were derived from the magma, not the wall rocks.

Hydrothermal deposits are not associated with all parts of a pluton. Where erosion has been deep enough to disclose the form of the pluton, the ore deposits appear to be clustered about the higher parts—the apex or "cupola." A few deposits lie in deep sags in the roof or along its steeper walls. Furthermore, in plutons so deeply eroded that no remnants of roof remain, ore bodies are rare. These relations suggest that the ore-depositing solutions were hot volatiles—residues from crystallization of the magma, concentrated in higher parts of the pluton by the upward convergence of the walls (Fig. 18-12).

Form of hydrothermal ore bodies   The commonest form of hydrothermal ore body is the **vein**. Unlike the cylindrical veins of animals and plants, mineral veins are generally tabular—hundreds if not thousands of times as long and high as they are thick. They may lie in any attitude; they pinch and swell, branch and swerve. Many occupy faults, as shown by offsets of geologic contacts across them. Some follow dikes or bedding surfaces, and still others occupy joints. The so-called "true fissure veins" differ sharply from their wall rocks, and generally break cleanly from them. These are so common in some districts that many of the mining laws of the United States are based on the fallacious assumption that all ore deposits are "true fissure veins"—an assumption that has proved highly remunerative

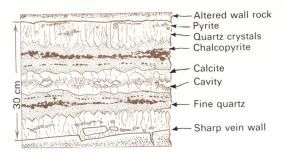

Figure 18-13
A section across a crustified vein.

for the legal profession and for some geological "expert witnesses," but one that hardly advances the mining industry.

Many veins whose geologic relations indicate that they formed at relatively shallow depths are filled with quartz, calcite, and other carbonate minerals plus a little feldspar, sulfides, and perhaps native gold. These minerals are arranged in bands (Fig. 18-13), in definite sequence inward from the walls. Unfilled cavities lined with well-formed crystals testify to the fact—already clear from the crust-like arrangement of the minerals—that the vein partly fills a formerly open channel and that the minerals were deposited in sequence from the passing solutions. These are **crustified veins**.

Other veins are not sharply separable from their walls. Microscopic study shows that the vein matter has replaced the wall rock without disturbing it, as in the contact metamorphic deposits. These **replacement veins** may pass along their trend into fissure veins.

Lodes are unusually thick veins or groups of veins, some scores of meters thick. Clearly, the walls of an opening so wide would have collapsed from the weight of the overlying rock, yet chunks of rock lying along the lower side of a dipping vein, and the crustified arrangement of well-formed crystals in some of the veins (Fig. 18-14) show that open space existed. Careful study of many lodes shows that the quartz along the walls has been broken and recemented with later quartz, commonly in several generations. The evidence of repeated rupture suggests that the fault occupied by the lode was repeatedly active during deposition of the quartz. Movement along an irregular fault would necessarily bring opposing bends into contact and leave openings

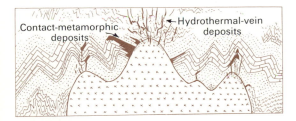

Figure 18-12
Concentration of ore deposits in and near the apex of a large pluton. [After B. S. Butler and G. F. Loughlin, 1913.]

422

Figure 18-14
Part of a crustified quartz lode from Grass Valley, California. [Photo by W. D. Johnston, Jr., U. S. Geological Survey.]

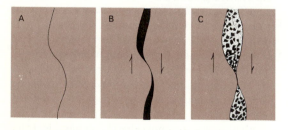

Figure 18-15
Growth of a lode in vein deposits along an irregular fault. (A) Fracture before movement. (B) Ore filling cavities after first movement. (C) Renewed movement and fracturing of old ore and emplacement of new.

between (Fig. 18-15). Renewed faulting after such open spaces were filled would produce new openings, to be filled in turn; in this way a lode would reach considerable widths, filled by a succession of hydrothermal emanations.

DISSEMINATED DEPOSITS

Other hydrothermal deposits are very irregularly shaped. Notable among these are the great so-called porphyry coppers, which supply by far the most of the world's copper, exemplified by the great mines of Bingham, Utah, Ely, Nevada, Morenci, Arizona, and Chuquicamata, Chile (the greatest copper deposit in the world). Most ore in these districts is disseminated along small, irregular fissures in porphyritic plutons, impregnated and partly replaced by very low-grade copper and iron sulfides carrying about 0.4 to 3 percent Cu. The rock surrounding and within the ore body is largely altered to clay and fine-grained mica, with a little epidote and chlorite. The ore is so intimately veined by sulfides, quartz, and, locally, K-feldspar, that it is hard to find an unbroken piece as big as a tennis ball in a mass comprising many thousand cubic meters.

The cause of the shattering of these huge masses of rock, so that they could be thoroughly impregnated with ore minerals, is unknown. That the shattering is confined to the ore body suggests the possibility that it was caused by the explosion into steam of magmatic water (which was excluded from the structure of all the crystallizing minerals except the micas), so that the residual magma became supersaturated with water, and exploded. Another possibility is that the fracturing was caused by rise and fall of magmatic pressure as various vents to the surface opened and closed, causing a sort of churning in the pluton. But such variations in internal pressure must be common to nearly all plutons, whereas the minute fracturing in the porphyry ore deposits seems limited to the mineralized parts.

Very different disseminated deposits are those exemplified by the lead ores of southeastern

Missouri and the zinc-lead deposits of the Tri-State district (Oklahoma, Missouri, and Kansas) and of Poland. These ores form irregular and indefinite "runs" in limestone (Fig. 18-16). The ores are sulfides that replace the carbonate wall rocks, and although the pattern of the ore bodies shows some control by faults and joints, there are no distinct veins. The sulfides are accompanied by silica, deposited as chert, and by dolomite and calcite, mostly well crystallized. These minerals were apparently introduced in solutions that permeated the entire rock but deposited their ores selectively in certain beds and along certain trends or channels, perhaps because of some obscure variations in permeability or chemical composition.

SUMMARY

The hydrothermal deposits consist of rocks that are on the one extreme almost indistinguishable from contact-metamorphic deposits and on the other almost identical with hot-spring deposits.

Observations of fumaroles and hot springs show that the minerals so formed change with the fall in temperature; the earlier-formed minerals are dissolved, and new minerals form. Although we have nowhere found in a single district a complete series of ore deposits filling the range from hot spring to contact metamorphic, parts and gradations of such a series are apparent in many parts of the world. Studies of many such partial series has made possible a tentative arrangement of ore deposits in a reasonably systematic sequence. This sequence probably reflects changes in composition, temperature, and pressure of magmatic solutions as they pass through and react with the minerals of the wall rocks and cool as they approach the surface. Where conditions favorable to the deposition of one or more ore minerals have remained fairly constant and the volume of solution has been large, ore bodies have formed. Where conditions changed too rapidly, or the volume of solution was too small, no ore body was formed, though small quantities of ore minerals were deposited. "Gold is where you find it"—but it has many times been proved that identifiable geologic factors have controlled the deposition of the ores and thus aid in the finding of new ore bodies or hidden extensions of those already known.

SEDIMENTARY DEPOSITS

Many economic deposits have been concentrated by sedimentation. Among those already discussed are the clay used in ceramics and cement; sand and gravel used in concrete; limestone for building stone, mortar, or cement; dolomite, rock salt, gypsum, and potassium salts. The total value of these deposits is very large—in the United States about twice as great each year as the value of all metals mined. The value of each depends chiefly on the amount of impurities in the sedimentary rock, on the cost of mining or quarrying, and on the distance to market.

Some sedimentary rocks, many of wide extent, are metallic ores. Among the most notable is the Kupferschiefer of Germany (Upper Permian) a bed less than 50 cm thick but extending over an area of many hundred square kilometers, consisting of copper, zinc, and a dozen other metal sulfides accompanied by much vegetable debris. It is interpreted as a shallow sea deposit where, in a reducing environment of rotting vegetation, sulfidizing bacteria precipitated the metals that were supplied in acid solutions from weathering mineral deposits on the desert shores.

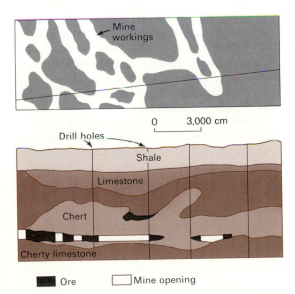

Figure 18-16
Map (top) and section (bottom) showing the irregular distribution of ore and silicification of limestone in a Missouri lead mine.

A comparable deposit of Precambrian age is that of the Mount Isa mine, in Queensland, Australia. Here a stratified ore body at least 5 km long, 1 to 2 km wide, and several meters thick lies rigidly parallel to overlying and underlying tuff beds throughout the mine, with no signs of any "feeding fissures." It is clearly not a hydrothermal deposit, for no replacement could be so faithfully strata-bound. It is likely, too, that the great ore deposits of the Broken Hill district of New South Wales are of sedimentary origin, but the rocks are so highly metamorphosed that this is less certain.

The iron ores of Alabama, Newfoundland, northeastern France, and parts of England are also examples of sedimentary deposits. The fossiliferous Alabama ores of Silurian age are of low grade (25 to 40 percent iron) but attain a thickness of 6 m and have been mined for nearly 25 km along the strike and to a depth of more than 600 m. The ore beds are generally sharply distinct, but locally they grade into sandstone. In these ores, hematite cements the rock and coats and replaces some of the fossils and sand grains. These show, by the abrasion of the hematite-covered surfaces, that the iron coating was formed before the agitation of the grains was stopped by their burial. Many hematite granules are flattened, suggesting that they were soft and were squeezed during the consolidation of the rock. Fossils made of calcite are undeformed. These facts suggest that the present hematite granules were once jelly-like aggregates of iron rust deposited on nuclei of clastic grains. It was presumably a hydrous oxide from which the water was driven by heat and pressure after burial. Such low-grade deposits can only be worked because the calcite content acts as a flux in the blast furnace and coking coal is close at hand.

Placers

We have repeatedly noted that during normal weathering, feldspars and other silicates tend to decompose to clays, so weakly coherent as to break up with the slightest transportation. Quartz, however, is normally quite stable and so is concentrated in the stream beds of a granitic region, constituting most of the sand on the stream bed, even though it makes up only 10 or 20 percent of the bedrock. Such minerals as magnetite, chromite, diamond, gold, and cassiterite (SnO_2), the chief source of tin, are also chemically stable in most climates, and since they are denser than quartz, tend to fall to the bottom of a stream or accumulate on the riffles of a sluice box, while quartz of the same grain size is carried on. Where streams from regions that supply such minerals reach the sea, the heavy minerals are further concentrated by the waves. This is the origin of the "black sands" of the Oregon coast (chromite), of the raised beaches of Nome, Alaska (gold), of the monazite (thorium) sands of the Malabar coast of India, of the ilmenite (titanium) sands of Florida, of the diamond beaches of Southwest Africa, and of the great tin-dredging grounds of Indonesia. In all these localities, minerals that form only very minor parts of the inland rocks have been concentrated because of their density.

MINERAL EXPLORATION

Exploration for economic deposits of the sedimentary rocks is based mainly on accurate geologic mapping and careful, systematic sampling. But prospecting for disseminated or vein deposits is much more difficult and complex. The first step, and this has been rather completely done by the prospectors of preceding generations, even in the most remote parts of the world, is the "panning" of stream gravels in search for placers or for minerals known to be characteristic associates of valuable minerals. Thus a particular variety of garnet is common in the diamond pipes of South Africa; a systematic search for this resistant and much more abundant mineral is the first step in searching for other diamond-bearing pipes. Every iron-mining area has been thoroughly surveyed magnetically for possible extensions. Detailed gravity surveys have located chromite bodies in the serpentine belt of Cuba.

The search for disseminated ores now entails not only geologic mapping in search of favorable environments, as judged by similarities with known ore occurrences, but considerable chemical investigation as well. Highly sensitive chemical tests have been devised, allowing quantitative determination of many valuable elements in such dilution as a few parts per billion in rocks ($0.000,000,00x$ percent). Stream sediments are sampled in the more promising areas, and followed upstream if encouraging values are found. Where the stream values fall off, the area from which the metal came is probably on one side or the other of the stream; the soils and vegetation on either hand are then sampled and followed up the slopes. In this manner very large deposits of gold have recently been found in Nevada and of zinc and lead in New Brunswick.

"Locoweed" concentrates selenium from the soil; because selenium is a constant companion of uranium in the Colorado Plateau, the presence of dense stands of locoweed has been successfully used to find uranium ores. Once the outcrop of a mineral deposit is found, very detailed mapping is done to identify the controls, whether faults, fissures, schistose partings, or igneous contacts. Every ore body is geologically controlled; the problem is to identify and extrapolate these controls so as to facilitate discovery of more ore.

With the depletion of our richest and most readily found ore bodies, it is inevitable that still more extensive geologic studies and the application of still more sensitive chemical and geophysical tests will be needed to supply the ever-expanding industrial needs of the world.

Questions

1. Why did Holland and Denmark, which suffered greatly during World War II, oppose the suggestion to abolish the commercial production of coal from the Ruhr area in Germany despite their fear of renewed German aggression?

2. Why, in view of the great mineral endowment of the United States, did Congress authorize "stockpiling" of certain minerals as a defense measure?

3. From the description of placer deposits, explain how they can be used as guides to ore in bedrock. Why does a prospector carry a "pan"?

4. In view of the nonreplenishable nature of mineral deposits, what are the advantages and disadvantages of a tariff on minerals?

5. Although a "fault trap" was not described in the text, draw a cross section of an oil reservoir formed by a fault trap in a gently dipping sequence of sediments.

6. Why do the oil fields of the Persian Gulf region have fewer wells per square mile than those of East Texas? Does this reflect geologic or other conditions?

7. Determination of the depth of relatively recent unconsolidated river-deposited sediments in the estuary of the Congo River is important for the solution of the problem of submarine canyons (Chapter 16). What methods now routine in oil finding might be used to determine this depth at enough points to define the form of the bedrock surface? Which method would be quickest?

8. Sandstones used for making glass must consist of almost pure quartz. How is such purity attained by natural processes?

9. Why are placer deposits usually the first to be discovered in a gold-mining region?

10. Pyrite is common in many deposits of copper, lead, zinc, and gold. Why is a conspicuously brown-stained outcrop often a guide to a buried ore deposit?

Suggested Readings

American Geographical Society, *World Atlas of Petroleum*. New York, 1950.

Bituminous Coal Institute, *Bituminous Coal Facts and Figures*. Washington, D.C., 1948.

Flawn, Peter T., *Mineral Resources*. Chicago: Rand McNally, 1967.

Lovering, T. S., *Minerals in World Affairs*. New York: Prentice-Hall, 1943.

Rickard, T. A., *Man and Metals*. New York: McGraw-Hill, 1932.

U.S. Geological Survey and Bureau of Mines, *The Mineral Resources of the United States*. Washington, D.C.: Public Affairs Press, 1948.

Sutulov, Alexander, *Minerals in World Affairs*. Salt Lake City: University of Utah Press, 1972.

CHAPTER 19

The Earth as a Planet –
Speculations on Its Early History

Since the earliest writing—and doubtless long before—man has speculated about the relations of the earth to the cosmos. It is pointless to review the various ideas suggested before Copernicus, for until his discoveries no scientific tools existed to test them. Today, though great advances in all sciences enable us to place somewhat closer limitations on the possibilities, the origin and early history of the earth remain among the most uncertain, as well as the most fascinating, of earth problems. Such a book as this can only sketch in broad outline the kinds of reasoning employed, list some of the results obtained, and outline a few of the speculations favored by modern studies; a full discussion would not only extend the book unconscionably but would demand far more background in the supporting sciences than is appropriate to the level of this work.

THE SOLAR SYSTEM AND THE EARTH'S PLACE IN IT

The recognition of the solar system was the great achievement of the brilliant Polish cleric Copernicus (1473–1543). With only crude observational facilities and no telescopes, years of patient observations enabled him to show that the earth and its sister planets revolve around the sun: the earth is not the center of the universe, as was taught by Aristotle (384–322 B.C.), whose erroneous dogmatism ruled the scholarly world for centuries.

Inspired by Copernicus, Tycho Brahe (1546–1600), astronomer to the Danish Court, using sextants as long as 10 meters, charted the wanderings of the moon and planets for many years. From the occasional retrograde motion of some of the planets against the background of the stars, he was able to show that their orbits are outside the earth's (Fig. 19-1). On his death,

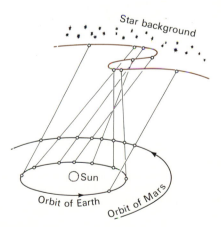

Figure 19-1
Diagram illustrating how any planet whose orbit is external to the earth's will at times appear to have a retrograde motion.

Brahe bequeathed his hard-won data to a young German mathematician, Johannes Kepler (1571–1630), who, after years of patient study, succeeded in reducing the huge mass of observations to rational rules, known ever since as Kepler's Laws:

1. All planets, including the earth, move in elliptical orbits, with the sun at one of the foci.
2. A line connecting the sun and a planet sweeps out equal areas in the plane of the orbit in equal periods of time (Fig. 19-2).
3. The square of the time, T, of a planet's revolution is proportional to the cube of its average distance, D, from the sun. In other words, the greater the average distance from the sun, the slower the revolution, but the ratio of T^2 to D^3 is the same for all planets.

Kepler's Laws, together with the brilliant work of Galileo (1564–1642) on the acceleration of gravity, furnished the background for Isaac Newton's deduction of the Law of Universal Gravitation. The moon follows an elliptical path about the earth, and the earth an elliptical path about the sun because, although the inertia of each tends to keep it moving in a momentarily tangential straight path, gravity tends to make it fall toward its principal (Fig. 19-3).

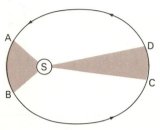

Figure 19-2
Illustration of Kepler's first two laws: first, that the orbit is an ellipse with the sun at one focus; second, that a line joining sun and planet sweeps out equal areas in equal times. Area SAB equals area SCD; the planet moves fastest when closest to the sun and slowest when farthest away.

TELESCOPES AND SPECTROSCOPES

Along with these advances in theory, a great new astronomical tool was developed. A Dutch spectacle maker, Hans Lippershey, about 1608, accidentally discovered that two lenses—one magnifying, the other reducing—when properly arranged, magnify distant objects.

Galileo learned of this; never having seen the Lippershey invention, he promptly developed his own telescope. His first (1609) magnified only three diameters, but he later attained a magnification of 33 diameters. One of the first objects he examined was the moon, discovering its magnificent craters (Fig. 19-4). He found the Milky Way to be a huge disc-shaped mass of stars "so numerous as to be almost beyond belief." He noted the spots on the sun, observed their motion, and found the sun's period of rotation. He discovered the four largest of Jupiter's thirteen moons and interpreted the rings on Saturn as bulges. He followed the phases of Venus, comparable to the familiar phases of the moon. Copernicus' sun-centered solar system

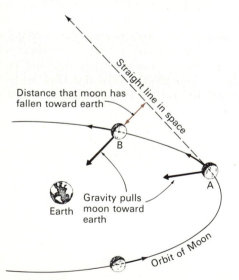

Figure 19-3
Gravitational force causes the moon to fall toward the earth from the course it would have if its inertia could maintain straight-line travel.

Figure 19-4 *(facing page)*
The moon at nearly full phase. The dark plains (maria) are floods of basaltic lava, in contrast to the highly cratered uplands (terrae). [Photo by Hale Observatories.]

was therefore no longer merely a simpler way to explain the planetary motions; it became a demonstrated fact, confirmed by the finding that Jupiter's moons revolve around Jupiter and not around the earth.

But man does not easily relinquish the self-flattering doctrine that he stands at the center of a universe created for his sole benefit. Though a devout churchman, Galileo was haled before the Inquisition and forced to retract publicly his support of the Copernican system as being contrary to the Bible. He continued privately in his convictions—"but it does move," he is said to have quietly remarked after his public recantation. His works were placed on the Index of Prohibited Books, where they remained for nearly 200 years. The University of Salamanca was still teaching that the earth is central to the universe in 1828. Galileo spent the last eight years of his life in house arrest. Though denied his telescope, he was permitted to work further in dynamics, a fundamental contribution underlying Newton's epochal discoveries. He died the year Newton was born.

Deviations of Saturn's orbit from the strict path required by Kepler's laws could only be explained by the gravitational attraction of another planet: close study of the portion of the sky where such a planet would have to be situated led to the discovery of Uranus (1781). Neptune (1846) and Pluto (1930) were discovered in turn by their gravitational distortions of the orbits of the next inner planets.

The next fundamental advance in astronomy was the application of **spectroscopy** to the study of the stars and planets. A spectrum of light is of course readily seen in a rainbow or in the common experiment in high school physics, in which sunlight is passed through a triangular glass prism and refracted into its component colors (Fig. 19-5). Light travels slower through glass than through air, and violet light, with shorter wave-

length, travels slower than red, with longer. We have already noted, in Chapter 10, how the solar spectrum extends to wavelengths both shorter and longer than those of the visible spectrum. Two further discoveries were needed before spectroscopy became valuable in astronomy.

Fraunhofer Lines

A Bavarian optician, Joseph von Fraunhofer, discovered in 1814 that the spectrum of sunlight contains sharply defined black lines: no light comes through the prism at certain wavelengths. Sunlight reflected from the moon shows black lines at precisely the same wavelengths. Fifty years later the German chemist R. W. Bunsen and his physicist colleague G. R. Kirchoff found that all gases at low pressure produce characteristic bright spectral lines when heated to incandescence. These **emission spectra** are each specific and characteristic of a particular gas. Some were found to yield bright lines at precisely the same wavelengths as some of the dark lines found by Fraunhofer. White light passed through a cool gas yields black lines—**absorption spectra**—at the same wavelengths as the emission spectra of the same gas. This means that the Fraunhofer lines in the solar spectrum are due to absorption by some gases in the outer atmosphere of the sun, or by the earth's atmosphere. Each element can be identified by its specific spectral lines.

If we drop some common table salt into a gas flame, it gives off a strong yellow light: incandescent sodium has strong emission spectra in the visible yellow. Other elements have other specific spectra. Kirchoff and Bunsen discovered the hitherto unknown elements rubidium and lithium by their characteristic spectra, and helium was identified in the solar atmosphere before it was found in our own. By their spectra, elements can be identified in stars many light years distant, if we can focus the light on a sufficiently sensitive spectroscope. For this the glass prism will not serve.

A great step forward was the invention, in 1878, of the **diffraction spectroscope**, by R. W. Wood, an American physicist, who devised a method of scratching almost incredibly fine parallel lines on a polished metal surface curved in such a way that it produces not only an enormously magnified spectrum but one in which lines of all wavelengths are equally in focus. This instrument per-

Figure 19-5
A diagram of a glass prism, showing how it refracts light to form a spectrum.

mits recognition of any element by its spectral signature. Molecules and ions can also be identified.

The emission and absorption spectra of an element are caused by the abrupt jump of electrons in their orbital shells from one energy level to another. The wavelengths that each element radiates and absorbs are unique because the nucleus of each element has a unique charge, and thus a unique input of energy is necessary to bring about a higher energy level for the electron involved (absorption spectrum). The same amount of energy is radiated in the emission spectrum.

Spectroscopes attached to the great reflecting telescopes enable us to photograph an incandescent object in the light of a single spectral line. From the displacement of certain spectral lines toward slightly lower wavelengths (Doppler effect), it has been determined that certain stars are approaching us; the solar system is traveling toward the constellation Hercules at many kilometers a second. On the other hand, the spectral shifts toward longer wavelengths in the vast majority of stars and galaxies (assemblages of millions of stars like our Milky Way) indicate that they are retreating at rates that increase with increasing distance from us. The universe seems to be expanding in all directions, although certain phenomena among the radio stars have caused some astronomers to seek another explanation for the "red shift." These matters are not further discussed, as they are far afield from geology.

Within our field of interest, though, and of vital importance in trying to understand the origin and evolution of the earth, is the fact that the spectroscope makes possible realistic appraisals of the chemical composition of the sun, of the distant stars, and of the atmospheres of such planets as have them. The estimates of the chemical compositions of the distant stars are not wild guesses but very respectable approximations. Yet, before the astronauts landed on the moon, we knew nothing of its composition, for it shines only by light reflected from the sun or the earth. Nor have we yet found a way to obtain the signatures of elements buried in the earth's core and mantle.

Radio Telescopes and Stellar Distances

Though versatile, optical telescopes and spectroscopes make use of only a part of the electromagnetic spectrum (Fig. 19-6). Most wavelengths

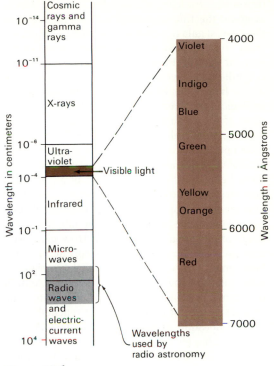

Figure 19-6
The electromagnetic spectrum. The fraction available to optical telescopes is shown enlarged. The diffraction spectroscope can cover the range between ultraviolet and microwaves. Radio astronomy makes use of the range shown by shading in the microwave–radio-wave area.

in the sun's radiation are shielded from our instruments by the earth's upper atmosphere and by its magnetic field—a happy circumstance for us, because exposure to radiation of certain ultraviolet wavelengths is very harmful and indeed lethal if long continued.

But radio waves of certain restricted frequencies can penetrate the atmosphere, and although radio astronomy is in its infancy, it has already found that the vast dust clouds in space, though opaque to the telescope, are transparent to radio waves emitted by hydrogen atoms. The radio signals have enabled radio astronomers to chart the "arms" of the Milky Way galaxy and to prove that we live in one of the arms of a spiral nebula about a third of the way from the center, as the telescopic astronomers had long suspected.

What an immensity of space we know! The light of the North Star that we see tonight began its journey, traveling at a speed of 299,776 km/sec at about the time of Columbus; that from one of the nearest galaxies comparable with our own, the great spiral nebula in Andromeda, started on its way far back in Precambrian time, for the nebula is 2 billion light-years distant. The demonstration of the immensity of space and of the duration of geologic time has wholly revolutionized man's view of the earth's place in the cosmos—a view held by nearly all as recently as 300 years ago. Concepts that then seemed ridiculous or even vicious heresies are now demonstrated facts. Far from being at the center of the universe, man occupies a relatively small planet orbiting an average-sized star in a galaxy of millions of stars —a galaxy that is itself only one of millions within range of our greatest telescopes. How many yet-unknown galaxies and solar systems lie beyond? How many stars are accompanied by planets? How many of the planets are habitable? And, of the habitable planets, has life developed on them, and what stage of evolution has it reached?

Consideration of the Law of Faunal Assemblages (Chapter 6) and of the vast parade of life, barely touched on in Appendix V, should suffice to show that extraterrestrial life closely comparable to the earth's is so improbable as to be virtually impossible. Evolution on earth

has been controlled by a vast succession of environmental changes so complex that precise parallelism of the course of evolution on any other planet with that on our own seems out of the question. Yet of course the existence of terrestrial life makes it almost certain that life also exists elsewhere and has doubtless evolved into organisms uniquely adapted to the available ecological niches, whether at all similar to those of the earth or not. It is safe to say we will never know; the time involved in communication with even the very near stars is so great that a question asked will not receive an answer in the lifetime of the questioner.

ORIGIN OF THE STARS

The origin of stars and planets is a fascinating subject of study, and there are nearly as many theories as there have been students. We review only a few of the currently popular schemes.

Figure 19-7 summarizes present knowledge of the distribution of the principal elements in the stars and dark clouds of outer space; in the earth's crust, oceans, and atmosphere; and in a living organism. The striking difference between the earth and the cosmos is at once apparent: the

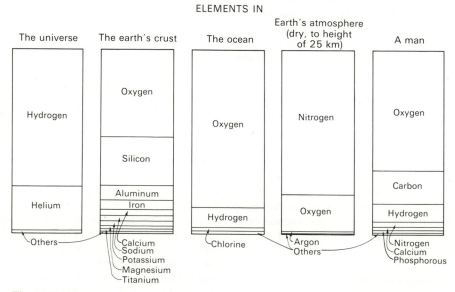

Figure 19-7
The distribution of the principal elements in the stars, in the earth's crust, ocean, and atmosphere, and in a living organism.

universe as a whole is composed almost entirely of the two lightest elements, hydrogen and helium. To be sure, nearly all the elements known on earth have also been detected in the turbulent atmosphere of the sun, but the most abundant earthly elements form almost infinitesimal parts of the sun and other stars. Hydrogen and helium are the overwhelmingly dominant components of stars, nebulae, and cosmic dust.

Physicists and chemists generally agree that hydrogen is the primary building block from which all other elements are ultimately formed. Its transformation into the next element, helium, is a complex process that was first suggested by Hans Bethe of Cornell University in 1938 as responsible for the sun's energy. The transformation begins at a temperature of about 10 million degrees; in simplified form, 4 hydrogen atoms (with 4 electrons) combine to form 1 helium atom (only 2 electrons), with the mass of the other 2 electrons being converted into the driving energy of the sun's furnace. This thermonuclear reaction is of the same general nature as that occurring in the hydrogen bomb. In larger and hotter stars, whose temperatures are thousands or even millions of times greater than the sun's, more complex thermonuclear reactions produce heavier and heavier elements in abundances that decrease, though not at a regular rate, with increase in atomic number.

Theoretical physicists suggest that during the evolution of extremely large stars, the reactions may become so intense that the star explodes, forming an extraordinarily bright **nova**. The enormous pressures of such explosions scatter stellar debris as clouds of dust and gas for tremendous distances. The great pressures and high temperatures combine to form some of the heaviest of the elements. Several *novae* have been recorded in history, though of course the actual explosions took place millions of years ago; the light of the explosions is just now reaching our eyes. Generally the intensely bright novae diminish rapidly in brilliance, some to complete extinction within a few months or years. The widely dispersed particles soon (on an astronomical time scale!) cool to form dark clouds.

Of course no one knows how a new star is formed; a currently popular theory among astronomers and physicists is that the cold dark clouds in space, derived from eruptions of novae, are slowly driven into more compact masses by light pressure from the surrounding galaxies. With increasing density, their own gravitational attraction tends further to condense them. As the dust clouds are by no means of uniform density, but are highly turbulent, their contraction would not be by simple centripetal motions of all particles: chance aggregations of higher density would deflect the motions of nearby particles, just as the attractions of the outer planets disturb the orbit of Saturn. The contraction would not yield a spherical aggregate of the whole cloud but a disc-like agglomeration, rotating generally in the plane of the largest aggregates. The shape would approximate that of nearly all the bright galaxies, including our own, and of the only planetary system we know. All parts of the collapsing cloud would eventually swing about their common center of gravity in a disc-shaped spiral.

As the cloud shrinks under self-gravitation and light pressure, it heats up and, if the mass is sufficient, it eventually reaches a temperature high enough for thermonuclear reactions to begin; the mass becomes luminous, and a star is born. In some clouds more than one concentration of mass is large enough to become a star; there are many double stars in our galaxy.

ORIGIN OF THE EARTH

The currently most popular hypothesis of the earth's origin is a direct sequel to this hypothesis of star formation. As the dust cloud parental to our solar system began to contract, certain scattered accumulations of matter grew at the expense of the more rarefied volumes between them. If massive enough these might also develop into stars, but in our system none became sufficiently massive (though Jupiter is hot enough to radiate three times as much heat as it receives from the sun). These aggregates became the **protoplanets,** composed largely of hydrogen and helium gases and dust, with some solid bodies resembling the various kinds of meteorites.

In our solar system (as we know it), the first protoplanet to form would naturally be parental to Pluto, most distant from the sun, and so the coolest. The next possible stable accumulation of gas, dust, and meteorites would have to be far enough away from Pluto to remain separate from it; the same would hold true for each of the planets that formed in succession inward from Pluto to

Mercury. Tidal attractions of these masses undoubtedly disrupted thousands of smaller bodies that failed to reach greater sizes; almost certainly the asteroids between Mars and Jupiter are residual from such breakups.

Because of the randomness of collisions with dust particles, meteorites, and larger aggregates that the protoplanets were gathering to themselves, their originally elongated orbits became nearer and nearer to circular, though none quite attained circularity. This is one of the most striking properties of the planetary orbits, and of prime importance in making the earth a habitable planet, for it assures that the solar radiation received throughout the year, and hence the surface temperature, varies but little from season to season.

The evolution of the protoplanets would be greatly affected by their masses (which determine their gravitational attraction), and by their distances from the sun (which determine the amount of radiation and solar wind they receive). Jupiter, by far the largest of the planets, is far enough away from the sun and large enough to have retained a composition comparable to that of the sun, largely hydrogen and helium with small amounts of the other elements. The inner planets, so much nearer the sun and with much less mass, are subject to far higher radiation. The gravitational attraction of the earth was so slight that the **solar wind**—the emanations of ions and photons shooting outward from the sun—swept away all the gas on and around the primitive earth, leaving nothing but solid meteorites and other solid accumulations to aggregate into our planet.

This conclusion is supported by a comparison of the abundances of the inert gases on the earth with their abundances in the sun and other stars as revealed by the spectroscope. We ignore helium, the lightest of the inert gases, because it is still escaping from the upper atmosphere, and the amount present depends on the release of radiogenic helium during erosion.

But the heavier gases, neon, argon, krypton, and xenon are too heavy to escape from the earth now, and their abundances should record the amount present from the beginning. Careful studies by Harrison Brown, an American geochemist, have shown that all these inert gases are far less abundant (compared with silicon) on the earth than they are in the sun, and that the lighter of them are more impoverished than the heavier, in direct order. If, in accord with the suggestion outlined above, the earth was originally composed of gases and small solid bodies of average solar composition, it has lost through diffusion and impact of the solar wind a truly stupendous mass of both hydrogen and helium—a mass many times that of the earth as we know it. In addition, huge volumes of all the other light gases have also escaped; what remains is only a minute fraction of their original abundance. The "fractionation factor" of the inert gases—that is, the abundance of the element compared with that of silicon divided by the terrestrial ratio to silicon—is shown in Table 19-1 and in graphic form in Figure 19-8.

The table suggests that these noble gases—so called because they never form compounds in nature and are always in the gaseous state—have

Table 19-1
Estimates of maximum and minimum ratios of rare gases to silicon on earth; their solar ratios to silicon, and their fractionation factors.

Gas	Atomic weight	Estimated terrestrial ratio to silicon (atoms per 10^4 atoms of silicon)		Cosmic ratio to silicon	Fractionation factor	
		Max	Min		Max	Min
Neon	20.2	8.1×10^{-10}	1.1×10^{-10}	37,000	$10^{10.5}$	$10^{9.7}$
Argon	39.9	2.5×10^{-7}	5.5×10^{-8}	1.000	$10^{8.8}$	$10^{8.1}$
Krypton	83.8	4.4×10^{-11}	6.0×10^{-12}	0.87	$10^{7.2}$	$10^{6.3}$
Xenon	131.3	3.6×10^{-12}	4.9×10^{-13}	0.015	$10^{6.8}$	$10^{5.7}$

SOURCE: Data from Harrison Brown, 1952.
NOTE: Fractionation factors are solar abundances compared with that of silicon divided by the maximum and minimum estimates of terrestrial abundance ratios.

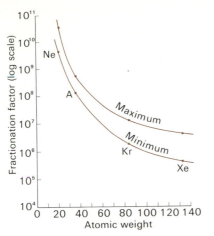

Figure 19-8
Fractionation factors of the noble gases. Note the logarithmic scale. Maximum fractionation above, minimum below, based on estimates of terrestrial abundance.

these molecules could not have been present as gases. This is shown by the fact that water is more than 10^7 times as abundant as one would expect if it had been present as water vapor throughout the time of depletion. Nitrogen is more than 3000 times as abundant, and oxygen, despite the fact that huge tonnages have been consumed by weathering in the long course of time, is still present (*in the atmosphere alone—ignoring that locked up in the sedimentary rocks as iron rust*) in 340 times the abundance that would have been expected had it always been gaseous. Carbon dioxide is more than 4000 times as abundant as the maximum expected.

If this hypothesis is correct, then these components of the present atmosphere were not there at the time of the great solar wind; they must have been combined, the oxygen and water perhaps as hydrous silicates, the nitrogen as nitrides or nitrates, the carbon dioxide as carbonates or else dissociated, and the carbon as carbides and graphite. In other words, the earth at that time had no atmosphere and no water vapor available to form the seas. *Both the atmosphere and oceans have formed from emanations from the interior of the solid earth*, as will be discussed from several other viewpoints later in this chapter.

The above hypothesis is widely but by no means universally accepted. Some distinguished scholars are impressed with the fact that the chondrites, one class of meteorites, contain rare gases in about the same proportions as the earth and that their mean composition is not far from the inferred composition of the earth as a whole (Chapter 8). If the earth formed by assemblage of great volumes of chondrites and metallic meteorites, one might expect just what is found,

been stripped away from the earth, or protoearth, apparently by the solar wind, in roughly inverse ratio to their atomic weights, the lighter elements being more depleted than the heavier. If water, nitrogen, oxygen, and carbon dioxide had been in gaseous form at the same time and the solar wind thus acted to strip them from the earth, one would expect that, as with the noble gases, their fractionation factors would depend on their molecular weights, listed in Table 19-2.

Instead of having been swept away in roughly inverse proportion to their atomic weights, the major components of today's atmosphere and ocean are in proportions that show no consistent dependence on their molecular weights. If the noble gases were swept away by the solar wind,

Table 19-2
Expectable fractionation factors for the present atmospheric components, had they been present as gases throughout earth history.

Molecules	Molecular weight	Expected fractionation factors
H_2O	18.02	Between 1×10^{10} and 1×10^{11}
N_2	28.02	Between 6×10^8 and 3×10^9
O_2	32	Between 3×10^8 and 1.5×10^9
CO_2	44.01	Between 5×10^7 and 2.5×10^8

SOURCE: Date from Harrison Brown, 1952.

Table 19-3
Some members of the solar system.

Object	Mean diameter, kilometers	Mean distance from sun (kilometers)	Mass Earth = 1	Density	Period of rotation
Sun	1,382,000		331,950	1.4	25 days
Mercury	44,640	57,584,000	0.05	6.1	58.67 days
Venus	12,160	107,520,000	0.81	5.06	243 ± .05 days
Earth	12,660	148,800,000	1.0	5.52	23 hr 56 min
Moon	3456	148,800,000	0.012	3.34	27 d 32 min
Mars	6720	229,560,000	0.11	4.12	24 hr 37 min
Jupiter	138,900	773,280,000	318.4	1.35	9 hr 56 min
Saturn	114,400	1,417,900,000	95.3	0.71	10 hr 40 min
Uranus	47,000	2,852,000,000	14.5	1.56	10 hr 48 min
Neptune	44,000	4,470,000,000	17.2	2.29	15 hr 48 min
Pluto	10,000?	5,870,000,000	0.8?	?	6.5 da ± 0.3 da

noble metal ratios and all, in which case there is no need to postulate an internal origin for earth's atmosphere and seas. We will note later that several features of the earth's history are difficult or impossible to reconcile with this view, hence we favor the majority view. But in science, mere wide acceptance is no evidence of truth!

The hypothesis of planetary origin just outlined gives a partial explanation for the differences in planetary densities and masses shown in Table 19-3: the four inner planets all have densities far higher than the outer five. It is suggested that the solar wind has stripped the inner four of nearly all the gas present in their protoplanets and perhaps even of some nongaseous but lighter particles. This mechanism, however, can hardly account for the anomalously low density of Saturn or the anomalously high density of Neptune. Least satisfactory of all, though, is the great discrepancy in density between the earth and the moon—far greater than the differences in self-compression caused by their differences in mass. Certainly the moon cannot contain iron in anything like the earth's proportion. Sir George Darwin more than half a century ago tried to explain this by assuming that the moon was thrown off the young earth by rapid rotation, leaving the Pacific basin as the scar whence it was torn away. Some of the strong geological arguments against this have already been mentioned in Chapter 8.

However uncertain the processes involved in forming the solar system, the systematic differences between the inner and outer planets must be significant. These are tabulated in Table 19-3 and illustrated in Figure 19-9.

The theory of origin of the planetary system just outlined requires many amendments, not merely because of the irregularities just mentioned but for many others also: the axis of Uranus is tilted away from the normal to the plane of the orbit, not the few degrees of the other planets, but more than 80°; several of the satellites revolve in orbits opposite to those of their principals, and there are many other irregularities. The theory, despite its incompleteness, is widely accepted as the best current approximation. Because of their historic interest, and because the current theory has borrowed from both of them, two older theories are briefly outlined, though they have been discredited as seemingly fatal objections have confronted them.

The **Nebular Hypothesis** was proposed by the German philosopher Immanual Kant in 1775, and was worked out in considerable detail by the French mathematician Pierre Laplace about 40 years later. They postulated a slowly revolving dust cloud, which, as it shrank because of gravitational attraction, speeded up its rotation, just as a skater spins faster when he pulls his arms closer to his body. The centifugal force built up by the rotation was so great that rings comparable to the rings of Saturn were successively left behind as the dust cloud shrank. The material of each ring was thought to aggregate into a planet; the residual mass, the sun, became heated by infalling material and self-compression and became incandescent.

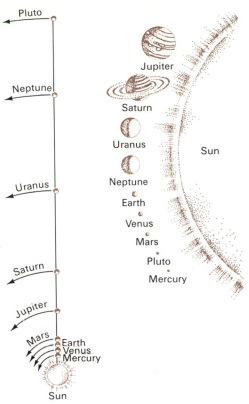

Figure 19-9
Diagrammatic representation of the sun and planets:
right, according to size; left, according to distance
from the sun.

lision of the sun with another star. The close
approach created enormous tides and wrenched
from the sun a long sausage-shaped filament of
matter that broke up into separate bodies that
whirled around the sun in a spiral. They suggested
that the ejected matter quickly cooled into solid
fragments ranging in size from dust particles to
large **planetesimals**.

Their greater gravitational attraction enabled
the large planetesimals to sweep up the smaller.
The pock-marked surface of the moon (which has
neither air nor water to erode away its scars of
impact) was considered evidence of such a process (Fig. 19-4). The accreting planets would at
first be small cold bodies unable to hold an atmosphere, but, as they grew, some could retain
volatiles and form atmospheres and oceans, and
gravitational compaction might heat even a
rather small body to the point where volcanism
could begin.

Later analysis has indicated that the likelihood
of such a "grazing" passage of two stars producing the ejection of a filament is remote. Again, the
small contribution of the sun to the system's
angular momentum is a difficulty. Most modern
guesses about planetary origin, have, however,
retained the idea of planetesimals and "cold"
accumulations of the planets, as first suggested
in this hypothesis.

Two objections to the scheme were early
pointed out: (1) the thin rings of dust would dis-
sipate into space, or at least not aggregate into
single planets any more than have Saturn's rings
and the belt of microplanets, the **asteroids**, that
orbit the sun between the paths of Mars and
Jupiter; (2) the sun should be spinning very much
faster than it is instead of loafing along at spin
rate that contributes only about 2 percent of the
angular momentum of the system as a whole.
Though this theory was popular for nearly a century, it has now been generally abandoned.

The **Planetesimal Hypothesis** was first suggested
by Georges Buffon, a French astronomer, in 1749,
but it was mainly developed early in the present
century by T. C. Chamberlin, geologist, and F. R.
Moulton, astronomer, of the University of Chicago and was further evolved by Harold Jeffreys,
geophysicist, and James Jeans, astronomer, of
Cambridge University. They assumed a near col-

Origin of the Earth's Core

Most current hypotheses of planetary origin
assume that the earth accumulated as a relatively
cool body, an assumption supported by its retention of water, carbon dioxide, nitrogen, sulfur,
chlorine, and the volatile metal mercury. These
are all much more abundant on the earth than they
are in the sun; had the earth been wholly molten,
all of them, including mercury, with an atomic
weight of 80, should be less abundant, as is
xenon, with nearly as heavy an atom. Had the
earth ever been molten throughout, its crust
should be much more uniform than it is, for layered intrusives, stratified by gravitative differentiation, should have crystallized throughout the
mantle. We saw in Chapter 8 that the mantle is
solid to depths of 2900 km but is so poor a conductor of heat that even in 4 billion years no heat
can have escaped by conduction from depths less
than half as great. Some students reject this argument on the ground that much heat can have been

carried away by radiation and convection. Laboratory studies do not, however, suggest that molten silicates are transparent to heat, though there is strong suggestion of active convection in the mantle even today. But even effective convection seems inadequate to account for solification of the mantle to great depths.

Still another argument for an earth that has never been molten throughout is the wide variation from place to place of the concentration of ore minerals; had there been a well-mixed molten globe, copper deposits should be as numerous and as rich in Kansas or Brazil, proportional to area, as those in Zambia, Chile, or Arizona, but obviously this is not so. For these and other reasons, most students, though not all, think it highly unlikely that the molten earth postulated by Buffon and Lord Rayleigh ever existed.

How, then, could the earth have developed a liquid core, in which are concentrated nickel, iron, silicon, and sulfur, making up more than 32 percent of the earth's mass? Some geophysicists think the core is still growing; iron melts at considerably lower temperatures than the high-pressure modifications of olivine, pyroxene, and garnet—the supposedly dominant components of the mantle. Thus, even though the mantle as a whole is solid, it may contain interstitial films of molten metals. Because of their greater density, these films are continually working their way downward as tidal stresses distort the mantle, so that the core may still be growing. Such an increasing concentration of mass toward the center tends to increase the speed of rotation, thus counteracting the tendency of the tides to slow it down.

Other students take a less uniformitarian view. They point out that since the earth was born it has lost much radioactivity. Four billion years ago the earth contained more than sixty times as much ^{235}U and nearly twice as much ^{238}U as it does now, and there were perhaps several radioactive elements of far shorter half-lives also evolving heat, though they are now so weakly radioactive as to have escaped notice. Thus the mantle might have heated up to the point of being molten even though the crust was not: the iron sank to the core in an almost catastrophically short time, and the convection that accompanied the cooling of the mantle produced a gigantic overturn, dragging all the sial to one side to form the land hemisphere and leaving the Pacific Basin.

We have seen in Chapter 8 that the floor of the Pacific is much too young to be thus explained, for whatever the condition of the mantle 4 billion years ago, it is certain that it was solid at so recent a time as the Jurassic.

Whatever the true history may have been, at the time the earth had accumulated most of its mass—the end of the astronomical history and beginning of its geological evolution—it had no atmosphere, no ocean, no sedimentary rocks. Its surface was doubtless littered by large and small fragments of meteorites, like the present-day surface of the moon. All the present atmosphere and oceans were sweated out of the interior by volcanic processes, and all the sedimentary rocks that today cover most of the continents were produced—partly by pyroclastic eruptions, but mainly by weathering, erosion, and sedimentation after the oceans and atmosphere had developed. This conclusion, long assumed by geochemists, seems strongly supported by the arguments that follow.

THE ORIGIN OF THE OCEAN

Were the earth born with neither atmosphere nor ocean, as is strongly suggested by the fractionation of the noble gases as compared with those of the present atmosphere and seas, they must have developed from the earth's interior. They could not possibly have been derived as was, say, the sodium in the sea, by mere erosion of the "average igneous rock."

The arguments for this have been cogently presented by the American geologist W. W. Rubey. He arbitrarily assumed that weathering and erosion can proceed in the absence of water and carbon dioxide. He then computed the volume of water and carbon dioxide that could have been freed from the "average igneous rock" during the erosion required to supply all the sedimentary rocks of the earth's crust. He found that even accepting the highest estimate of sedimentary volume and thus the greatest volume of igneous rocks from which to derive them, such volatile substances as water, carbon dioxide, nitrogen, chlorine, and sulfur are present in far greater quantities than could have been so derived.

His assumption was, of course, an impossible one; without water and carbon dioxide, the complex weathering reactions that produce the clays and release the metallic ions when feldspars are decomposed could not go on. So there had to be an ocean and an atmosphere before any weathering could take place. The impossible assumption, however, was not without value, for it focused attention on the problem posed by the existence of a very large ocean and atmosphere at present, though both were almost surely absent from the early earth.

We pointed out in Chapter 17 that data on both the abundance and average composition of the sedimentary rocks are weak and that the volume of igneous rocks eroded to produce them may have been underestimated by an order of magnitude. Despite this serious deficiency in our knowledge, it has for several decades been clear that whichever volume of igneous rocks (within these limits) one chooses as the source of the sedimentary rocks, the volatile substances enumerated above are far more abundant in the atmosphere and oceans alone (not counting the vast quantities locked up in the sedimentary rocks) than could have been supplied simply by erosion of the igneous rocks. The discrepancy is truly prodigious, as we shall see. Two semi-independent estimates have been made of the amount of these "excess volatiles" (Table 19-4).

Poldervaart used many of Rubey's data, but made separate computations. He did not specifi-

cally compute the excess volatiles, but as the table shows, except for his estimate of carbon dioxide (nearly $2\frac{1}{2}$ times Rubey's), he was in close agreement. Should Poldervaart's estimate of CO_2 prove the better, it serves only to strengthen Rubey's argument further that the ocean has grown during geologic time by emanations from within the earth.

The argument is broadly as follows: if all the "excess volatiles"—those that could not have been derived from erosion of pre-existing igneous rock—were in gaseous form, that is, in an atmosphere either at high temperature or in an ocean-atmosphere association like today's, the fluid or fluids would be under very high pressure and very strongly acid. If all were in the atmosphere, the atmospheric pressure at the earth's surface would have been somewhere between 200 and 500 times that of the present. At such concentrations the strongly acidic solutions would rapidly corrode any igneous rock or meteorite with which it came in contact. All the bases, major and minor, aluminum, calcium, magnesium, potassium, sodium, iron, and any other, would be taken into solution. For reasons we review later, in discussing the primitive atmosphere, little if any free oxygen was present in either ocean or atmosphere. Thus it is a gross oversimplification (but one that on analysis proves immaterial) to assume

Table 19-4
Estimated quantities (in units of 10^{14} metric tons) of volatile substances now at or near the earth's surface, together with the amounts of these substances that could have been derived by weathering of the average igneous rock.

Volatile substance	Source*	H_2O	Total as CO_2	Cl	N	S
Present in atmosphere, in surface and ground water, and in organisms	R	14,600	1.5	276	39	13
	P	14,430	2	310	40	15
Buried in ancient sedimentary rocks	R	2100	920	30	4	15
	P	2010	2280	30	5	15
Totals	R	16,700	921	306	43	28
	P	16,440	2282	340	45	30
Supplied by weathering of average igneous rock (maximum estimate)	R	130	11	5	0.6	6
Difference—that is, the "excess volatiles" unaccounted for by rock weathering	R	16,600	910	300	42	22

*R; estimated by W. W. Rubey, 1951. P; estimated by A. Poldervaart, 1955.

that the materials carried to the sea in early geologic history were in the same proportions as those found today in streams draining igneous rocks. The four main bases in such modern streams are present in roughly the following proportions:

Ca	53
Mg	8
Na	25
K	14
	100 percent

In addition, large quantities of aluminum and iron are now, and doubtless were in the past, carried to the sea in solution, along with many other elements. At first the water would have been too acid to permit calcite to be deposited, but when it was neutralized, calcite would be an early mineral to form and to draw on the reservoir of CO_2.

Over the past $3\frac{1}{2}$ billion years, evolution has undoubtedly modified the tolerance of many organisms to changes in salinity, concentration of carbon dioxide, nitrogen, sulfur, and many other chemicals, but within the span of the metazoan record, from the beginning of the Cambrian onward, these modifications seem to have been small. Few forms of life can exist under a pressure of carbon dioxide more than a few times that of the present atmosphere (0.03 percent of one atmosphere). Certain mosses can live under pressures of as much as one atmosphere of CO_2, and we may follow Rubey in assuming that life could begin when the pressure of CO_2 reached this level.

If we accept this level as fixing the time of origin of plant life and hence the beginning of photosynthesis, to which virtually all the atmospheric oxygen is due (Fig. 19-10), the sea water (on the hypothesis of an original atmosphere) would then still be slightly acidic, salinity would have been twice that of the present ocean, and the amount of bases dissolved in order to bring about these conditions would be fully as much as the most generous estimates of rock weathered through all of geologic time. All this at a time when more than half the CO_2 postulated as present in the primitive atmosphere was either still there or dissolved in the ocean! Under this artificial assumption that all the excess volatiles were in the primitive atmosphere, prodigious quantities of limestone would have had to be

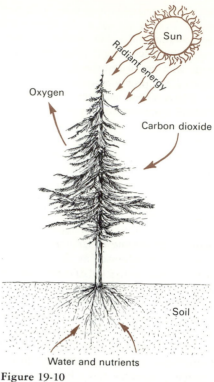

Figure 19-10
Photosynthesis utilizes carbon dioxide from the air and releases oxygen to it.

deposited before the carbon dioxide pressure could have fallen to as little as one atmosphere; the early Precambrian of every continent should be rich in limestone and dolomite.

It is of course true that our record of early Precambrian sedimentary rocks is extremely slender, but such samples as we have are, if anything, less rich in carbonate rocks than are the rocks of later times. This hypothesis also requires the solution of more sodium—from the same minerals that supplied the calcium, even before the beginning of life—than can reasonably be inferred for all of geologic time. The conclusion thus seems inescapable that *the primitive ocean could not have contained all or even a large fraction of the excess volatiles now present.* That the ocean and atmosphere could have been fully developed early in earth history seems impossible in view of the very persuasive evidence that the earth was born without either. Other wholly independent arguments tend also to show the gradual formation of the ocean throughout geologic time.

Gradual Accumulation of the Ocean

Let us assume, for the moment, not only the uniform action of physical and chemical laws throughout geologic time, but also that other conditions were relatively uniform—that is, that only a small fraction of the excess volatiles were ever present in the atmosphere at any one time, as is true today (in other words, that all the volatiles have been added to the atmosphere slowly from the earth's interior throughout geologic time). If, for example, the pressure of carbon dioxide never exceeded one atmosphere, carbonates could have begun precipitating when the atmosphere and ocean contained less than 10 percent of the total volatiles. The total atmospheric pressure, instead of being hundreds of times as great as at present, would have been only 10 percent higher, the salinity of the sea about that of today, and its acidity only slightly higher, according to Rubey's computations.

On this hypothesis, life could perhaps have evolved in the very early Precambrian (Figure 6-1) without the earlier deposition of vast thicknesses of limestone on the sea floor. If, at this time, not enough free oxygen had yet accumulated to oxidize hydrogen sulfide, Rubey estimated that not more than 6 percent of the total excess volatiles could have been present in this early atmosphere and ocean; the volume of the ocean would have been much smaller than at present.

The balance between the positive and negative ions in sea water is very delicate. A decrease of only 1 percent of the sodium—all other elements remaining unchanged—would render the sea highly acidic, fatal to many if not most organisms. The same holds for carbon dioxide, whose relations to the other components must lie within very narrow limits indeed to avoid fatal consequences on a massive scale to both flora and fauna. Despite Cuvier's conclusions (Chapter 6), the stratigraphic record contains no evidence whatever of such wholesale massive extinctions. If we assume, then, that the ocean has grown slowly throughout geologic time, we must find a mechanism whereby water and carbon dioxide—and when examined in detail, other of the excess volatiles—are simultaneously added to the atmosphere and ocean in closely consistent ratios.

Today, CO_2 is being added to the atmosphere and ocean in three ways: (1) by rock weathering, (2) by decay and artificial combustion of carbon-bearing materials, and (3) by emanations from volcanoes and hot springs. If we ignore the last two, we find the loss of CO_2 in deposition is far greater than the gain from weathering. Weathering of sedimentary rocks furnishes roughly equivalent amounts of CO_2 and Ca on the average, but weathering of igneous rocks uses up great quantities of CO_2 in balancing the calcium released (see Table 19-4 and Chapter 10). Rough calculations by several workers indicate that the amount of carbon dioxide removed from the ocean to maintain this balance would, *in only a few hundred thousand years,* render the ocean so strongly alkaline—the abundance of OH^- ions would rise so high—that brucite, a rare mineral of composition $Mg(OH_2)$, would be the first stable precipitate from the sea, rather than calcite. Earlier in geologic history, this condition would have arisen within a very brief time, geologically speaking, unless another source of CO_2 can be found, for of course combustion added trivial amounts before the beginning of the industrial age. And in the early Precambrian, before the development of abundant plant life, even decay must have been a trivial source.

The fact is that no brucite-bearing strata have ever been found; we must conclude that the ratio of CO_2 to water has not strayed far from its present value at any time in recognizable geologic history. Neither brucite-bearing strata nor massive biologic extinctions record even relatively mild aberrations. We are driven to seek another source; it is clearly the third source in the preceding paragraph: volcanic emanations. This was suggested as long ago as 1894 by the distinguished Swedish geologist A. G. Högbom, and has been widely accepted ever since. Let us look into the possibilities.

Similarity of Magmatic Gases and Excess Volatiles

Rubey gathered analyses of gases from volcanoes, igneous rocks, and hot springs; his data are summarized in Table 19-5.

Obviously the various analyses differ greatly, and some more recent analyses differ still more. Nevertheless, when all the modifying factors, such as dilution by ground water, organic activity, reworking of sedimentary components, and many others are considered, the similarity is close enough to make it probable that the "excess volatiles" have indeed been released by

Table 19-5

Composition of gases from volcanoes, igneous rocks, and hot springs, and of "excess volatiles" in the atmosphere, hydrosphere, and sedimentary rocks

Gas	Kilauea and Mauna Loa	Basalt and diabase	Obsidian, andesite, and granite	Fumaroles, steam wells, and geysers	Excess volatiles
H_2O	57.8	69.1	85.6	99.4	92.8
Total C as CO_2	23.5	16.8	5.7	0.33	5.1
S_2	12.6	3.3	0.7	0.03	0.13
N_2	5.7	2.6	1.7	0.05	0.24

SOURCE: Median analyses of major components, calculated to weight percent; after W. W. Rubey, 1951.

igneous activity within the earth. They have reached the surface either directly in lava (already shown to be inadequate as the sole source), by fumarolic activity, or via ground water from hot springs. Furthermore, the volcanic release after the evolution of life must have been in nearly uniform proportions; otherwise, the effects of drastic changes in the ratio of CO_2 should be recorded in the rocks.

Though the proportions must have been roughly uniform, the rate was probably not. The early earth contained more than 60 times as much ^{235}U and twice as much ^{238}U as at present; volcanism must surely have been considerably more active then, so that the ocean would have grown faster in early Precambrian time than later.

That hot springs and volcanism have been responsible for the growth of the ocean in geologic time has been questioned by some students on the ground that the oxygen of igneous rocks is richer in the heavy isotope ^{18}O than are hot-spring waters. Rubey has pointed out, however, that only one percent of hot-spring waters need be juvenile in order to supply the amount needed. Isotopic analysis is not yet competent to exclude this possibility (see Table 7-3).

ORIGIN OF THE ATMOSPHERE

Thus "excess volatiles" must have been added to the atmosphere and oceans in roughly constant proportions throughout geologic time. Among them, of course, is the dominant component of the atmosphere, nitrogen (78.05 percent), the extremely important minor constituent, carbon dioxide (0.03 percent), and variable amounts of water vapor. The question remains as to the source or sources of the other constituents: oxygen (20.99 percent), argon (0.94 percent), and helium, neon, krypton, xenon, and hydrogen, which together make up only 0.01 percent. Whence came they?

The abundance of oxygen is the most striking aspect of the atmosphere, for it is not, like nitrogen, a component of volcanic emanations. Furthermore, many processes are operating to consume it: oxidation of metals at the earth's surface, respiration of animals, combustion of carbon compounds. A little is released in the high atmosphere through dissociation of water by solar radiation. Because of its low density and the high temperatures prevailing in the upper atmosphere, most of the hydrogen produced in this way escapes to outer space. Much of the oxygen remains in the upper atmosphere in the form of ozone molecules (O_3), whose absorption of some of the most lethal wavelengths in the sun's spectrum is of crucial importance to life. But some of the oxygen is carried by atmospheric turbulence down to the earth's surface. Though this amount is trivial compared to that produced by photosynthesis, it was of prime importance in the earth's early history, for it was then the only logical source of the oxygen necessary for the beginning of life. By far the greatest source of oxygen today—and it must be continually produced to make up for the vast quantities consumed—is photosynthesis by green plants. In this process the plants use the sun's radiation to decompose carbon dioxide and water, thus building hydrocarbons into their structures and releasing oxygen to the air (Fig. 19-10).

Helium is the major noble gas in the atmosphere; it is slowly escaping to space, but is being replaced by helium formed during radioactive disintegration of uranium and thorium and re-leased during erosion. Small amounts of helium and other noble gases may be released from the mantle and crust by volcanism. A little hydrogen has also been detected in volcanic emanations.

THE ORIGIN OF LIFE

The classical philosophers of Miletus (600 B.C.) and the Neoplatonists, such as St. Augustine, of the early Christian era, thought that life is an inherent property of all matter, and that the germs of life are present throughout the Universe—a concept held by some philosophers to this day. During the Middle Ages grotesque creatures were fancied as living on the other planets and even on the sun. With the growth of the biologic sciences, and especially since the work of Pasteur, it has become obvious that this simplistic view cannot be true. Special conditions, varying from one life group to another, are necessary for reproduction and growth, and under many conditions no life at all is possible. The turbulent atmosphere of the sun is composed wholly of atoms, ions, and electrons; clearly no organism can exist there.

Professor George Wald of Harvard University has shown that carbon, hydrogen, and oxygen are essential to all life (silicon cannot substitute for carbon except in unicellular organisms, and there only as inert capacitors, despite the fact that the two elements are otherwise much alike). Unless these three elements are available, no life even remotely resembling that on earth is possible. Yet, although the environment controls the possibilities of life, life itself also modifies the environment, as is demonstrated by the process of photosynthesis by plants.

The environmental conditions for the origin of life on earth depended quite obviously on the earth's earlier history. It may be assumed that volcanism began early but that at first the volatile emanations were swept away by the solar wind of the young sun. As the solar radiation diminished, some volatiles, such as water and carbon dioxide, were retained in sufficient quantity to act as a partial radiation shield, and an anoxic atmosphere began to form. The subsequent physical and chemical processes during the evolution of the atmosphere have been penetratingly studied by the American geophysicists L. V. Berkner

and L. C. Marshall, on whose work the following condensed account is largely based.

As already noted, water vapor is subject to dissociation into its constituent oxygen and hydrogen by solar radiation in the upper atmosphere. The earth is not massive enough to retain the hydrogen, but some of the much heavier oxygen reaches the ground in rains and atmospheric turbulence. Only a little oxygen can accumulate in this way, and most would be quickly used up in oxidizing many of the minerals of exposed rocks. Oxygen generation would be somewhat self-controlled because molecular oxygen (O_2) absorbs the same wavelengths that control the dissociation of water vapor. After accumulation of as little as 0.1 percent of the present atmospheric content of O_2, the oxygen-releasing dissociation of water vapor would be almost halted.

Berkner and Marshall concluded that at such low oxygen concentrations, much of the O_2 would disintegrate under ultraviolet radiation into atomic O and ozone (O_3), both of which are far more active chemically than molecular oxygen, O_2. Thus, even though the oxygen content would remain extremely low, the atmosphere would be oxidizing; the abundant "red beds" of the Precambrian do not necessarily indicate an oxygen-rich atmosphere, as has commonly been thought. In fact, Berkner, Marshall, and other students think that the lowest organisms—and therefore presumably the earliest to evolve—could have been poisoned by too much oxygen. Independent evidence of an oxygen-poor atmosphere is the association of water-worn pebbles of uraninite and pyrite in the middle Precambrian placer deposits of the Witwatersrand gold field of South Africa. Uraninite is readily disintegrated in an oxygen-rich atmosphere, and pyrite, as it so readily oxidizes, is unknown in recent stream gravels. No placer uraninites are known from rocks less than 2 billion years old.

The early atmosphere was thus primarily one of nitrogen and carbon dioxide, though the content of water vapor would continually increase after the atmosphere became dense enough to permit liquid water to accumulate in the primitive lakes and seas.

In order for life to form, three chemical requirements must be met: (1) hydrocarbons must be present, (2) no free hydrogen may be present, and (3) water must be available. Under present conditions, virtually all the hydrocarbons on earth are formed by living organisms, chiefly through photosynthesis. A few bacteria, themselves composed of hydrocarbons, can convert carbon dioxide directly into organic compounds by oxidizing sulfur, iron, or nitrogen. But, before the rise of bacteria, how did the hydrocarbons necessary for the *beginning* of life become available? Carbides of metals, particularly of iron, nickel, and cobalt, are common in meteorites, though very rare on earth. The primitive earth must have contained at least some of them. When these carbides react with water, they form hydrocarbons of inorganic origin. Methane (CH_4) is an important constituent of the atmospheres of Jupiter, Saturn, and Neptune, where it is almost certainly of inorganic origin. The primitive earth should have lost most of any original methane content, but some might remain enclosed in solid accumulations in the early history of the planet. So at least some hydrocarbons of nonbiogenic origin may have been available early in earth history or have been developed by reaction of carbides with water.

We have repeatedly cited the availability of water from volcanic processes; what about the third essential, absence of free hydrogen? Except for the small amount in the upper atmosphere, escaping to space after dissociation of water vapor, the earth could retain little free hydrogen after its early degassing. What it did produce was quickly combined with other volcanic products to form ammonia, hydrogen sulfide, and hydrochloric acid.

Accordingly, fairly early in its evolution, the earth possessed the chemical properties suitable for life. How did life's complex association of compounds with the property of reproduction arise? Small amounts of oxygen were available to oxidize the simpler organic compounds into such more complex substances as alcohols, aldehydes, and other constituents. No one, of course, knows the details of the process. Experiments in which electric discharges have been passed through organic compounds of very primitive type have yielded amino acids—compounds that in nature are found only in organisms. Perhaps lightning strokes formed some similar compounds in the early seas or in water-saturated soil.

According to the Russian biochemist A. I. Oparin and his astronomer colleague V. Fesenkov, who have given much effort to studying the preconditions of life, small organic molecules would remain diffuse, but if by chance large molecules analogous to a virus were formed, they would tend to aggregate. If modern proteins are dispersed in a suitable medium, and those of different composition are mixed in equal amounts, one variety always coagulates into larger segregations than the other; it grows by absorbing part of the material making up the other.

But this process does not go on indefinitely. Eventually the large aggregates split into fragments, additional growth of which depends on the availability of additional coagulants in the solution. We see, in the growth of one variety of protein accumulating at the expense of another, a kind of natural selection. The larger bodies that grow and split leave their fragments in competition for whatever salts are useful for growth; here again, is an opportunity for natural selection to operate. Eventually a complex developed with the capacity to select just the right constituents from the watery solution to enable it to grow and split into fragments that themselves had the power of regeneration. Thus the process of fermentation began, and plant life had come into being. The time required for this evolution may well have been many hundred million years. If the earth is indeed about 4500 million years old it may have required more than a billion years between then and about 3100 m.y. ago, when the oldest fossil algae and bacteria yet found were incorporated into the rocks of the Fig Tree Series of Swaziland—a span of time more than twice as long as that since the beginning of the Cambrian.

Blue-green algae were among the earliest organisms. The basis for their recognition is of course not the identification of specific algal cells in these oldest known fossiliferous rocks, but the preservation at many stratigraphic levels dating from 3100 m.y. on of cabbage-shaped or corrugated, brain-shaped structures of calcite, dolomite, and silica, similar to the structures built by present-day algal colonies.

One of the problems that confronted early life was the very high incidence of ultraviolet light during the long period before the oxygen content of the atmosphere sufficed to produce the ozone screen in the upper atmosphere that now protects us from most of these nearly lethal rays. Even now a day at the seashore or in a snowfield can convince one of their intensity; though the oxygen content of the air was only 0.1 percent of that of today, the rays would have killed any organism fully exposed to them.

It seems likely, then, that the earliest organisms formed in damp places in the soil or on the sea floor at depths great enough to screen out most of these rays. The modern blue-green algae are photosynthesizers, but in some families the mechanism seems highly diffuse and more primitive than in most plants. Perhaps these are less modified than others from the primitive species.

With the beginning of photosynthesis, the amount of oxygen increased slowly over several hundred million years as the plants gradually evolved more and more efficient mechanisms. Berkner and Marshall estimate that it was not until about the beginning of the Cambrian that the atmosphere contained as much as 1 percent of its present oxygen content; others place the date a few hundred million years earlier. It was not until photosynthesis developed that animal evolution was possible. Professor Cloud of the University of California, Santa Barbara, has pointed out that photosynthetic plants provide local sources of enrichment of oxygen, so that animal life might arise in isolated areas long before the atmosphere and oceans had accumulated enough free oxygen to support long travel for animals. Perhaps many local sites for natural selection may account for the wide diversity of life in the earliest Cambrian faunas, for these contain representatives of nearly every invertebrate phylum, implying a long evolutionary history (Appendix V).

Because of this, many paleontologists think that Berkner and Marshall should have put the time at which oxygen became adequate for migratory life considerably earlier than the beginning of the Cambrian.

With the flourishing of vegetation, the oxygen content continued to increase, facilitating the migration of both plant and animal life. Some have thought that the heavy external armor of so many early organisms—even the earliest fish of Ordovician time—was a protective mechanism against ultraviolet radiation. Certainly the bryozoa of today are far less massive than the early Paleozoic ones, and no modern fish is truly armored.

Thus as life evolved, it not only necessarily adjusted to the environment, it fundamentally modified the environment. "The present is indeed the key to the past," but only in the sense that the same physical laws prevailed as do today; surely the oceans of today are far larger than those of the distant past, and the composition of the atmosphere is radically different from that of the early Precambrian.

Questions

1. In terms of the ocean's ultimate origin, what is wrong with the statement that "the water of the oceans comes from the rivers?"

2. Explain why sea water is salty. Why does the atmosphere contain little or no hydrogen whereas a chemical analysis of sea water shows about 11 percent, by weight, of this element?

3. What is meant by "excess volatiles?" Excess over what? How does this jibe with the statement that the protoearth lost its volatiles to the solar wind?

4. What is the most abundant element in the universe? How do we know this?

5. Refer to Figure 19-7.

   a) Explain why there is more oxygen in the ocean than the atmosphere. Is this free oxygen?

b) Where is the phosphorous concentrated in a man? The calcium?

c) Why is helium recorded in only the first column?

d) What changes would you make in column 4 to approximate an early Precambrian atmosphere?

6. Suppose the sun's light were suddenly turned off, as we turn off an electric light. How long would it be before we knew about it?

7. Granting that a space ship exists that could go to a planet in Andromeda, and that you are an astronaut, why would it be impossible for you to tell us what was there?

8. Permafrost is found on earth in the polar regions, and pictures taken during the 1965 flyby of Mars suggest the possibility that permafrost composed of both ice and frozen CO_2 may be present on that planet. Why do we not have CO_2 permafrost on earth?

9. Armored fish of the early Paleozoic lived in shallow water and along mud flats. Why might they be better adapted to the risks of their environment than they would be if they lived in similar waters of today: Consider both physical and biological risks.

Suggested Readings

Brancazio, P. J., and A. G. W. Cameron (eds.), *The Origin and Evolution of Atmospheres and Oceans.* New York: John Wiley & Sons, 1964.

Fanning, A. E., *Planets, Stars and Galaxies* (rev. by Menzel, D. H.). New York: Dover Publications, 1966.

Faul, Henry, *Ages of Rocks, Planets, and Stars.* New York: McGraw-Hill, 1966.

Kuiper, G. P. (ed.), *The Atmospheres of the Earth and Planets* (2nd ed.). Chicago: University of Chicago Press, 1952.

Oparin, A. I., and V. Fesenkov, *Life in the Universe.* New York: Twayne Publishers, 1961.

Seaborg, G. T., and E. G. Valens, *Elements of the Universe.* New York: E. P. Dutton, 1965.

Symposium on the Evolution of the Earth's Atmosphere. Proceedings of the National Academy of Sciences v. 53, no. 6, 1965, pp. 1169–1226.

Scientific American Offprints

210. William A. Fowler, "The Origin of the Elements" (September 1956).

250. Gart Westerhout, "The Radio Galaxy" (August 1959).

253. John H. Reynolds, "The Age of the Elements in the Solar System" (November 1960).

833. Harold C. Urey, "The Origin of the Earth" (October 1952).

Appendixes

Maps and Mapping

Maps are the shorthand summary used by the student of the earth in presenting his data and observations. The earth's crust is complex. The intricate patterns of land and sea, the forms of hills and valleys, the labyrinths that men have dug in mining—all are so complex in form that a true picture of them cannot be given in words alone. A map, however, condenses the findings about them into intelligible form.

MAP SCALES

A blueprint of a machine part or a dress pattern may be thought of as a map. Most of these are **full-scale** maps on which a centimeter represents a centimeter on the object it portrays.

Few geographic and geologic maps, however, are full size. Most are drawings to scale. In such **reduced-scale** maps, a centimeter on the map may correspond to 10, 100 or 1,000,000 centimeters on the ground, depending on the scale of reduction used by the map maker. If he decides to reduce the length of objects on the map to $\frac{1}{10}$ of their true length on the ground he plots on a $\frac{1}{10}$ scale. The fraction is the ratio of reduction and simply means that 1 cm on the map equals 10 cm on the ground. Most of the newer maps of the *Topographic Atlas of the United States,* prepared by the United States Geological Survey, are drawn on a scale of $\frac{1}{24,000}$—1 cm on the map corresponds with 24,000 cm or 240 m on the ground.

LIMITATIONS OF MAPS

On a full-scale drawing, it is possible to show, for example, the head of a nail 1 mm across in full size. If the nail were to be correctly represented on a $\frac{1}{200}$ scale, it would have to be drawn as a point only $\frac{1}{200}$ mm across, a point far too small to be seen. Thus if the nail is to be shown at all on the $\frac{1}{200}$ scale it must be shown diagrammatically only. The position might be indicated but the size must be greatly exaggerated if the nail is to be seen. This limitation of reduced scale maps must always be kept in mind by the map user.

All maps are generalizations, drawn to perform a particular service. All represent selections of data chosen to emphasize a particular point, and these are often exaggerated in relation to other features. A navigator's chart emphasizes the features important to navigation—for example, shoals and shallow rocks are emphasized more than deep-water features of similar size; a good road map stresses highway junctions, and in doing so, may distort the distances between them.

The maps of Seattle Harbor (Fig. I-1) illustrate one effect of map scale; details such as the docks in Elliot Bay can be shown only on the larger-scale map.

Whatever the scale, limitations in drawing and printing make it almost impossible for maps to be accurate to better than 0.2 mm in the location of points. It is difficult to make a legible pencil mark narrower than this. The scale of a map of

450

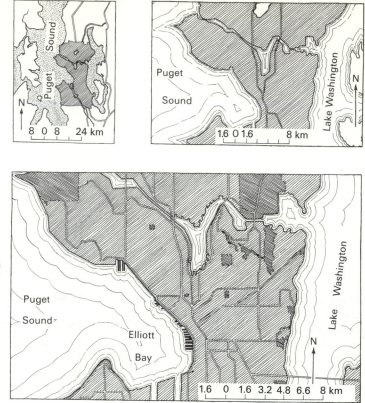

Figure I-1
Maps of Seattle Harbor on different scales. Note that the smaller-scale maps show much less detail, although all contain about the same number of lines per square centimeter. [After maps of the U. S. Geological Survey and of the Seattle Chamber of Commerce.]

the United States that could be printed on this page would be about $\frac{1}{25,000,000}$; on it points less than 12 km apart can not be distinguished without distortion. A thin line representing the Mississippi River on such a map would scale at least 6.5 km in width.

Maps of the earth—the summaries of geographical knowledge—have still another limitation: they must depict the curved surface of the earth on a flat surface. This cannot be done without distorting either the distances or the angular relations between points. Most maps are compromises between these distortions. Note the differences in the maps shown in Figure I-2.

TOPOGRAPHIC MAPS

The maps described thus far may be called **planimetric maps**; they show the relative positions of points but do not indicate their elevations above

or below sea level. The **relief** of an area is the difference in altitude between the highest and lowest points within it. Although, by skillful shading, a so-called **relief map** can give an impression of the relative steepness of the slopes in an area, such a map cannot be used to determine accurately the actual differences in elevation. It is impossible to read height accurately from such a map.

To meet this difficulty, geodesists have devised **topographic maps**, which are designed to show the elevations as well as the positions of points. They portray the three-dimensional form of the land surface—its **topography**.

A topographic map depicts a three-dimensional surface—one having length, breadth, and varying height above a reference plane or **datum** (usually mean sea level)—on a two-dimensional piece of paper. On such a map, lines called **contours** are drawn to portray the intersections of the ground surface with a series of horizontal

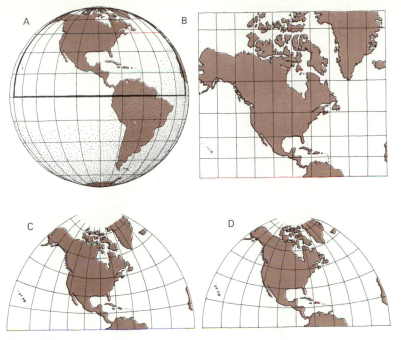

Figure I-2
The part of the earth outlined in A is reproduced in three common map projections: B, the Mercator; C, the stereographic; and D, the polyconic. By comparing the size and shape of the longitude and latitude grid, it is clear that each projection distorts the surface in varying degrees. The distortion is unavoidable in transferring a spherical surface to a flat one.

planes at definite intervals above (or below) the datum plane (Fig. I-3).

There are many different methods of making topographic maps and their actual making is a quite complex process. To illustrate the principles, we have chosen the "Plane Table Method," a method still widely used in making geologic maps even though most topographic maps are now made from aerial photographs.

The first step in the preparation of a topographic map by any method is to acquire both **horizontal** and **vertical control** for the measurements. To attain this double control, we first establish the position of a point on the earth's surface (*a*) by its latitude and longitude and (*b*) by its altitude with respect to sea level. The more points determined, the better our control. If we knew the elevation, latitude, and longitude of many points on the surface, we could construct the map; the decision of how many we must determine for a given area depends on the scale and

contour interval of the map, and the relief of the area.

Once the horizontal and vertical control (position and elevation) of one point, and the direction of the north-south line through it have been established, we can quickly determine many other points by a process called **triangulation.**

The first step in triangulation is the selection and measurement of a **base line.** The base line is a straight line from the point for which we have established control to another point. Each end of the base line is marked by a stake holding a flag (Fig. I-4), or by some other suitable marker, and the distance between them is measured carefully with a steel tape.

The accuracy of the whole map depends on the base line; therefore, as a check, the measurement is generally repeated. After the base line has been measured, it must also be carefully plotted on the plane table sheet (paper on which the map is to be constructed) in accordance with

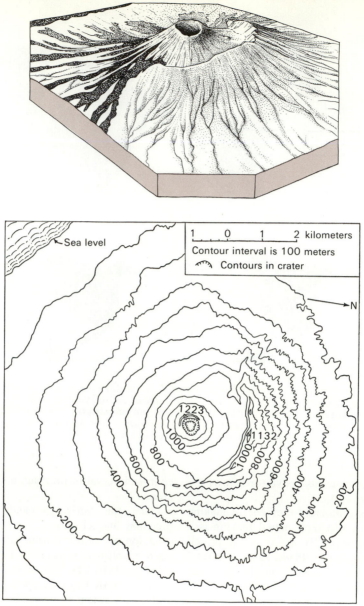

Figure I-3
Relief model and topographic (contour) map of the volcano Vesuvius.
[After Il Vesuvio sheet, Instituto Geografico Militaire.]

the reduced scale chosen for the map. Actually, the plotted length of the base line determines the scale, for if there are errors either in its measurement or in its plotting, they will be carried throughout the map.

Once the base line has been plotted, a **plane table** (essentially a drawing board mounted on a tripod) is set up over one end of the base line. Then the edge of an **alidade** (a telescope with a ruler as its base) is placed along the line drawn on the plane table sheet to represent the base line (Fig. I-4). The plane table, carrying the alidade, is rotated until the telescope points directly at the flag on the other end of the base line, and then

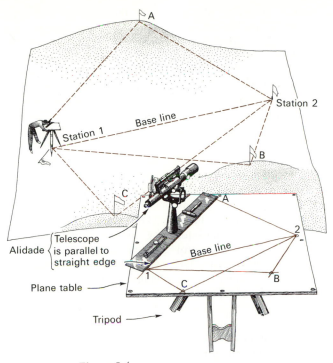

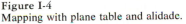

Figure I-4
Mapping with plane table and alidade.

clamped firmly in position. It is now correctly oriented, since the base line on the ground and the plotted base line on the plane table sheet have exactly the same trend (azimuth) in relation to true north.

With the table still clamped in this position, the telescope is pointed successively toward each of several other flags, such as A, B, C in Figure I-4, or other marked points that are visible, and lines are drawn along the edge of the alidade to indicate the directions of these lines of sight. The plane table is then taken to the other end of the base line, oriented in the same manner by sighting back to the first flag, and the process of sighting upon and drawing lines toward each point visible from this location is repeated. The point of intersection of the two lines of sight toward an object, one line of sight having been drawn from the first end of the base line and the other from the second end, marks the true position of that object on the reduced-scale map (Fig. I-4). This point and the two ends of the base line form the apexes of a triangle. The new point can then be used, just as if it were one end of the base line,

to determine the position of additional points, extending the system of triangles (triangulation net) within a given area.

A surveyor's transit or theodolite can be used in triangulation instead of a plane table and alidade. With a transit, the horizontal angle from the base line to the point to be determined is carefully measured at each end of the base line. Knowing the two angles, and having measured one side (the base line) of the triangle, we can compute the position of the apex of the triangle by trigonometry and plot it on the map.

Vertical control is established in the same way, and can be done at the same time as the horizontal triangulation. We establish the elevation of one end of our base line from some point of known elevation. Then, with the elevation of one point in our triangulation system established, we can readily compute the elevation of a second station. We have already determined the distance between the two stations by triangulation. The only other measurement we need for the computation is the vertical angle between the horizon and the line of sight to the station. The alidade is equipped with a vertical arc (Fig. I-4) for making this measurement. Thus, the vertical control is extended throughout the triangulation net, and the elevations of many points are determined.

By these and other auxiliary techniques (not described here) for obtaining horizontal and vertical control, we locate enough additional points to allow the topographer to sketch the contours in their proper relation to the ground surface as determined by the positions and elevations of the many points thus located and plotted. The map that results from this procedure is, we repeat, a generalization. Many factors other than the already mentioned baseline determinations affect its accuracy—for example, the number and spacing of the control points, trees that obscure the ground forms, the skill of the topographer, and the amount of time he has at his disposal for study of the shape of the land surface. Topographical engineers rate a map as excellent if, on testing it, they find not more than 10 percent of the elevations in error by an amount more than half of the contour interval.

Recently, great strides have been made in preparing topographic maps from photographs taken from airplanes. This saves much time formerly spent in surveying on the ground. The basic principles are, nonetheless, the same as those employed in making maps on the ground and, for the basic horizontal and vertical control, a preliminary ground survey is still necessary.

Good topographic maps are available for relatively few parts of the land surface of the earth. Less than half the area of the United States has been mapped on a scale that permits drawing contours with intervals as small as one hundred feet. Poland and several of the "backward Balkans" have much better map coverage. For large parts of the earth's surface, we have only crude maps.

HYDROGRAPHIC MAPS

Hydrographic maps are topographic maps of the sea floor. They not only depict the shorelines of the water bodies but, by soundings (measurements of depth made by vessels at sea), they also show something of the topography of the bottom.

In the construction of maps of the sea floor, soundings were formerly made by measuring the length of a rope or wire paid out until it reached bottom. As a measurement in the deep part of the ocean would require several hours, it is not surprising that relatively few such soundings were made, except at shallow depth near the coasts.

Since World War I sonic sounding has superseded measurement of depth by wire or rope. In sonic sounding the time required for a sound signal to travel from a ship to the sea bottom and rebound is measured, and the depth is then calculated from the speed of sound in sea water. This method has been refined in the past decade by the development of the **precision profiler**, which sends out a very sharply focused sound wave so that a very narrow area on the sea floor is involved in the reflection. The sound intensity is enough to penetrate the sediments on the sea floor to considerable depths so that sub-bottom rock structures also can be recorded under favorable conditions. By sonic sounding it is now an easy matter for a ship to chart a continuous record of the depths traversed while it is under way. The position of the ship is determined to within a distance of a few hundred meters by radio signals from shore stations.

Although sonic sounding has greatly increased our knowledge of the ocean floor, the vastness of the sea, the lack of interest of many navigators in obtaining detailed information of this kind from little-traveled sea lanes, and the cost of operating a vessel for surveying purposes alone, still conspire to prevent more than a mere sampling of the topography of the ocean floor.

The hydrographer, compared with the topographer, is severely handicapped, for he is unable to see the sea bottom and therefore cannot choose the most suitable points to use for control in mapping. A series of points of equal depth can be connected by a contour line in several ways, but obviously only one such contour line represents the actual form of the sea floor. On land the topographer can see the topographic forms and sketch between his points accordingly; the hydrographer must get additional control or else make an interpretation which will probably be inaccurate in minor details, and may be seriously inaccurate.

Identification of Minerals

The laboratory techniques in most common use today for the identification of minerals are noted here.

PETROGRAPHIC ANALYSIS

Petrographic analysis is the most frequently used method for the precise identification of both minerals and rocks. A small piece of the substance to be identified is ground with abrasives on a revolving plate until it is 0.03 mm thick — much thinner than a sheet of paper. It is then mounted between thin glass slides. This **thin section** can then be examined under the petrographic microscope. In a thin section most minerals are transparent, or nearly so, and the optical properties which distinguish different minerals can be readily measured.

An alternative petrographic method is to crush the mineral to powder, place the powder in a drop of liquid of known optical properties on a glass slide, cover with a thin glass plate, and examine the fragments immersed in the liquid under the petrographic microscope.

X-RAY ANALYSIS

As explained in Chapter 3, it is possible by means of X-rays to work out the internal structure of a mineral — the geometric arrangement of the ions or atoms within it. Since the internal structure is the most distinctive characteristic of a mineral, X-ray analysis is one of the most fundamental methods of mineral identification.

CHEMICAL ANALYSIS

A chemical analysis, or even a qualitative chemical test for some particular element, will generally help to identify an unknown mineral, although even a complete chemical analysis may fail to establish the identity of some. Some distinct minerals, diamond and graphite for example, have identical chemical compositions and so cannot be distinguished chemically. Furthermore, most minerals are highly insoluble silicates, difficult to treat by standard chemical procedures which require dissolving the substance to be analyzed. Most minerals are also

456

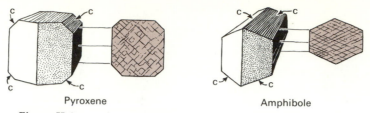

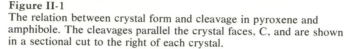

Pyroxene   Amphibole

Figure II-1
The relation between crystal form and cleavage in pyroxene and amphibole. The cleavages parallel the crystal faces, C, and are shown in a sectional cut to the right of each crystal.

"solid solutions" whose compositions vary widely. For these reasons standard chemical procedures are little used in ordinary mineral identification, though they may be employed in special kinds of research on minerals.

As supplements to petrographic and other methods, however, a few special chemical techniques have proved useful in mineral identification. Many minerals that are too opaque to be readily identified by ordinary petrographic methods, can be identified by simple chemical tests made on the surface of the thin section or on a polished piece of the mineral while it is being examined under the miscroscope.

The spectroscope is widely used to detect elements that may be present in small amounts in a mineral. Its use requires that the mineral be heated in an arc until it vaporizes.

In the last twenty years many highly refined analytical methods have been developed for analysis of minerals and rocks: emission spectroscopy, neutron-activation, X-ray fluorescence, and many others; these are beyond the scope of this book and are not further discussed.

DETERMINATION BY PHYSICAL PROPERTIES

The common rock-forming minerals, and also many of the rarer minerals of economic value, can usually be identified without special instruments by a careful study of their physical properties. This method suffices for recognition of the minerals listed at the end of this appendix. The more important physical properties are given here.

CLEAVAGE. Many minerals **cleave** (break) along *smooth planes* controlled by the internal structure of the crystal (Figs. II-1 and 3-11). Some minerals—mica, for example—have only one cleavage and can be split into countless thin flakes, all of which are parallel to one another; many minerals have two cleavages; others have three or more. Broken fragments of these minerals have characteristic shapes, which aid in identifying the mineral, because the number of cleavages and the angles between them are characteristic for a particular mineral.

FRACTURE. Many minerals fracture irregularly instead of cleaving along smooth planes. Such rough fragments are less readily identified than cleavage fragments, but some minerals, of which quartz is an example, usually break with characteristic curved surfaces (**conchoidal fracture**). Others have a splintery or **fibrous fracture** that helps to distinguish them.

FORM. As mentioned previously, minerals tend to crystallize into definite, characteristically shaped crystals, bounded by smooth planes called **crystal faces**. When crystal faces are present, their shapes and interfacial angles are diagnostic (Figs. 3-3 and II-2), but many minerals occur in shapeless, granular forms, or in crystals so small that the crystal faces are not visible. In some minerals, crystal faces are parallel to cleavage surfaces; in others, they are not. For this reason, they should, of course, be carefully distinguished from cleavage surfaces. The distinction is not difficult if one remembers that crystal faces appear only on the outside of the crystal, whereas cleavage surfaces appear only on the broken or cracked fragments of a crystal.

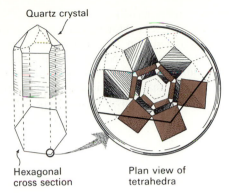

Quartz crystal

Hexagonal
cross section

Plan view of
tetrahedra

Figure II-2
The crystal form of quartz and its internal structure.
The internal structure consists of silicon tetrahedra
closely linked together because each oxygen atom
serves as the corner for two tetrahedra. The linked
tetrahedra lie in spirals around screw axes,
producing a hexagonal basal pattern of the tetrahedra,
which is also reflected in the external shape of the
crystal. In the enlarged plan view shown at the *right*,
the linked tetrahedra are indicated for one screw axis.
[Adapted from models by L. Pauling and by P. Niggli.]

COLOR. All specimens of some minerals, such
as magnetite and galena, have a constant or
uniform color; but others, such as quartz and
calcite, are variable in color because of pigments
that may be present as impurities. Even in
minerals with constant intrinsic color, alteration
of the surface through exposure to air and
moisture may change the surface color. Nevertheless, the color of a freshly broken surface
may be diagnostic, and even the color of the
altered surface film aids in identifying some
minerals.

STREAK. The color of the powdered mineral—
which is called the **streak** because it is easily
obtained as a "chalk mark" by rubbing the
mineral against a piece of unglazed porcelain—is
more constant and, for some minerals, more
helpful in identification than the color of the
mineral in a larger mass. The streak of a mineral
may be similar to, or it may be entirely different
from, the color of the mineral itself: silvery-gray
galena gives a silvery-gray streak; but both the
black and red varieties of hematite show a characteristic brownish-red streak that is very helpful
in identifying the mineral.

LUSTER. The **luster** of a mineral refers to the
way ordinary light is reflected from its surfaces.
Metallic luster is like that of polished metals;
vitreous luster is like that of glass; adamantine
like that of diamond. Other self-explanatory
terms used to describe luster are resinous, silky,
pearly, and dull or earthy.

HARDNESS. The relative hardness of two different minerals can be determined by pushing
a pointed corner of one firmly across the flat
surface of the other. If the mineral with the point
is harder, it will scratch or cut the other. Laboratory tests of the hardness of minerals are usually
recorded in terms of a **scale of hardness** ranging
from 1 to 10. Each number refers to the hardness
of a specific mineral, ten of which, arranged in
order of increasing hardness, constitute the scale.

1.	talc	6.	orthoclase
2.	gypsum	7.	quartz
3.	calcite	8.	topaz
4.	fluorite	9.	corundum
5.	apatite	10.	diamond

When specimens to make up this series are not
available, it is convenient to know that the steel
of a pocketknife is about $5\frac{1}{2}$ in this scale, a copper
penny $3\frac{1}{2}$, and the thumbnail about $2\frac{1}{2}$.

SPECIFIC GRAVITY. The **specific gravity**, or density, of a mineral is given by the formula:

$$\text{Sp. Gr.} = \frac{\text{(wt. mineral in air)}}{\text{(wt. mineral in air)} - \text{(wt. mineral in water)}}$$

Specific gravity, therefore, is stated as a number
indicating the ratio of the weight of the substance
to that of an equal volume of water. Specific
gravity is easily determined by an ordinary spring
scale for large specimens or a sensitive micro-
balance for smaller ones. The specific gravity of
different minerals varies greatly.

With a little practice, moderate differences
in specific gravity can be detected from the
"heft" of a moderate-sized specimen held in
the hand. Quartz, with a specific gravity of
2.65, may be used as the standard of comparison. Gypsum (2.2 to 2.4) would then be called
light; olivine (3.2 to 3.6) heavy; magnetite (5.0
to 5.2) very heavy.

OTHER PROPERTIES.   Many other physical properties are useful in identifying minerals. Some minerals are attracted by a magnet, others are not; some conduct electricity better than others; some "fluoresce," or glow in various colors when ultraviolet light is played on them; some are characterized by fine striations, or "twinning lines," on certain cleavage surfaces; others have different striations—"growth striations"—on certain crystal faces. Minerals also differ in fusibility, in solubility, in their reactions to simple chemicals—such as bubbling when dilute hydrochloric acid is applied to them; and in many other ways, but the major physical properties listed above will suffice to identify the more common kinds.

List of Minerals

The grouping of the minerals in the list that follows is not alphabetical but is based on similarities in chemical composition, physical properties, associations, or uses. The first group includes *common carbonates, sulfates, chlorides, and oxides*. These minerals are found mainly in the sedimentary rocks, though some of them (quartz and magnetite, for example) are abundant in many igneous and metamorphic rocks as well. The second group, *common rock-forming silicates*, includes chiefly minerals that crystallized in igneous and metamorphic rocks, but which may also be found as clastic particles in sedimentary rocks. Kaolinite, included in this group as a representative of the abundant clay minerals, is chiefly a weathering product. The third group, important *ore minerals*, are much less common, but are listed because of their importance as sources of valuable metals or other commercial materials.

In the tables those properties most useful in sight identification are italicized. Some minerals have more than one name; the less common names are given in parentheses. The chemical name and chemical formula follow the mineral name.

Common carbonates, sulfates, chlorides, and oxides.

Mineral	Form	Cleavage	Hardness	Sp. Gr.	Other properties
Calcite. Calcium carbonate, $CaCO_3$.	"Dog-tooth" or flat crystals showing excellent cleavages; granular, showing cleavages; also masses too fine grained to show cleavages distinctly.	*Three highly perfect cleavages at oblique angles*, yielding rhomb-shaped fragments (Fig. 3-7).	3	2.72	Commonly colorless, white, or yellow but may be any color owing to impurities. Transparent to opaque, transparent varieties showing *strong, double refraction* (e.g., 1 dot seen through calcite appears as 2). Vitreous to dull luster. *Effervesces readily in cold dilute hydrochloric acid.*
Dolomite. Calcium magnesium carbonate, $CaMg(CO_3)_2$.	*Rhomb-faced crystals showing good cleavage;* also in fine-grained masses.	*Three perfect cleavages* at oblique angles, as in calcite.	3.5–4	2.9	Variable in color, but commonly white. Transparent to translucent. Vitreous to pearly luster. *Powder will effervesce slowly in cold dilute hydrochloric acid, but coarse crystals will not.*
Gypsum. Hydrous calcium sulfate, $CaSO_4 \cdot 2H_2O$.	Tabular crystals, and cleavable, granular, fibrous, or earthy masses.	*One perfect cleavage, yielding thin, flexible folia;* 2 other much less perfect cleavages.	2	2.2–2.4	Colorless or white, but may be other colors when impure. Transparent to opaque. Luster vitreous to pearly or silky. Cleavage flakes flexible but *not elastic* like those of mica.
Halite. (Rock salt.) Sodium chloride, NaCl.	Cubic crystals (Fig. 3-10), in granular masses.	Excellent cubic cleavage (3 cleavages mutually at right angles).	2–2.5	2.1	*Colorless to white,* but of other colors when impure. The color may be unevenly distributed through the crystal. Transparent to translucent. Vitreous luster. *Salty taste.*
Opal. Hydrous silica, with 3% to 12% water, $SiO_2 n H_2O$. Does not have a definite geometric internal structure, hence is a mineraloid, not a true mineral.	*Amorphous.* Commonly in veins or irregular masses showing a banded structure. May be earthy.	None; *conchoidal fracture.*	5.0–6.5	2.1–2.3	Color highly variable, often in wavy or banded patterns. Translucent or opaque. *Somewhat waxy luster.*
Chalcedony. (Cryptocrystalline quartz.) Silicon dioxide, SiO_2.	Crystals too fine to be visible; some are conspicuously banded, or in masses.	None; *conchoidal fracture.*	6–6.5	2.6	Color commonly white or light gray, but may be any color owing to impurities. Distinguished from opal by *dull* or *clouded* luster.
Quartz. (Rock crystal.) Silicon dioxide, SiO_2.	*Six-sided prismatic crystals,* terminated by 6-sided triangular faces; also massive (Fig. II-2).	None or very poor; *conchoidal fracture.*	7	2.65	Commonly *colorless* or white, but may be yellow, pink, amethyst, smoky-translucent brown, or even black. *Transparent* to opaque. *Vitreous to greasy luster.*

Common carbonates, sulfates, chlorides, and oxides (*continued*).

Mineral	Form	Cleavage	Hardness	Sp. Gr.	Other properties
Magnetite. A combination of ferric and ferrous oxides, Fe_3O_4.	Well-formed, 8-faced crystals, more commonly in compact aggregates, disseminated grains, or loose grains in sand.	None; conchoidal or uneven fractures; may show a rough parting resembling cleavage.	5.5–6.5	5.0–5.2	*Black.* Opaque. Metallic to submetallic luster. *Black streak. Strongly attracted by a magnet.* Magnetite is an important iron ore.
Hematite. Ferric iron oxide, Fe_2O_3.	Highly varied, compact, granular, fibrous, or earthy, micaceous; rarely in well-formed crystals.	None, but fibrous or micaceous specimens may show parting resembling cleavage; splintery to uneven fracture.	5–6.5	4.9–5.3	Steel-gray, reddish-brown, red, or iron-black in color. Metallic to earthy luster. *Characteristic brownish-red streak.* Hematite is the most important iron ore.
"Limonite." Microscopic study shows that the material called limonite is not a single mineral. Most "limonite" is a very finely crystalline variety of the mineral **Goethite** containing absorbed water. Hydrous ferric oxide with minor amounts of other elements, roughly $Fe_2O_3 \cdot H_2O$.	Compact or earthy masses; may show radially fibrous structure.	None; conchoidal or earthy fracture.	1–5.5	3.4–4.0	*Yellow,* brown, or black in color. Dull earthy luster, which distinguishes it from hematite. *Characteristic yellow-brown streak.* A common iron ore.
Ice. Hydrogen oxide, H_2O.	Irregular grains; lacelike flakes with hexagonal symmetry; massive.	None; conchoidal fracture.	1.5	0.9	Colorless, white or blue. Vitreous luster. *Melts at 0°C., so is liquid at room temperature. Low specific gravity.*

Common rock-forming silicates.

Mineral	Form	Cleavage	Hardness	Sp. Gr.	Other properties
Potassium Feldspar. (Orthoclase, microcline, and sanidine.) Potassium aluminum silicate, $KAlSi_3O_8$.	Boxlike crystals (Fig. 3-3,B); massive.	One perfect and 1 good cleavage, making an angle of 90°.	6	2.5–2.6	Commonly *white, gray, pink,* or *pale yellow;* rarely colorless. Commonly opaque but may be transparent in volcanic rocks. Vitreous. Pearly luster on better cleavage. *Distinguished from plagioclase by absence of striations.*
Plagioclase feldspar. (Sodalime feldspars.) A solid solution group of sodium calcium aluminum silicates, $NaAlSi_3O_8$ to $CaAl_2Si_2O_8$.	In well-formed crystals and in cleavable or granular masses.	*Two good cleavages nearly at right angles* (86°). May be poor in some volcanic rocks.	6–6.5	2.6–2.7	Commonly *white* or *gray,* but may be other colors. Some gray varieties show a play of colors called *opalescence.* Transparent in some volcanic rocks. Vitreous to pearly luster. Distinguished from orthoclase by the presence on the *better cleavage surface of fine parallel lines or striations.*
Muscovite. (White mica; isinglass.) A complex potassium aluminum silicate, $KAl_3Si_3O_{10}(OH)_2$ approximately, but varying.	Thin, scalelike crystals and scaly, foliated aggregates.	*Perfect in one direction, yielding very thin, transparent, flexible scales.*	2–3	2.8–3.1	*Colorless,* but may be gray, green, or light brown in thick pieces. *Transparent to translucent.* Pearly to vitreous luster.
Biotite. (Black mica.) A complex silicate of potassium, iron, aluminum, and magnesium, variable in composition but approximately $K(Mg,Fe)_3AlSi_3O_{10}(OH)_2$.	Thin, scalelike crystals, commonly 6-sided, and in scaly, foliated masses.	*Perfect in one direction, yielding thin, flexible scales.*	2.5–3	2.7–3.2	Black to dark brown. Translucent to opaque. Pearly to vitreous luster. White to greenish streak.
Pyroxene. A solid-solution group of silicates. Chiefly silicates of calcium, magnesium, and iron, with varying amounts of other elements. The commonest varieties are *augite* and *hypersthene.*	Commonly in short, 8-sided, prismatic crystals; *the angle between alternate faces nearly 90°.* Also as compact masses and disseminated grains.	*Two cleavages at nearly 90°* (Fig.II-1). Cleavage not always well developed; in some specimens, conchoidal or uneven fracture.	5–6	3.2–3.6	Commonly greenish to black in color. Vitreous to dull luster. Gray-green streak. Distinguished from amphibole by the *right-angle cleavage, 8-sided crystals,* and by the fact that most crystals are short and stout, rather than long, thin prisms, as in amphibole (Fig. II-1).
Amphibole. A group of complex, solid-solution silicates, chiefly of calcium, magnesium, iron, and aluminum. Similar to pyroxene in composition, but containing a little hydroxyl (OH^-) ion. The commonest of the many varieties of amphibole is *hornblende.*	*Long, prismatic, 6-sided crystals;* also in fibrous or irregular masses of interlocking crystals and in disseminated grains.	*Two good cleavages meeting at angles of 56° and 124°* (Fig. II-1).	5–6	2.9–3.2	Color black to light green; or even colorless. Opaque. *Highly vitreous luster on cleavage surfaces.* Distinguished from pyroxene by the *difference in cleavage angle* and in crystal form. Amphibole also has much better cleavage and higher luster than pyroxene.

Common rock-forming silicates (continued).

Mineral	Form	Cleavage	Hardness	Sp. Gr.	Other properties
Olivine. Magnesium iron silicate, $(Fe, Mg)_2SiO_4$.	Commonly in small, glassy grains and granular aggregates.	So poor that it is rarely seen; *conchoidal fracture.*	6.5–7	3.2–3.6	*Various shades of green,* also yellowish; opalescent and brownish when slightly altered. *Transparent to translucent. Vitreous luster.* Resembles quartz in small fragments but has *characteristic greenish color,* unless altered.
Garnet. A group of solid solution silicates having variable proportions of different metallic elements. The most common variety contains calcium, iron and aluminum, but garnets may contain many other elements.	Commonly in well-formed *equidimensional crystals* (Fig. 3-3, C) but also massive and granular.	None; *conchoidal or uneven fracture.*	6.5–7.5	3.4–4.3	Commonly *red, brown, or yellow,* but may be other colors. Translucent to opaque. *Resinous to vitreous luster.*
Sillimanite. (Fibrolite.) Aluminum silicate, Al_2SiO_5.	*In long slender crystals,* or fibrous.	Parallel to length, but rarely noticeable.	6–7	3.2	Gray, white, greenish-gray, or colorless. *Slender prismatic crystals or in a felted mass of fibers.* Streak white or colorless.
Kyanite. (Disthene.) Aluminum silicate, Al_2SiO_5.	Long, *bladelike crystals.*	One perfect, and one poor cleavage, both parallel to length of crystals; and a crude parting across the crystals.	4–7	3.5–3.7	Colorless, white, or a *distinctive pale blue color.* Can be scratched by knife parallel to cleavage, but is harder than steel across cleavage.
Staurolite. Iron-aluminum silicate, $Fe(OH)_2(Al_2SiO_5)_2$.	*Stubby prismatic crystals,* and in *cross-shaped twins.*	Poor and inconspicuous.	7–7.5	3.7	Red-brown or yellowish-brown to brownish black. Generally in well-shaped crystals larger than the minerals of the matrix enclosing them.
Epidote. A complex group of calcium, iron, aluminum silicates, $Ca_2(Al,Fe)_3(SiO_4)_3(OH)$.	*Short, 6-sided crystals* or radiate crystal groups (Fig. 3-3, A) and in granular or compact masses.	One good cleavage; in some specimens, a second poorer cleavage at an angle of 115° with the first.	6–7	3.4	Characteristic *yellowish-green (pistachio green)* color. Vitreous luster.
Chlorite. A complex group of hydrous magnesium aluminum silicates containing iron and other elements in small amounts.	Commonly in *foliated or scaly masses;* may occur in tabular, 6-sided crystals resembling mica.	*One perfect cleavage,* yielding thin, flexible, but inelastic, scales.	1–2.5	2.6–3.0	*Grass-green to blackish-green color.* Translucent to opaque. Greenish streak. Vitreous luster. *Very easily disintegrated.*

Mineral	Form	Cleavage	Hardness	Sp. Gr.	Other properties
Serpentine. A complex group of hydrous magnesium silicates, roughly $H_4Mg_3Si_2O_9$.	Foliated or fibrous, usually massive.	Commonly only one cleavage, but may be in prisms. Fracture usually conchoidal or splintery.	2.5–4	2.5–2.65	*Feels smooth*, or even *greasy*. Color *leek-green* to *blackish-green* but varying to brownish-red, yellow, etc. *Luster resinous* to *greasy*. Translucent to opaque. Streak white.
Talc. Hydrous magnesium silicate, $Mg_3(OH)_2Si_4O_{10}$.	In tiny *foliated scales* and soft compact masses.	*One perfect cleavage*, forming thin scales and shreds.	1	2.8	White or silvery white to apple green. *Very soft*, with a *greasy feel*. Pearly luster on cleavage surfaces.
Kaolinite. Hydrous aluminum silicate, $H_4Al_2SiO_9$. Representative of the 3 or 4 similar minerals common in clays.	Commonly in soft, compact, *earthy masses*.	Crystals always so small that cleavage is invisible without microscope.	1–2	2.2–2.6	White color, but may be stained by impurities. *Greasy feel. Adheres to the tongue*, and *becomes plastic when moistened.* "*Clay-like*" odor when breathed upon.

Important ore minerals (*See also iron ores listed on p. 460*).

Mineral	Form	Cleavage	Hardness	Sp. Gr.	Other properties
Galena. Lead sulfide, PbS.	Cubic crystals common, but mostly in coarse to fine granular masses.	*Perfect cubic cleavage* (three cleavages mutually at right angles).	2.5	7.3–7.6	*Silvery-gray color.* Metallic luster. Silvery-gray color. Lead-gray streak. Chief ore of lead.
Sphalerite. Zinc sulfide (nearly always containing a little iron), ZnS.	Crystals common, but chiefly in fine to coarse-granular masses.	*Six highly perfect cleavages at 60° to one another.*	3.5–4	3.9–4.2	Color ranges from white to black but is commonly *yellowish-brown. Translucent* to opaque. *Resinous* to *adamantine luster.* Streak white, pale yellow or brown. Most important ore of zinc.
Pyrite. ("Fool's gold.") Iron sulfide, FeS_2.	*Well-formed crystals*, commonly cubic, with striated faces (Fig. 3-3,D); also granular masses.	None; uneven fracture.	6–6.5	4.9–5.2	Pale *brassy-yellow color*; may tarnish brown. Opaque. Metallic luster. Greenish-black or brownish-black streak. Brittle. Not a source of iron, but used in the manufacture of sulfuric acid. Commonly associated with ores of several different metals.

Important ore minerals (*See also the iron ores listed on p. 460*) (*continued*).

Mineral	Form	Cleavage	Hardness	Sp. Gr.	Other properties
Chalcopyrite. Copper iron sulfide, $CuFeS_2$.	Compact or disseminated masses, rarely in wedge-shaped crystals.	None; uneven fracture.	3.5–4	4.1–4.3	Brassy to *golden-yellow. Tarnishes* to blue purple, and reddish iridescent films. Greenish-black streak. Distinguished from pyrite by deeper yellow color and softness. A common copper ore.
Chalcocite. (Copper glance.) Cuprous sulfide, Cu_2S.	Massive, rarely in crystals of roughly hexagonal shape. May be tarnished and stained to blue and green.	Indistinct, rarely observed.	2.5–3	5.5–5.8	Blackish-gray to steel gray, commonly *tarnished to green or blue.* Dark gray streak. *Very heavy.* Metallic luster. An important ore of copper.
Copper. (Native copper.) An element, Cu.	*Twisted and distorted leaves and wirelike forms; flattened or rounded grains.*	None.	2.5–3	8.8–8.9	*Characteristic copper color,* but commonly stained green. *Highly ductile* and malleable. Excellent conductor of heat and electricity. *Very heavy.*
Gold. An element, Au.	Massive or in thin plates; also in flattened grains or scales; distinct crystals very rare.	None.	2.5–3	15.6–19.3	*Characteristic gold-yellow color and streak. Rarely in crystals. Extremely heavy.* Very malleable and ductile.
Silver. An element, Ag.	In flattened grains and scales; rarely in wirelike forms, or in irregular needlelike crystals.	None.	2.5–3	10–11	*Color and streak are silvery-white,* but may be tarnished gray or black. *Highly ductile and malleable. Very heavy.* Mirrorlike metallic luster on untarnished surfaces.
Cassiterite. Tin dioxide, SnO_2.	Well-formed, 4-sided prismatic crystals terminated by pyramids; 2 crystals may be intergrown to form knee-shaped twins; also as rounded pebbles in stream gravels.	None; curved to irregular fracture.	6–7	7	*Brown to black.* Adamantine luster. White to pale-yellow streak. Chief ore of tin.
Uraninite. (Pitchblende.) Uranium oxide, UO_2 to U_3O_8.	Regular 8-sided or cubic crystals; massive.	None; fracture uneven to conchoidal.	5–6	6.5–10	Color black to brownish-black. Luster submetallic, pitchlike, or dull. Chief mineral source of uranium, radium, etc.
Carnotite. Potassium uranyl vanadate, $K_2(UO_2)_2(VO_4)_2 \cdot 8H_2O$.	Earthy powder.	Not apparent.	Very soft	4.1 approx.	*Brilliant canary-yellow color.* An ore of vanadium and uranium.

Identification of Rocks

The classification of rocks given in this Appendix is a field classification based on features that can be seen without the aid of the petrographic microscope. Much more elaborate classifications have been built with the aid of this instrument and other laboratory techniques. Nearly all such classifications, however, have been made by expanding and adding varietal subdivisions to the major rock classes listed in the tables below. Therefore the field classification represents a broad framework into which more elaborate subdivisions can be fitted.

The field classification is based primarily upon the texture and mineral composition of the rock. Remember that rocks grade into one another, and hence some properties of an individual specimen you may be examining are likely to fall between those listed as typical of two major rock classes. To use the tables and lists below as guides in identifying an unknown rock specimen the student must be thoroughly familiar with the common rock-forming minerals listed in the Mineral Table of Appendix II. He must also have clearly in mind the basic distinctions between sedimentary, igneous, and metamorphic rocks and the general range of textures found in them.

This fundamental information is given in Chapter 4. The common rock textures are here summarized in glossary form for easy reference.

COMMON TEXTURES OF SEDIMENTARY ROCKS

Differences in *the nature of the constituent particles,* and in *how they are bound together* determine the texture of a sedimentary rock.

Clastic (Greek, "broken"). Composed of broken and worn fragments of pre-existing minerals, rock particles, or shells that have been cemented together. Further distinctions can be made on the *size* of the particles, and on the amount of *rounding* by wear of the individual fragments.

Organic. Composed of accumulations of organic debris (shells, plant remains, bones, etc.) in which the individual organic particles are so well preserved (not notably broken and worn) that organic features dominate the texture of the rock.

Crystalline. Composed of crystals precipitated from solution and therefore tightly interlocked by mutual interpenetration during growth. The rock owes its coherence to this interlocking of crystals, instead of to the presence of a cement as in the clastic and organic textures.

COMMON TEXTURE OF IGNEOUS ROCKS

Differences in the *degree of crystallinity,* and in the *size of the crystals* determine the texture of an igneous rock. Both of these factors are controlled primarily by *rate of cooling,* though the chemical composition of the magma and its content of volatile materials play roles.

BASIC TEXTURES

Pyroclastic (Greek, "broken by fire"). Composed of slivers of volcanic glass, bits of frothy pumice, phenocrysts, and broken fragments of volcanic rock deposited together. The glass slivers and pumice may be largely altered to clay. Pyroclastic rocks are the products of volcanic explosions or of pyroclastic flows.

Glassy. Composed almost entirely of massive or streaky volcanic glass. Small phenocrysts of feldspar or other minerals may be scattered through the glass. The glass may be frothy, filled with minute bubbles, forming a *pumiceous glassy* texture.

Aphanitic (Greek, "invisible"). Composed chiefly of tiny crystals (less than 0.5 millimeter in diameter), with or without a glassy residue between the crystals. The crystals are mere specks, large enough to be seen but too small to identify without the aid of the microscope. Their presence gives the rock a stony or dull luster in contrast to the glassy (vitreous) luster of rocks with glassy texture. Most lava flows have aphanitic texture; in some, flow has aligned the tiny mineral grains, giving a streaky or flow-banded appearance.

Granular (Latin, "a grain"). Composed of crystals that are large enough to be seen and identified without the aid of lens or microscope. In different rocks the average size may vary from about 0.5 millimeter to more than 1 centimeter in diameter, but the common granular rocks such as granite have grains averaging from 3 to 5 millimeters in size.

MODIFYING TEXTURE

Porphyritic. Composed of two widely different sizes of minerals, giving a spotted appearance. Because porphyritic texture is most common in small intrusive bodies or in lavas, it has been attributed to a *change in the rate of cooling while the magma was crystallizing.* The inferred process is explained as follows: A large body of magma underground may cool to the temperature at which one or more minerals begin to crystallize. Because cooling is slow, the crystals of these minerals grow to considerable size. If, when the magma is perhaps half crystallized, a fissure opens in the roof of the chamber, some of the magma with its suspended crystals may escape to form a lava flow at the surface. The still-liquid portion of the magma quickly freezes at the surface of the ground and surrounds the large crystals, called *phenocrysts,* with a *groundmass* of aphanitic crystals. The phenocrysts were formed underground, the aphanitic groundmass at the surface. Such a lava has a *porphyritic aphanitic* texture. The adjective porphyritic is used to modify the prevailing texture of the groundmass. Rocks with *porphyritic granular* texture—large crystals in a granular groundmass of finer grain—are common in intrusive bodies. *Porphyritic glassy* texture appears in some lava flows, and in the pumice fragments of pyroclastic rocks. Rarely, conditions other than a change in the rate of cooling may produce porphyritic rocks.

COMMON TEXTURES OF METAMORPHIC ROCK

Difference in the *orientation,* or alignment, of the crystals and in the *size* of the crystals determine the texture of a metamorphic rock. There are two general textural groups: *foliated textures* (Latin, "leafy"), in which platy or leaflike minerals such as mica or chlorite are nearly all aligned parallel to one another so that the rock splits readily along the well-oriented, nearly parallel cleavages of its constituent mineral particles, and *nonfoliated,* composed either of equidimensional minerals or of randomly oriented platy minerals, so that the rock breaks into angular particles.

BASIC TEXTURES

Gneissose (from Greek, "banded rock"). Coarsely foliated; individual folia are 1 millimeter or more, even several centimeters, thick. The folia may be straight, pancake-like, or wavy and crenulated. They commonly differ in composition; feldspars, for example, may alternate with dark minerals. Mineral grains are coarse, easily identified.

Schistose (Greek, "easily cleaved"). Finely foliated, forming thin parallel bands along which the rock splits readily. Individual minerals are distinctly visible. The minerals are mainly platy or rodlike—chiefly mica, chlorite, and amphibole. Equidimensional minerals like feldspar, garnet, and pyroxene may be present but are not abundant.

Slaty (from Old High German, "to split"). Very fine foliation, producing almost rigidly parallel planes of easy splitting due to the nearly perfect parallelism of microscopic and ultramicroscopic crystals of platy minerals, chiefly mica.

Granoblastic (Greek, "sprouting grains"). Unfoliated or only faintly foliated. Composed of mutually interpenetrating mineral grains that have crystallized simultaneously. Minerals are large enough to be easily identified without the microscope, and are chiefly equidimensional kinds such as feldspar, quartz, garnet, and pyroxene. Corresponds roughly to the granular texture of igneous rocks.

Hornfelsic (German, "hornlike rock"). Unfoliated. Mineral grains commonly microscopic or ultramicroscopic, though a few may be visible. Breaks into sharply angular pieces with curved fracture surfaces.

HOW TO USE THE ROCK TABLES AND LISTS

After carefully examining a specimen of rock, but before referring to the tables and lists in this Appendix, the student should ask himself the following three basic questions, and if in doubt about the answers, should refer back to the material in Chapter 4, Appendix II, and the glossary of textures just given.

1. What is the texture of the rock? (Glossary above)
2. Of what minerals is it composed? (Appendix II)
3. Is it an igneous, sedimentary, or metamorphic rock? (Chapter 4)

Once this basic information is worked out, turn to the appropriate rock table (Sedimentary, p. 468; Igneous, p. 469; Metamorphic, p. 475) and find the rock's name, then check against the description of the rock in the appropriate rock list.

COMMON SEDIMENTARY ROCKS

CONGLOMERATE. Conglomerate is cemented gravel. Gravel is an unconsolidated deposit composed chiefly of rounded pebbles. The pebbles may be of any kind of rock or mineral and of all sizes. Most conglomerates, especially those deposited by streams, have much sand and other fine material filling the spaces between the pebbles. Some cleanly washed beach conglomerates contain little sand.

BRECCIA. Sedimentary breccias resemble conglomerate except that most of their fragments are angular instead of rounded. They commonly grade into conglomerates. Since their constituent fragments have been little worn, however, it is apparent that the components of breccia underwent relatively less transportation and wear before they were deposited. There are many kinds of breccias other than sedimentary breccias. Volcanic breccias, as well as sedimentary breccias, are described in this appendix; glacial breccias in Chapter 13; and fault breccias in Chapter 5.

SANDSTONE. Sandstone is cemented sand. Sand, by definition, consists of particles from 2 millimeters to $\frac{1}{16}$ millimeter in diameter. Sandstones commonly grade into either shale or conglomerate. Three general varieties of sandstone are recognized:

Quartz sandstone is composed mainly of the mineral quartz. Most sand is chiefly quartz but

Table III-1
Sedimentary rocks.

CLASTIC SEDIMENTARY ROCKS

Consolidated rock	Chief mineral or rock components	Original unconsolidated debris	Diameter of fragments
Conglomerate	Quartz, and rock fragments	Gravel (rounded pebbles)	More than 2 mm
Breccia	Rock fragments	Rubble (angular fragments)	
Sandstone	. . .	Sand	2 to $\frac{1}{16}$ mm
Quartz Sandstone	Quartz	Quartz-rich sand	
Arkose	Quartz and feldspar	Feldspar-rich sand	
Graywacke	Quartz, feldspar, clay, rock fragments, volcanic debris	"Dirty sand," with clay and rock fragments	
Siltstone	Quartz, clays	Silt	Less than $\frac{1}{16}$ mm; more than $\frac{1}{256}$ mm
Shale	Clay minerals, quartz	Mud, clay, and silt	Less than $\frac{1}{256}$ mm
Clastic Limestone	Calcite	Broken and rounded shells and calcite grains	Variable

ORGANIC AND CHEMICAL SEDIMENTARY ROCKS

Consolidated rock	Chief mineral or rock components	Original nature of material	Chemical composition of dominant material
Limestone	Calcite	Shells; chemical and organic precipitates	$CaCO_3$
Dolomite	Dolomite	Limestone, or unconsolidated calcareous ooze, altered by solutions	$CaMg(CO_3)_2$
Peat and Coal	Organic materials	Plant fragments	C, plus compounds of C, H, O
Chert	Opal, chalcedony	Siliceous shells and chemical precipitates	SiO_2 and $SiO_2 n H_2O$
Evaporites, or Salt Deposits	Halite, gypsum, anhydrite	Evaporation residues from the ocean or saline lakes	Varied, chiefly NaCl and $CaSO_4 \cdot 2H_2O$

contains small amounts of many other minerals and even small particles of rock.

Arkose is a feldspar-rich sandstone. It may contain nearly as many particles of partly weathered feldspar as of quartz, or even more. Most arkoses have been formed by the rapid erosion of coarse feldspar-rich rocks such as granites and gneisses, and the rapid deposition of this eroded debris before the feldspar has had time to weather completely into clay.

Graywacke is a cemented "dirty sand" containing clay and rock fragments in addition to quartz and feldspar. Many graywackes contain much pyroclastic debris in various stages of weathering and decomposition; others are crowded with bits of slate, greenstone, or other metamorphic rocks; and still others are rich in ferromagnesian minerals. All contain appreciable amounts of clay. Graywackes are commonly dark gray, dark green, or even black. Like arkose, they indicate rapid erosion and deposition without much chemical weathering.

Sand, the original material that is cemented into sandstone, accumulates in many different environments. Some sand is deposited by streams; some is heaped up in dunes by the wind; some is spread out by waves and currents along beaches or in the shallow water of the continental shelves; some is washed by turbidity currents down steep submarine slopes onto the deep-sea floor.

SHALE. Shale is hardened mud. Mud is a complex mixture of very small mineral particles less than $\frac{1}{256}$ mm in diameter (chiefly clay), and coarser grains, called silt, from $\frac{1}{256}$ to $\frac{1}{16}$ mm in diameter. Shale frequently contains small bits of organic matter.

The predominant minerals in shale are the hydrous aluminum silicates called clay minerals, but most shales also contain appreciable amounts of mica, quartz, and other minerals. Shale splits readily along closely spaced planes, parallel or nearly parallel to the stratification. Some rocks of similar grain size and composition show little layering and break into small angular blocks: these are more correctly called *mudstone;* when of appropriate grain size, *siltstone.*

Shales accumulate in many different environments. As the main load brought down to the sea by great rivers is mud and fine sand, it is not surprising that shale is the most abundant marine sedimentary rock. Mud deposited in deltas, on lake bottoms, and on plains along sluggish rivers may also harden into shale.

Many shales are black, some because they contain large amounts of carbon-rich organic matter in various stages of decomposition, some because of the precipitation of black iron sulfide (FeS_x) by sulfur bacteria. The iron sulfide may later crystallize into pyrite (FeS_2), forming small brass-colored crystals sprinkled through the rock. Many blue-green, dark gray, gray-green, or purplish-red mudstones owe their color to decomposed volcanic material.

LIMESTONE. Limestone is composed almost entirely of calcium carbonate ($CaCO_3$), chiefly as the mineral calcite, though aragonite (which is also $CaCO_3$ but with a different crystalline form) may be plentiful.

Organic limestones are common rocks, and occur in great variety because of the many kinds of shells from which they are formed. Among the most common are: *coral limestone*, which contains a framework of coralline deposits but also includes the shells of many other animals, especially foraminifers, molluscs, and gastropods; *algal limestone*, made largely of deposits of calcite precipitated by algae and bacteria; *foraminiferal limestone*, composed chiefly of the tiny shells of foraminifers; *coquina*, composed mostly of the coarse shells of molluscs; and *chalk*, which consists largely of almost ultramicroscopic blades and spines from *coccoliths*, the tests of minute algae.

Clastic limestones are composed of broken and worn fragments of shells or of crystals of calcite. The white sands of the Florida Keys are made up largely of calcite grains worn from shells and organic limestones.

Chemically precipitated limestone is also forming today in shallow warm seas, in hot springs, and in saline lakes. The role of inorganic precipitation is, however, difficult to separate from that of biochemical and organic agents. Very fine-grained, flour-like, white *calcareous ooze* (calcareous means calcite-rich) is abundant in parts of the southwest Pacific and on the shallow Bahama Banks of the Atlantic. Some of this ooze consists of microscopic shells, but much of that in the Bahamas consists of tiny spines and crystals of aragonite and calcite, perhaps in part precipitated inorganically, or else precipitated from sea water as a result of the life processes of such microorganisms as algae and bacteria.

Limestone deposited from hot springs is coarsely crystalline, and commonly full of small irregular holes stained yellow or red by iron oxides. Such limestone is called *travertine.*

Limestones differ greatly in texture and color depending on the size of the shells or crystals composing them and the impurities they contain. Some black limestones are rich in hydrocarbons from the partially decayed bodies of organisms, as shown by the strong, fetid odor they give off when freshly broken. Most limestones, however, are light colored and contain many fossils.

DOLOMITE. Dolomite rock is composed chiefly of dolomite, the mineral of the same name. Dolomite resembles limestone, and also grades into it, by changes in the amount of calcite in the rock. Chemical and microscopic tests are generally necessary to determine the relative amounts of the minerals calcite and dolomite in the rock.

Most dolomite appears to result from alteration of limestone or its parental calcareous ooze by magnesia-bearing solutions. The alteration that formed most dolomite is thought to have taken place during slow deposition, by the action of the magnesium ions in sea water on calcareous ooze or other calcareous deposits. Some limestone, however, changed to dolomite long after it was deposited and consolidated.

Dolomite has rarely, perhaps never, been deposited directly as a precipitated sediment.

FINE-GRAINED SILICEOUS ROCKS. Rocks composed almost entirely of fine-grained silica are common, but they rarely form large masses. Many different kinds of siliceous (siliceous means silica-rich) sedimentary rocks have been described and named, but the most common is *chert,* a hard rock with grains so fine that a broken surface appears uniform and lustrous.

Chert nodules, many resembling a knobby potato in shape and size, are common in limestone and dolomite. Dark-colored chert nodules are often called *flint.* Chert also appears as distinct beds and as thin, wedgelike, discontinuous layers. Beds of chert are commonly associated with volcanic deposits.

The microscope shows that some cherts are made up largely of spines or lacelike shells of silica (opal) secreted by microscopic animals and plants. In other cherts, siliceous fossils are rare or absent, but siliceous shells may have been partly dissolved and reprecipitated as structureless silica during cementation. Abundant undissolved siliceous shells usually make the rock porous and light in weight. An example is *diatomite,* a white rock composed almost entirely of the siliceous shells of microscopic plants called diatoms.

Not all fine-grained siliceous rocks are of organic origin. Some are believed to have precipitated around silica-bearing submarine hot springs. Many have been formed by the replacement of wood, limestone, shale, or other materials by silica-bearing solutions. *Petrified wood* is a familiar example.

PEAT AND COAL.   Peat and coal are not common sedimentary rocks but their economic importance justifies their mention here.

Peat is an aggregate of slightly decomposed plant remains. It can be seen in process of accumulation in swamps and shallow lakes in temperate climates and even spreading over steep hillsides in wet sub-arctic regions. Coal is the result of compression and more thorough decomposition of the plant material in ancient peat bogs which were buried under later sediments. Coals grade from *lignite,* which differs little from peat, through *bituminous* to *anthracite,* which may contain 90 percent or more of carbon. From evidence obtained in mines and by geologic mapping, we infer that the grade of the coal depends largely on the depth to which it has been buried (i.e., the pressure and heat to which it has been subjected). The nature of the original plant material may also affect the variety of coal.

EVAPORITES OR SALT DEPOSITS.   Evaporites vary in mineral composition and texture. They are now being formed by the evaporation of landlocked masses of sea water, as at the Rann of Kutch in northwest India, and in saline lakes like Great Salt Lake. When sea water evaporates completely many different salts are precipitated from it, but *rock salt* (halite, $NaCl$) is the most abundant. In nature, however, calcium sulfate, which occurs both as a hydrated form, *rock gypsum* ($CaSO_4 2H_2O$), and as the anhydrous mineral called *anhydrite* ($CaSO_4$), is much more common than rock salt. Gypsum separates out early in the process of evaporation and will, therefore, accumulate in quantity from water bodies that are not saline enough to precipitate halite. Rock gypsum, accompanied by little or no rock salt, is abundant in the Paris Basin of France, in the Dakotas, and elsewhere. Thick beds of rock salt, accompanied by gypsum and anhydrite, are found in Utah, Texas, New Mexico, Saskatchewan, Germany, Iran, India and many other areas.

In a few places where relatively complete evaporation of sea water has occurred, deposits of potassium salts and other valuable, latecrystallizing minerals are found. Many rare and useful mineral products such as potash, salsoda, borax, nitrates, sodium sulfate, and epsom salts are recovered from salt deposits formed by the evaporation of ancient desert lakes. Commercial deposits of sulfur—presumably formed by reduction of sulfates by hydrocarbons—locally accompany evaporites.

COMMON IGNEOUS ROCKS

VOLCANIC TUFF.   Volcanic tuff is a fine-grained pyroclastic deposit composed of fragments less than 4 mm in diameter. Most of the fragments are volcanic glass, either microscopic slivers called *shards* or frothy bits of *pumice.* Other common constituents are broken phrenocrysts and fragments of solidified lava. Pieces of the basement rock on which the volcano rests may also be present.

Pumice and other kinds of glass fragments have been seen to form by the explosive disruption of sticky lava highly charged with gases. Evidently the gas pressure increases until it exceeds the containing pressure on the magma; then the pent-

Table III-2
Igneous rocks.

TEXTURES	PREDOMINANT MINERALS			
	Feldspar and quartz	Feldspar predominates (no quartz)	Ferromagnesian minerals and feldspar (no quartz)	Ferromagnesian minerals (no quartz or feldspar)
PYROCLASTIC	**Volcanic tuff** (fragments up to 4 mm in diameter) **Volcanic Breccia** (fragments more than 4 mm in diameter)			Rocks of the texture and composition represented by this part of the table are rare.
GLASSY	**Obsidian** (if massive glass) **Pumice** (if a glass froth)		**Basalt Glass**	
APHANITIC (generally porphyritic-aphanitic)	**Rhyolite** and **Dacite**	**Andesite**	**Basalt**	
GRANULAR	**Granite** (potassium feldspar predominates) and **Granodiorite** (plagioclase feldspar predominates)	**Diorite**	**Gabbro** **Dolerite** or **Diabase** (if fine grained)	**Peridotite** (with both olivine and pyroxene) **Pyroxenite** (with pyroxene only) **Serpentine** (with altered olivine and pyroxene)

INCREASING GRAIN SIZE →

← DECREASING SILICA CONTENT →

up gases separate into bubbles, causing the lava to expand tremendously and to froth. Upon breaking out to the surface, the froth disrupts further into a cloud of glass fragments and pumice which may be blown high into the air in a great volcanic explosion, or may froth forth more quietly and roll down the slope of the volcano as a *pyroclastic flow*, or "glowing avalanche."

The fragments from a volcanic explosion may be cemented together in the same way as the fragments of a sedimentary rock, forming an ordinary volcanic tuff. The component particles deposited by many pyroclastic flows, however, when viewed under the microscope, show flattening and collapse of the bits of frothy pumice and glass shards upon one another as if the rock had been welded—stuck together under its own weight while sticky and partly melted. Such *welded tuffs* are common products of rhyolitic and dacitic volcanoes. They are often confused with rhyolite and dacite lavas because of the close similarity of the welded fragmental matrix to the flow-banded aphanitic texture of lava flows.

VOLCANIC BRECCIA. Volcanic breccia is composed dominantly of fragments more than 4 mm in diameter. In general, fragments of lava are more abundant than in tuff; glass slivers and pumice may be scarce. *Scoria* (see p. 40) is abundant in some breccias. The scoria may form large angular blocks, streamlined bombs 2.5 to 15 cm long shaped into cigarlike or teardrop forms by flying through the air while still molten, or small bits of frothy lava less than an inch in diameter, called *lapilli*.

Some volcanic breccias are formed like the tuffs, but many are products of volcanic mudflows. Heavy rains falling on the steep slopes of a volcanic cone have been seen to set great avalanche-like slides of unconsolidated pyroclastic debris in motion. Other mudflows are formed by eruption clouds falling into rivers, or onto snowfields and glaciers, or by explosive eruptions through crater lakes. The water-soaked volcanic debris may travel for many miles down stream valleys.

OBSIDIAN. Obsidian is natural glass, formed chiefly from magmas of rhyolitic, dacitic, or andesitic composition. It is lustrous and breaks with a curved fracture. Most obsidians are black because of sparsely disseminated grains of magnetite and ferromagnesian minerals, but they may be red or brown from the oxidation of iron by hot magmatic gases. Thin pieces of obsidian are almost transparent.

Obsidian forms lava flows and rounded domes above volcanic vents. It also is found as thin selvages along the edges of intrusions, and, rarely, makes small intrusive masses. Most intrusive obsidians have a dull, pitch-like luster, and are called *pitchstone*.

PUMICE.   Pumice is obsidian froth, characteristically light-gray to white and crowded with tiny bubbles. The bubbles are so numerous that pumice will float on water. Pumice is abundant as fragments in tuffs and breccias. It also may form distinct flows, or more commonly, it caps flows of obsidian or rhyolite, and grades downward into the unfrothed lava beneath.

BASALT GLASS.   Basalt glass is a jet-black natural glass formed by chilling of basaltic magma. Unlike obsidian, it is not noticeably transparent on thin edges. Basalt glass has never been found in large flows like those of obsidian; on this fact is based the inference that basalt magma crystallizes much more readily than rhyolite. Basalt glass forms thin crusts on the surfaces of lava flows, small fragments in volcanic breccia, and thin contact selvages in volcanic necks and dikes. Breccias of basalt glass form in abundance when basalt magma is extruded into water and quickly quenched. These may quickly alter to a yellow mineraloid called palagonite.

RHYOLITE.   Rhyolite has an aphanitic groundmass generally peppered with phenocrysts of quartz and potassium feldspar. The color of rhyolite ranges widely, but generally is white or light yellow, brown, or red. Most rhyolites are flow banded; that is, they show streaky irregular layers that were formed by the flowing of the sticky, almost congealed magma.

Dacite is like rhyolite except that plagioclase predominates instead of potassium feldspar. It bears the same relation to rhyolite that granodiorite does to granite (see below).

Rhyolite and dacite are found in lava flows and as small intrusions.

ANDESITE.   Andesite is an aphanitic rock, generally porphyritic, that resembles dacite but contains no quartz. Plagioclase feldspar is the most common phenocryst, but pyroxene, amphibole, or biotite may appear. Most ande-

sites are flow banded, though not so conspicuously as rhyolites. Andesites range from white to black, but most are dark gray or greenish gray.

Andesite is abundant as lava flows and as fragments in volcanic breccias, tuffs, and mudflows. Glacier-clad andesite volcanoes tower above mountain ranges such as the Andes (from which the name), the Cascades, and the Carpathians. Andesite also forms small intrusive masses.

BASALT.   Basalt is a black medium-gray aphanitic rock. Most basalts are nonporphyritic, but some contain phenocrysts of plagioclase and olivine.

Basalt is the world's most abundant lava and is very widespread, forming great lava plateaus that cover thousands of square miles in the northwestern United States, India, and elsewhere. It is the chief constituent of the isolated oceanic islands. Although it typically forms lava flows, basalt is also common in small intrusive masses.

GRANITE.   Granite, characterized by a granular texture, has feldspar and quartz as its two most abundant minerals, and in consequence most granite is light colored. Biotite or hornblende, or both, are also present in most granite.

Technically, the term *granite* is reserved for those granular quartz-bearing igneous rocks that have potassium feldspar as the chief mineral. Those in which plagioclase predominates are called *granodiorite*. (Compare rhyolite and dacite above.) Granodiorite can usually be distinguished from granite by the fine striations that characterize one cleavage surface of plagioclase.

Geologic mapping shows that great quantities of granite and granodiorite are present in the earth's crust. They form large intrusive masses along the cores of many mountain ranges and in other areas where deep erosion has occurred, such as northeastern Canada, the Scandinavian region, and eastern Brazil. They are typical continental rocks; the Seychelles Islands of the Indian Ocean and Iceland are the only oceanic islands on which they occur in large masses.

Some granites are of metamorphic instead of igneous origin. (Chapters 4 and 9.)

DIORITE.   Diorite is a granular rock composed of plagioclase and lesser amounts of ferromagnesian minerals. The most common ferromagnesian minerals are hornblende, biotite, and pyroxene. In general, diorite masses are much smaller than those of granite or granodiorite.

GABBRO.   Gabbro is a granular rock composed chiefly of plagioclase and pyroxene commonly with small amounts of other ferromagnesian minerals, especially olivine. If ferromagnesian minerals predominate over the plagioclase so that the rock is dark-colored, it is generally correct to call it gabbro, though the microscopic distinction from diorite rests on the composition of the plagioclase, a character not determinable with the unaided eye.

Gabbro is widely distributed in both large and small masses. Dikes and thin sills of fine-grained gabbro are especially common. In most of these small intrusions, the mineral grains are so small that they are barely recognizable without the aid of the microscope. Such gabbros, intermediate in grain size between basalt and normal gabbro, are called *dolerite*. (Some geologists prefer the name *diabase*.)

PERIDOTITE, PYROXENITE, AND SERPENTINE. Granular rocks composed almost entirely of ferromagnesian minerals and without feldspar are common in some areas. If the rock contains olivine as a conspicuous constituent, it is called *peridotite*; if it is made up almost wholly of pyroxenes, it is called *pyroxenite*.

Olivine is a very unstable mineral, easily altered to a mixture of greenish hydrous minerals. Some varieties of pyroxene also alter easily. These alterations probably occur soon after consolidation of the magma and are caused by the hot gases and solutions that escape from the crystallizing peridotite or perhaps from nearby granite or gabbro masses. Such altered peridotites and pyroxenites are called *serpentine*. Because serpentine is composed almost entirely of secondary minerals which did not solidify directly out of the magma, it is often classed as a metamorphic rock instead of an igneous rock. Nearly all plutonic igneous rocks, however, show some features that suggest alteration and "working over" by hot gases during the last stages of crystallization, although most are not modified as much as serpentine.

Serpentine forms sills, dikes, and other small intrusive masses.

PORPHYRY.   This ancient term is used rather indefinitely. It may be applied to porphyritic-textured, fine-grained intrusive igneous rocks in which phenocrysts constitute 25 percent or more of the volume. The ground-mass may be either coarse-grained, aphanitic, or fine-grained granular. The name of the rock whose composition and texture fit the groundmass part of the rock is usually prefixed to the word porphyry. Thus *diorite* porphyry has a fine-grained granular ground-mass and contains abundant phenocrysts of plagioclase and perhaps some ferromagnesian mineral. Andesite porphyry is similar except that the groundmass is aphanitic.

The noun "porphyry," as distinguished from the adjective "porphyritic," should not be applied to porphyritic rocks with a coarse granular groundmass or to porphyritic lava containing abundant glass. The former should be called porphyritic diorite and the latter porphyritic andesite if they have the same composition as diorite and andesite.

Granite porphyry, granodiorite porphyry, and diorite porphyry form many dikes near granite and granodorite masses. Rhyolite porphyry, dacite porphyry, and andesite porphyry are common in volcanic plugs and other small intrusive masses.

METAMORPHIC ROCKS

HORNFELS.   Hard, unfoliated, very fine-grained rock which breaks into sharp angular pieces. In many hornfelses traces of original structures such as stratification, flow banding, or slaty cleavage can be seen, but the rock will not break along them. The mineral composition is highly variable, and grains are, in general, too small to be recognizable without a microscope.

Hornfels is formed by the partial or complete recrystallization, near an igneous intrusion, of such fine-grained rocks as shale, shaly limestone, slate, chlorite schists, tuff, and lavas.

QUARTZITE.   Very hard, sugary-textured granoblastic rock, composed predominantly of interlocking quartz grains. Unlike most sandstones, quartzite breaks across the grains, not around them. Colors range from white through pale buff to pink, red, brown, and black, but most quartzite is light colored.

Quartzite is formed by the metamorphism of quartz sandstone. It is a widely distributed metamorphic rock.

Sandstone with a cement of silica (sedimentary "quartzite" is difficult to tell from metamorphic quartzite since both break across the grains. Distinction by use of the petrographic microscope is usually not difficult, for the cement can be readily distinguished from the original sand grains. Metamorphic quartzite can also be distinguished from silica-cemented sandstone by the rocks associated with it in the field, for true quartzite is associated with other metamorphic rocks, and sandstone with other sedimentary rocks.

MARBLE. Granoblastic, fine- to coarse-grained rock composed chiefly of calcite or dolomite or both. Many marbles show a streaky alteration of light and dark patches; others show brecciated structures healed by veinlets of calcite.

Marble is formed by the metamorphism of limestone and dolomite; if from dolomite, it commonly contains magnesium-bearing silicates such as pyroxene, amphibole, and serpentine.

TACTITE. Granoblastic, but variable in texture, grain size, and mineral composition. Tactite is rich in silicates of calcium, iron, and magnesium—amphibole, pyroxene, garnet, and epidote. It occurs in many areas where limestone or dolomite has been invaded by granite or granodiorite. From this it is inferred that fluids escaping from the congealing magma have carried into the limestone large quantities of silica, iron, and other substances that combined with the calcite and dolomite to form new minerals. Ores of iron, copper, tungsten, and other metals may be associated with these rocks.

AMPHIBOLITE. Granoblastic, rocks consisting chiefly of plagioclase and amphibole. Garnet, quartz, and epidote may be present in small quantities. Amphibolites have been formed by the metamorphism of basalt, gabbro, and rocks of similar composition; some are derived from impure dolomite.

GRANULITE. Granoblastic-textured, medium- to coarse-grained rock consisting chiefly of feldspars, pyroxenes, and garnet, but commonly containing small amounts of many other minerals such as quartz, kyanite, and staurolite. Most show an indefinite streakiness or a faint foliation. The feldspar may show a fine mottling when viewed under a lens or microscope. Granulites are formed by the high-grade metamorphism of shale, graywacke, and many kinds of igneous rock.

SLATE AND PHYLLITE. Very fine-grained, exceptionally well-foliated rocks. Because of their excellent foliation, they split into thin sheets. Mineral grains are too small to be identified without the microscope or X-rays. Slate is dull on cleavage surfaces; phyllite is shiny and coarser grained, containing some mineral grains large enough to be identified by the eye. Slate and, to a lesser extent, phyllite, commonly show remnants of sedimentary features such as stratification, pebbles, and fossils.

Slate and phyllite are abundant. Most were formed by the metamorphism of shale, but others are derived from tuffs or other fine-grained rocks.

CHLORITE SCHIST OR GREENSCHIST. Green, very fine-grained, schistose to slaty rock. Most are soft, greasy, and easily pulverized, composed of chlorite, plagioclase, and epidote—all except chlorite generally are in grains too small to identify. Remnants of original volcanic structures such as phenocrysts and scoria may be present.

Chlorite schists are common. They are often called *greenschist* or, if poorly foliated, *greenstone*, from the color of the chlorite. Most have formed by metamorphism of basalt or andesite and their corresponding tuffs, but some have been derived from dolomitic shale, gabbro, and other ferromagnesian rocks.

MICA SCHIST. Schistose rock composed chiefly of muscovite, quartz, and biotite in varying proportions; any one of these minerals may predominate. The most common varieties are rich in muscovite.

Mica schist is one of the most abundant metamorphic rocks. Like slate, most has been formed from shales and tuffs, although some derives from arkose, shaly sandstone, rhyolite, or other rocks. It represents more intense metamorphism than slate.

AMPHIBOLE SCHIST. Schistose rock, composed chiefly of amphibole and plagioclase, with varying amounts of garnet, quartz, or biotite. It is a common metamorphic derivative of basalt, gabbro, chlorite schist, and related rocks.

GNEISS. Coarse-grained gneissose rock with distinct layers or lenses of different minerals. Mineral composition is variable, but feldspar is especially abundant. Other minerals common in gneiss are quartz, amphibole, garnet, and mica.

Gneisses are among the most plentiful metamorphic rocks. They may be derived from many different rocks—granite, granodiorite, shale, rhyolite, diorite, slate, and schist, among others.

MIGMATITE. Migmatites are highly complex rocks (see Chapter 9 and Fig. 9-42). In general, they are intimate small-scale mixtures of igneous and metamorphic rocks, characterized by a roughly banded or veined appearance. They are widespread, especially near large granite masses. Their mineral composition is complex and highly variable, but most contain abundant feldspar and quartz, and smaller amounts of biotite, and amphibole.

Table III-3
Metamorphic rocks.

Name	Texture	Commonly derived from	Chief minerals
UNFOLIATED OR FAINTLY FOLIATED			
Hornfels	Hornfelsic	Any fine-grained rock	Highly variable
Quartzite	Granoblastic, fine grained	Sandstone	Quartz
Marble	Granoblastic	Limestone, dolomite	Calcite, magnesium and calcium silicates
Tactite	Granoblastic, but coarse and variable	Limestone or dolomite plus magmatic emanations	Varied; chiefly silicates of iron, calcium, and magnesium, such as garnet, epidote, pyroxene, amphibole
Amphibolite	Granoblastic	Basalt, gabbro, tuff	Hornblende and plagioclase, minor garnet and quartz
Granulite	Granoblastic	Shale, graywacke, or igneous rocks	Feldspar, pyroxene, garnet, kyanite, and other silicates
FOLIATED			
Slate (and **Phyllite**)	Slaty	Shale, tuff	Mica, quartz
Chlorite schist	Schistose to slaty	Basalt, andesite, tuff	Chlorite, plagioclase, epidote
Mica schist	Schistose	Shale, tuff, rhyolite	Muscovite, quartz, biotite
Amphibole schist	Schistose	Basalt, andesite, gabbro, tuff	Amphibole, plagioclase
Gneiss	Gneissose	Granite, shale, diorite, mica schist, rhyolite, etc.	Feldspar, quartz, mica, amphibole, garnet, etc.
Migmatite	Coarsely banded, highly variable	Mixtures of igneous and metamorphic rocks	Feldspar, amphibole, quartz, biotite

Chemical Data

Table IV-1
The atomic numbers, symbols, and names of the elements.

Atomic number	Symbol	Element	Atomic number	Symbol	Element
1	H	Hydrogen	30	Zn	Zinc
2	He	Helium	31	Ga	Gallium
3	Li	Lithium	32	Ge	Germanium
4	Be	Beryllium	33	As	Arsenic
5	B	Boron	34	Se	Selenium
6	C	Carbon	35	Br	Bromine
7	N	Nitrogen	36	Kr	Krypton
8	O	Oxygen	37	Rb	Rubidium
9	F	Fluorine	38	Sr	Strontium
10	Ne	Neon	39	Y	Yttrium
11	Na	Sodium	40	Zr	Zirconium
12	Mg	Magnesium	41	Nb	Niobium
13	Al	Aluminum	42	Mo	Molybdenum
14	Si	Silicon	43	Tc	Technetium
15	P	Phosphorus	44	Ru	Ruthenium
16	S	Sulfur	45	Rh	Rhodium
17	Cl	Chlorine	46	Pd	Palladium
18	Ar	Argon	47	Ag	Silver
19	K	Potassium	48	Cd	Cadmium
20	Ca	Calcium	49	In	Indium
21	Sc	Scandium	50	Sn	Tin
22	Ti	Titanium	51	Sb	Antimony
23	V	Vanadium	52	Te	Tellurium
24	Cr	Chromium	53	I	Iodine
25	Mn	Manganese	54	Xe	Xenon
26	Fe	Iron	55	Cs	Cesium
27	Co	Cobalt	56	Ba	Barium
28	Ni	Nickel	57	La	Lanthanum
29	Cu	Copper	58	Ce	Cerium

Table VI-1 (continued).

Atomic number	Symbol	Element	Atomic number	Symbol	Element
59	Pr	Praseodymium	81	Tl	Thallium
60	Nd	Neodymium	82	Pb	Lead
61	Pm	Promethium	83	Bi	Bismuth
62	Sm	Samarium	84	Po	Polonium
63	Eu	Europium	85	At	Astatine
64	Gd	Gadolinium	86	Rn	Radon
65	Tb	Terbium	87	Fr	Francium
66	Dy	Dysprosium	88	Ra	Radium
67	Ho	Holmium	89	Ac	Actinium
68	Er	Erbium	90	Th	Thorium
69	Tm	Thulium	91	Pa	Protactinium
70	Yb	Ytterbium	92	U	Uranium
71	Lu	Lutetium	93	Np	Neptunium
72	Hf	Hafnium	94	Pu	Plutonium
73	Ta	Tantalum	95	Am	Americium
74	W	Tungsten	96	Cm	Curium
75	Re	Rhenium	97	Bk	Berkelium
76	Os	Osmium	98	Cf	Californium
77	Ir	Iridium	99	E	Einsteinium
78	Pt	Platinum	100	Fm	Fermium
79	Au	Gold	101	My	Mendelevium
80	Hg	Mercury	102	No	Nobelium

Table IV-2
Chemical composition of the earth's crust, ocean, and atmosphere.

Element	Rocky crust (a)	Rocky crust (b)	Ocean	Atmosphere (dry air to height of 25 kilometers)
O	46.6%	43.8	85.79%	21.0%
Si	27.7	27.0	—	—
Al	8.1	10.0	—	—
Fe	5.0	5.9	—	—
Ca	3.6	5.1	—	—
Na	2.8	2.2	1.14	—
K	2.6	1.7	—	—
Mg	2.1	3.2	0.14	—
Ti	0.4	1.0	—	—
H	0.14		10.67	—
Cl	0.03		2.07	—
N	0.005		—	78.1
A	—		—	0.9
CO_2	—		—	0.03 (variable)

NOTES AND SOURCES:
(a) As estimated from composition of exposed rocks. (After B. Mason, 1952.)
(b) As estimated from seismic properties of the crust correlated with chemical compositions of rocks having similar elastic properties. (After L. C. Pakiser and R. Robinson, 1966.)

Table IV-3
Average chemical compositions of igneous rocks and sedimentary rocks.

Constituent	Igneous rocks	Sedimentary rocks
SiO_2	59.14%	57.95%
TiO_2	1.05	0.57
Al_2O_3	15.34	13.39
Fe_2O_3	3.08	3.47
FeO	3.80	2.08
MgO	3.49	2.65
CaO	5.08	5.89
Na_2O	3.84	1.13
K_2O	3.13	2.86
H_2O	1.15	3.23
P_2O_5	0.30	0.13
CO_2	0.10	5.38
SO_3	–	0.54
BaO	0.06	–
C	–	0.66
Total	99.56	99.93

NOTES AND SOURCES: The compositions in the table above are based on 5159 analyses of igneous rocks compiled by F. W. Clarke, and on selected analyses of sedimentary rocks compiled by C. K. Leith and W. J. Mead. The sedimentary rocks have been weighted in the proportions of 82 percent shale, 12 percent sandstone and 6 percent limestone. The compositions are reported as *oxides*, which is the conventional system for reporting data on the composition of rocks and minerals.

Table IV-4
Chemical composition of dissolved solids in river water and in the sea.

Ion	River water (weighted average)	Sea water
CO_3^{--}	35.15%	0.41 (HCO_3^-)%
SO_4^{--}	12.14	7.68
Cl^-	5.68	55.04
NO_3^-	0.90	–
Ca^{++}	20.39	1.15
Mg^{++}	3.41	3.69
Na^+	5.79	30.62
K^+	2.12	1.10
$(Fe, Al)_2O_3$	2.75	–
SiO_2	11.67	–
Sr^{++}, H_3BO_3, Br^-	–	0.31
Total	100.00	100.00

SOURCE: After F. W. Clarke.

Fossils

Life is virtually ubiquitous over the earth, from high in the air, where pollen and single-celled organisms have been carried by the winds, to the deepest sunless trenches of the oceans. As we noted in Chapter 6, the more ancient the strata, the less do the fossils contained in them resemble living forms; our contemporary organisms, like ourselves, are the modified survivors — the best adjusted to present environments — of an almost incredibly great number of species, most long vanished from the earth. The study of living organisms is the province of the biologist; the study of their mostly unsuccessful ancestors or former competitors is the occupation of many paleontologists. Anything but a most cursory treatment is far beyond the scope of this book, but we feel that at least a sample of the diverse life-forms of the past should be presented.

To the zoologist, fossils are invaluable as a record of the evolution of life; for the paleontological geologist, in addition to the evolutionary record, they furnish clues to past environments and through the correlation of strata by short-lived forms of life provide basic data for the history of the earth. The ecologic niche of an individual modern relative cannot alone be evidence that its fossil analogues occupied the same environment. For example, in the Early Eocene (London Clay), the Nipa palm (now living in the Indo-Malayan area) and the mollusk *Astarte* lived in association. Most of the other fossils in the London Clay are related to modern tropical forms as is the Nipa palm, but all of the living members of the genus *Astarte* are found only in cold water and nowhere within thousands of kilometers of the present habitat of the palm. Obviously the *Astarte* line has adapted to a very different environment today than this genus occupied in the early Eocene. Though an individual genus may thus modify its living habits, associations comprised of many genera are not likely to. Faunal and floral *groups*, then, offer trustworthy guides to habitats of the geologic past, though, of course, less trustworthy the more remote the past.

THE PLANT AND ANIMAL KINGDOMS

The two great divisions of life are the plant and animal kingdoms. Most plants, as noted in Chapter 19, are able to synthesize organic matter from simple inorganic substances in the presence of light; animals live only on organic matter so produced, or on each other. Other living things, such as the bacteria, differ from both plants and animals in neither performing photosynthesis nor integrating organic compounds from others. Instead, they are the main agents of decay, breaking down complex organic matter into water, carbon dioxide, and simple salts. Without them, life could not have persisted to the present, for all the carbon in the atmosphere would long since have been combined in organic matter. As we saw (Fig. 6-1), bacteria are known from very old rocks of the Lake Superior country, and, in fact, have been identified in rocks of the Fig Tree Series of Swaziland, South Africa, known to be more than three billion years old. They probably evolved along with the blue-green algae as the earliest organisms on earth.

The plant kingdom has existed for many million more years than the animal kingdom, but its fossil residues are much less resistant to geologic processes, and it is only in rocks of Silurian and younger age that plant fossils useful in correlation are recognizable. For this reason, we consider the animal kingdom first, as its fossils are generally more abundant and therefore more useful even in the younger rocks.

The Animal Kingdom

Since the time of the great Swedish naturalist Linnaeus (1707–1778), biologists have classified the organisms of both plant and animal kingdoms in a hierarchy of his devising. The basic unit in the Linnaean System is the species, which may be defined as a population whose members may differ widely among themselves — as, for example, the Great Dane and the Mexican Hairless dogs — but are connected, through intermediate varieties, into an interbreeding group. Species closely similar but not normally interbreeding, such as dogs, wolves, and coyotes, are grouped as a single genus.

In identifying an individual in this system, both the generic and specific names, in latinized form, are used: the domestic dog is *Canis familiaris*, the coyote *Canis latrans*. The generic name is always capitalized, the specific name never, even when it has been derived from a proper name. Both are invariably printed in italics, generally followed by the name of the individual who first described and defined the species.

The genera are in turn grouped with similar ones into families, the families into orders, etc. The hierarchy of the Linnaean System is as follows:

> Kingdom
> Phylum
> Class
> Order
> Family
> Genus
> Species

As noted in Chapter 6, the history of life is a history of successions of species, some long-lived, some very short-lived, so that assemblages of the faunas and floras have continually changed through time by natural selection of those species best suited to a contemporary environment and elimination of those less fit. New species arose through successive mutations that tended better to fit a creature to its environment; deleterious mutations led to prompt elimination of the line in which they took place. Even without unfavorable mutations a species might be eliminated by the competition of a newly evolved, more efficient competitor or by its own inability to adjust to environmental changes clearly indicated in the geologic record. Thus random mutations and environmental changes have enabled natural selection not only to produce the remarkably divergent species of fossil and living organisms, but also to permit the adaptation of some form or other to almost every environment on or near the earth's surface, in every water body, and through much of the atmosphere.

ANIMAL FOSSILS

The Great Auk became extinct in 1844, the passenger pigeon soon after. The aurochs, Irish elk, mammoth, and mastodon, painted on many a cave wall by Pleistocene men, did not survive that epoch. The one-toed horse did not evolve until the Pleistocene, but his ancestry is well documented back to the Eocene, with characteristic intermediate forms at various spans of the Tertiary. The geologic time scale has been developed by intercomparisons of the fossil assemblages of known stratigraphic succession. Such intercomparisons show that the several classes of the animal kingdom are represented by fossils during the time intervals shown in Table V-1, though, of course, no single species survived even a small fraction of the time its class existed.

Though, as the table shows, many Classes have long pedigrees, it should not be overlooked that at the next lower rung of the Linnaean ladder, most of the Orders are far more limited in their time span. The orders Blastoidea, Cystoidea, Ophiocistoidea, and Heterostela among the Echinodermata, for example, are all limited to the Paleozoic; they have since been extinct. All of the Ammonoidea, among the stratigraphically most

Table V-1
Life spans of some classes of the animal kingdom.

Phylum	Class	Life span
Protozoa (one-celled)	Foraminifera	Cambrian to Holocene
	Radiolaria	Cambrian (Precambrian?) to Holocene
Porifera (sponges)	several varieties	Cambrian to Holocene
Coelenterata (corals and related forms)	Hydrozoa	Cambrian? to Holocene
	Anthozoa	Ordovician to Holocene
Brachiopoda (lamp-shells)	Inarticulata	Cambrian to Holocene
	Articulata	Cambrian to Holocene
Bryozoa ("moss-corals")	Phylactolaemata	Cretaceous to Holocene
	Gymnolaemata	Ordovician to Holocene
Annelida (worms)	several varieties	Precambrian to Holocene
Arthropoda	Trilobita (trilobites)	Cambrian through Permian
	Crustacea (lobsters, crabs)	Cambrian to Holocene
	Arachnida (spiders, scorpions)	Cambrian to Holocene
	Insecta (insects)	Devonian to Holocene
Molluska	Gastropoda (snails)	Cambrian to Holocene
	Pelecypoda (clams)	Cambrian to Holocene
	Cephalopoda (squids, nautilids)	Cambrian to Holocene
Echinodermata (echinoids)	Pelmatozoa (sea lilies)	Cambrian to Holocene
	Eleutherozoa (star fish)	Cambrian to Holocene
Chordata	Graptolithina (graptolites)	Cambrian to Carboniferous
	Vertebrata (vertebrates)	Ordovician to Holocene

useful orders of the Cephalopoda because they swarmed in the seas of the late Paleozoic and Mesozoic, had died out by the end of the Cretaceous. Many other examples could be given. The earliest mammals are of Triassic age, the earliest birds of Jurassic, both descended from reptilian stock.

We cannot, of course, in a book of this size offer a systematic coverage of paleontology; all we can do is to offer a few examples of the kind of material that has been used in the compilation of the geologic time scale, and is still used as the primary basis of correlation of the stratified rocks.

Protozoa

Figure 17-25 illustrates a representative collection of Foraminifera from the ocean floor off Central America. Although Foraminifera have been reported from rocks as old as Cambrian, their record is very limited until early in the Carboniferous. They abounded during Cretaceous time and fragments of their minute skeletons are common constituents of the widespread chalks that gave the name to the system. A genus of Foraminifera whose members were much larger, some several centimeters across, and coin-shaped, the *Nummulites*, was very prominent in the Eocene and Oligocene; the Pyramids of Egypt are faced with limestone mainly composed of their shells. In later Tertiary and Quaternary time these larger species disappeared, and few foraminifers of the living genera exceed a millimeter or two in length; many are much smaller. The Radiolaria, also minute unicellular organisms (Fig. V-1), have tests composed of silica rather than of calcium carbonate as foraminiferal shells are. They have been reported from the Precambrian rocks of Brittany but there is some question as to the correctness of the identification. Certainly they have persisted ever since the beginning of the Cambrian. The Radiolaria and the Foraminifera have been much studied during the past few decades, because their small size commonly enables them to escape being ground up during rotary drilling operations (as are most larger fossils). Foraminifers especially have proved useful in subsurface correlations of strata from one oil well to another.

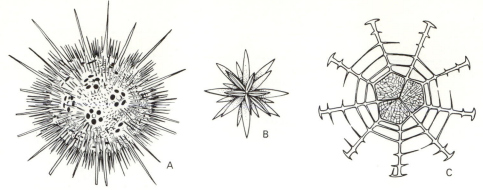

Figure V-1
Some representative Radiolaria. (A) *Haeckeliana* (Holocene). (B) *Zygacantha* (Holocene). (C) *Tetraphormis* (Cretaceous to Holocene). [After A. S. Campbell *in* R. C. Moore (ed.). *Treatise on Invertebrate Paleontology,* Part D, Geological Society of America and University of Kansas Press, 1954.]

Porifera

The stratigraphically most useful class of the Porifera is the Archaeocyathidae (Fig. V-2) considered an independent phylum by some students; these conical creatures were confined to the Cambrian, and since they are readily recognized, serve as particularly valuable "guide fossils." Although sponges have been recognized in all later systems, most of them are so simple and so readily disintegrated that they are of little value stratigraphically.

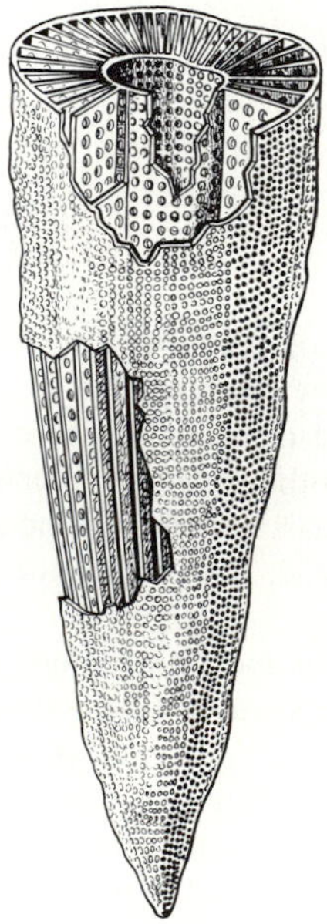

Coelenterata

The Hydrozoa, represented today by such creatures as the sea anemone and jellyfish, are known by traces from the early Paleozoic, but are so sparse and generally poorly preserved that they are of little value below the Cenozoic. In the Cenozoic they are represented by ancestors of the living "staghorn" corals and by the Millepora, or stinging corals, which form colonies an inch or two thick, and grow in wall-like patterns as minor constituents of living coral reefs.

Much more important as stratigraphic aids are the Anthozoa, or true corals. Of these, the subclass Tetracoralla, or horn corals, were very prominent from the Ordovician through the Paleozoic (Fig. V-3, A, B). Other members of this subclass formed chains, such as *Halysites* (Fig. V-3,C), or compact masses, such as *Favosites*

Figure V-2
Syringocnema, a Lower and Middle Cambrian archaeocyathid. [After V. J. Okulitch, *in* R. C. Moore (ed.), *Treatise on Invertebrate Paleontology*, Part E, Geological Society of America and University of Kansas Press, 1955.]

(Fig. V-3,D); these were especially abundant in Silurian and Devonian time and became extinct at the close of the Paleozoic.

The subclass Hexacoralla arose during the Mesozoic and is now the dominant variety. These

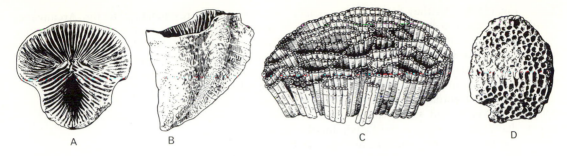

Figure V-3
(A and B) *Streptelasma trilobatum* Whiteaves, a Late Ordovician horn coral. [After R. J. Ross, Jr., U. S. Geological Survey, 1957.] (C) *Halysites catenularius* Edwards and Haime, a Late Ordovician chain coral. [After R. G. Creadick, 1941, via H. W. Shimer and R. R. Schrock, *Index Fossils of North America*, M.I.T. Press, 1944.] (D) *Favosites prolificus* Billings, a Late Ordovician compact coral. [After R. J. Ross, Jr., U. S. Geological Survey, 1957.]

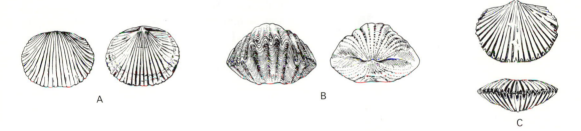

Figure V-4
(A) *Dinorthis occidentalis* Okulitch, Late Ordovician, Montana. (B) *Lepidocyclus gigas* Wang, Late Ordovician, Montana. (C) *Austinella whitfieldi* Winchell and Schuchert, Middle or Late Ordovician, Idaho [Representative Ordovician brachiopods, after R. J. Ross, Jr., U. S. Geological Survey, 1957.]

organisms have radial symmetry and are either colonial or solitary. The shallow-water colonial forms are confined to warm seas, but many solitary forms live in very deep and cold environments.

Brachiopoda

Brachiopods are solitary marine animals commonly anchored by a stalk. The phylum is in decline, represented during the Cenozoic by less than a third of the number of genera that lived in the Ordovician. They are relatively small shelled creatures, generally about 25 mm or so across; the largest ever found was only about 30 cm wide. They are bivalves; in the Class Inarticulata, the valves are held together by muscles alone. The Class Articulata has toothed hinges as well as

muscles for this service. One inarticulate brachiopod, *Lingula*, has survived from the Ordovician to the present day with little change in form—perhaps the longest lived genus known. Most inarticulate brachiopods have phosphatic shells, whereas all the articulate forms have calcareous shells. Figure V-4 illustrates some articulate species from the Late Ordovician of Montana and from the Middle or Late Ordovician of Idaho.

Bryozoa

Bryozoa are extremely small animals but they, like the corals, commonly grow into fantastically complex colonies attached to the sea floor. Many colonies are branching, others have leaf-like forms, but all are characterized by minute tubules and perforations in which the individual animals lived (Fig.V-5,A). As they grow, some colonies change form, even so drastically that broken fragments of fossils, now known from perfect matching of the fractured ends to have belonged to a single colony, earlier had been mistakenly assigned not merely to different species but to different genera! The identification of most bryozoa is a matter for experts, but some readily recognized genera have proved very useful "guide fossils." For example, the genus *Archimedes,* so-called because of its resemblance to the Archimedes screw used for raising water in many primitive cultures, is a guide to the Mississippian of the mid-continental United States, though its range extends into the early Pennsylvanian of the western Cordillera (Fig. V-5,B).

Annelida

Although annelid worms are known from both Precambrian strata and younger rocks, their soft parts are so poorly preserved that they are almost useless as stratigraphic guides. A few lived in protective calcareous tubes: thus *Serpula*, though a long lived genus, is useful in certain stratigraphic investigations. Its characteristic coiled tubes are practicable markers in the Cenozoic of California. Worm marking and trails formed in soft sediments are very common, and are similarly useful in some places, even though these rather vague markings are not demonstrably restricted to strata of one age.

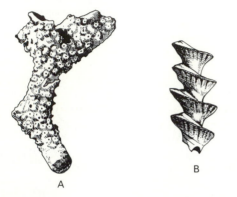

Figure V-5
(A) *Hallopora ramosa* d'Orbigny, from the Ordovician of Ohio. [After C. O. Dunbar, *Historical Geology*, John Wiley & Sons, 1960.] (B) *Archimedes wortheni* Hall, a common bryozoan in the Mississippian of Missouri. [After H. W. Shimer and R. R. Shrock, *Index Fossils of North America*, M.I.T. Press, 1944.]

Arthropoda

The phylum Arthropoda contains, in the class Insecta alone, many times as many genera as all the other phyla together. But the most important stratigraphically are the Trilobita, a class that

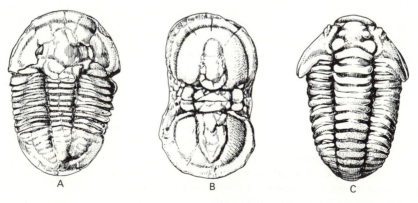

Figure V-6
(A) *Asaphiscus wheeleri* Meek. (B) *Peronopsis interstrictus* White. Both are Middle Cambrian trilobites from the House Range, Utah. [After A. R. Palmer, U. S. Geological Survey, 1954.] (C) *Flexicalymene retrorsa* Foerste, Lake Ordovician, Ohio. [After R. J. Ross, Jr., U. S. Geological Survey, 1967.]

Figure V-7
A fossil *Plecia* from the Green River Shale, Eocene, Wyoming. [After W. H. Bradley, U. S. Geological Survey.]

flourished in the Cambrian and Ordovician and then gradually lessened in abundance and finally died out in the Permian. Any fossil of this class is thus conclusive evidence of the Paleozoic age of the strata containing it; the genera are, of course, much more limited in range than the class as a whole and allow the recognition of many restricted stratigraphic zones, especially in the early Paleozoic. The name refers to the three longitudinal lobes that all these creatures have, a characteristic immediately definitive of the class. A few representative specimens are illustrated in Figure V-6.

Despite their prodigious numbers, both of individuals and of species, insects have left few fossil records, though some are preserved with great clarity (Fig. V-7). Some Carboniferous cockroaches were as long as 10 centimeters. Because of their rarity, insect fossils are not as valuable as those of many other classes.

Much more useful as stratigraphic guides are the minute bivalved Ostracoda, which are crustacea, distantly related to the modern crabs. Like the Foraminifera and Radiolaria, many ostracods are small enough to escape destruction

during well-drilling and thus are useful in oilfield correlations.

Although a few fossil spiders have been recognized, the most important arachnids, stratigraphically, are the eurypterids, giant water scorpions, some as long as 2 m. They flourished from Ordovician to Permian time, when their line died out.

Molluska

Among the most useful fossils are the mollusks: the gastropods (snails), pelecypods (clams), and cephalopods, all of which have flourished from the Cambrian to the present day, though many orders have had much shorter life spans. The Lamellibranchia (pelecypods) were not abundant until the late Paleozoic but are extremely useful in strata of that age and younger. Representative examples of some Miocene clams are illustrated in Figure V-8.

Gastropoda, the class to which the snails belong, were also not abundant until the Carboniferous, but their fossils, too, are both abundant in and useful for stratigraphic correlations of rocks of that and younger ages. Representatives of several varieties are illustrated in Figure V-9.

The Cephalopoda, represented by the present-day chambered nautilus of the Philippine Seas and by the more widespread octopus and squid, have been known in rocks of all ages from Cambrian on. The squid shell is internal and not well developed; the squid's only stratigraphically important fossil relatives are the belemnoids, which lived from the Mississippian to the Paleocene. Their fossil remains are the cigar-shaped internal shells, very dense and solid. They abound in the Cretaceous but are not good guide fossils because of their simplicity.

Far more important stratigraphically are extinct cephalopods that are related to the living nautilus: the rather similar Nautiloidea and the Ammonoidea. Both orders had chambered shells. As the animal—and the shell—grew, he successively vacated the smaller inner chambers, leaving them filled only with air and thus enabling the organism to float. All the chambers were connected by a narrow tube, the siphuncle. The classification of the nautiloids depends largely on the location of the siphuncle and the pattern of the junctions

486

Figure V-8
Some Miocene clams from Jamaica. (A) *Echinochama antiqua*. (B) *Codakia spinulosa*. (C) *Cardium (Trachycardium) waylandi*. (D) *Chione sawkinsi*. [After W. P. Woodring, U. S. Geological Survey, 1925.]

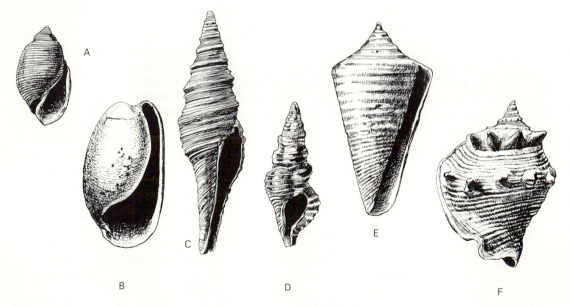

Figure V-9
Some Miocene snails from Jamaica. (A) *Acteon textilis*. (B) *Bulla vendryesiana*. (C) *Polystira barretti*. (D) *Carinodrillia bocatoroensis*. (E) *Conus (Leptoconus) planiliratus*. (F) *Strombus bifrons*. [After W. P. Woodring, U. S. Geological Survey, 1928.]

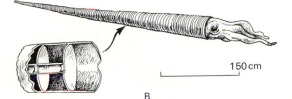

150 cm

A

B

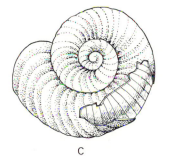

C

Figure V-10
(A) A nautiloid cephalopod, showing representative sutures. [After K. von Zittel, 1916.] (B) An orthoceratid of the Ordovician. [After C. R. Knight.] (C) *Plectoceras occidentale*, a loosely coiled Ordovician nautiloid. [After C. O. Dunbar, *Historical Geology*, John Wiley & Sons, 1960.]

(called sutures) of the partitions that formed living chambers with the wall of the outer shell. The nautiloids have been in existence since the Late Cambrian, but they are doubtless a dying class, represented now only by the single genus *Nautilus*. In the nautiloids the partitions between the successive living chambers are simply curved walls so that their sutures, too, are simple curves (Figure V-10,A). The early forms left mostly

straight conical shells. Some forms from the Ordovician are as much as 5 m long and 25 cm across the large end—they were doubtless the most formidable carnivores of their time (Fig. V-10,B). A few Ordovician nautiloids had loosely curved shells (Fig. V-10,C).

The Ammonoidea became differentiated from the nautiloids in Devonian time and died out at the end of the Cretaceous. They offer some of the most accurate stratigraphic correlations for the span of time in which they lived, as they were abundant, evolved rapidly, and their floating dead bodies were widely distributed by currents. They are differentiated from the nautiloids by more complex septa between the chambers; accordingly the sutures where the septa join the shell are more complex. Sutures of the goniatites, which lived in the late Paleozoic, are rather simple multiple curves (Fig. V-11,A), but in the later evolving ammonites they become complex, indeed (Fig. V-11,B).

Most of the ammonites were coiled rather tightly, but some were less tightly coiled, and others though evidently coiled at immature stages, became nearly straight when adult. Examples are shown in Figure V-11,C,D, and E.

Echinodermata

Echinoderms of both great classes, Pelmatozoa and Eleutherozoa, are found as fossils in rocks from the Cambrian to the present. The Pelmatozoa, represented by living sea lilies, are attached to the sea floor, some by stalks many feet high. Crinoids, another form of Pelmatozoa, were abundant in Paleozoic times. Many limestones contain so many segments of crinoid stalks as to be called crinoidal limestones. A representative Mississippian crinoid is illustrated in Figure V-12,A.

The Eleutherozoa, or mobile echinoids, occur as fossils in rocks from Cambrian to present, but were unimportant stratigraphically until the Mississippian. Since then starfishes and many other varieties have flourished. Echinoids of the kind represented by modern sand dollars have proven widely useful in Cenozoic stratigraphy. A Pliocene form is illustrated in Figure V-12,B.

488

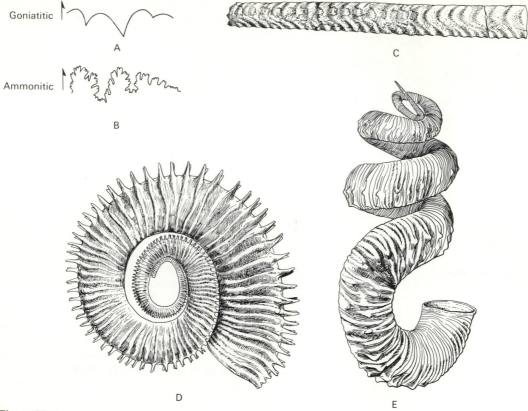

Figure V-11
(A) Representative goniatite sutures. (B) Representative ammonite sutures. The left-hand end of the patterns corresponds to the middle of the whorl; the sutures were symmetrical about this line. (C) *Baculites thomi* Reeside. (D) *Exiteloceras jenneyi* Whitfield. (E) *Didymoceras nebrascence* Meek and Mayden. [C, D, and E are all Late Cretaceous species, courtesy of W. A. Cobban, U. S. Geological Survey.]

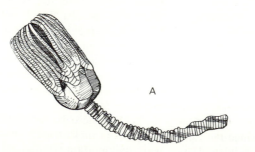

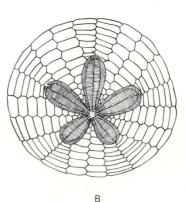

Figure V-12
(A) *Platycrinites hemisphericus*, a Mississippian crinoid. [After C. O. Dunbar, *Historical Geology*, John Wiley & Sons, 1960.] (B) *Dendraster eccentricus*, a sea urchin living since the Pliocene. [After U. S. Grant and L. G. Hertlein, *Western American Cenozoic Echinoidea*, Univ. of California Press, 1938.]

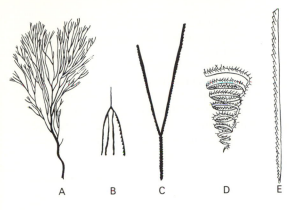

Figure V-13
(A) *Dendrograptus fruticosus* Hall, Lower
Orovician, Quebec. (B) *Tetragraptus fruticosus* Hall,
Lower Ordovician, Quebec. (C) *Dicranograptus
ramosus longicaulis* Elles and Wood, Middle
Ordovician, Scotland. (D) *Monograptus turriculatus*
Barrande, Lower Silurian, Bohemia. (E) *Monograptus
dubius* Seuss, Upper Silurian, England. [After O. M.
B. Bulman, *in* R. C. Moore (ed.), *Treatise on
Invertebrate Paleontology*, Part V, Geological Society
of America and University of Kansas Press, 1955.]

Chordata

GRAPTOLITHINA. It was long thought that the
earliest chordates were the primitive fish of the
Ordovician, but in the late 1930's unusually well-
preserved specimens of Graptolithina—a class
hitherto assigned to a much more primitive
evolutionary level—were recognized to be
chordates. The earliest graptolithinas may be of
Middle Cambrian age; by the late Cambrian they
were widespread. The dendroids, an early form,
resemble small ferns (Fig. V-13,A); their much
more useful relatives, the graptolites, flourished
from the Early Ordovician to the beginning of the
Devonian, when they died out. Most graptolites
have been flattened between the bedding surfaces
of shales. They look like pencil marks (hence the
name, from the Greek for "written rock"). Those
few that have been found as three-dimensional
fossils in limestone or chert permit something to
be learned of their anatomy, including the fact
that they had notochords, a characteristic feature of chordates.

Most graptolites were very small creatures,
rarely more than a few centimeters long, though
some as long as 50 cm have been found in Silurian

strata. They are thought to have floated freely in
the sea, supported by tiny air sacs, and because
this facilitated wide distribution, they are unusually valuable for long distance correlations. A
few examples are illustrated in Figure V-13,B,C,
D, and E.

VERTEBRATA. We know no living descendants of
the graptolites; it seems most unlikely that they
are in any way connected with vertebrate evolution, despite their relatively advanced nervous
systems. From what phylum, then, did the vertebrates evolve? Some students have thought,
because amphibians resemble eurypterids in
form, that an arachnid stem might be ancestral.
Physiologically, however, the eurypterids differ
fundamentally from even the most primitive vertebrates. Others have thought the worms might be
ancestral to the vertebrates because some worms
have principal nerves approaching the mammalian spinal cord in form, but here too there are
fundamental physiological difficulties.

Perhaps the evolution of the vertebrates took
place in an ecological setting different from that of
any vertebrate fossil yet discovered. More likely
our difficulties in tracing vertebrate ancestry are
because evolution was by way of soft-bodied
forms which were not preserved as fossils. At any
rate, some authorities, notably Professor A. S.
Romer of Harvard University, whose general
ideas we are here following, think it likely that the
vertebrates had an echinoid ancestry: our distant
cousins the sea lilies are closer relatives than our
other cousins, the scorpions and the worms of
today!

Some bone fragments in the Ordovician of
Colorado show that vertebrates existed then, but
the earliest vertebrate fossils well-enough preserved for sound analysis are Late Silurian. These
were small fish: jawless bottom-dwellers, mud-
eaters with suckers rather than teeth, heavily
armored (presumably against the giant carnivorous eurypterids of the time, but possibly as
protection against lethal solar radiation). Because
of their bony armor, these primitive fish are called
ostracoderms, from the Greek word for "bone
skinned" (Fig. V-14,A). Though differing tremendously in shape and in armor from any existing
lamprey, they were physiologically so similar that
they are considered ancestral both to the lamprey
and to all higher vertebrates.

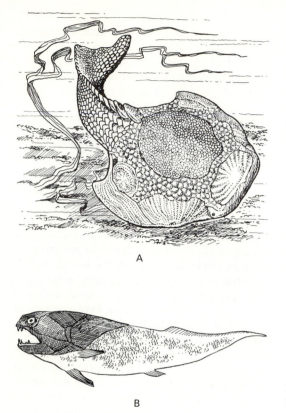

A

B

Figure V-14
(A) An ostracoderm, *Psammolepis*. [After Mark and
Bystrow, courtesy of Alfred S. Romer, Harvard
University.] (B) A giant Late Devonian placoderm,
Dinichthys, some of which attained a length of 9 meters.
[After Obruchev, 1964, courtesy of Alfred S. Romer,
Harvard University.]

The kidney structures of the lampreys and all
higher vertebrates suggest very strongly that they
evolved from a freshwater ancestor, which may,
at first, seem difficult to reconcile with the marine
ancestry of all crinoids and of phyla from which
vertebrates might have evolved. Because osmotic
pressures through organic membranes must be
balanced with those of the environment, the operation of the kidneys to separate excess salt from
the blood stream is essential to life in a fresh-
water environment. Though this function is most
disadvantageous to marine life, all fish, even salt-
water dwellers, retain kidneys that operate like
our own; they have developed auxiliary organs to
counter the kidney's activity. It seems most im-

probable that a marine organism without a fresh-
water ancestry would have developed both anti-
thetic operations.

During the Devonian, fishes multiplied and
diversified so rapidly that the period is sometimes
referred to as the Age of Fishes. Some abandoned
their freshwater habitat and took to the sea. The
first fishes with jaws evolved, and by the end of
the period all four of the major groups of fish now
known had been established:

1. the Agnatha, represented by the ostracoderms and the modern lamprey;

2. the Placodermi, the first jawed fishes, which
had armored heads and grew to great size. These
were undoubtedly the most ferocious carnivores
of the Late Devonian but the line died out completely by the end of the Paleozoic (Fig. V-14,B).

3. the Chondrichthyes, the ancestors of the
modern sharks and other cartilagenous fish such
as the rays and skates; and

4. the Osteichthyes, or bony fish, ancestral to
nearly all living vertebrates.

We can devote no space to the evolution of the
first three of these groups; we pass directly to the
bony fishes from which all other vertebrates and
most modern fish have descended. And of the
bony fishes, too, we will omit further discussion of
the great group from which modern fish arose, and
refer only to a second, very much smaller group,
the fleshy-finned fishes—very unsuccessful as
fishes, but rather important as the ancestors of the
amphibians and of all land vertebrates, including
ourselves.

Fleshy-finned fish The fleshy-finned fish divided, during the Devonian, into two groups: the
lungfish and the lobe-finned fish. The lungfish
developed the ability to breathe air through a
nostril in the top of its head. They were fresh-
water fish, and as the Devonian was a time of
widespread deserts (Chapter 17), they adapted to
a life of intermittent water-supply. When the
stream or lake home dried up in summer, they
could dig themselves holes in the muddy bottom,
wrap themselves in mud, and survive by breathing air until the waters returned with the next
rainy season. Like hibernating bears of cold countries they were wholly inactive while dormant.
Descendants of the early lungfish, with much the

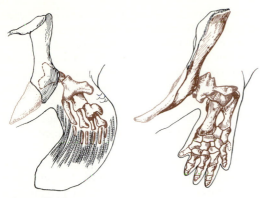

Figure V-15
A comparison between the fin of a Late Devonian lobe-finned fish, *Eusthenopteron foordi*, and the limb of a primitive amphibian, *Euryops*, of late Paleozoic age. [After W. K. Gregory, via A. S. Romer, *The Vertebrate Story*. Copyright 1959 by The University of Chicago Press.]

same habits, live today in the upper Nile drainage, in Queensland, Australia, and in the Gran Chaco of Paraguay. Despite their partial adaptation to land life, their fins were never modified to permit walking; the land vertebrates were not derived from this subdivision of the fleshy-finned fish, but from their cousins, the lobe-finned fishes.

The lobe-finned fishes also developed lungs but at the same time they had a tremendous advantage over the ancestors of the surviving lungfish in that their lobe-like fins grew more sturdy and more muscular. They probably did not have any advantage as far as food-seeking or protection was concerned, but when the droughts came, they were able to move, awkwardly and slowly, no doubt, from one waterhole to another, adapting their diet, perhaps to the dormant lungfishes or possibly to primitive land plants. Their jaws did not become so specialized as those of the lungfish, whose diet of mollusks led them to develop shell-cracking modifications of their jaws. By the end of the Devonian, the lobe-finned fishes had been so modified that, somewhat like the modern salamander, they lived much of their lives on land: the Amphibia had evolved (Fig. V-15).

The ancestry of the primitive amphibians is unquestionable: the head bones, the armor over their heads and foreparts, their teeth and even their vestigial tails are closely similar to those of their ancestors, the lobe-finned fish.

Reptilia Although the amphibians were the first land vertebrates and thrived in the late Paleozoic, their modern amphibian descendants are few: the salamanders, frogs, toads, and a few others. They were succeeded by their much modified descendants, the reptiles, dominant in the Mesozoic—the Age of Reptiles.

Of course, no one really knows all the factors in what was doubtless a very complex story, but one obvious weakness in the life scheme of the amphibians is the necessity for them to deposit their eggs in water. Here both egg and young are vulnerable to attack by any number of animals. The reptiles developed the greatly advantageous attribute of laying eggs with shells, which could be hatched in the sand. Because the shelled eggs contain yolks adequate for nourishment of the embryo—as amphibian eggs do not—the young, when hatched, are further developed and hence less vulnerable to attack in infancy than amphibian young. Whatever the reason, the stratigraphic record is unmistakable: by the end of the Permian the Amphibia were all but eliminated and the reptiles, in fantastic variety and modes of life, dominated the land. Soon after, they dominated the sea also, to which some groups returned.

The earliest of the "stem" reptiles, from which all others evolved, developed in the Pennsylvanian, when the amphibians were at their height. Like their direct ancestors, the amphibians, and their more remote ancestors, the bony fishes, they had solid skulls without lateral openings. But early in their history some descendants developed openings in the sides of the skull, facilitating the contraction of the jaw muscles. By the patterns and numbers of these lateral skull openings paleontologists have classified the reptiles and are able to trace the various lineages. We will not go into the details of the classification; all of the main branches of the reptiles were well established by the end of the Permian. Of these, one subclass developed toward mammalian characters and is ancestral to mammals; another toward the plesiosaurs and ichthyosaurs, which returned to the sea, and of which no living representative is known today; still another evolved along the line of the crocodiles. It was ancestral to dinosaurs, the "ruling reptiles" of the Mesozoic, which also died out at the end of that era, but are represented today by their cousins the birds and the mammals.

Every reader of this book will know how many

kinds of dinosaurs looked, from museums, and cartoons and science fiction. We shall trace only the reptilian family tree that leads to the birds and mammals.

Birds Although, among the many diverging lines of "ruling reptiles" several took on the outward shape of running birds such as ostriches, and still other reptiles actually became fliers, none of these variants is ancestral to the birds we know. The birds sprang from the basic stock of the "ruling reptiles" and not from any variant bird-like forms otherwise recognized. Indeed, both the running and the flying bird-like reptiles were flourishing long after the earliest feathered birds (known from the Upper Jurassic) were well established.

Mammalian Evolution

Early in the history of the "stem" reptiles, in the late Carboniferous, an aberrant branch arose with skull modifications and tooth arrangement that turned from the normal reptilian scheme toward that of the mammals. These carnivorous offshoots of the reptile line looked a lot like lizards, but were structurally quite distinct. Their Permian and Triassic descendants took more and more the ultimate form of mammals both in skull structure and in the arrangement of the limbs, which changed from the sprawling appendages of a lizard to upright supports growing under the body. Beasts of this kind dominated the late Permian and earliest Triassic, but then arose the "ruling reptiles"; by the end of the Triassic the mammal-like reptiles were extinct.

But they had left behind the first mammalian stem—animals that still laid eggs like their reptilian ancestors (and the living Australian platypus), but were in all other respects truly mammals. The placental mammals (whose young are attached to the mother's womb by a placenta) had probably arisen by the Late Jurassic, though the evidence is extremely slender: it must be based on similarities in such things as teeth and jawbones rather than on flesh and soft parts, which are not preserved. The creatures were tiny, many no bigger than mice, and carnivorous, probably subsisting mainly on insects and worms. Many, perhaps all, were tree-dwellers. They were far outnumbered by the marsupials, whose newborn young are carried in the mother's pouch.

It was not until the end of the Cretaceous, when the great reptiles died out, that the the time came for the great mammalian expansion that made the Cenozoic the Age of Mammals. This expansion, with amazing speed, filled virtually every conceivable ecologic niche. Both marsupials and placental mammals took part in it, though eventually the placentals, with their great advantage in protection of the young, took the dominant place. By Eocene time porpoises and gigantic whales had occupied the seas, and flying mammals, the bats, were fully at home in the air. Our own line, the primates, retained its arboreal habitat well into the Tertiary, though by Miocene time a likely ancestor of man, *Proconsul,* whose remains have been found near Lake Victoria, had lost some of the stronger indicators of tree life.

PLANT FOSSILS

Of the many classifications of the plant kingdom, we regard the simplest as adequate to our purpose and present it in Table V-2.

Thallophyta

BACTERIA. We have noted that the Fig Tree Series of Swaziland has yielded bacterial fossils more than 3 billion years old. Some much younger ones from the Lake Superior region are illustrated in Figure 6-1. While of great interest as throwing light on the antiquity of life on the earth, these organisms are too rarely preserved to be useful in correlations.

DIATOMACEAE. Though these minute unicellular organisms enclosed in siliceous tests seem so simple that one would think they had evolved very early, their oldest fossils yet found are of Jurassic age. This flora is so diversified, however, as to suggest a much longer history. Possibly their ancestors did not secrete silica tests, or perhaps their tests have been dissolved from the older rocks, for although their tests are of silica, the silica is hydrated and is much more soluble than most forms of silica.

The Jurassic species were marine; the earliest terrestrial diatoms are of Miocene age. Within the last forty years, largely through the work of K. E. Lohman, of the U.S. Geological Survey,

Table V-2
Life spans of some members of the plant kingdom.

Division	Subdivision	Class	Life span
Thallophyta (nonvascular plants)	Bacteria*		Precambrian to Holocene
	Diatomacea*		Jurassic to Holocene
	Algae (photosynthetic)		Precambrian to Holocene
	Fungi (degenerate, non-photosynthetic)		Precambrian to Holocene
Bryophyta (liverworts and mosses)			Silurian to Holocene
Pteridophyta (spore-bear-ing vascular plants)			Silurian to Holocene
	Psilophytales (Psylopsida)		Devonian
	Lycopodiales (Lycopsida)		Silurian to Holocene
	Articulatales (Sphenopsida)		Devonian to Holocene
	Filicales (Pteropsida)		Late Devonian to Holocene
Spermophyta (seed-bearing plants)			Carboniferous to Holocene
	Gymnospermae ("naked seeds")		Carboniferous to Holocene
		Pteridospermae (seed-ferns)	Carboniferous to Jurassic
		Cycadales (cycads)	Triassic to Holocene
		Bennettitales (cycad-like but with flowerlike shoots)	Triassic to Cretaceous
		Ginkgoales (ginkgoes and re-lated forms)	Triassic to Holocene
		Coniferales (pines, cypresses, and related forms)	Carboniferous to Holocene
	Angiospermae (flowering plants, pro-tected seeds)		Triassic to Holocene
		Monocotyledonae (grasses, lilies, palms)	Triassic to Holocene
		Dicotyledonae (beans, peas, most plants)	Cretaceous to Holocene

*In many classifications, these organisms are excluded from both animal and plant kingdoms, and along with a few other microorganisms set apart as Protista.

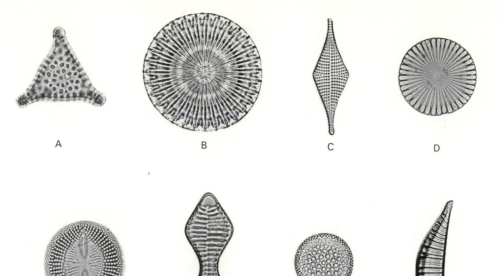

Figure V-16
Some representative diatoms. (A) *Trinacria aries* Witt, Late Cretaceous, California, marine. (B) *Lepidodiscus elegans* Witt, Eocene, Russia, marine. (C) *Raphoneis angularis* Lohman, Pliocene only, California, marine. (D) *Melosira clavigera* Grunow, mid-Miocene to Lower Pliocene, California, marine. (E) *Raphidodiscus marylandicus* Christian, middle Miocene only, very widespread, marine. (F) *Rhabdonema valdelatum* Tempere and Brun, Miocene to Pleistocene, marine. (G) *Coscinodiscus elegans* Greville, Eocene in Barbados, to Miocene in California, marine. (H) *Rhopalodia gibberula* (Ehrenberg) Miller, Pliocene to Recent, very widespread, commonly in hot springs. [Photos by K. E. Lohman, U. S. Geological Survey.]

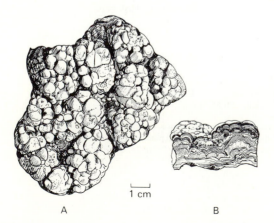

Figure V-17
An algal colony from the Green River Shale, Eocene, Wyoming. (A) Top view. (B) Sectional view. [After W. H. Bradley, U. S. Geological Survey.]

they have become extremely useful horizon markers in the Cretaceous and Tertiary of the Pacific States. Some representative specimens are illustrated in Figure V-16.

ALGAE. As mentioned in Chapter 19 cabbage-shaped heads and columns formed by algae have been found in strata of all ages from far back in the Precambrian to the present. Although they have been successfully used in correlations on a gross scale, they are not sufficiently differentiated to be particularly valuable. A representative Eocene freshwater algal colony is shown in Figure V-17.

Much more useful than the colonial algae are the minute discoasters and coccoliths, which are fragments of floating algae. Their small size—most are at the very limit of optical microscopes and must be studied by phase contrast or electron microscopes—permits a virtually worldwide distribution by marine currents; identical forms have been found in Eocene rocks from Iran, New Zealand, Britain, California, Japan, and Italy. Despite the difficulty of their study, they are rapidly becoming extremely useful guide fossils—the coccoliths for rocks of Jurassic and younger age, the discoasters for the Cenozoic. Representative specimens are illustrated in Figure V-18.

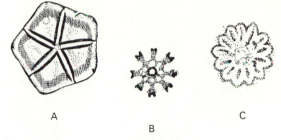

A

B

C

Figure V-18
Some microscopic planktonic organisms. (A) *Braarudosphaera bigelowi*, a Paleocene coccolith. (B) *Discoaster septemradiatus*, Middle Eocene. (C) *Discoaster helianthus*, Paleocene. All are from California [After A. O. Woodford, 1965. Specimens from M. N. Bramlette, University of California, San Diego.]

Bryophyta

The bryophytes include the existing liverworts and mosses, and though they have been found sporadically in rocks of Silurian and all later ages, they are of little use stratigraphically; their preservation is usually very poor.

Pteridophyta

PSILOPHYTALES. The oldest land plants were very simple ones perhaps ancestral to (or perhaps paralleling the ancestry of) the living *Psilotum*. They are found in the Silurian of Australia and in the Middle Devonian of Germany, Scotland, and eastern Canada. These were extremely simple rush-like plants that rose vertically from a horizontal rhizome (Fig. V-19). There were many representatives in the Upper Devo-

Figure V-19
One of the earliest land plants, *Rhynia*, from the Old Red of Scotland. [After Kidston and Lang, *Trans. Royal Society Edinburgh* v. 52, 1921–1922.]

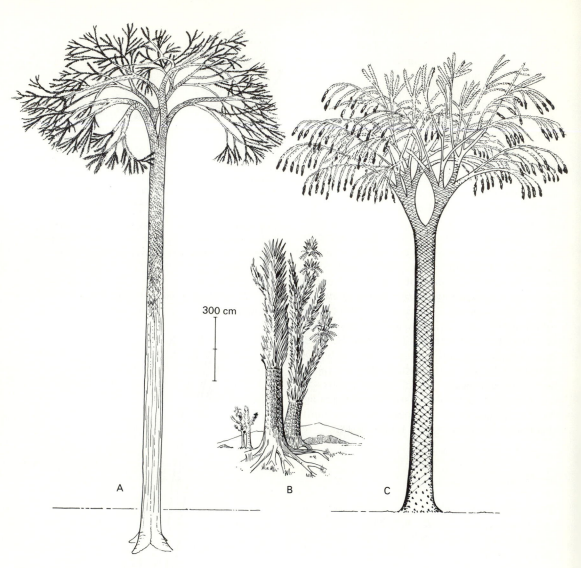

300 cm

A   B   C

Figure V-20
Some lycopsids of the Carboniferous coal swamps. (A) *Lepidodendron.* [After a drawing by Edward Valliamy, in *Plant Life Through the Ages,* by A. C. Seward, Cambridge University Press, 1932.] (B) *Sigillaria.* (C) The bark of *Lepidondendron,* showing leaf scars on the trunk. [B and C after C. O. Dunbar, *Historical Geology,* John Wiley & Sons, 1960.]

nian, some with trunks more than a meter in diameter. Though it was once thought that all higher plants were derived from these simple plants, this is now seriously questioned. Other classes were probably developing at the same time.

LYCOPODIA. The living lycopsids (club mosses), though widely distributed over the earth, are few compared with their abundant varieties in the Devonian and Carboniferous, when they dominated the flora of the coal swamps. Two genera, especially, *Lepidodendron* and *Sigillaria,* grew to

Figure V-21
One of the great contributors to the Carboniferous coal swamps, *Calamites*. [Modified from Arthur Cronquist, *Introductory Botany*, Harper & Row, 1961.]

immense heights of as much as 40 or 50 m (Fig. V-20). The line became much reduced in Triassic and later time.

Articulatales The only living descendant of the once abundant trees of Carboniferous time belonging to this group is the common *Equisetum* or scouring rush, so-called because of the silica spicules on its jointed stems. The group was represented in the Carboniferous by great trees more than 30 cm in diameter, the most common being *Calamites* (Fig. V-21). Few forms survived the Triassic period.

FILICALES.   The Filicales include the ferns, the only division of the pteridophytes that is still

prominent. They dominated the Late Devonian floras and produced tree-sized genera during the Carboniferous. An Eocene fern from one of the many fossil forests of Yellowstone Park is shown in Figure V-22.

Spermatophyta

GYMNOSPERMAE.   Of the seed-bearing plants, the gymnosperms, whose seeds are more or less exposed, have the longer lineage, but they are surely losing out to their cousins, the angiosperms, whose seeds are better protected.

Pteridospermae The seed ferns originated in the Late Devonian and became important components of the Carboniferous flora, especially in the Gondwanaland areas, characterized by *Glossopteris* and *Gangamopteris* as the dominant genera (Fig. V-23). This class died out by Jurassic time.

Cycadales The cycads, now chiefly southern hemisphere inhabitants, became very widespread in the Triassic and are found as fossils from Siberia to Virginia. They dominated many

Figure V-22
An Eocene fern from Yellowstone Park, Wyoming. [After Erling Dorf, Princeton Univ.]

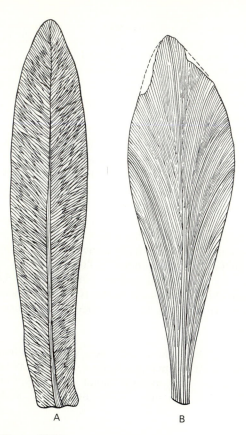

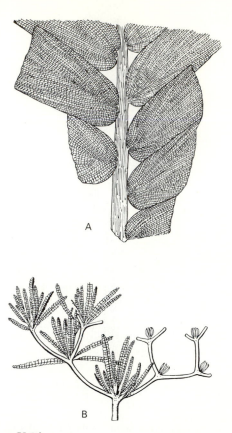

Figure V-23
Leaves of the seed ferns. (A) *Glossopteris*. (B)
Gangamopteris. Both are from the late Paleozoic
Gondwanaland flora. [After E. A. N. Arber, 1905.]

Figure V-24
(A) Fragment of a cycad leaf (*Sphenozamites
rogersiana*) from the Triassic of Virginia. [After
M. Fontaine, U. S. Geological Survey.] (B) Restoration
of *Wielandiella*, a Triassic cycad of Sweden. [After
A. G. Nathorst, 1910.]

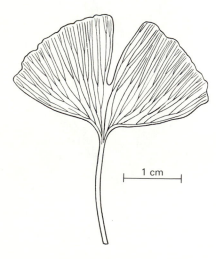

Figure V-25
A leaf of *Ginkgo biloba*, the only surviving
representative of the Ginkgoales. [After
Erling Dorf, Princeton Univ.]

1 cm

Jurassic floras but are now much reduced. Figure V-24 represents some Triassic forms. Most
living forms have stumpy pine-like trunks to
which the leaves cling for many years; they have
few branches and a crown of large leaves that
reminds the layman of palms. The Bennettitales
were very similar trees, but carried their seeds in
exposed shoots resembling flowers.

Ginkgoales The ginkgoes arose in the Triassic and became very widespread, with many genera, during the Cretaceous. Just one species survives and that, perhaps only because it is cultivated for its grace and foliage. A leaf of the modern genus, *Ginkgo biloba*, is shown in Figure V-25.

Coniferales The conifers rose in the Carboniferous and are there represented by one of the important trees of the coal forests, *Cordaites*. (Some paleobotanists classify Cordaites as a separate class, but all agree it was some kind of a proto-conifer). It is illustrated in Figure V-26. The great Eocene fossil forests of the Yellowstone Park have preserved trunks by the thousand and leaves by the million of conifers ancestral to the modern redwoods. Some trunks and a leaf fragment are illustrated in Figure V-27.

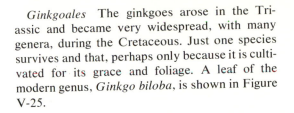

300 cm

Figure V-26
A *Cordaites* from the Pennsylvanian. [After C. O. Dunbar, *Historical Geology*, John Wiley & Sons, 1960.]

Figure V-27
Petrified trunks and a leaf fragment from one of the petrified Eocene forests of Yellowstone Park. Many of the trees were ancestral to the living redwoods. [From Erling Dorf, Princeton Univ.]

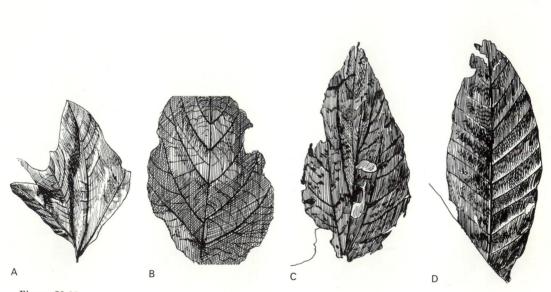

Figure V-28
A succession of fossil forests (Eocene) were buried beneath falls of volcanic ash. Lamar River canyon in northeastern Yellowstone Park. [After Erling Dorf, Princeton Univ.]

A   B   C   D

Figure V-29
Some leaves from the fossil forests of Yellowstone Park. (A) Leaf of an extinct sycamore. (B) Leaf of an extinct grape. (C) Leaf of a tree related to the rare Chinese katsura tree. (D) A meliosoma leaf, whose nearest living relatives are restricted to tropical and subtropical forests. [After Erling Dorf, Princeton Univ.]

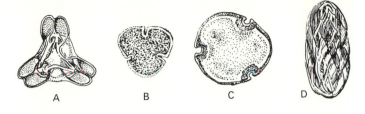

Figure V-30
Some representative pollen grains, greatly enlarged. (A) *Oculopolis*,
Cretaceous of Tennessee, a plant with no close relatives living.
(B and C) *Tilia pollenites*, Tertiary of Kentucky, a relative of the
living linden tree. (D) *Ephedra pollenites*, Cretaceous of Kentucky,
a relative of the living "Mormon tea." [After Robert Tschudy,
U. S. Geological Survey.]

ANGIOSPERMAE. The angiosperms are so familiar that they require no illustration. Nonetheless, it seems worthwhile to illustrate a tremendous wealth of fossil trees, 27 successive layers of
forests, some with trees estimated to be more than
1000 years old, each in turn buried beneath ash
from a nearby volcano. Both the succession of
forests and some representative fossil leaves are
shown in Figures V-28 and V-29.

One of the relatively recent developments of
paleobotany is the growing emphasis on the use
of pollen in correlation. The pollen of many plants
is highly characteristic, and as the spore is both
small and highly resistant chemically, it is commonly preserved when larger fossils such as
leaves and invertebrates are destroyed. Pollen
spores are therefore of increasing use and value in
stratigraphy. A few representative specimens are
illustrated in Figure V-30.

Page references to important concepts and
technical terms are given in **boldface** type.

Zittel, K. von, 487
Zone
  aeration, **310**
  almandine, 200
  Benioff, **134**
  benthic, **362**
  biotite, 200
  capillary saturation, **310**
  chlorite, 200
  contact, 200
  fracture, **137**

  life, **162**
  Low-Velocity, **111**
  metamorphic, **200**
  ophiolitic, 136
  pelagic, **362**
  rift, **158**, 164
  shadow, **114**
  sillimanite, 200
  staurolite, 200
  strand, **383**
  subduction, **131**–137